Non-Convent
Energy Sou
and
Utilisation

(Energy Engineering)

For Students of B.E. / B. Tech, also useful for Competitive Examinations

---

Er. R.K. RAJPUT

M.E. (Heat Power Engg.) Hons., Gold Medalist; Grad. (Mech. Engg. & Elect. Engg.);
M.I.E. (India), C.E. (India)
M.S.E.S.I.; M.I.S.T.E.

Recipient of:
"Best Teacher (Academic) Award"
"Distinguished Author Award"
"Jawahar Lal Nehru Memorial Gold Medal"
for an outstanding research paper
(Institution of Engineers–India)
Man of Achievement Award
Principal (Formerly):
● Thapar Polytechnic College, Patiala
● Punjab College of Information Technology, Patiala

S. CHAND
PUBLISHING

S Chand And Company Limited
(ISO 9001 Certified Company)

S Chand And Company Limited
(ISO 9001 Certified Company)

Head Office: Block B-1, House No. D-1, Ground Floor, Mohan Co-operative Industrial Estate, New Delhi – 110 044 | Phone: 011-66672000

Registered Office: A-27, 2nd Floor, Mohan Co-operative Industrial Estate, New Delhi – 110 044 Phone: 011-49731800

www.**schandpublishing.com**; e-mail: **info@schandpublishing.com**

Branches

Chennai	:	Ph: 23632120; chennai@schandpublishing.com
Guwahati	:	Ph: 2738811, 2735640; guwahati@schandpublishing.com
Hyderabad	:	Ph: 40186018; hyderabad@schandpublishing.com
Jalandhar	:	Ph: 4645630; jalandhar@schandpublishing.com
Kolkata	:	Ph: 23357458, 23353914; kolkata@schandpublishing.com
Lucknow	:	Ph: 4003633; lucknow@schandpublishing.com
Mumbai	:	Ph: 25000297; mumbai@schandpublishing.com
Patna	:	Ph: 2260011; patna@schandpublishing.com

First Edition 2012
Reprint 2013
Second Revised Edition 2014
Reprints 2014, 2016, 2018, 2020, 2021

Reprint 2022

ISBN: 978-81-219-3971-3   **Product Code:** H3NCU64ENVS10ENAB14O

PRINTED IN INDIA

By Vikas Publishing House Private Limited, Plot 20/4, Site-IV, Industrial Area Sahibabad, Ghaziabad – 201 010 and Published by S Chand And Company Limited, A-27, 2nd Floor, Mohan Co-operative Industrial Estate, New Delhi – 110 044.

PREFACE TO THE SECOND EDITION

I take the pleasure in presenting the Second Edition of this book. The warm reception which the previous edition and reprints have received is a matter of much satisfaction to me.

In this edition, the book has been thoroughly revised and a new Section on **"SHORT ANSWER QUESTIONS"** has been added to make the book still more useful to the students.

I am greatly indebted to the management and the editorial team of S. Chand & Company Pvt. Ltd. for all help and support in publication of this edition.

The constructive suggestions for improvement of the book are most welcome.

Er. R.K. RAJPUT
(Author)

PREFACE TO THE SECOND EDITION

The primary purpose of writing this book is to make available to the student community a book which deals *exhaustively* and *systematically* various topics on **"Non-conventional, Renewable and Conventional Energy sources and systems."** The subject matter has been prepared in lucid, direct and easily understandable style, with simple diagrams and worked out examples wherever necessary.

The book consists of *16 chapters*, namely: **1.** *Introduction to energy sources*; **2.** *Principles of solar radiation*; **3.** *Solar energy collectors*; **4.** *Solar energy storage and applications*; **5.** *Wind energy*; **6.** *Bioenergy (Energy from biomass)*; **7.** *Geothermal energy*; **8.** *Ocean energy*; **9.** *Direct energy conversion*; **10.** *Hydrogen energy*; **11.** *Nuclear power plant*; **12.** *Steam thermal power plants*; **13.** *Diesel engine power plant*; **14.** *Gas turbine power plants*; **15.** *Hydro-electric power plants*; **16.** *Environmental aspects of energy generation and pollution control*.

At the end of each chapter *Highlights, Theoretical Questions, Unsolved examples* (wherever necessary) have been added to make this treatise a *complete comprehensive book on the subject*.

This book will prove to be of immense use to the students preparing for Engineering undergraduate and competitive Examinations.

The author's thanks are due to his wife Ramesh Rajput, without whose cooperation and encouragement, this book would have never materialised.

In the end, the author wishes to express his sincere thanks to Sh. Navin Joshi (Executive Vice President), S. Chand and Company Pvt. Ltd. and Editorial Staff for their unstinted support for bringing out this book in such a good shape and in a short span of time.

Although every care has been taken to make the book free of error, yet the author shall be obliged if errors present are brought to his notice. Any constructive criticism/suggestions will be always most welcome.

Er. R.K. Rajput
(Author)

CONTENTS

6. BIO ENERGY (ENERGY FROM BIOMASS)   154–188

15. HYDRO-ELECTRIC POWER PLANTS

INTRODUCTION TO SI UNITS AND CONVERSION FACTORS

A. INTRODUCTION TO SI UNITS

SI, the international system of units are divided into three classes :

1. Base unit
2. Derived units
3. Supplementary units.

From the scientific point of view division of SI unit into these classes is to a certain extent arbitrary, because it is not essential to the physics of the subject. Nevertheless the General Conference, considering the advantages of a single, practical, world-wide system for international relations, for teaching and for scientific work, decided to base the international system on choice of six well-defined units given in Table 1 below :

Table 1. SI Base Units

Quantity	Name	Symbol
length	metre	m
mass	kilogram	kg
time	second	s
electric current	ampere	A
thermodynamic temperature	kelvin	K
luminous intensity	candela	cd
amount of substance	mole	mol

The second class of SI units contains derived units, *i.e.*, units which can be formed by combining base units according to the algebraic relations linking the corresponding quantities. Several of these algebraic expressions in terms of base units can be replaced by special names and symbols can themselves be used to form other derived units.

Derived units may, therefore be classified under three headings. Some of them are given in Tables 2, 3 and 4.

Table 2. Examples of SI Derived Units Expressed in Terms of Base Units

Quantity	SI Unit	
	Name	Symbol
area	square metre	m^2
volume	cubic metre	m^3
speed, velocity	metre per second	m/s
acceleration	metre per second squared	m/s^2
density, mass density	kilogram per cubic metre	kg/m^3
concentration (of amount of substance)	mole per cubic metre	mol/m^3
activity (radioactive)	1 per second	s^{-1}
specific volume	cubic metre per kilogram	m^3/kg
luminance	candela per square metre	cd/m^2

Table 3. Derived Units with Special Names

Quantity	SI Unit			
	Name	Symbol	Expression in terms of other units	Expression in terms of SI base units
frequency	hertz	Hz	–	s^{-1}
force	newton	N	–	$m.kg.s^{-2}$
pressure	pascal	Pa	N/m^2	$m^{-1}.kg.s^{-2}$
energy, work, quantity of heat power	joule	J	N.m	$m^2.kg.s^{-2}$
radiant flux, quantity of electricity	watt	W	J/s	$m^2.kg.s^{-3}$
electric charge	coloumb	C	A.s	s.A
electric tension, electric potential	volt	V	W/A	$m^2.kg.s^{-3}.A^{-1}$
capacitance	farad	F	C/V	$m^{-2}.kg^{-1}.s^4$
electric resistance	ohm	Ω	V/A	$m^{-2}.kg.s^{-3}.A^{-2}$
conductance	siemens	S	A/V	$m^{-2}.kg^{-1}.s^3.A^{-1}$
magnetic flux	weber	Wb	VS	$m^2.kg.s^{-2}.A^{-1}$
magnetic flux density	tesla	T	Wb/m^2	$kg.s^{-2}.A^{-1}$
inductance	henry	H	Wb/A	$m^2.kg.s^{-2}.A^{-2}$
luminous flux	lumen	lm	–	cd.sr
illuminance	lux	lx	–	$m^{-2}.cd.sr$

Table 4. Examples of SI Derived Units Expressed by Means of Special Names

Quantity	SI Unit		
	Name	Symbol	Expression in terms of SI base units
dynamic viscosity	pascal second	Pa-s	$m^{-1}.kg.s^{-1}$
moment of force	metre newton	N.m	$m^{-2}.kg.s^{-2}$
surface tension	newton per metre	N/m	$kg.s^{-2}$
heat flux density, irradiance	watt per square metre	W/m^2	$kg.s^{-3}$
heat capacity, entropy	joule per kelvin	J/K	$m^2.kg.s^{-2}.K^{-1}$
specific heat capacity, specific entropy	joule per kilogram/kelvin	J/(kg.K)	$m^2.s^{-2}.K^{-1}$
specific energy	joule per kilogram	J/kg	$m^2.s^{-2}$
thermal conductivity	watt per metre kelvin	W/(m.K)	$m.kg.s^{-3}.K^{-1}$
energy density	joule per cubic metre	J/m^3	$m^{-1}.kg.s^{-2}$
electric field strength	volt per metre	V/m	$m.kg.s^{-3}.A^{-1}$
electric charge density	coloumb per cubic metre	C/m^3	$m^{-3}.s.A$
electric flux density	coloumb per square metre	C/m^2	$m^{-2}.s.A$
permitivity	farad per metre	F/m	$m^{-3}.kg^{-1}.s^4.$
current density	ampere per square metre	A/m^2	–
magnetic field strength	ampere per metre	A/m	–
permeability	henry per metre	H/m	$m.kg.s^{-2}.A^{-2}$
molar energy	joule per mole	J/mol	$m^2.kg.s^{-2}mol^{-1}$
molar heat capacity	joule per mole kelvin	J/(mol.K)	$m^2.kg.s^{-2}.K^{-1}.mol^{-1}$

The SI units assigned to third class called "Supplementary units" may be regarded either as base units or as derived units. Refer to Table 5 and Table 6.

Table 5. SI Supplementary Units

Quantity	SI Units	
	Name	Symbol
plane angle	radian	rad
solid angle	steradian	sr

Table 6. Examples of SI Derived Units Formed by Using Supplementary Units

Quantity	SI Units	
	Name	Symbol
angular velocity	radian per second	rad/s
angular acceleration	radian per second squared	rad/s^2
radiant intensity	watt per steradian	W/sr
radiance	watt per square metre steradian	$W-m^{-2}.sr^{-1}$

Table 7. SI Prefixes

Factor	Prefix	Symbol	Factor	Prefix	Symbol
10^{12}	tera	T	10^{-1}	deci	d
10^{9}	giga	G	10^{-2}	centi	c
10^{6}	mega	M	10^{-3}	milli	m
10^{3}	kilo	k	10^{-6}	micro	μ
10^{2}	hecto	h	10^{-9}	nano	n
10^{1}	deca	da	10^{-12}	pico	p
			10^{-15}	fasnot	f
			10^{-18}	atto	a

B. DEFINITIONS

The SI seven base units and two supplementary units are defined below :

(i) **Metre.** The metre is the length equal to 1,650, 763.73 wavelengths in vacuum of the radiation corresponding to the transition between the levels $2p_{10}$ and $5d_5$ of the krypton-86 atom.

(ii) **Kilogram.** One kilogram is equal to the mass of the international prototype of the kilogram.

(iii) **Second.** The second is the duration of 9,192,631,770 periods of the radiation corresponding to the transition between the two hyperfine levels of the ground state of the Cesium-133 atom.

(iv) **Ampere.** One ampere is that constant current which, if maintained in two straight parallel conductors of infinite length, of negligible circular cross-section and placed one metre apart in vacuum, would produce between these conductors a force equal to 2×10^{-7} newton per metre of length.

(v) **Kelvin.** The kelvin is the fraction $\dfrac{1}{273.16}$ of thermodynamic temperature of the triple point of water.

(vi) **Candela.** The candela is the luminous intensity, in the perpendicular direction, of a surface of a $\dfrac{1}{600,000}$ square metre of a black body at a temperature of freezing platinum under a pressure of 101, 325 newton per square metre.

(vii) **Mole.** The mole is the amount of substance of a system which contains as many elementary entities as there are atoms in 0.012 kg, carbon 12. When the mole is used, the elementary entities must be specified and may be atoms, molecules, ions, electrons, other particles or specified groups of such particles.

(viii) **Radian.** The radian is the plane angle between two radii of a circle that cut off on the circle an arc equal in length to the radius.

(ix) **Steradian.** The steradian is the solid angle which having its vertex in centre of a sphere, cuts off an area of the surface of the sphere equal to that of a square with sides of length equal to the radius of the sphere.

Some Other Definitions

Newton. The newton (N) is a derived unit of force and is defined as the unit of force which when acting on a mass of 1 kilogram gives it an acceleration of one metre per second per second. Since acceleration due to gravity equals 9.81 m/s^2, kilogram force equals 9.81 newtons.

Joule. The joule (J) is a derived unit of energy, work or quantity of heat and is defined as the work done when a force of one newton acts so as to cause a displacement of one metre. Energy is defined as the capacity to do work. A unit of energy in nuclear physics is the electron volt (eV) which is defined as the energy gained by an electron in rising through a potential difference of one volt.

$$1 \text{ eV} = 1.6021 \times 10^{-19} \text{ J}.$$

Watt. The watt (W) is a unit of power (*i.e.*, rate of doing work)

$$\text{Power in watts} = \frac{\text{work (or energy) in joules}}{\text{time in seconds}}$$

Thus 1 watt equals 1 Joule/sec.

1 kilo watt-hour (kWh) = 1000 watt-hours = 3600000 joules.

Coulomb. The coulomb (C) is the derived unit of charge. It is defined as *the quantity of electricity passing a given point in a circuit when a current of 1 A is maintained for 1 second.*

$$Q = I.t$$

where Q = charge in coulombs,

   I = current in amperes, and

   t = time in second.

1 coulomb represents 6.24×10^{18} electrons.

Ohm. The ohm (Ω) is the unit of electric resistance and is defined as *the resistance in which a constant current of 1 A generates heat at the rate of 1 watt.*

Siemen. The siemen is a unit of electric conductance (*i.e.*, reciprocal of resistance). If a circuit has a resistance of 5 ohms, its conductance is 0.2 siemens. A more commonly used name for siemen is *mho* ($\mho$).

Volt. The volt is a unit of potential difference and electromotive force. It is defined as *the difference of potential across a resistance of 1 ohm carrying a current of 1 ampere.*

Hertz. The hertz (Hz) is a unit of frequency. 1 Hz = 1 cycle per second.

Horse-power. It is a practical unit of mechanical output. BHP (British horse power or brake horse power) equals 746 watts. The metric horse power equals 735.5 watts. To avoid confusion between BHP and metric horse power, the mechanical output of machines in SI units, is expressed in watts or kilowatts.

C. SALIENT FEATURES OF SI UNITS

The salient features of SI units are as follows :

1. It is a coherent system of units, *i.e.*, product of quotient of any two base quantities results in a unit resultant quantity. For example, unit length divided by unit time gives unit velocity.

2. It is a rationalised system of units, applicable to both, magnetism and electricity.

3. It is a non-gravitational system of units. It clearly distinguishes between the units of mass and weight (force) which are kilogram and newton respectively.

4. All the units of the system can be derived from the base and supplementary units.

5. The decimal relationship between units of same quantity makes possible to express any small or large quantity as a power of 10.

6. For any quantity, there is one and only one SI unit. For example, joule is the unit of energy of all forms such as mechanical, heat, chemical, electrical and nuclear. However, kWh will also continue to be used as unit of electrical energy.

Advantages of SI Units :

1. Units for many different quantities are related through a series of simple and basic relationship.

2. Being an absolute system, it avoids the use of factor 'g' i.e., acceleration due to gravity in several expressions in physics and engineering which had been a nuisance in all numericals in physics and engineering.

3. Being a rationalised system, it ensures all the advantages of rationalised MKSA system in the fields of electricity, magnetism, electrical engineering and electronics.

4. Joule is the only sole unit of energy of all forms and watt is the sole unit of power hence a lot of labour is saved in calculations.

5. It is a coherent system of units and involves only decimal co-efficients. Hence it is very convenient and quick system for calculations.

6. In electricity, all the practical units like volt, ohm, ampere, henry, farad, coulomb, joule and watt accepted in industry and laboratories all over the world for well over a century have become absolute in their own right in the SI system, without the need for any more practical units.

Disadvantages :

1. The non-SI time units 'minute' and 'hour' will still continue to be used until the clocks and watches are all changed to kilo seconds and mega seconds etc.

2. The base unit kilogram (kg) includes a prefix, which creates an ambiguity in the use of multipliers with gram.

3. SI units for energy, power and pressure (i.e., joule, watt and pascal) are too small to be expressed in science and technology, and, therefore, in such cases the use of larger units, such as MJ, kW, kPa, will have to be made.

4. There are difficulties with regard to developing new SI units for apparent and reactive energy while joule is the accepted unit for active energy in SI systems.

D. CONVERSION FACTORS

1. **Force :**

$$1 \text{ newton} = \text{kg-m/sec}^2 = 0.012 \text{ kgf}$$
$$1 \text{ kgf} = 9.81 \text{ N}$$

2. **Pressure :**

$$1 \text{ bar} = 750.06 \text{ mm Hg} = 0.9869 \text{ atm} = 10^5 \text{ N/m}^2$$
$$= 10^3 \text{ kg/m-sec}^2$$
$$1 \text{ N/m}^2 = 1 \text{ pascal} = 10^{-5} \text{ bar} = 10^{-2} \text{ kg/m-sec}^2$$
$$= 1.01325 \times 10^5 \text{ N/m}^2$$

3. **Work, Energy or Heat :**

$$1 \text{ joule} = 1 \text{ newton metre} = 1 \text{ watt-sec}$$
$$= 2.7778 \times 10^{-7} \text{ kW-h} = 0.239 \text{ cal}$$
$$= 0.239 \times 10^{-3} \text{ kcal}$$
$$1 \text{ cal} = 4.184 \text{ joule} = 1.1622 \times 10^{-6} \text{ kWh}$$
$$1 \text{ kcal} = 4.184 \times 10^3 \text{ joule} = 427 \text{ kgf m}$$
$$= 1.1622 \times 10^{-3} \text{ kWh}$$

$$1 \text{ kWh} = 8.6042 \times 10^5 \text{ cal} = 860.42 \text{ kcal}$$
$$= 3.6 \times 10^6 \text{ joule}$$
$$1 \text{ kgf.m} = \left(\frac{1}{427}\right) \text{kcal} = 9.81 \text{ joules}$$

4. Power :

$$1 \text{ watt} = 1 \text{ joule/sec} = 0.86 \text{ kcal/h}$$
$$1 \text{ h.p.} = 75 \text{ m kgf/sec} = 0.1757 \text{ kcal/sec} = 735.5 \text{ watt}$$
$$1 \text{ kW} = 1000 \text{ watts} = 860 \text{ kcal/h}$$

5. Specific heat :

$$1 \text{ kcal/kg-}°\text{K} = 0.4184 \text{ joules/kg-K}$$

6. Thermal conductivity :

$$1 \text{ watt/m-K} = 0.8598 \text{ kcal/h-m-}°\text{C}$$
$$1 \text{ kcal/h-m-}°\text{C} = 1.16123 \text{ watt/m-K} = 1.16123 \text{ joules/s-m-K}.$$

7. Heat transfer co-efficient :

$$1 \text{ watt/m}^2\text{-K} = 0.86 \text{ kcal/m}^2\text{-h-}°\text{C}$$
$$1 \text{ kcal/m}^2\text{-h-}°\text{C} = 1.163 \text{ watt/m}^2\text{-K}.$$

The following conversion factors may be used to convert the quantities in non-SI units into SI units.

To convert	To	Multiply by
angstroms	m	10^{-10}
atmospheres	kg/m^2	10332
bars	kg/m^2	1.02×10^4
Btu	joules	1054.8
Btu	kWh	2.928×10^{-4}
circular mils	m^2	5.067×10^{-10}
cubic feet	m^3	0.02831
dynes	newtons	10^{-5}
ergs	joules	10^{-7}
ergs	kWh	0.2778×10^{-13}
feet	m	0.3048
foot-pounds	joules	1.356
foot-pounds	kg-m	0.1383
gauss	tesla	10^{-4}
grams (force)	newton	9.807×10^{-3}
horse power (metric)	watts	735.5
lines/sq. inch	tesla	1.55×10^{-5}
Maxwell	webers	10^{-8}
mho	siemens	1
micron	metre	10^{-6}
miles	km	1.609
mils	cm	2.54×10^{-3}
poundals	newton	0.1383
pounds	kilogram	0.454

pounds (force)	newtons	0.448
pounds/sq. ft.	N/m²	47.878
pounds/sq. inch	N/m²	6894.43

E. IMPORTANT ENGINEERING CONSTANTS AND EXPRESSIONS IN S.I. UNITS

Engineering Constants and Expressions	M.K.S. System	SI Units
1. Value of g_0	9.81 kg-m/kgf-sec²	1 kg-m/N-sec²
2. Universal gas constant	848 kgf-m/kg mole-°K	848 × 9.81 = 8341 J/kg-mole-°C ($\because$ 1 kgf-m=9.81 joules)
3. Gas constant (R)	29.27 kgf-m/kg-°K	$\dfrac{8341}{29}$ = 287 joules/kg-K
4. Specific heat (for air)	for air c_v = 0.17 kcal/kg-°K c_v = 0.24 kcal/kg-°K	for air c_v = 0.17 × 4.184 = 0.71128 kJ/kg-K c_p = 0.24 × 4.184 = 1 kJ/kg-K
5. Flow through nozzle-Exist velocity (C_2)	$91.5\sqrt{U}$ where U is in kcal	$44.7\sqrt{U}$ U where U is in kJ
6. Refrigeration 1 ton	= 50 kcal/min	= 210 kJ/min
7. **Heat transfer** Stefan Boltzmann Law is given by :	$Q = \sigma T^4$ kcal/m²-h where $\sigma = 4.9 \times 10^{-8}$ kcal/h-m²-°K⁴	$Q = \sigma T^4$ watts/m²-h where $\sigma = 5.67 \times 10^{-8}$ W/m²K⁴

F. DIMENSIONS OF QUANTITIES

Different units can be represented dimensionally in terms of units of length L, mass M, time T and current I. The dimensions can be derived as under :

1. Velocity = length/time = $L/T = LT^{-1}$
2. Acceleration = velocity/time = $LT^{-1}/T = LT^{-2}$
3. Force = mass × acceleration = MLT^{-2}
4. Charge (coulomb) = current × time = IT
5. Work or energy = force × distance = ML^2T^{-2}
6. EMF or potential = work/charge
   = $ML^2T^{-2}/IT = ML^2I^{-1}T^{-3}$
7. Power = work/time = $ML^2T^{-2}/T = ML^2T^{-3}$
8. Current density = current/area = $I/L^2 = IL^{-2}$
9. Resistance = emf/current = $ML^2I^{-1}T^{-3}/I = ML^2I^{-2}T^{-3}$
10. Electric flux density = electric flux or charge/area = $IT/L^2 = ITL^{-2}$
11. MMF = current × number of turns = I
12. Conductance = I/resistance = $1/ML^2I^{-2}T^{-3} = I^2T^3M^{-1}L^{-2}$

13. Electric field intensity = volt/metre = $ML^2I^{-1}T^{-3}/L = MLI^{-1}T^{-3}$

14. Resistivity = $\dfrac{\text{resistant} \times \text{area}}{\text{length}}$

$$= (M^2I^{-2}T^{-3})(L^2)/L$$
$$= ML^3I^{-2}T^{-3}$$

15. Magnetic field intensity (H) = MMF/length
$$= I/L = IL^{-1}$$

16. Magnetic flux = emf × time = $(ML^2I^{-1}T^{-3})(T) = ML^2I^{-1}T^{-2}$

17. Magnetic flux intensity = magnetic flux/area
$$= (ML^2I^{-1}T^{-2})/L^2 = MI^{-1}T^{-2}$$

18. Impedence = emf/current = $ML^2I^{-2}T^{-3}$

19. Admittance = 1/impedence = $I^2T^3M^{-1}L^{-2}$

20. Inductance = magnetic flux/current
$$= ML^2T^{-2}I^{-1}/I = ML^2T^{-2}I^{-2}$$

21. Capacitance = electric charge/potential
$$= IT/ML^2T^{-3}I^{-1} = M^{-1}L^{-2}T^4I^2$$

CHAPTER 1

INTRODUCTION TO ENERGY SOURCES

1.1 Energy – Energy terms – Characteristics of energy – Energy and thermodynamics – Energy parameters – Energy planning – Energy audit – Electrical energy and power – Cogeneration; 1.2 Classification of energy; 1.3 Energy resources; 1.4 Non-conventional energy sources – Solar energy – Wind energy/power – Energy from biomass and biogas – Ocean energy – Wave energy – Tidal energy/ power – Geothermal energy – Hydrogen energy – Thermoelectric power – Fuel cell – MHD generator; 1.5 Renewable and non-renewable energy sources – Renewable energy sources – Non-renewable energy sources; 1.6 Alternative energy sources; 1.7 Energy scenario in Indian context; 1.8 Electricity generation from non-conventional energy sources; 1.9 Impact of energy sources on environment; 1.10 Fuels – Classification of fuels – Solid fuels – Liquid fuels – Gaseous fuels – Calorific or heating values of fuels. *Highlights – Theoretical Questions.*

1.1 ENERGY

1.1.1. Energy Terms

- **Energy:** *It is the capability to produce motion, force, work, change in shape, change in form etc.*

Energy exists in several forms such as:

— Chemical energy

— Nuclear energy

— Mechanical energy

— Electrical energy

— Internal energy

— Bio-energy in vegetables and animal bodies

— Thermal energy etc.

- **Energy science:** It focusses attention on the *'energy'* and *'energy transformations'* involved in the various other branches of science to National economy and civilization.

- **Energy technology:** *It is the applied part of energy sciences for work and processes, useful to human society, nations and individuals.*

— Energy technologies deal with plants and processes involved in the energy transformation and *analysis* of the *useful energy (exergy)* and *worthless energy (anergy).*

1

— Energy technology co-relates various sciences and technologies.

— Energy technology deals with the complete *energy route and its steps* such as: (*i*) Exploration of energy resources; discovery of new resources; (*ii*) Extraction or tapping of renewable or growing of bio-farms; (*iii*) processing; (*iv*) Intermediate storage; (*v*) Transportation/Transmission; (*vi*) Reprocessing; (*vii*) Intermediate stage; (*viii*) Distribution; (*ix*) Supply; (*x*) Utilisation, conservation, receiving.

1.1.2. Characteristics of Energy

Energy possesses the following *characteristics*:

1. It can be stored.

2. It can neither be created nor destroyed.

3. It is available is several forms.

4. It does not have absolute value.

5. It is associated with a potential. Free flow of energy takes place only from a higher potential to a lower potential.

6. It can be transported from one system to other system or from one place to another place.

7. The energy is measured in Nm or in joules.

The forms of energy are graded as per their availability or energy content.

● The total mass and energy in the closed system remains unchanged (as per law of conservation of energy).

1.1.3. Energy and Thermodynamics

"Thermodynamics" is a branch of energy which deals with conversion of heat into work or vice versa:

— More than 30 per cent energy conversion processes involve *thermodynamics,* while more than 30 per cent energy conversion processes involve *electromagnetic energy* and more than 30 per cent involve *chemical energy.*

In most of the energy conversion processes, First law and Second law of thermodynamics are applicable:

● *First law* of thermodynamics relates to *conservation of energy* and throws light on concept of internal energy.

● *Second law* of thermodynamics indicates the limit of *converting heat into work* and introduces the principle of increase of entropy. Following statements are based on this law:

— Spontaneous processes are irreversible.

— The internal energy of the environment is worthless for obtaining useful work.

— All forms of energy are not identical with reference to useful work.

— Every energy conversion process has certain *'losses'*.

1.1.4. Energy Parameters

In order to conserve fuel, it is imperative to adopt measures for maximising economic development with minimum energy consumption.

1. *Energy intensity:*

The term *'energy intensity'* is defined as *energy consumption per unit of Gross National Product (GNP).*

When the per unit energy consumption for the production of energy intensive raw materials, like steel and aluminium, is reduced, there may be a marginal fall in energy (GNP ratio) with continuation of the downward trend.

Developed countries have reduced *'energy intensity'*, resulting in less energy consumption and at the same time achieving higher production.

2. *Energy-GDP elasticity:*

It is defined as the *percentage growth in energy requirement for 1% growth in GDP.*

The *lower* the value of elasticity, the *higher* is the overall efficiency.

- The value of elasticity for the developed countries ranges from 0.8 to 1.0 whereas for India it is about 1.2.

1.1.5. Energy Planning

It is an essential management *tool that decides various activities in advance with reference to resources and time frame.* This includes forecasts, budget, infrastructure, technology, planning etc. The energy policies are framed for the purpose of energy planning and to be followed by the higher to lower hierarchy level.

Energy planning includes the following **steps:**

(*i*) To collect data.

(*ii*) To evaluate trends.

(*iii*) To determine demand.

(*iv*) To determine availability of resources.

(*v*) To plan entire energy route for each sector

   — Exploration/Extraction/Conversion

   — Processing/By product/Cleaning

   — Storage/Transport or Transmission

   — Distribution/Supply.

(*vi*) To evaluate economic viability and decide tariff/rates.

(*vii*) To formulate short-term/mid-term/long-term plans.

1.1.6. Energy Audit

"Energy audit" is *an official survey/study of energy consumption/processing/supplying aspects related with an organisation, system, process, plant, equipment.*

The *objectives* of the 'Energy Audit' are to recommend "steps" to be taken by the management for:

(*i*) Improving the energy efficiencies,

(*ii*) Reducing the energy costs, and

(*iii*) Improving the productivity without sacrificing quality, standard of living/comforts and environmental balance.

The Energy Audit is officially recommended by the *Management* and is carried out by the Energy Audit Group headed by the Energy Auditor.

Energy audit is usually carried out in following *three stages* within certain agreed time frame:

1. Simple Walk – through energy audit.

2. Intermediate Energy audit.

3. Comprehensive/Exhaustive energy audit.

The *procedure of 'Energy Auditing'* is dictated by the *size, complexity* and *recurring energy costs of the plant.*

For energy intensive processes/plants, thorough comprehensive energy audit and high investments in Energy Conservation Measures are justified.

1.1.7. Electrical Energy and Power

1.1.7.1. Electrical Energy

It is an essential gradient for the industrial and all-round development of any country. It is preferred due to the following *advantages:*

(*i*) Can be generated centrally in bulk.

(*ii*) Can be easily and economically transported from one place to another over long distances.

(*iii*) Losses in transport are minimum.

(*iv*) Can be easily sub-divided.

(*v*) Can be adapted easily and efficiently to domestic and mechanical work.

Electrical energy is obtained, conventionally, by *conversion from fossil fuels (coal, oil, natural gas), the nuclear and hydro sources.* Heat energy released by burning fossil fuels or by fusion of nuclear material is converted to electricity by first converting heat energy to the mechanical form through a thermocycle and then converting mechanical energy through generators to the electrical form. *Thermo-cycle is basically a low efficiency process— highest efficiencies for modern large size plants range up to 40%,* while smaller plants may have considerably lower efficiencies. The *earth has fixed non-replenishable resources of fossil fuels and nuclear materials. Hydro-energy,* though replenishable, is also *limited in terms of power.*

In view of the ever increasing per capita energy consumption and exponentially rising population, the earth's non-replenishable fuel resources are not likely to last for a long time. Thus a coordinated world-wide action plan is, therefore, necessary to ensure that energy supply to humanity at large is assured for a long time and at low economic cost. The following *factors* need to be considered and actions to be taken accordingly: (*i*) *Energy consumption curtailment;* (*ii*) *To initiate concerted efforts to develop alternative sources of energy including unconventional sources like solar, tidal, geothermal energy etc. ;* (*iii*) *Recycling of nuclear wastes;* (*iv*) *Development and application of anti-pollution technologies.*

Decentralised and Dispersed generation:

☞ • **Decentralised generation:** It covers a local energy source to generate electric power for distribution to consumers in particular area. These may be *mini/ microlevel hydel* or *wind turbine units.*

"*Keylong*", the district headquarters of Lahaul (H.P.) was electrified in 1964 by (2 × 50) kW hydel units.

"Şunderbans" in West Bengal was not accessible to grid power but was electrified during 1997 by solar power 410 kW by SPV modules, and biomass-based power plant of (5 × 100) kW.

☞ • **Dispersed generation:** It refers to the use of generating units of *less than 25 kW output* to serve individual .homes, business and defence installation in remote areas.

Examples: Diesel generators, solar PV installations, kiosk type mini hydro-plants, fuel cells and wind generators etc.

1.1.7.2. Power

Any physical unit of energy when divided by a unit of time automatically becomes a *unit of power.* However, it is in connection with the mechanical and electrical forms of energy that the term *"power"* is generally used. The rate of production or consumption of heat energy and, to a certain extent, of radiation energy is not ordinarily thought of as power. *Power is primarily associated with mechanical work and electrical energy.* Therefore, **power** can be defined as the *rate of flow of energy* and can state that a *power plant is a unit built for production and delivery of a flow of mechanical and electrical energy.*

In common usage, a machine or assemblage of equipment that produces and delivers a flow of mechanical or electrical energy is a **power plant.** Hence, an internal combustion engine is a power plant, a water wheel is a power plant, etc. However, what we generally mean by the term is that assemblage of equipment, permanently located on some chosen site which receives raw energy in the form of a substance capable of being operated on in such a way as to produce electrical energy for delivery from the power plant.

1.1.8. Cogeneration

In a **cogeneration system**, *mechanical work is converted into electrical energy in an "electrical generator",* and the *discharge heat,* which would otherwise be dispersed to the environment, is *utilised in an "industrial process"* or in other ways. The *net result is an overall increase in the efficiency of fuel utilisation.*

Cogeneration is *the simultaneous generation of electricity and steam (or heat) in a single power plant.*

It is *highly energy efficient* and is especially suitable for sugar mills, textile, paper, fertilizer and crude oil refining industries.

Cogeneration is *advisable* for industries and municipalities if they can produce electricity cheaper, or more conveniently than brought from a utility. It is *not* usually used by large utilities which tend to produce *electricity only.*

Cogeneration (from energy resource point of view) is *beneficial only* if it *saves primary energy* when compared with separate generation of electricity and steam or heat.

Cogeneration of heat and electricity can be dealt with in the following *two ways:*

1. Topping cycle. In this mode, fuel is burnt to generate electric power and the *discharged heat from the turbine is supplied as 'process heat'.* The requirements of process steam pressure vary widely between 0.5 bar and 40 bar.

• This cycle can *provide true savings in primary energy.*

2. Bottoming cycle. In this mode, *fuel is consumed to process heat,* and *waste heat is then utilised for power generation.*

1.2. CLASSIFICATION OF ENERGY

Energy may be *classified* as follows:

Energy:

1. *Stored in earth:*

   (*i*) Chemically bonded:

      (*a*) Oil

      (*b*) Gas

      (*c*) Coal

   (*ii*) Geotherm

   (*iii*) Atomic:

      (*a*) Fission

      (*b*) Fusion (futuristic)

2. *Continually received by earth:*

   ● Solar insolation

      (*a*) Ocean temp. difference

      (*b*) Tidal

      (*c*) Hydro

         — Irrigation

         — Other benefits (flood control etc.)

         — Hydroelectric (no thermal limits)

      (*d*) Wind

         — Wind mill generator

      (*e*) Direct

         — PVC

         — Concentrator—Steam turbine

Water input

↓

Thermocycle

—Steam turbo-generator

—Gas turbo-generator-combined cycle

—M.H.D. combined cycle

Conservation

— Cogeneration

— High thermal generation

— Low trans. loss

— High efficiency motors

— Curb wasteful use

1.3 ENERGY RESOURCES

The various sources of energy can be *classified* as follows:

A. 1. **Commercial (or Conventional) energy sources:**

   (*i*) Coal

   (*ii*) Lignite

   (*iii*) Oil and natural gas

   (*iv*) Hydroelectric

   (*v*) Nuclear fuels.

These sources form the basis of industrial, agricultural transport and commercial development in the modern world. In the industrialised countries, commercialised fuels are predominant source not only for economic production, but also for many household tasks of general population.

2. **Renewable energy sources:**
   (*i*) Solar photo-voltaic
   (*ii*) Wind
   (*iii*) Hydrogen fuel-cell.
3. **New sources of energy:**

Most prominent *new sources of energy* as identified by UN are:
   (*i*) Tidal energy
   (*ii*) Ocean waves
   (*iii*) OTEC (Ocean Thermal Energy Conversion)
   (*iv*) Geothermal energy
   (*v*) Peat
   (*vi*) Tar sand
   (*vii*) Oil shales
   (*viii*) Coal tar
   (*ix*) Draught animals
   (*x*) Agricultural residues etc.

- *Coal, oil, gas, uranium and hydro* are commonly known as **"commercial" or "conventional energy sources"**. These represent about 92% of the total energy used in the world.
- *Firewood, animal dung and agricultural waste* etc. are called as **non-commercial energy sources.** These represent about 8% of the total energy used in the world.

As per Planning Commission of India, the geographical distribution of various energy resources available in the country are given in Table 1.1

Table 1.1. Primary Commercial Energy Resources

Region of India	Coal (*Bt*)	Lignite (*Bt*)	Crude oil (*Mt*)	Natural gas (*BCM*)	Hydropower (*TWH*)
Northern	1.06	2.51	0.03	0.00	225.00
Western	56.90	1.87	519.47	516.42	31.40
Southern	15.46	30.38	45.84	80.94	61.80
Eastern	146.67	0.00	2.19	0.29	42.50
North-Eastern	0.89	0.00	166.17	152.00	239.30
Total	220.98	34.76	733.70	749.65	600.00

Bt = Billion tonnes; *Mt* = Million tonnes; *BCM* = Billion cubic metres; *TWH* = Trillion Watt hours.

The production of commercial primary energy resources is shown in Table 1.2.

Table 1.2. Production of Commercial Energy Sources

Source of energy	Unit	Production Periods					
		1960-61	1970-71	1980-81	1990-91	2001-02	2006-07
Coal	Mt	55.67	72.95	114.01	211.73	325.65	405.00
Lignite	Mt	0.05	3.39	4.80	14.07	24.30	55.96
Crude oil	Mt	0.45	6.82	10.51	33.02	32.03	33.97
Natural gas	BCM	—	1.44	2.35	1.79	29.69	37.62
Hydro Power	BkWh	7.84	25.25	46.54	71.66	82.80	103.49
Nuclear Power	BkWh	—	2.42	3.00	6.14	16.92	19.30

B. The energy sources *can also be classified* as follows:

1. **Primary energy sources.** These sources are obtained from environment.

   Examples: Fossil fuels, Solar energy, Hydro energy and Tidal energy.

   These resources can further be classified as:

   (*a*) (*i*) **Conventional energy sources:**

   Examples: Thermal power and hydel power.

   (*ii*) **Non-conventional energy sources:**

   Examples: Wind energy, Geothermal energy, Solar energy and Tidal energy.

   (*b*) (*i*) **Renewable:** These sources are being *continuously produced in nature and are inexhaustible.*

   Examples: Wood, Wind energy, Biomass, Biogas, Solar energy etc.

   (*ii*) **Non-renewable:** These are *finite and exhaustible.*

   Examples: Coal, petroleum etc.

2. **Secondary energy resources.** These resources do not occur in nature but are *derived from primary energy resources.*

   Examples: Electrical energy from coal burning, H_2 obtained from hydrolysis of H_2O.

Fossile fuel as a conventional energy source:

Some of the fossil fuels are discussed briefly below:

1. **Coal.** It is a conventional energy source and is formed due to conversion of vegetable matter. It is composed of mainly *carbon* and *hydrocarbons*. It is found in Jharkhand, UP, MP, Bihar etc. in India.

Use of coal:

1. It is used to generate electricity. Power plants use coal for heating the water to generate steam which runs the turbines to generate electricity.

2. It is heated in a furnace to make coke, which is used to smelt iron for making steel.

3. The heat obtained from coal is used by various industries in making plastics, tar, synthetic fibre, etc.

Environmental problems:

(*i*) Due to combustion of coal, CO_2 is produced which is responsible for causing *global warming.*

(*ii*) Coal also produces SO_2 which is a cause for *acid rain.*

2. **Natural gas.** It is one of the fossil fuels and is formed by decomposition of remains of dead animals and plants buried under the earth. It is mainly composed of methane (CH_4) with small amount of propane and ethane. When refined, it is colourless and odorless, but can be *burned to release large amount of energy.*

- *It is the cleanest fossil fuel.*

Merits

1. It has a high calorific value and it burns without smoke.

2. It can be easily transported through pipelines.

Uses

1. It is used in thermal power plants for generating electricity.

2. It is used as domestic and industrial fuel.

Reserves and production of "petroleum" and "natural gas" in India with problem areas:

Almost 40 per cent of the energy needs of the world, are met by oil. The rising prices of oil has brought a considerable strain to the economy of world, more so in the case of the developing countries that do not possess oil resources enough for their own consumption.

With today's consumption and a resource amount of 2.5×10^5 million tonnes of oil, it is estimated that it may suffice for about 100 years unless more oil is discovered. As such, the world must start thinking of a change from a world economy dominated by oil.

Petroleum. India is not particularly rich in petroleum reserves. Our fuel oils are produced by *refining petroleum or crude oil*. The potential oil-bearing areas are located in Assam, Tripura, Manipur, West Bengal, Ganga Valley, Punjab, Himachal Pradesh, Kutch, eastern and western coastal areas (in Tamil Nadu, Andhra Pradesh and Kerala), Andaman and Nicobar Islands, Lakshadweep, and in the continental shelves adjoining these areas.

Gas. Gas is *incompletely* utilised at present and huge quantities are burnt off in the oil production process because of the non-availability of ready market. The reason may be the high transportation cost of the gas. Transporting gas is *costlier* than transporting oil. Large reserves are estimated to be located in inaccessible areas.

Gaseous fuels can be *classified* as: (*i*) Gases of *fixed composition* such as acetylene, ethylene, methane etc; (*ii*) Composite industrial gases such as producer gas, coke oven gas, blast furnace gas etc.

Note. Energy cannot be economically stored in electrical form in large quantities. Energy in large quantities is stored in conventional forms (Hydro-reservoirs, coal stocks, fuel stocks, nuclear fuel stocks). Electrical energy is generated, transmitted and utilised almost simultaneously without intermediate storage in electrical form. Hence a large *electrical network* is formed *to pool up electrical energy* available from various generating stations and to distribute to various consumers over the large geographical area. Consumers draw power as per their load requirement (*e.g.* lighting, heating, mechanical drives etc.)

1.4 NON-CONVENTIONAL ENERGY SOURCES

A plenty of energy is needed to sustain industrial growth and agricultural production. The existing sources of energy such as coal, oil, uranium etc. may not be adequate to meet the ever increasing energy demands. These conventional sources of energy are also depleting and may be exhausted at the end of the century or beginning of the next century. Consequently sincere and untiring efforts shall have to be made by the scientists and engineers in exploring the possibilities of harnessing energy from several non-conventional energy sources. The various non-conventional energy sources are as follows:

(*i*) Solar energy

(*ii*) Wind energy

(*iii*) Energy from biomass and biogas

(*iv*) Ocean thermal energy conversion

(*v*) Tidal energy

(*vi*) Geothermal energy

(*vii*) Hydrogen energy

(*viii*) Fuel cells

(*ix*) Magneto-hydro-dynamic generator

(*x*) Thermionic converter

(*xi*) Thermo-electric power.

Advantages of non-conventional energy sources:

The leading advantages of non-conventional energy sources are:

1. They do not pollute the atmosphere.
2. They are available in large quantities.
3. They are well suited for decentralised use.

According to energy experts the non-conventional energy sources can be used with advantage for *power generation* as well as other applications in a large number of locations and situations in our country.

- The non-conventional energy programme was initiated in *India* in *1983-84*, managed and implemented by MNES (Ministry of Non-conventional Energy Sources), Govt. of India.

The estimated potential of non-conventional energy resources in India is as given below:

Category	Estimated potential (MW)
1. Wind power	45,195
2. Biomass power	16,881
3. Small hydro	15,000
4. Cogeneration Bagasse	5,000
5. Waste energy	2,700
6. Solar power (Grid)	2,533

Brief description of important non-conventional energy sources:

1.4.1. Solar Energy

On this planet, human life and all other forms of life are completely dependent on the daily flow of solar energy. The production of food and all other life-support systems of the natural environment are dependent on the sun.

- **Solar energy** travels in small particles called *photons*. Converting even a part of the solar energy at even a very low efficiency can result in a *far more* energy that could conceivably be harnessed or utilised for power generation.

- The amount of solar energy is expressed in *"solar constant"*. The solar constant is the total energy that falls on a unit area exposed normally to the rays of the sun, at the average sun-earth distance.

The most accepted value of solar constant is *1.353 kW/m²*. A number of scattering and absorption processes in the atmosphere *reduce* the maximum heat flux reaching the earth's surface to around *1 kW/m²*.

The heat flux reaches earth's surface by two modes: (*i*) Direct (*ii*) Diffuse. It is the *only direct heat energy* which can be collected through a *"collector"*. The *ratio* of direct to totally heat energy varies from place to place and depends on atmospheric conditions like dust, smoke, water vapour and other suspended matter. The *ratio varies between 0.64* and *0.88* according to different investigators.

Since the altitude of the sun and length of day vary with the season, the solar energy received on a summer day is *many times* the energy received on a winter day. As a result the *total energy for most of the areas in plains in India is around 6000 MJ/m² per year*.

Advantages:

1. It is a renewable source of energy.

2. Free of cost.

3. Non-polluting source of energy.

Disadvantages:

1. *Low efficiency.*

2. It is of intermittent type in nature, so for night hours this energy is not available, and as such, *storage is required.*

☞ **Impact on environment:**

1. Solar thermal system may pose a *health hazard* because of the careless disposal of the heat transfer fluids (*e.g.* glycol nitrates and sulphates; CFCs and aromatic alcohols) used.

2. Solar photovoltaic modules pose disposal problems owing to the presence of arsenic and cadmium.

3. The total system comprising solar power generator with accessories contain several pollutants.

4. Solar reflectors cause hazard to eyesight.

1.4.2. Wind Energy

Man has been served by the power from winds for many centuries but the total amount of energy generated in this manner is *small.* The expense of installation and variability of operation have tended to *limit* the use of the windmill to intermittent services where *its variable output has no serious disadvantage.* The principal services of this nature are the *pumping of water into storage tanks and the charging of storage batteries.*

- Windmill power equipment may be *classified* as follows:

    1. *The multi-bladed turbine wheel.* This is the foremost type in use and its efficiency is about 10 per cent of the kinetic energy of the wind passing through it.

    2. *The high-speed propeller type.*

    3. *The rotor.*

- The propeller and rotor types are *suitable for the generation of electrical energy,* as both of them possess the ability to start in very low winds. The *Propeller type is more likely to be used in small units* such as the driving of small battery charging generators, whereas the *rotor,* which is rarely, seen, is more practical *for large installations,* even of several hundred kilowatts capacity.

- In India, the wind velocity along coastline has a range 10-16 kmph and a survey of wind power has revealed that wind power is capable of exploitation for pumping water from deep wells or for generating small amounts of electric energy.

Modern windmills are capable of working on velocities as low as 3-7 kmph while *maximum efficiency is attained at 10-12 kmph.*

- A normal working life of 20 to 25 years is estimated for windmills.

- The great advantage of this source of energy is that *no operator is needed and no maintenance and repairs are necessary for long intervals.*

Merits/Characteristics of wind power/energy. Some characteristics of wind energy are given below:

1. No fuel provision and transport are required in wind energy systems.

2. It is a renewable source of energy.

3. Wind power systems are non-polluting.

4. Wind power systems, up to a few kW, are less costly, but on a large scale, costs can be competitive with conventional electricity. Lower costs can be achieved by mass production.

Demerits/Problems associated with wind energy:

1. Wind energy systems are *noisy in operation.*
2. *Large areas* are needed to install *wind farms* for electrical power generators.
3. Wind energy available is *dilute and fluctuating in nature.* Because of dilute form, conversion machines have to be necessarily large.
4. Wind energy *needs storage means* because of its irregularity.

☞ **Impact on environment:**

1. The development of wind farm in a forest area needs *cutting of trees* leading to *environmental degradation.*
2. The environment is degraded due to noise pollution caused by wind turbines.
3. Interference of large wind turbines with television signals (through reflection).
4. Visual intrusion of wind turbines gives negative public response on the existing landscape.

1.4.3. Energy from Biomass and Biogas

Biomass. *Green plants trap solar energy through the process of "photosynthesis" and convert it into organic matter, known as **biomass**.*

Wood, charcoal, agricultural waste produce the bioenergy after burning; cowdung, garbage are aerobically decomposed to obtain the energy.

Dried animal dung or cattle dung cakes are used directly as fuels in rural areas but it produces smoke and has low efficiency of burning.

Biogas: Biogas is formed due to the *decomposition of organic waste matter.* During decomposition of organic matter, the gases such as carbon dioxide, hydrogen and hydrogen sulphide are formed.

The organic waste is generally animal dung, plant waste etc. These waste products contain carbohydrates, proteins, which are broken down by bacteria in *absence* of oxygen caerobic conditions.

Advantages:

1. Continuous supply of energy.
2. Renewable in nature.
3. Cheap in cost.

Disadvantages:

1. Power generating units are huge and bulky.
2. Biogas generation depends on temperature, therefore, in water or cold areas like J & K additional source of energy is required.

☞ **Impact on environment:**

1. Domestic use of biomass in rural areas creates *air pollution.*
2. A large scale energy-crop plantation is *water consuming with increased use of pesticides and fertilizers, causing water pollution and flooding.*
3. The production of biomass on large scale and its harvesting accelerates *soil erosion and loss of nutrients.*

1.4.4. Ocean Energy

India is having large potential of ocean thermal energy which could be of the order of about 50,000 MW.

Ocean Thermal Energy Conversion (OTEC) plants convert the heat in the ocean into *electrical energy* with the help of *temperature difference*. The large temperature difference between *warm* surface sea water (28-30°C) and *cold* deep sea water (5-12°C) is used to generate electricity with the help of ocean thermal energy conversion system.

☞ **Impact on environment:**
1. OTEC plant creates *adverse impacts on marine environment* since the massive flow of water disturbs thermal balance, changes salinity gradient and turbidity.
2. The leakage of ammonia, used as a working fluid in closed cycle OTEC system, may cause much *damage to the ocean ecosystem*.

1.4.5. Wave Energy

The ocean waves are caused by wind, which in turn is caused by uneven heating and subsequent cooling of earth's crust and rotation of the earth.

The most of the sea surface in the form of wind waves forms a source of energy. Floating propellers are placed in shallow waters, near the shores, and due to motion of the waves the propellers also get the motion and this kinetic energy can be used to drive turbines.

The harnessing of wave energy requires the development of special power conversion devices.

Advantages:
1. The wave energy is a *cheap and inexhaustible* source of energy.
2. Wave-power devices, unlike solar or wind devices, *do not use up large land masses.*
3. It is *pollution-free.*
4. A staggered array of power devices can *produce* electricity, *protect* coastlines from the destructive action of waves, *minimise* erosion and even *help create* artificial harbours.

Limitations:
1. Wave lacks dependability.
2. There is a scarcity of accessible sites of large wave activity.
3. Economic factors like capital investment, cost of maintenance, repair and replacement hinder the development.

1.4.6. Tidal Energy/Power

The rise and fall of tides offers a means for storing water at the rise and discharging the same at fall. Of course the head of water available under such cases is very low but with increased catchment area considerable amounts of power can be generated at a negligible cost.

● The use of tides for electric power generation is practical in a few favourably situated sites where the geography of an inlet of bay favours the construction of a large scale hydroelectric plant. To harness the tides, a dam would be built across the mouth of the bay in which large gates and low head hydraulic turbines would be installed. At the time of high tide the gates are opened and after storing water in the tidal basin the gates are closed. After the tide has receded, there is a working hydraulic head between the basin water and open sea/ocean and the water is allowed to flow back to the sea through water turbines installed in the dam. With this type of arrangement, the generation of electric power is *not continuous*. However by using reversible water turbine the turbine can be run continuously as shown in Fig. 1.1.

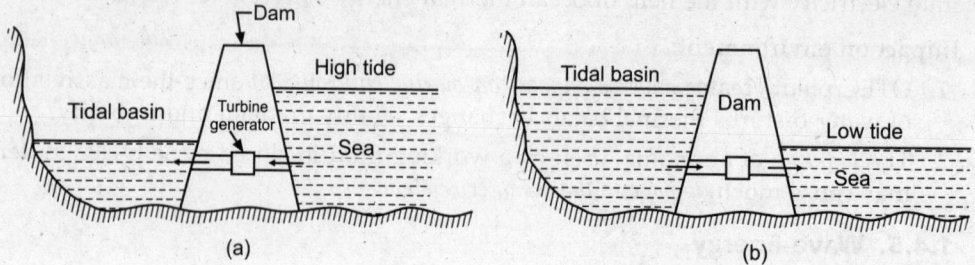

Fig. 1.1. Generation of power by tides.

1.4.7. Geothermal Energy

In many places on the earth *natural steam escapes from surface vents*. Such natural steam wells suggest the possibility of tapping terrestrial heat (or geothermal energy) in this form and using it for the development of power. Unfortunately, the locations where the steam-producing sub-strata seem to be fairly close to the surface are far removed from centres of civilization where the power could be usefully employed. Nevertheless, there are probably many places where, although no natural steam vent or hot springs are showing, deep drillings might tap a source of underground steam. The cost of such explorations and the great likelihood of an unsuccessful conclusion are not very conductive to exploitation of this source of energy.

There are two ways of *electric power production from geothermal energy:*

(*i*) Heat energy is transferred to a working fluid which operates the power cycle. This may be particularly useful at places of fresh volcanic activity where the molten interior mass of earth vents to the surface through fissures and substantially high temperatures, such as between 450 to 550°C can be found. By embedding coil of pipes and sending water through them steam can be raised.

(*ii*) The hot geothermal water and/or steam is used to operate the turbines directly. From the well-head the steam is transmitted by pipelines up to 1 m in diameter over distances up to about 3 km to the power station. Water separators are usually employed to separate moisture and solid particles from steam.

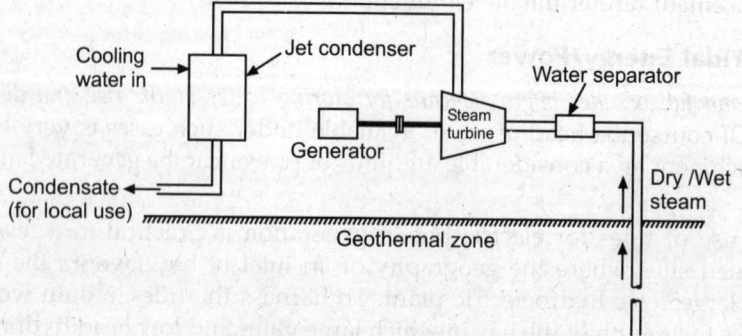

Fig. 1.2. Geothermal power plant.

Presently, only steam coming out of the ground is used to generate electricity, the hot water is discarded, because it contains as much as 30% dissolved salts and minerals, and these cause serious rust damage to the turbine. The water, however, contains more than 1/3rd of the available thermal energy.

Advantages:

1. It is almost free from the pollution.
2. It is a cheap and clean source of energy.

Disadvantages:

1. The drilling operations result in the noise pollution.
2. Air pollution results in case of release of gases like H_2S, NH_3 present in the steam waste.

☞ **Impact on environment:**

1. Gases escape into the atmosphere and *drop down as acid rain*.
2. The soil and water are *polluted by the chemicals like sulphates, chlorides and carbonates of lead, arsenic*.
3. Owing to the discharge of waste hot water, rivers are infected and consequently drinking water, farming and fisheries are adversely affected.
4. The exhausts, blow downs and centrifugal separation cause *noise pollution*.

1.4.8. Hydrogen Energy

Hydrogen energy is a non-conventional energy source. Hydrogen is considered as an *alternative future source of energy*. It has a tremendous potential because it can be produced from water which is available in abundance in nature. Hydrogen atoms in the core of sun combine to form helium atoms which is called as *fusion reaction*. It gives radiant energy which sustains the life on the earth.

Hydrogen can be separated from water by means of electrical energy. It can also be obtained from fossil fuels.

Advantages:

1. Its burning is non-polluting.
2. Hydrogen energy has a very high energy content.

Applications:

1. It is used for generating electricity for domestic appliances
2. It is utilised in automobiles.
3. It is employed for industrial uses

1.4.9. Thermo-electric Power

According to *Seebeck effect*, when the two ends of a loop of two dissimilar metals are held at different temperatures, an *electromaotive force is developed and the current flows in loop*. This method, by selection of suitable materials, can also be used for *power generation*. This method involves *low initial cost and negligible operating cost*.

1.4.10. Fuel Cell

When an *electric current is passed through a dilute solution of an acid or an alkali* by means of two platinum electrodes, *hydrogen* is produced at the *cathode* and oxygen is evolved at *aviode*. If this process is *reversed* by removing the power supply and connecting the two electrodes through a suitable resistance, the presence of hydrogen at one electrode and oxygen at the other *will produce a small current in the external circuit,* water being produced as a by-product. This *reverse process of electrolysis is the essence of the "fuel cell technology" as, the chemical energy stored in hydrogen and oxygen have been combined to produce electricity.*

- A fuel cell does not have moving components and as such, it is *quieter* and *requires less maintenance and attention* in operation.
- Fuel cells *convert* chemical energy directly to electrical energy at room temperatures.
- These cells are *very efficient* and are *not subject to Carnot limitation.*

1.4.11. Magneto-Hydro-Dynamic (MHD) Generator

The MHD working principle is based on *Faraday's law of electromagnetic induction which states that change in magnetic field induces an electric field in any conductor located in the magnetic field.* This electric field while acting on the free charges in the conductor causes a current to flow in the conductor. In *"MHD generator",* an ionised gas is used as a conductor. *If such ionised gas is passed at a high velocity through a powerful magnetic field, then current is generated and can be extracted by placing electrodes in a suitable position in the stream.* It produces *D.C. power directly.*

In MHD generation, all kinds of heat sources like coal, gas, oil, solar etc. can be used. MHD systems are of two types: (*i*) Open cycle system; and (*ii*) Close cycle system.

1.5 RENEWABLE AND NON-RENEWABLE ENERGY SOURCES

1.5.1. Renewable (Non-Conventional) Energy Sources

Renewable energy sources include *both 'direct'* solar radiation intercepted by collectors (*e.g.* solar and flat-plate thermal cells) and *'indirect'* solar energy such as wind, *hydropower, ocean energy* and *biomass resources* that can be managed in a sustainable manner. *Geothermal* is considered renewable because the *resource is unlimited.*

Advantages:

The *advantages* of renewable energy sources are:

1. These energy sources recur in nature and are *inexhaustible.*
2. The power plants using renewable sources of energy *do not have any fuel cost* and hence their *running cost is negligible.*
3. As renewables have low energy density, there is *more or less no pollution or ecological balance problem.*
4. These energy sources can help to *save foreign exchange* and *generate local employment* (since most of the devices and plants used with these sources of energy are simple in design and construction, having been made from local materials, local skills and by local people).
5. These are *more site specific* and are employed for local processing and application, their economic and technological losses of transmission and distribution being *nil.*
6. Since conversion technology tends to be flexible and modular, renewable energy can usually be *rapidly deployed.*

Demerits/Limitations:

1. Owing to the low energy density of renewable energy sources large size plants are required, and as such the *cost of delivered energy is increased.*

2. These energy sources are *intermittent and also lack dependability*.

3. The user of these sources of energy has to *make huge additional investment* before deriving any benefit from it (whereas in case of conventional energy sources, the processing cost has traditionally been borne by large industries which borrow money from a bank and then charge the customer for each unit of energy used).

4. These energy sources, due to their low energy density, have *low operating temperatures leading to "low efficiencies"*.

5. Since the renewable energy plants have low operational efficiency, the heat rejections are large which *cause thermal pollution*.

6. These energy sources are *energy-intensive*.

● If *broadly interpreted*, the *definition of* renewable resources also *includes the chemical energy stored in food and nonfuel plant products and even the energy in air used to dry materials or to cool, heat and ventilate the interiors of buildings*.

From an *"operational point of view"*, the correct way to treat renewable energy is *as a means to reduce the demand for conventional energy forms*. Thus, in performing economic and financial analyses, there is *no real distinction* between renewable energy technologies and those designed to improve the efficiency of conventional energy use.

Another point is that *cost-effective approaches to energy efficiency* – ranging from no-or-low cost measures (*e.g.* reducing excess air in boilers, shutting down equipment when not needed) to systems requiring moderate capital investment, such as heat recuperators, boiler replacements or cogeneration units – can improve the financial and economic feasibility of renewable as well as conventional energy systems. *Improvements in the efficiency of energy use* can be teamed with a variety of energy supply technologies, and this fact must be recognised when assessing the relative economics of renewable and conventional energy systems.

Usefulness of renewable energy resources:

Any analysis of the usefulness of renewable resources in developing countries must consider the following *basic facts*:

1. The renewable energy resources and conversion systems are *technically capable* of meeting many of the power and fuel needs of a modern technological civilisation, from small-scale, decentralised uses to large-scale urban and industrial concentration.

2. Although renewable technologies are *economically competitive with fossil fuels in their ability to provide electricity, mechanical power, thermal energy and liquid fuels*, such technologies have not yet been deployed internationally, and the primary obstacles to their further development are institutional.

3. *Worthwhile and widespread deployment* of solar energy systems for the production of electric power, thermal energy and liquid fuels will require *advanced materials and concepts to be competitive with conventional options*.

In fact, the systems for conversion and storage for many of the renewable sources will only be commercially feasible after *sophisticated research and development* – surprisingly analogous to that now taking place in the computer and information fields.

Barriers in the implementation of renewable energy systems:

There are a number of obstacles to the effective deployment and widespread diffusion of renewable energy systems; among these are:

1. *Inadequate documentation and evaluation of past experience*, a *paucity* of validated field performance data and a *lack* of clear priorities for future work.

2. *Weak or non-existent institutions and policies* to finance and commercialise renewable energy systems. With regard to *energy planning*, separate and completely uncoordinated organisations are often responsible for petroleum, electricity, coal, foresty, fuelwood, renewable resources and conservation.

3. *Technical and economic uncertainties* in several renewable energy systems; *high economic and financial costs* for some systems in comparison with conventional supply options and energy-efficient measures.

4. *Skeptical attitudes* towards renewable energy systems on the part of energy planners and a lack of qualified personnel to design, manufacture, market, operate and maintain such systems.

5. *Inadequate donor coordination* in renewable energy assistance activities, with little or no information exchange on successful and unsuccessful projects.

The following points may be mentioned in this connection:

1. The energy demand is increasing by leaps and bounds due to rapid industrialisation and population growth, and hence the conventional source of energy will not be sufficient to meet the growing demand.

2. Conventional sources *except hydro* are non-renewable and are bound to finish up one day.

3. Conventional sources (fossil fuels, nuclear) also cause pollution, thereby their use degrades the environment.

4. Large hydro resources affect wild life, cause deforestation and pose various social problems.

Owing to *these reasons* it has become important *to explore and develop non-conventional energy sources to reduce too much dependence on conventional resources.*

☞ *The "renewable energy technologies" are better than most conventional energy technologies in the following ways:*

1. Can be produced in *large numbers* and introduced quickly.

2. Can often be built on, or close to the site where the energy is required, this minimises transmission costs.

3. Can be matched in scale to the need, and can deliver energy of the quality that is required for a specific task thus reducing the need to use premium fuels or electricity to provide low grade forms of energy such as hot water.

4. Although there are physical and environmental risks associated with the construction and operation of renewable energy technologies, as there are with all energy conversion systems, they tend to be *relatively modest* by comparison with those associated with fossil fuels or nuclear fuels.

5. *Increased flexibility and security of supply due to availability of diversity of systems.*

1.5.2. Non-Renewable Energy Sources

The non-renewable energy sources are those which *do not get replenished after their consumption e.g.* coal once get burnt is consumed without replacement of the same (fossil fuels, nuclear fission fuels).

1.6 ALTERNATIVE ENERGY SOURCES

There are the sources which are *non-traditional.* They are *alternatives to the conventional energy sources.*

The demarcation between conventional and non-conventional is *not rigid. To day non-conventional becomes conventional after a few decades.*

- Concern for the *environment* due to ever-increasing use of fuels and rapid depletion of natural resources has led to development of *alternative sources of energy* which are *"renewable"* and *"environmental friendly"*.

1.7 ENERGY SCENARIO IN INDIAN CONTEXT

For all socio-economic activities, *"energy"* is a *primary source.* The consumption of energy of a nation is considered as an index of development. All sectors such as industry, transport, telecommunication, agriculture, household services etc. depend on energy.

With the increase in population, the 'energy demand' is increasing and the energy sources are becoming scarce. The main sources of energy in India are *coal, oil,* and *water.* Commercial consumption of energy is mostly from *coal* and *petroleum* and the other sources are *natural gas* and *water.*

The maximum portion of energy available from traditional energy sources such as fuelwood, agricultural waste, animal residue is used by the industrial sector. A large amount of petroleum products is used by the transport sector. As a result of modernisation, the consumption of energy by agricultural sector has also grown rapidly.

Due to growing population, it becomes necessary to look for alternate sources of energy to meet the future demands of the growing economy.

Energy position in India. The total power generation capacity in India in 1947 was only 1360 MW and in 1991 it grew to 65,000 MW, of which 45,000 MW (69%) was generated in thermal plants. Table 1.3, shows the power generating capacity by different types of generating plants.

Table 1.3. Indian Generation Capacity (MW)

Type of plant	1991	8th Plan 1997	9th Plan 2002	10th Plan 2007
Thermal	45,000	28,000	32,000	58,000
Hydro	18,500	18,700	26,000	23,000
Nuclear	1,500	1,320	2,880	—
Others	—	38,000	61,000	81,000

Table 1.4. below exhibits the renewable energy potential and installed capacity in India (2009) :

Table 1.4. Renewable Energy Potential and Installed capacity in India (2009)

S.No.	Source	Estimated potential	Installed capacity or number
1.	*Wind power*	45,000 MW	9756 MW
2.	*Wind pumps*	—	1284 Nos.
3.	*Small hydro (up to 25 MW)*	15,000 MW	2345 MW
4.	*Solar photovoltaic power plants*	50 MW/sq km	110 MW
5.	*Biomass power*	16,000 MW	638 MW
6.	*Biomass gasifier*	—	87 MW

7.	Biogas plants	12 million	3.9 million
8.	Bagasse cogeneration	5000 MW	1034 MW
9.	Waste to energy (i) Municipal solid waste (ii) Industrial waste	 1700 MW 1000 MW	 23.70 MW 35.21 MW

1.8 ELECTRICITY GENERATION FROM NON-CONVENTIONAL ENERGY SOURCES

It has been widely recognised that the fossil fuels and other conventional resources, presently used in generation of electrical energy, *may not be either sufficient or suitable* to keep pace with the ever increasing world demand for electrical energy. The prospects for meeting this demand and avoiding a crisis in supply would be *improved if new and alternative energy sources could be developed.*

The *important non-conventional electricity sources* are:

(i) Magneto hydrodynamic systems

(ii) Solar electric power plants

(iii) Photovoltaic cells

(iv) Fuel cells

(v) Wind energy

(vi) Geothermal energy

(vii) Tidal-energy

(viii) Ocean thermal energy

(ix) Organic wastes, biogas, rice straw etc.

The details are given in the chapters to follow.

☞ 1.9 IMPACT OF ENERGY SOURCES ON ENVIRONMENT

A *high quality environment* is one:

(i) which offers the most favourable living conditions for people of diverse interests;

(ii) which is conducive to good health and well-being of inhabitants, in which all biological variables are intact and healthy and in which a diverse and stable biotic community is maintained.

- A normal constituent of environment is *energy*. A permanent flow of energy controlled by nature takes place from sun to environment. This energy is used by humans, animals, insects etc.

  Man, by using this solar energy for food, hydropower, fuel etc., *participates harmoniously in the natural flow of energy through the environment.* In addition ,man is *also* using sources of energy which are limited, non-renewable and. not included in the normal flow of energy through the environment. By doing so, *man is polluting the environment:*

- An *important environmental aspect of energy generation* is the **"air pollution"**. Therefore, it is imperative to *limit the extent of air pollution* to maintain the air quality at a reasonable and acceptable level and protect public health and welfare.

- For improving our present level of economy and standard of living, it is necessary to *have continued expansion of electric power facilities.*

The environmental aspects of power plants are gaining more and more importance. It is possible to build new power plants with little detrimental environmental effects if sufficient thought, planning and study are given to the potential problems before the project is designed and constructed. The engineers and environmentalists, through mutual cooperation, may find a solution which is best from the point of engineering as well as ecology.

Note: For details, refer to chapter on *"Environmental Aspects of Energy Generation and Pollution Control (Chapter-16).*

1.10   FUELS

Fuels may be *"chemical"* or *"nuclear"*. Here we shall consider chemical fuels only.

A *chemical fuel* is a substance which releases heat energy on combustion. The principal combustible elements of each fuel are *carbon* and *hydrogen*. Though *sulphur* is a combustible element too, but its presence in the fuel is considered to be *undesirable*.

1.10.1. Classification of Fuels

Fuels can be *classified* according to whether:

1. They *occur in nature* called *'primary fuels'* or are *prepared* called *'secondary fuels'*;

2. They are *in solid, liquid or gaseous state.*

The detailed classification of fuels can be given in a summary form as below:

Type of fuel	Natural (Primary)	Prepared (Secondary)
Solid	Wood	Coke
	Peat	Charcoal
	Lignite coal	Briquettes.
Liquid	Petroleum	Gasoline
		Kerosene
		Fuel oil
		Alcohol
		Benzol
		Shale oil.
Gaseous	Natural gas	Petroleum gas
		Producer gas
		Coal gas
		Coke-oven gas
		Blast furnace gas
		Carburetted gas
		Sewer gas.

1.10.2. Solid Fuels

Coal. Its main constituents are carbon, hydrogen, oxygen, nitrogen, sulphur, moisture and ash. Coal passes through different *stages* during its formation from vegetation. These stages are enumerated and discussed below:

Plant debris-Peat-Lignite-Brown coal-Sub-bituminous coal-Bituminous coal-Semi-bituminous coal-Semi-anthracite coal-Anthracite coal-graphite.

Peat. It is the first stage in the formation of coal from wood. It contains huge amount of moisture and therefore it is dried for about 1 to 2 months before it is put to use. It is used as a domestic fuel in Europe and for power generation in Russia. In India it does not come in the categories of good fuels.

Lignites and brown coals. These are intermediate stages between peat and coal. They have a woody or often a clay-like appearance associated with high moisture, high ash and low heat contents. Lignites are usually amorphous in character and impose transport difficulties as they break easily. They burn with a smoky flame. Some of this type are suitable for local use only.

Bituminous coal. It burns with long yellow and smoky flames and has high percentages of volatile matter. The average calorific value of bituminous coal is about 31350 kJ/kg. It may be of two types, namely, *caking* or *non-caking*.

Semi-bituminous coal. It is softer than the anthracite. It burns with a very small amount of smoke. It contains 15 to 20 per cent volatile matter and has a tendency to break into small sizes during storage or transportation.

Semi-anthracite. It has less fixed carbon and less lustre as compared to true anthracite and gives out longer and more luminous flames when burnt.

Anthracite. It is very hard coal and has a shining black lustre. It ignites slowly unless the furnace temperature is high. It is non-caking and has high percentage of fixed carbon. It burns either with very short blue flames or without flames. The calorific value of this fuel is high to the tune of 35500 kJ/kg and as such is *very suitable for steam generation.*

Wood charcoal. It is obtained by destructive distillation of wood. During the process the volatile matter and water are expelled. The physical properties of the residue (charcoal) however, depends upon the rate of heating and temperature.

Coke. It consists of carbon, mineral matter with about 2% sulphur and small quantities of hydrogen, nitrogen and phosphorus. It is solid residue left after the destructive distillation of certain kinds of coals. It is smokeless and clear fuel and can be produced by several processes. It is mainly used in blast furnace to produce heat and at the same time to reduce the iron ore.

Briquettes. These are prepared from fine coal or coke by compressing the material under high pressure.

Analysis of coal:

The following two types of analyses is done on the coal:

1. Proximate analysis.

2. Ultimate analysis.

1. **Proximate analysis.** In this analysis, *individual elements are not determined; only the percentage of moisture, volatile matters, fixed carbon and ash are determined.*

   Example. Moisture = 4.5%, volatile matter = 5.5%, fixed carbon = 20.5%.

   This type of analysis is easily done and is for *commercial purposes* only.

2. **Ultimate analysis.** In the ultimate analysis, the percentages of various elements are determined.

   Example. Carbon = 90%, hydrogen = 2%, oxygen = 4%, nitrogen = 1%, sulphur = 15% and ash = 1.5%.

This type of analysis is useful for *combustion calculations.*

Properties of coal:

Important properties of coal are given below:

1. Energy content or heating value.
2. Sulphur content.
3. Burning characteristics.
4. Grindability.
5. Weatherability.
6. Ash softening temperature.

A *good coal should have:*

(*i*) Low ash content and high calorific value.

(*ii*) Small percentage of sulphur (less than 1%).

(*iii*) Good burning characteristics (*i.e.* should burn freely without agitation) so that combustion will be complete.

(*iv*) High grindability index (in case of ball mill grinding).

(*v*) High weatherability.

Ranking of coal:

ASME and ASTM have accepted a specification based on the fixed carbon and heating value of the mineral matter free analysis.

- **Higher ranking** is done on the basis of fixed carbon percentage (*dry basis*).
- **Lower ranking** is done on the heating value on the *moist basis*.

    Example: 62% C and a calorific value of 5000 kcal/kg is ranked as (62—500) rank.

Rank is an inherent property of the fuel depending upon its relative progression in the classification process.

Grading of coal:

Grading is done on the following basis:

(*i*) Size

(*ii*) Heating value

(*iii*) Ash content

(*iv*) Ash softening temperature

(*v*) Sulphur content.

Example: A grade written as 5–10 cm, 5000-A8-F24-S1.6 indicate the coal as having:

— a size of 5–10 cm,

— heating value of 5000 kcal/kg,

— 8 to 10% ash,

— ash softening temperature of 2400–2590°F, and

— a sulphur content of 1.4 to 1.6%.

A rank and grade of a coal gives a complete report of the material. Thus the following are the rank and grade of the coal described above:

(62—500), 5—10 cm, 500-A8-F24-S1.6.

1.10.3. Liquid Fuels

The chief source of liquid fuels is petroleum which is obtained from wells under the earth's crust. These fuels have proved *more advantageous in comparison to sold fuels* in the following respects:

Advantages:

1. Require less space for storage.
2. Higher calorific value.
3. Easy control of consumption.
4. Staff economy.
5. Absence of danger from spontaneous combustion.
6. Easy handling and transportation.
7. Cleanliness.
8. No ash problem.
9. Non-deterioration of the oil in storage.

Petroleum. There are different opinions regarding the origin of petroleum. However, now it is accepted that petroleum has originated probably from organic matter like fish and plant life etc., by bacterial action or by their distillation under pressure and heat. It consists of a mixture of gases, liquids and solid hydrocarbons with small amounts of nitrogen and sulphur compounds. In India the main sources of petroleum are Assam and Gujarat.

Heavy fuel oil or crude oil is imported and then refined at different refineries. The refining of crude oil supplies the most important product called *petrol*. Petrol can also be made by polymerization of refinery gases.

Other liquid fuels are kerosene, fuel oils, colloidal fuels and alcohol.

The following table gives *composition* of some common liquid fuels used in terms of the elements in weight percentage.

Fuel	Carbon	Hydrogen	Sulphur	Ash
Petrol	85.5	14.4	0.1	—
Benzene	91.7	8.0	0.3	—
Kerosene	86.3	13.6	0.1	—
Diesel oil	86.3	12.8	0.9	—
Light fuel oil	86.2	12.4	1.4	—
Heavy fuel oil	88.3	9.5	1.2	1.0

Important properties of liquid fuels:

  (1) Specific gravity   (2) Flash point   (3) Fire point

  (4) Volatility   (5) Pour point   (6) Viscosity

  (7) Carbon residue   (8) Octane number   (9) Cetane number

(10) Corrosive property   (11) Ash content   (12) Gum content

(13) Heating value   (14) Sulphur content.

The requisite properties vary from device to device which uses the fuel to generate power. For example, *higher the octane number, higher can be the compression ratio and the thermal efficiency will be higher.* Similarly, *the cetane number of a diesel oil should be as high as possible.*

In general the liquid fuels should have:

  (*i*) Low ash content   (*ii*) High heating value

(*iii*) Low gum content   (*iv*) Less corrosive tendency

  (*v*) Low sulphur content   (*vi*) Low pour point.

Viscosity and other properties vary from purpose to purpose to which the fuel is employed.

1.10.4. Gaseous Fuels

Natural gas. The main constituents of natural gas are methane (CH_4) and ethane (C_2H_6). It has calorific value nearly 21000 kJ/m^3. Natural gas is used alternately or simultaneously with oil for internal combustion engines.

Coal gas. This gas mainly consists of hydrogen, carbon monoxide and hydrocarbons. It is prepared by carbonisation of coal. It finds its use in boilers and sometimes used for commercial purposes.

Coke-oven gas. It is obtained during the production of coke by heating the bituminous coal. The volatile content of coal is driven off by heating and major portion of this gas is utilised in heating the ovens. This gas must be thoroughly filtered before using in gas engines.

Blast furnace gas. It is obtained from smelting operation in which air is forced through layers of coke and iron ore, the example being that of pig iron manufacture where this gas is produced as by product and contains about 20% carbon monoxide (CO). After filtering it may be blended with richer gas or used in gas engines directly. The heating value of this gas is very low.

Producer gas. It results from the partial oxidation of coal, coke or peat when they are burnt with an insufficient quantity of air. It is produced in specially designed retorts. It has low heating value and in general is suitable for large installations. It is also used in steel industry for firing open hearth furnaces.

Water or Illuminating gas. It is produced by blowing steam into white hot coke or coal. The decomposition of steam takes place liberating free hydrogen and oxygen in the steam combines with carbon to form carbon monoxide according to the reaction:

$$C + H_2O \longrightarrow CO + H_2$$

The gas composition varies as the hydrogen content if the coal is used.

Sewer gas. It is obtained from sewage disposal vats in which fermentation and decay occur. It consists of mainly marsh gas (CH_4) and is collected at large disposal plants. It works as a fuel for gas engines which in turn drive the plant pumps and agitators.

Gaseous fuels are becoming popular because of following *advantages* they possess:

Advantages:
1. Better control of combustion.
2. Much less excess air is needed for complete combustion.
3. Economy in fuel and more efficiency of furnace operation.
4. Easy maintenance of oxidizing or reducing atmosphere.
5. Cleanliness.
6. No problem of storage if the supply is available from public supply line.
7. The distribution of gaseous fuels even over a wide area is easy through the pipe lines and as such handling of the fuel is altogether eliminated. Gaseous fuels give economy of heat and produce higher temperatures as they can be preheated in regenerative furnaces and thus heat from hot flue gases can be recovered.

Important properties of gaseous fuels:
1. Heating value or calorific value.

2. Viscosity.
3. Specific gravity.
4. Density.
5. Diffusibility.

Typical composition of some gaseous fuels is given below:

Fuel	H_2	CO	CH_4	C_2H_4	C_2H_6	C_4H_8	O_2	CO_2	N_2
Natural gas	—	1	93	—	3	—	—	—	3
Coal gas	53.6	9.0	25	—	—	3	0.4	3	6
Blast furnace gas	2	27	—	—	—	—	—	11	—

1.10.5. Calorific or Heating Values of Fuels

The **calorific value** *of the fuel is defined as the energy liberated by the complete oxidation of a unit mass or volume of a fuel.* It is expressed in kJ/kg for solid and liquid fuels and kJ/m³ for gases.

Fuels which contain hydrogen have two calorific values, the *higher* and the *lower*. The *'lower calorific value'* is the heat liberated per kg of fuel after deducting the heat necessary to vaporise the steam, formed from hydrogen. The *'higher or gross calorific value' of the fuel is one indicated by a constant-volume calorimeter in which the steam is condensed and the heat of vapour is recovered.*

The lower or net calorific value is obtained by subtracting latent heat of water vapour from gross calorific value. In other words, the relation between Lower Calorific Value (L.C.V.) and Higher Calorific Value (H.C.V.) can be expressed in the following way:

$$\text{L.C.V.} = (\text{H.C.V.} - 2465\, m_w) \qquad \qquad ...(1.1)$$

where m_w is the mass of water vapour produced by combustion of 1 kg of fuel and 2465 kJ/kg is the latent heat corresponding to standard temperature (saturation) of 15°C.

> **In MKS units:**
>
> $$\text{L.C.V.} = (\text{H.C.V.} - 588.76\, m_w)$$
>
> where m_w is the mass of water vapour produced by combustion of 1 kg of fuel and 588.76 is the latent heat value in kcal as read from steam tables for 1 kg of water vapour

Dulong's formula (Solid/liquid fuels). Dulong suggested a formula for the calculation of the calorific value of the solid or liquid fuels from their chemical composition which is as given below:

Gross calorific value,

$$\text{H.C.V.} = \frac{1}{100}\left[33800\,C + 144000\left(H - \frac{O}{8} \right) + 9270\,S \right] \text{ kJ/kg} \quad ...(1.2)$$

> **In MKS units :**
>
> $$\text{H.C.V.} = \frac{1}{100}\left[8080\,C + 34500\left(H - \frac{O}{8} \right) + 2240\,S \text{ kcal/kg} \right]$$

where C, H, O and S are carbon, hydrogen, oxygen and sulphur in percentages respectively in 100 kg of fuel. In the above formula, the oxygen is assumed to be in combination with hydrogen and only extra surplus hydrogen supplies the necessary heat.

HIGHLIGHTS

1. *Energy* is the capability to produce motion, force, work, change in shape, change in form etc.
2. The various sources of energy are:
   (*i*) *Commercial primary energy resources:* Coal, lignite, oil and natural gas, hydro-electric and nuclear fuels.
   (*ii*) *Renewable energy resources:* Solar photovoltanic; Wind; Hydrogen fuel-cell.
   (*iii*) *New sources of energy.*
3. Coal, oil, gas, uranium and hydro are known as *'commercial'* or *'conventional'* *energy sources.*
   Firewood, animal dung and agricultural waste etc. are called as *non-commercial energy* sources.
4. The various *non-conventional energy sources* are:
   (*i*) Solar energy; (*ii*) Wind energy; (*iii*) Energy from biomass and biogas; (*iv*) Ocean thermal energy conversion; (*v*) Tidal energy; (*vi*) Geothermal energy; (*vii*) Hydrogen energy; (*viii*) Fuel cells (*ix*) Magneto-hydrodynamic generator; (*x*) Thermionic converter; (*xi*) Thermoelectric power.

THEORETICAL QUESTIONS

1. Explain briefly the following:
   (*i*) Energy; (*ii*) Energy technology; (*iii*) Energy planning.
2. Describe commercial and non-commercial energy sources.
3. What are conventional and non-conventional energy sources? Describe the fossil fuel as a conventional energy source.
4. Write short note on Decentralised and Dispersed generation.
5. Discuss the priamry and secondary energy sources. Also describe the future of non-conventional energy sources in India.
6. Describe the energy position of India.
7. Give brief review of various sources of renewable energy.
8. Describe the various non-conventional energy sources relevant to India.
9. Discuss the merits and demerits of the various renewable technologies developed in India.
10. Describe the fossil fuel as a conventional energy source.
11. Discuss the possibility of exploiting the non-conventional energy in India.
12. Discuss briefly any two of the following non-conventional energy sources:
    (*i*) Solar energy   (*ii*) Wind energy
    (*iii*) Geothermal energy   (*iv*) Energy from biomass and biogas.
13. Explain briefly the impact of conventional sources of energy on environment.
14. What is a chemical fuel? How does it differ from a nuclear fuel?
15. How are chemical fuels classified?
16. Explain briefly the following solid fuels:
    Lignites and brown coals, bituminous coal, and coke.
17. List the advantages of liquid fuels.
18. Describe the following gaseous fuels:
    Coal gas, Coke-oven gas, Blast furnace gas and Producer gas.
19. What are the advantages of gaseous fuels?

CHAPTER

2

PRINCIPLES OF RADIATION

2.1 Solar energy - General aspects – Sun and earth- Solar energy – an introduction – Advantages, disadvantages and applications of solar energy; **2.2 Solar Energy terms and definitions.** – Solar radiation – Solar constant (I_{sc}) – Cloudy index – Solar radiation geometry – Solar day-length, sunrise and sunset – Local solar time (LST) or Local apparent time (LAT) – Apparent motion of sun; **2.3 Measurement of solar radiation** – Pyramometer-Pyrhetiometer-Sunshine recorder; **2.4 Solar radiation data** – General aspects – Solar radiation data for India – Solar insolation; **2.5 Estimation of average solar radiation; 2.6 Solar radiation on an inclined surface.** *Highlights – Theoretical Questions – Unsolved Examples.*

2.1. SOLAR ENERGY–GENERAL ASPECTS

2.1.1. Sun and Earth

Sun:

- It is a sphere of very hot gases and is largest members of the solar system.
- The diameter of the sun is 1.39×10^6 km.
- The distance between 'sun' and 'earth' is 1.50×10^8 km.
- It completes its one rotation in four weeks when observed from earth. But the equator of the 'sun' takes 27 days and polar regions takes about 30 days for each rotation.
- The heat generation is mainly due to various kinds of fusion reactions but most of the energy is released in which hydrogen (*i.e.*, four protons) combine to form helium. An *effective* black body temperature of sun is 5577 K.

The fusion reaction is as follows:

$$4(_1H^1) \rightarrow \,_2He^4 + 26.3 \text{ MeV}$$

This energy is produced in the interior of the solar sphere and transmitted out by the radiation into system.

Net energy radiated, $\quad E = \varepsilon\sigma T_s^4$

where ε = Emissitivity of surface, σ = Stefan's Boltzmann constant, and T_s = Effective black body surface temperature of sun.

Earth:

- It is almost round in shape and has a diameter of 1.27×10^4 km.
- Its real shape is a sphere flattened at the poles and buldged in the plane normal to the poles.

- The earth's *inner core* is a solid mass made of *iron and nickel* and the next outer core is melted state of iron and nickel. The *outermost portion* is made of *rocks*.

- The existence of blue green algae indicates beginning of photosynthesis at least 3×10^9 years ago. As a result of photosynthesis, the level of O_2 and O_3 is increased in the atmosphere which block the ultra violet (UV) solar radiation coming from the 'sun'. Half the earth is lit by the sunlight at a time. It reflects one-third of the sunlight that falls on it, is known as *earth's albedo*.

- The length of days and nights keep changing because the earth is spinning about its axis which is inclined at an angle of 23.5°.

2.1.2. Solar Energy – An Introduction

The sun emits radiant energy as a *spectrum* corresponding to a 'black body' at a temperature of about 5500°C of which *only a small amount is intercepted by the earth*. Solar radiation is absorbed in the atmosphere and at the earth surface at a rate of 10.3×10^6 W.

The solar irradiance just outside the atmosphere is about 1353 W/m². Because solar radiation is *attenuated* as it travels through the atmosphere, the total power falling on horizontal surface, known as the global irradiance, achieves a *maximum of about* 1000 W/m² (*i.e.,* 1kW/m²) at sea level.

Global irradiance is actually made up of two components.

(*i*) *"Direct beam radiation" from the sun* and

(*ii*) *"Diffuse radiation" from the sky* (radiation that has been *scattered* by the atmosphere).

- The amount of radiation received varies throughout the day as the *path of solar radiation* through the atmosphere *lengthens* and *shortens*. For the same reason, seasonal and *lattitudal variations* can cause the total solar energy received (known as **insolation** or **solar irradiation**) to range from an average of 2 MJ/m²/day (or 0.55 kWh/m²/day) in a northern winter to an average of 20 MJ/m²/day (or 5.55 kWh/m²/day) in the tropical regions of the world.

— The *diffuse energy* may amount to only 15–20 percent of global irradiance on a clear day and 100 percent on a *cloudy day*.

- The solar energy *variability* is important in *system design* and *economics*. Unlike conventional fossil fuel technologies, the performance of solar systems can vary markedly from one location to another. Consequently, to design a system to convert solar energy, one *must have data on the solar radiation received at a particular site*, preferably on a month-to-month basis.

  If the system's size is calculated from the yearly amount of solar energy, it may not provide energy output in months of low insolation. *For convenience, the monthly solar radiation is usually expressed in terms of the daily average irradiation for the month; e.g., MJ/m²/day*. Most available radiation data are for global irradiation; this is the starting point for assessing a site. *If possible, data should be obtained from the nearest available meterological station, and due allowance made for any localized micro-climate.*

- Solar radiation can be converted to other useful forms of energy, principally: (*i*) **Heat:** This can be used directly to heat or distil water or to dry crops. The relatively simple conversion can be carried out by means of a variety of solar thermal collectors. (*ii*) **Mechanical or Electrical power:** These two forms, which are easily and efficiently interconvertible, can serve a variety of end-uses, including water pumping, lighting and refrigeration.

However, the *energy conversion technology is much more complex than that of heat production*. Conversion can be achieved by *two* completely different routes:

(*i*) Solar thermodynamics.

(*ii*) Solar photovoltaic.

☞ ● *Conditions for utilization of solar energy, in India, are favourable* since for nearly six months of the year sunshine is uninterrupted during the day, while in the other six months cloudy weather and rain provide conditions suitable for water power. Thus, a *coordination of solar energy with water power can provide a workable plan for most places in India.*

Following *renewable energy sources find their origin in 'Sun'.*

(*i*) Wind	(*ii*) Ocean thermal
(*iii*) Ocean wave	(*iv*) Ocean tide
(*v*) Geothermal	(*vi*) Biomass
(*vii*) Organic chemicals	(*viii*) Fossil fuels.

2.1.3. Advantages, Disadvantages and Applications of Solar Energy

Following are the advantages, disadvantages and applications of solar energy:

Advantages:

1. It is *clean, cheap* and *abundantly available.*

2. It is *re-usable* source of energy.

3. It is *eco-friendly* (*i.e.,* pollution free)

4. It *decreases* green house gas emissions.

Disadvantages:

1. *High capital cost* due to requirement of large area.

2. *Limited* to sunshine hours.

3. *Need of tracking* due to change in position of sun.

4. There is a need of *storage.*

Applications:

Solar energy is used in:

(*i*) Solar cooling;

(*ii*) Solar water heating;

(*iii*) Solar distillation;

(*iv*) Solar pumping;

(*v*) Electric power generation.

● **Solar energy conversion systems and their applications:**

1. *Passive heating systems:*

    Low temperature ($t < 150°C$): Cooling; Residential heating; Water heating; Drying; Biomass energy processes; Energy conservation of conventional non-renewables; Green-houses.

2. *Solar thermal Systems:*

    Medium temperature ($150°C < t > 300°C$): Process heat supply; Hot-water; Steam supply, Heat for chemical industry; Desalination plants.

3. *Solar thermal systems:*

    High temperature ($t > 300°C$): High temperature steam for industry; Electrical power generation.

4. *Solar to electrical energy conversion by PV systems:*

Very small *mV*, *mW* applications; small low voltage, low wattage applications; Medium voltage and medium power applications in kW range upto about 350 kW; Extremely useful for remote, stand-alone applications.

5. *Solar-diesel hybrid system:*

Stand-alone power plants rated 1 kW to 350 kW for remote applications, farms, villages, off-shore, mountain, desert *etc.*

6. *Solar central receiver thermal power plants:*

Feed power into electrical network, range 1 MW to 200 MW.

2.2 SOLAR ENERGY TERMS AND DEFINITIONS

2.2.1. Solar Radiation

Solar radiation is the *energy radiated by the sun.*

— The *radiated energy received on earth surface is called* **Solar irradiation**.

— *Solar radiation received on a flat horizontal surface on earth* is called **Solar insolation**.

The solar radiation is of the following two types:

1. *Extraterrestrial solar radiation:*

The *intensity of sun's radiation outside the earth's atmosphere* is called *"extraterrestrial"* and has no *diffuse components.*

Extraterrestrial radiation is the *measure* of solar radiation that would be received in the *absence of atmosphere.*

2. *Terrestrial solar radiation:*

The *radiation received on the earth surface* is called *"terrestrial radiation"* and is nearly 70 percent of extraterrestrial radiation.

Spectral distribution of solar radiation:

Light rays radiated from the sun are in the form of *electromagnetic waves* in infrared, visible and ultraviolet frequence bands. The *frequency spectrum* of solar light is a *graph of wavelength against power density,* shown in Fig. 2.1.

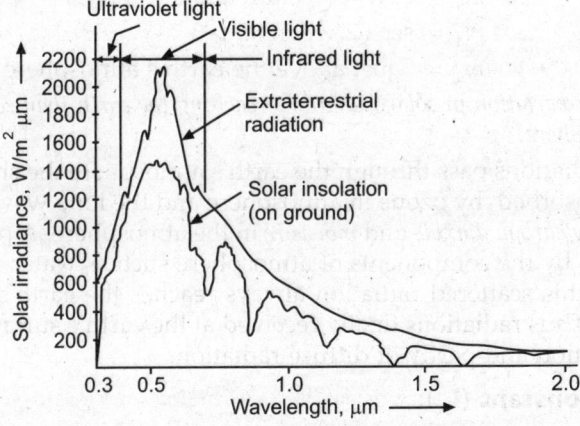

Fig. 2.1. Solar spectrum.

The solar spectrum has the following *three basic levels:*

(*i*) *"Infrared band"* with wavelengths *too long* for response by human eye (frequency range: 4×10^{14} to 7.5×10^{10} Hz) *wavelengths* : between 0.75 micron and 1.95 microns (1 micro, $\mu m = 10^{-6}$ m).

(ii) **Visible band:** *Frequency range:* 6×10^{16} to 7.69×10^{14} *Hz; Wave lengths:* Between 0.39 micron and 0.75 micron.

(iii) **Ultraviolet band:** *Frequency range:* 6×10^{16} to 7.5×10^{10} *Hz.*

Wavelengths: Between 0.005 micron to 0.39 micron.

Terms used in solar radiations:

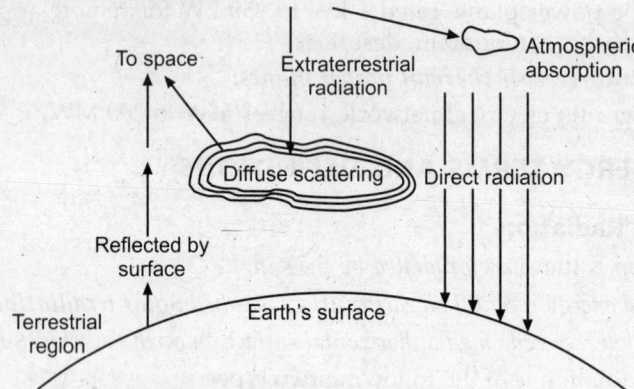

Fig. 2.2. Direct, diffuse and total solar radiations.

Refer to Fig. 2.2.

Beam (or direct) radiation (I_b). *Solar radiation received on the surface of earth without change in directions* is known as *"beam or direct radiation"*.

Diffuse radiation (I_d). *The solar radiation received from the sun after its direction has been changed by reflection and scattering by atmosphere* is known as *"diffuse radiation"*.

Total radiation (I_T). *The sum of beam and diffuse radiations intercepted at the surface of earth per unit area of location* is known as *"total radiation". It is also known as "Insolation".*

Mathematically: $I_T = I_b + I_d$ ...(2.1)

Airmass (m_a). *It is the path length of radiation through the atmosphere, considering the vertical path at level as unity.*

$m_a = 1$, when sun is at zenith (*i.e.,* directly above head).

$m_a = 2$, when zenith angle (θ_z) is 60°.

$m_a = \sec \theta_z$, when $m_a > 3$

$m_a \doteq 0$ just above the earth's atmosphere

- *"Reasons for variation in solar radiations reaching the earth than received on the outside of the atmosphere"*:

 As solar radiations pass through the earth's atmosphere the shortwave 'intraviolet rays' are 'absorbed' by ozone in atmosphere and the long wave *infrared waves* are 'absorbed' by *carbon dioxide* and *moisture* in the atmosphere. A portion of radiations is 'scattered' by the components of atmosphere such as water vapour and dust. A portion of this scattered radiation always reaches the earth's surface as 'diffuse radiation'. Thus radiations finally received at the earth's surface consists partly of beam radiation and partly of diffuse radiation.

2.2.2 Solar Constant (I_{sc}):

The *"solar constant" (I_{sc}) is the energy from the sun received on a unit area perpendicular to solar rays at the mean distance from the sun (1.5×10^8 km) outside the atmosphere.*

Solar constant is characterised by the following:

(i) It is constant and *not* affected by daily, seasonal, atmospheric condition, clarity of atmosphere etc.

(ii) It is on a *unit area on imaginary spherical surface around earth's atmosphere for mean distance between the sun and the earth.*

(iii) It is on surface *normal* to sun's rays. Sun rays are practically parallel (beam radiation).

(iv) It has a measured value of "*1353 W/m²*".

- Isc in terms of kJ/m². hour $= \dfrac{1353 \times 3600}{1000} = 4870.8$ kJ/m² hour

- The value of solar constant remains constant throughout the year. *However,* this value changes with location because earth-sun distance changes seasonally with time. The extraterrestrial relation observed on different days is known as *apparent extraterrestrial solar irradiance* and can be calculated on any of the year using the following relation:

$$I_0 = I_{sc}\left[1 + 0.033 \cos\left(\frac{360\,(n-2)}{365}\right)\right] \qquad \ldots(2.2)$$

Or,
$$I_0 = I_{sc}\left[1 + 0.033 \cos\left(\frac{360\,n}{365}\right)\right] \qquad \ldots[2.2\,(a)]$$

where,

$I_0 =$ Apparent extraterrestrial solar irradiance (W/m²),

$n =$ Number of days of the year counting January 1 as the first day of the year, and

$I_{sc} =$ Solar constant = 1353 W/m².

(The standard value of the solar constant based on experimental measurements is 1367 W/m² with accuracy of ±1.5%).

According to Eqn. (2.2), the apparent solar irradiance will be maximum during December last or first weak of January as the earth's centre is nearest to the sun during these days.

2.2.3. Clarity Index and Concentration Ratio

Clarity Index:

The ratio of radiation received on earth's horizontal surface over a given period to radiation on equal surface area beyond earth's atmosphere in direction perpendicular to the beam is called "*Clarity index*".

It depends upon the clarity of atmosphere for passage of solar beam radiation. Clarity index can be between 0.1 to 0.7.

Concentration ratio:

It is the *ratio of solar power per unit area of the concentrator surface* (kW/m²) *to power per unit area on the line focus or point focus* (kW/m²).

2.2.4. Solar Radiation Geometry

The various angles which are useful for conversion of beam radiation on the arbitrary surface are:

Refer to Fig 2.3:

(i) **Latitude angle (ϕ):**

The '*latitude of a place*' is the angle *subtended by the radial line joining the place to the centre of the earth, with the projection of the line on the equatorial plane.*

- The latitude is taken as *positive* for any location towards the 'northern hemisphere' and *negative* towards the 'southern hemisphere' *i.e.,* the latitude(s) at equator is 0^0 while at north and south poles are + 90° and – 90° respectively.

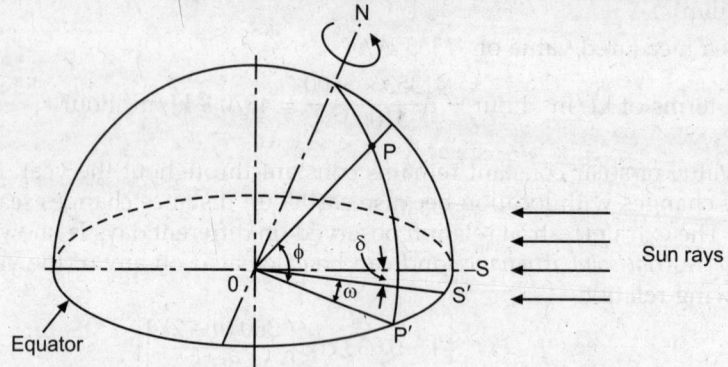

Fig. 2.3. Latitude, sun's declination δ and hour angle ω.

(*ii*) **Declination angle (δ):**

It is the angle made by the line *joining the centres of the sun and the earth with its projection on the equatorial plane.* This angle varies from a *maximum* value of +23.5° on *June 21* to *minimum* of –23.5° on *December 21*.

The declination (in degrees) for any day may be calculated from the approximate equation of *"Cooper".*

$$\delta = 23.45 \sin \left[\frac{360}{365} (284 + n) \right] \qquad \qquad ...(2.3)$$

where, *n* is the day of the year.

Fig. 2.4. shows the variation of declination angle.

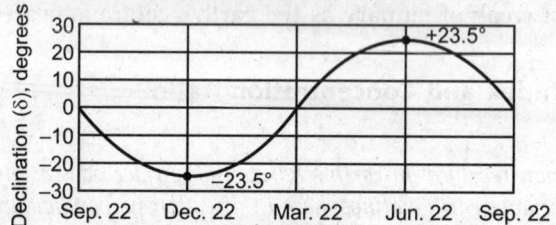

Fig. 2.4. Variation of declination angle.

(*iii*) **Hour angle (ω):**

It is angle through which the earth must be rotated to bring the meridian of the plane directly under the sun. In other words, it is the angular displacement of the sun, east or west of the local meridian, due to rotation of the earth on its axis at an angle of 15° per hour.

It is measured from noon based on the local solar time (LST) or local apparent time (LAT), being *positive in the morning* and *negative in the afternoon.* It *is the angle* measured in the earth's equilateral plane, between the projection $\overline{OP}$ and the projection of a line from the centre of the sun to the centre of the earth.

- At solar noon ω being zero and each hour angle equating 15° of longitude with "morning positive" and "afternoon negative" (e.g. ω = + 15° for 11.00 and ω = – 37.5° for 14.30), hour angle can be expressed as :

$$\omega = 15 (12 - LST)$$

(*iv*) **Altitude angle (α) or solar altitude:** Refer to Fig. 2.4.

It is a vertical angle between the projection of the sun rays on the horizontal plane and direction of the sunrays, passing through the point.

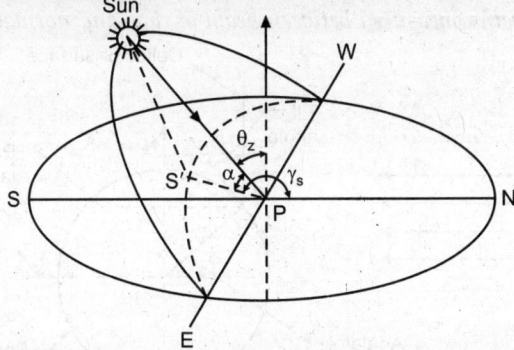

Fig. 2.5. Sun's altitude, zenith and solar azimuth angles.

(*v*) **Zenith angle (θ_z):**

It is a vertical angle between sun's rays and a line perpendicular to the horizontal plane through the point.

Mathematically, $\theta_Z = \dfrac{\pi}{2} - \alpha$ ...(2.4)

(*vi*) **Solar azimuth angle (γ_s):**

It is the polar angle (in degrees) along the horizontal east or west of north.

Or

It is a horizontal angle measured from north to the horizontal projection of the sun's rays. This angle is *positive* when measured *west wise*.

The following expressions hold good for angles θ_z and γ_s in terms of *basic angles* ϕ, δ and ω:

$$\cos \theta_z = \cos \phi \cos \omega \cos \delta + \sin \phi \sin \delta \qquad \text{...(2.5)}$$

$$\cos \gamma_s = \sec \alpha \, (\cos \phi \sin \delta - \cos \delta \sin \phi \cos \omega) \qquad \text{...(2.6)}$$

and, $\sin \gamma_s = \sec \alpha \cos \delta \sin \omega$ ...(2.7)

(*vii*) **Surface azimuth angle (γ):** Refer to Fig. 2.6.

It is the angle of deviation of the normal to the surface from the local meridian, the *zero point being south, east positive and west negative*.

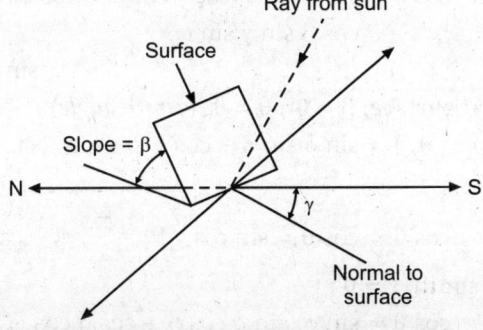

Fig. 2.6. Surface azimuth angle slope.

(*viii*) **Slope or tilt angle (β):**

It is the angle made by the plane surface with the horizontal.

It is taken to be *positive* for surfaces sloping towards the *south* and *negative* for surfaces sloping towards *north*.

(*ix*) **Incident angle (θ):**

It is the angle being measured between beam of rays and normal to the plane.

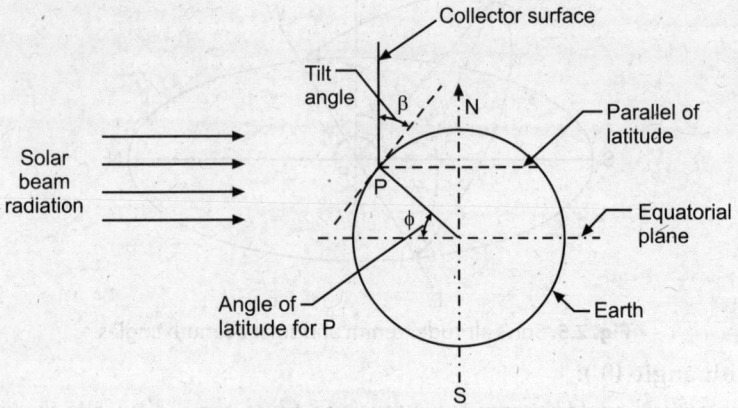

Fig. 2.7. Latitude angle (φ) and tilt angle (β).

Calculations of cos θ for any hour and any 'location':

The angle of incident θ is the *most significant parameter*. The power collected by the collector surface is *less* than the power available from the sun rays by *factor cosθ* as shown in Fig. 2.7. The angle of incident θ depends on the *position of sun* in the sky.

General equation for cos θ:

$$\cos \theta = [\sin \phi \, (\sin \delta \cos \beta + \cos \delta \cos \gamma \cos \omega \sin \beta)$$
$$+ \cos \phi \, (\cos \delta \cos \omega \cos \beta - \sin \delta \cos \gamma \sin \beta)$$
$$+ \cos \delta \sin \gamma \sin \omega \sin \beta] \qquad ...(2.8)$$

where,　　θ = Angle of incidence,

　　　　　ϕ = Angle of latitude,

　　　　　δ = Angle of declination,

　　　　　γ = Angle of surface azimuth, and

　　　　　β = Tilt angle (angle of slope).

1. For a **vertical surface; β = 90°:**

∴　　　　$\cos \theta = \sin \phi \cos \delta \cos \gamma \cos \omega - \cos \phi \sin \delta \cos \gamma +$
　　　　　$\cos \delta \sin \gamma \sin \omega$ 　　　　...(2.9)

(∵ sin 90° = 1 and cos 90° = 0)

2. For a **Horizontal surface; β = 0°; θ = θ_z** (*zenith angle*) :

∴　　$\cos \theta \, (= \cos \theta_z) = \sin \phi \sin \delta + \cos \phi \cos \delta \cos \omega.$ 　　...(2.10)

(∵ cos 0° = 1, sin 0° = 0)

Also,　　　$\cos \theta = \cos \theta_z = \sin \alpha \left(\because \theta_z = \dfrac{\pi}{2} - \alpha \right)$ 　　...(2.10(*a*))

3. **Surface facing south, γ = 0 :**

$$\cos \theta = \sin \phi \, (\sin \delta \cos \beta + \cos \delta \cos \omega \sin \omega)$$
$$+ \cos \phi \, (\cos \delta \cos \omega \cos \beta - \sin \delta \sin \beta)$$

∴　　$\cos \theta = \sin \delta \sin (\phi - \beta) + \cos \delta \cos \omega \cos (\phi - \beta)$ 　　...(2.11)

4. **Vertical surface facing south, β= 90°, γ = 0 :**

$$\cos \theta = \sin \phi \cos \delta \cos \omega - \cos \phi \sin \delta \qquad ...(2.12)$$

The *angle of incidence θ can also be expressed in terms of* θ_z (*zenith angle*) *as:*

$$\cos \theta = \cos \theta_z \cos \beta + \sin \theta_z \sin \beta \cos(\gamma_s - \gamma) \qquad ...(2.13)$$

$$\cos \gamma_s = \frac{(\cos \theta_z \sin \phi - \sin \delta)}{\sin \theta_z - \cos \phi} \qquad ...(2.14)$$

Fig. 2.8. shows the schematic representation for different angles.

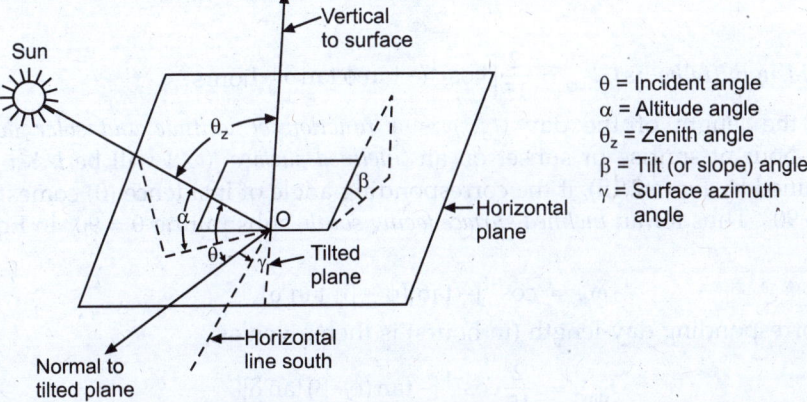

θ = Incident angle
α = Altitude angle
θ_z = Zenith angle
β = Tilt (or slope) angle
γ = Surface azimuth angle

Fig. 2.8. Schematic representation for different angles.

2.2.5. Solar Day-Length, Sunrise and Sunset

We make the following observations:

(i) During '*winter*' the sun rises late and sets early, the *day length is shorter*.

(ii) During '*summer*' sun rises early, sets late and *day length is longer*.

(iii) With the increase of the angle of latitude (from equator to north pole) the *difference* in day length between summer and winter becomes more and more prominent.

The '*sunrise hour*' '*sunset hour*' and '*day-length*' depend upon *latitude of the location* and *season and day in the year*.

The *expressions* are derived as follows:

For a horizontal surface on the ground Eqn. (2.10) is written as:

$$\cos \theta = \sin \phi \sin \delta + \cos \phi \cos\delta \cos \omega_s \qquad ...(2.15)$$

(For sunrise and sun set the hour angle is designated as ω_s)

The hour angle ω varies during the day.

At *sunrise*, as the sun light is *parallel* to the ground surface, therefore, angle of incidence $\theta = 90°$, $\cos \theta = 0$

Inserting this value in Eqn. (2.13), we get,

$$\cos 90° = \sin \phi \sin \delta + \cos \phi \cos \delta \cos \omega_s$$

or, $$\sin \phi \sin \delta = - \cos \phi \cos \delta \cos \omega_s$$

$$(\because \cos 90° = 0)$$

or, $$\cos \omega_s = - \frac{\sin \phi \sin \delta}{\cos \phi \cos \delta}$$

$$= - \tan \phi \tan \delta$$

i.e., $$\omega_s = \cos^{-1} (- \tan \phi \tan \delta) \qquad ...(2.16)$$

Day-length (t_{day}) in hours:

$$\text{Total day-length} = \omega_s + \omega_s$$
$$= 2\,\omega_s$$
$$= 2\cos^{-1}(-\tan\phi\tan\delta)$$

Since $15°$ of the hour angle is equivalent to 1 hour, hence $2\cos^{-1}(-\tan\phi\tan\delta)°$

corresponds to $\left[\dfrac{2\cos^{-1}(-\tan\phi\tan\delta)}{15}\right]$ hours.

$\therefore$ *Day-length (in hours)*, $t_{day} = \dfrac{2}{15}\left[\cos^{-1}(-\tan\phi\tan\delta)\right]$ hours ...(2.17)

Thus, the length of the day (t_{day}) is a *function of latitude and solar declination.* The angle hour at sunrise or sunset on an *'inclined surface'* (ω_{si}) will be *lesser* than the value obtained by Eqn. (2.15), if the corresponding angle of incidence (θ) comes out to be more than $90°$. Thus for an *inclined surface facing south*, substituting $\theta = 90°$ in Eqn. (2.11), we get:

$$\omega_{si} = \cos^{-1}[-\tan(\phi-\beta)\tan\delta] \qquad \qquad ...(2.18)$$

The corresponding day-length (in hours) is then given as:

$$t_{day} = \frac{2}{15}\cos^{-1}[-\tan(\phi-\beta)\tan\delta] \qquad \qquad ...(2.19)$$

2.2.6. Local Solar Time (LST) or Local Apparent Time (LAT)

It is the time used for calculating the hour angle. LST can be obtained from the standard time observed on a clock by applying the following *two corrections*:

(*i*) The correction which arises due to the *difference in longitude between a location and the meridian on which standard time is based.*

This correction has a magnitude of *4 minutes* for *every degree difference in longitude.*

(*ii*) This correction called the *equation of time correction* is due to the fact that earth's orbit and rate of rotation are subjected *to small perturbations.* This correction is *based on experimental observations.*

Hence,

Local Solar Time (LST) = Standard time $\pm$ 4 (Standard time longitude – longitude of location) + (Equation of time correction). ...(2.20)

The **+ ve** *sign is used for 'Western hemisphere' and* **– ve** *sign for 'Eastern hemisphere'.*

● In **India** *standard time is based on* "**82.5° E longitude**".

2.2.7. Apparent Motion of Sun

The *"apparent motion of sun"*, *caused by the rotation of the earth about its axis, changes the angle at which the direct component of light will strike the earth.* From a fixed location on earth, the sun *appears* to move throughout the sky. The position of the sun depends on the location of a point on earth, the time of day and the time of year.

This *apparent motion of the sun has a major impact on the amount of power received by a solar collector.* When the sun's rays are perpendicular to the absorbing surface, the power density of the surface is equal to the inside power density. However, as the angle between the sun and the absorbing surface *changes*, the intensity on the surface is *reduced*.

The angle between the sun and a fixed location on earth depends on the particular location (the longitude of the location), the time of year and the time of day. In addition, the time at which the sun rises and sets depends on the longitude of the location. Therefore, the complete modelling of the sun's angle to a fixed position on Earth requires the latitude, longitude, day of the year, and time of day.

Example 2.1. *Determine the sunset hour angle and day-length at a location latitude of 32°on March 30.*

Solution. *Given*: Latitude, $\phi = 32°$; Day the year = March 30

Sunset hour angle, ω_s:

Sunset hour angle is calculated by using the relation:

$$\omega_s = \cos^{-1}(-\tan\phi\tan\delta) \qquad ...[\text{Eqn. (2.16)}]$$

Let us first calculate the value of δ by using the relation:

$$\delta = 23.45\sin\left[\frac{360}{365}(284 + n)\right] \qquad ...[\text{Eqn. (2.3)}]$$

Here, n =number of days = 89 for Mach 30.

$$\left[n = \underset{Jan.}{31} + \underset{Feb.}{28} + \underset{Mar.}{30} = 89\right]$$

$\therefore$

$$\delta = 23.45\sin\left[\frac{360}{365}(284 + 89)\right] = 3.22°$$

Hence,

$$\omega_s = \cos^{-1}(-\tan 32°.\tan 3.22°) = \textbf{88.2° (Ans.)}$$

Day-length, t_{day} (hours):

Day-length is calculated by using the relation:

$$t_{day} = \frac{2}{15}\left[\cos^{-1}(-\tan\phi\tan\delta)\right]\text{hours}$$

$$= \frac{2}{15}\times 88.2° \text{ (already calculated)} = \textbf{12.78 hours (Ans.)}$$

Example 2.2. *Determine the day-length in hours at Delhi (latitude = 28.6°) on June 28 in a leap year.*

Solution. *Given:* $\phi = 28.6°$; Day of the leap year = June 28.

Day-length, t_{day} (hours):

Day length is given as:

$$t_{day} = \frac{2}{15}[\cos^{-1}(-\tan\phi\tan\delta)]\text{ hours} \qquad ...[\text{Eqn. (2.17)}]$$

Declination (δ) is found by using Cooper's equation:

$$\delta = 23.45\sin\left[\frac{360}{365}(284 + n)\right] \qquad ...[\text{Eqn. (2.3)}]$$

Here,

$$n = \text{number of days} = 180 \text{ for June 28.}$$

$$\left[n = \underset{Jan.}{31} + \underset{Feb.}{29} + \underset{Mar.}{31} + \underset{Apr.}{30} + \underset{May}{31} + \underset{June}{28} = 180\right]$$

$\therefore$

$$\delta = 23.45\sin\left[\frac{360}{265}(284 + 180)\right] = 23.24°$$

Substituting the various values in the above equation, we get :

$$t_{day} = \frac{2}{15}\left[\cos^{-1}(-\tan 28.6°. \tan 23.24°)\right]$$

$$= \frac{2}{15}\cos^{-1}(-0.234) = \textbf{13.8 hours} \textbf{ (Ans.)}$$

Example 2.3. *Calculate Local Apparent Time (LAT) and declination at a location latitude 24° 20′ N, longitude 77°30′E at 12.30 IST on July 24. Equation of time correction = – (1′ 06″).*

Solution. *Given:* Latitude = 24° 21′ N; Longitude = 77°30′E′;

$$IST = 12.30; \text{Day of the year} = \text{July 24.}$$

Local apparent time (LAT or LST):

We know that;

$$LAT = IST - 4 \text{ (Standard time longitude} - \text{longitude of location)}$$
$$+ \text{(Equation of time correction)} \qquad ...[\text{Eqn. (2.20)}]$$

Inserting the values in the above eqn., we get:

$$LAT = 12^h 30′ - 4(82°30′ - 77°30′) - (1′06″)$$
$$= 12^h 30′ - 4 \times 5 - 1′06″ = \textbf{12}^h \textbf{ 8′ 54″ (Ans.)}$$

(IST is the local civil time corresponding to **82.5° E longitude.**

Declination) δ:

Using Cooper's equation, we obtain:

$$\delta = 23.45\sin\left[\frac{360}{365}(284 + n)\right] \qquad ...[\text{Eqn. (2.3)}]$$

Here, n = number of days = 205 for July 24

$$\left[n = \underset{\text{Jan.}}{31} + \underset{\text{Feb.}}{28} + \underset{\text{Mar.}}{31} + \underset{\text{Apr.}}{30} + \underset{\text{May}}{31} + \underset{\text{June}}{30} + \underset{\text{July}}{24} = 205\right]$$

$$\therefore \qquad \delta = 23.45\sin\left[\frac{360}{365}(284 + 205)\right] = \textbf{19.82° (Ans.)}$$

Example 2.4. *Calculate the angle made by beam radiation with the normal to a flat plate collector, pointing the south location in New Delhi (27° 30′N, 76° 42′E) at 10.00 hour solar time on October 29. The collector is tilted at an angle of 35° with the horizontal. Also calculate the day-length.*

Solution. *Given:* $\phi = 27°30′ = 27.5°$; solar time on October 29 = 10.00 hour (LST). Angle of tilt, $\beta = 35°$.

Incident angle, θ :

Since the surface is facing south, $\gamma = 0$, therefore, the using following relation, we get:

$$\cos\theta = \sin\delta\sin(\phi - \beta) + \cos\delta\cos\omega\cos(\phi - \beta) \qquad ...[\text{Eqn. (2.11)}]$$

Let us first calculate the declination (δ) by using the relation:

$$\delta = 23.45\sin\left[\frac{360}{365}(284 + n)\right] \qquad ...[\text{Eqn. (2.3)}]$$

Here, n = number of days = 302 for October 29

$$[n = 31 + 28 + 31 + 30 + 31 + 30 + 31 + 31 + 30 + 29 = 302]$$

$$\therefore \qquad \delta = 23.45\sin\left[\frac{360}{365}(284 + 302)\right] = -14.43°$$

Also, hour angle, $\omega = 15\,(12 - LST) = 15\,(12 - 10) = 30°$

Inserting various values in the above eqn., we have:

$$\cos\theta = \sin(-14.43°)\sin(27.5° - 35°)$$
$$+ \cos(-14.43)\cos 30°\cos(27.5° - 35°)$$
$$= 0.0325 + 0.8315 = 0.864$$

$\therefore$ $\qquad\qquad\qquad \theta = \textbf{30.23° (Ans.)}$

Day-length, t_{day}:

$$t_{day} = \frac{2}{15}\Big[\cos^{-1}(-\tan\phi\,\tan\delta)\Big] \text{ hour} \qquad\qquad ...\text{Eqn.(2.17)}$$

$$= \frac{2}{15}\Big[\cos^{-1}\{-\tan 27.5°.\tan(-14.43°)\}\Big]$$

$$= \textbf{10.97 hours (Ans.)}$$

Example 2.5. *Calculate the angle made by beam radiation with normal to a flat plate collector on November 30, at 9.00 A.M. solar time for a location at 27°30' N. The collector is tilted at an angle of latitude plus 12°, with the horizontal and is pointing due south.*

Solution. *Given:* $\phi = 27°30' = 27.5°$; Day of the year = November 30;

Local Solar Time (LST) = 9.00 AM; Tilt angle, $\beta = 27.5 + 12 = 39.5°$.

Angle made by the beam radiation, θ:

Since collector is pointing due south, therefore, $\gamma = 0$.

In such a case, we shall use the following equation:

$$\cos\theta = \sin\delta\sin(\phi - \beta) + \cos\delta\cos\omega\cos(\phi - \beta) \qquad ...[\text{eq. (2.11)}]$$

Declination, δ:

In order to calculate declination (δ), using Cooper's equation, we get:

$$\delta = 23.45\sin\Big[\frac{360}{365}(284 + n)\Big]$$

(Here, $\qquad\qquad n$ = number of days = 334 for November 30).

$\therefore$ $\qquad\qquad \delta = 23.45\sin\Big[\frac{360}{365}(284 + 334)\Big] = -21.97°$

Also, $\qquad\qquad \omega = 15\,(12 - LST) = 15\,(12 - 9) = 45°$

Inserting the various values in the above eqn., we obtain:

$$\cos\theta = \sin(-21.97°)\sin(27.5° - 39.5°) + \cos(-21.97°)\cos 45°$$
$$\cos(27.5° - 39.5°)$$

$$= 0.0778 + 0.6414 = 0.7192$$

or, $\qquad\qquad \theta = \cos^{-1}(0.7192) = \textbf{44.01° (Ans.)}$

Example 2.6. *Calculate the sun's attitude zenith and solar azimuth angles, at 9.00 A.M. solar time on August 30 at latitude 25°N.*

Solution. *Given*: Latitude, $\phi = 25°$; Local Solar Time (LST) on August 30 = 9.00 A.M.

Zenith angle, θ_z:

We know that: Hour angle, $\omega = 15\,(12 - LST)$

$$= 15\,(12 - 9) = 45°$$

δ (declination) can be found by using the relation:

$$\delta = 23.45 \sin\left[\frac{360}{365}(284 + n)\right]$$

where, n = number of days = 242 for August 30

$$\left[n = \underset{\text{Jan.}}{31} + \underset{\text{Feb.}}{28} + \underset{\text{Mar.}}{31} + \underset{\text{Apr.}}{30} + \underset{\text{May}}{31} + \underset{\text{June}}{30} + \underset{\text{July}}{31} + \underset{\text{Aug.}}{30} = 242\right]$$

$\therefore$ $\delta = 23.45 \sin\left[\frac{360}{365}(284 + 242)\right] = 8.48°.$

In order to calculate θ_z, use the relation:

$$\cos \theta_z = \sin \phi \sin \delta + \cos \phi \cos \delta \cos \omega \qquad ...[\text{Eqn. (2.10)}]$$

$$= \sin 25°. \sin 8.48° + \cos 25°. \cos 8.48°. \cos 45°$$

$$= 0.0623 + 0.6338 = 0.6961$$

$\therefore$ $\theta_z = \cos^{-1}(0.6961) = \textbf{45.88° (Ans.)}$

Solar azimuth angle, γ_s:

Using the relation:

$$\cos \gamma_s = \frac{\cos \theta_z \sin \phi - \sin \delta}{\sin \theta_z . \cos \phi} \qquad ...[\text{Eqn. (2.14)}]$$

$$= \frac{\cos 45.88°. \sin 25° - \sin 8.48°}{\sin 45.88°. \cos 25°}$$

$$= \frac{(0.2942 - 0.1475)}{0.6506} = 0.2254$$

$\therefore$ $\gamma_s = \cos^{-1}(0.2254) = \textbf{76.97° (Ans.)}$

Example 2.7. *Calculate the sun's altitude and azimuth angle at 8.30 A.M. solar time on March 18 for a location at 35° N latitude.*

Solution: *Given* : Latitude, ϕ= 35° N; Day of the year = March 18; Local solar time (LST) = 8.30 A.M.

Sun's altitude angle, α:

ω (hour angle) is calculated from the relation:

$$\omega = 15(12 - \text{LST})$$

$$= 15(12 - 8.5) = 52.5°.$$

δ (declination) can be found by using the relation:

$$\delta = 23.45 \sin\left[\frac{360}{365}(284 + n)\right]$$

where n = number of days = 77 for March 18.

$$\left[n = \underset{\text{Jan.}}{31} + \underset{\text{Feb.}}{28} + \underset{\text{Mar.}}{18} = 77\right]$$

$\therefore$ $\delta = 23.45 \sin\left[\frac{360}{365}(284 + 77)\right] = -1.613°$

α (altitude angle) can be calculated from the relation:

$$\sin \alpha = \sin \phi \sin \delta + \cos \phi \cos \delta \cos \omega \qquad ...[\text{Eqn. (2.10)}]$$

$$= \sin 35° \sin(-1.613°) + \cos 35° \cos(-1.613°) \cos 52.5°$$

$$= -0.0161 + 0.4984 = 0.5145$$

$$\therefore \quad \alpha = \sin^{-1}(0.5145) = \textbf{30.96° (Ans.)}$$

Solar azimuth angle, γ_s:

Using the relation:

$$\sin \gamma_s = \sec \alpha . \cos \delta . \sin \omega \qquad\qquad ...[Eqn. (2.73)]$$

$$= \sec 30.96°. \cos(-1.613°) \sin 52.5°$$

$$= 0.9248$$

or, $$\gamma_s = \sin^{-1}(0.9248) = \textbf{67.64° (Ans.)}$$

2.3. MEASUREMENT OF SOLAR RADIATION

It is important to measure solar radiation, owing to the increasing number of solar heating and cooling applications, and the necessity for *accurate solar radiation data to predict performance.*

The following *three* devices are used for measuring the solar radiations.

1. Pyranometer; 2. Pyrheliometers;
3. Sunshine recorders.

2.3.1. Pyranometer

A **pyranometer** is *a device used to measure the "total hemispherical solar radiation".* The total solar radiation arriving at the outer edge of the atmosphere is called the *'solar constant'.*

The *working principle* of this instrument is that sensitive surface is exposed to total (beam, diffuse and reflected from the earth and surrounding) radiations.

The description of a pyranometer is given below:

Refer to Fig. 2.9.

Construction. It consists of a "black surface" which receives the beam as well diffuse radiations which rises heat. A *"glass dome"* prevents the loss of radiation received by the black surface. A *"thermopile"* is a temperature sensor, and consists of a number of thermocouples connected in series to increase the sensitivity. The *"supporting stand"* keeps the black surface in a proper position.

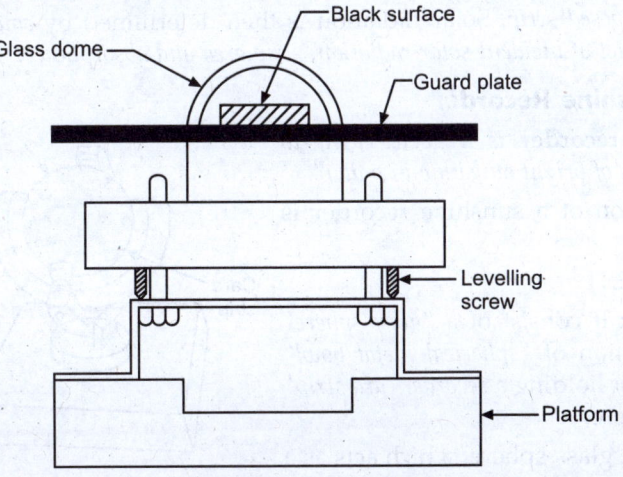

Fig. 2.9. Pyranometer.

Working: When the pyranometer is exposed to sun, it starts receiving the radiations. As a result, the surface temperature starts rising due to absorption of the radiation. The increase in the temperature of the absorbing surface is detected by the *thermopile*. The thermopile generates a *thermo emf* which is *proportional to the radiations absorbed;* this thermo emf is *calibrated in terms of the received radiations.* This will measure the global solar radiations.

2.3.2. Pyrheliometer

A **pyrheliometer** *is a device used to measure "beam or direct radiations".* It collimates the radiation to determine the beam intensity as a function of incident angle.

This instrument uses a collimated detector for measuring solar radiation from the sun and from a small portion of the sky around the sun at normal incidence.

The description of a thermoelectric type pyrheliometer is given below:

Refer to Fig. 2.10.

Construction: In this instrument, two identical blackened manganin strips A and B are arranged in such a way that either can be exposed to radiation at the base of *collimating tubes* by moving a *reversible shutter.*

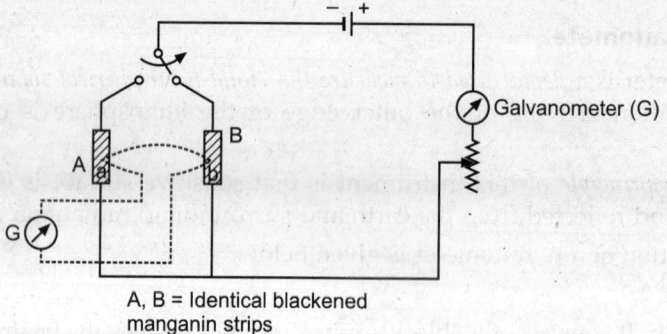

A, B = Identical blackened manganin strips

Fig. 2.10. Circuit diagram for the thermoelectric type pyrheliometer.

Working: One strip is placed in radiation and a current is passed through the *shaded strip* to heat it to the same temperature as the *exposed strip.* When there is *no difference in temperature,* the electrical energy supplied to shaded strip *must equal the solar radiation absorbed by the exposed strip.* Solar radiation is then determined by *equating the electrical energy to the product of incident solar radiation, strip area and absorptance.*

2.3.3. Sunshine Recorder

A **sunshine recorder** *is a device used to measure the "hours of bright sunshine in a day".*

The description of a sunshine recorder is given below:

Refer to Fig. 2.11.

Construction: It consist of a *"glass-sphere"* installed in a section of *"spherical metal bowl"* having grooves for holding a *recorder card strip"* and the glass sphere.

Working. The glass-sphere, which acts as a convex lens, focusses the sun's rays/beams to

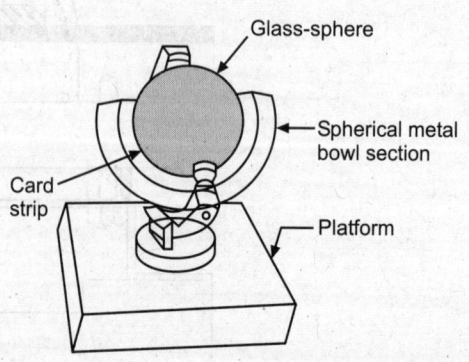

Fig. 2.11. Sunshine recorder.

a point on the card strip held in a groove in the spherical bowl mounted concentrically with the sphere.

Whenever there is a *bright sunshine, the image formed is intense enough to burn a spot on the card strip*. Through the day, the sun moves across the sky, the image moves along the strip. Thus a burnt space whose *length is proportional to the duration of sunshine* is obtained on the strip.

2.4. SOLAR RADIATION DATA

2.4.1. General Aspects

Solar radiation data should include the following informations:

(*i*) Whether they are *instantaneous* measurements or values *integrated* over some period of time (usually hour or days,) (*ii*) The *time* or *time period* of the measurements; (*iii*) Whether the measurements are of *beam, diffuse* or *total radiation* and the *instrument used*; (*iv*) The receiving surface *orientation* (usually horizontal, it may be inclined at a fixed slope or normal); (*v*) If averaged, the *period* over which they are averaged.

An instrument called **Solarimeter** is used to measure most of the data on solar radiation received on the surface of the earth. It gives *readings for instantaneous measurements at rate throughout the day for total radiation on a horizontal surface*. Integrating the plot of rate of energy received per unit area per unit time over a whole day gives the *"langleys"* of radiation received on a horizontal surface (the unit 'langley' has been adopted in honour of Samuel Langley who made the first measurement of the spectral distribution of the sun; 1 *langley = 1 cal/cm^2*).

- When data are *not* available, '**Maps**' can be used as a source of average radiation. **Charts** are also available for *clear day horizontal radiation for any period for any latitude*. **Tables** are also available for *hours of sunshine for various locations*.
- The solar radiation data is collected for various locations in the world on the basis of:
1. Solar power calculations with reference to the movement of the sun, latitude of the location etc.
2. Hourly measurements of solar radiation at the location and calculation of:
  (*i*) *"Daily average"* global radiation (H_{dg}) at the location for the month ($kJ/m^2.day$);
  (*ii*) *"Monthly average"* global radiation (H_{mg}) at the location for various months ($kJ/m^2. month$); (*iii*) *"Yearly average"* global radiation (H_{yg}) at the location for a few years ($kJ/m^2. year$).

The data in terms of kJ/m^2. day or kWh/m^2. day for various days/months/an year can be readily used for calculating:

(*i*) Available solar energy at the location;

(*ii*) Determining the surface area of the solar collectors;

(*iii*) Determining rating of solar plant.

2.4.2. Solar Radiation Data for India

- India is in the *"northern hemisphere"* within latitudes of 7° and 37.5° N.
- The average solar radiation values for India are between 12.5 and 22.7 MJ/m.^{2} day.
- The *peak solar radiation in India* occurs in some parts of Rajasthan and Gujrat and is equal to *25 MJ/m^2*.
- The solar radiation *reduces to about 60 percent* during monsoon months.

2.4.3. Solar Insolation

The **solar insolation** *is the solar radiation received on a flat horizontal surface at a particular location on earth at a particular instant of time.*

The *unit of solar radiation is* **W/m²**.

The *parameters of the solar insolation for a given flat horizontal surface are:*

(*i*) Daily variation (hour angle); (*ii*) Seasonal variation and geographical location of the particular surface; (*iii*) Atmospheric clarity; (*iv*) Shadows of trees, tall structures, adjacent solar panels etc.; (*v*) Degree of latitude for location; (*vi*) Surface area m²; (*vii*) Tilt angle.

2.5. ESTIMATION OF AVERAGE SOLAR RADIATION

"Angstrom's equation" **for average daily global radiation:**

The *expression* for the average radiation on a horizontal surface in terms of constants *a* and *b* and observed values of average length of solar days, as suggested by **Angstrom** (1924) is given by:

$$\frac{H_g}{H_c} = a + b\left(\frac{L_a}{L_m}\right) \qquad \qquad ...(2.21)$$

where, H_g = Monthly average of the daily global radiation on a horizontal surface at the location (kJ/m² day),

H_c = Monthly average of the daily global radiation on the same horizontal surface at the same location but on *clear* sky day (kJ/m². day),

a, b = Constants determined from various cities in the world by measurements,

L_a =   Average length of *solar day* for a particular month calculated/observed (hours), and

L_m = Length of the *longest solar day* in the month (hours).

Modified *Angstrom's equation:*

The modified Angstrom's equation is written by replacing H_c by H_0 in Eqn. (2.21) as:

$$\frac{H_g}{H_o} = a + b\left(\frac{L_a}{L_m}\right) \qquad \qquad ...(2.22)$$

where, H_o = Monthly average of the daily *extraterrestrial radiation*, which would fall on a horizontal surface at the location under consideration (kJ/m². day).

The expression for H_o is given by:

$$H_o = I_{sc}\left[1 + 0.033\cos\left(\frac{360n}{365}\right)\right]\int[(\sin\phi\sin\delta) + (\cos\phi\cos\delta\sin\omega)]\,dt \quad ...(2.23)$$

where, n = Number of days in the year,

I_{sc} = Solar constant = 4870.8 kJ/m² hour,

ϕ = Angle of latitude for the location,

δ = Angle of declination of equatorial plane, and

ω = Hour angle for local apparent time.

The integral $\int dt$ in Eqn. (2.23) is in terms of hours. It can be changed to angle $\int d\omega$ in radians.

Now, $1 \text{ hour} = 15° = \dfrac{15\pi}{180} \text{ rad.}$

$\therefore$ $dt = \dfrac{180}{15\pi}d\omega = \dfrac{12}{\pi}d\omega$

Inserting the value of dt in Eqn. (2.23), and integrating between *sunrise* $(-\omega_s)$ to sunset $(+\omega_s)$ for total day light period, we get:

$$H_o = \frac{12}{\pi} I_{sc} \left[1 + 0.003 \cos \left(\frac{360}{365} n \right) \right] \int\limits_{-\omega_s}^{+\omega_s} (\sin \phi \sin \delta + \cos \phi \cos \delta \cos \omega) d\omega \quad ...(2.24)$$

By solving the integral, we get

$$H_o = \frac{24}{\pi} I_{sc} \left[\left\{ 1 + 0.033 \cos \left(\frac{360}{365} n \right) \right\} (\omega_s \sin \phi \sin \delta + \cos \phi \cos \delta \sin \omega_s) \right] kJ/m^2. \, day$$

(where, ω_s is in radian) ...(2.25)

From Eqn. (2.25), the value of H_o is calculated in terms of kJ/m^2. day for a *clear sky day*, for a *flat surface location*.

Hence, the value of H_g (Daily global radiation-monthly average) for a horizontal surface at the location is then calculated by using the values of H_o, a, b in Eqn. 2.22.

- H_o *can be obtained from charts or can be calculated* by the *following* **Empirical relation:**

$$H_o = \frac{24}{\pi} I_{sc} \left[\left\{ 1 + 0.033 \cos \left(\frac{360}{365} n \right) \right\} \left(\cos \phi \cos \delta \sin \omega_s + \frac{2\pi\omega_s}{360} \sin \phi \sin \delta \right) \right] ...(2.26)$$

where, I_{sc} = Solar constant per hour,

        n = Day of the year, and

        ω_s = Sunrise hour angle (*degrees*).

The '*declination* δ' *can be obtained from* '*Cooper equation*' and the sunrise hour angle (*i.e.,* ω_s) from the following equation:

$$\omega_s = -\tan \phi \tan \delta$$

Example 2.8. *The following observations were made in Bhopal during the mouth of March:*

Average length of the day = 8.4 hours; Longest day during the month = 9 hours; Angstrom's constants for Bhopal: a = 0.27, b = 0.50; Solar radiation per day for a clear day = 2100 J/m². day. Calculate the average daily global radiation.

Solution.   *Given:* L_a = 8.4 hours, L_m = 9 hours; a = 0.27; b = 0.50;

     H_c = 2100 J/m^2. day.

Average daily global radiation, H_g:

Angstrom's equation is written as:

$$\frac{H_g}{H_c} = a + b \left(\frac{L_a}{L_m} \right) \quad ...[Eqn. \, (2.21)]$$

or, $$H_g = H_c \left[a + b \left(\frac{L_a}{L_m} \right) \right]$$

Inserting the various values in the above eqn., we get:

$$H_g = 2100 \left[0.27 + 0.50 \left(\frac{8.4}{9.0} \right) \right] = \textbf{1547 kJ/m}^2\textbf{.day (Ans.)}$$

Example 2.9. *Calculate the average value of solar radiation on a horizontal surface located in Ahmedabad (22°00'N, 73° 10'E), for May 28. Average solar day hours are 10.5 hours. Angstrom's constants are: a = 0.28, b = 0.48.*

Solution: *Given*: ϕ = 22°, Average solar day hours L_a = 10.5 hours; Angstrom's constants: a = 0.28, b = 0.48.

Monthly average of daily global radiation, H_g:

We know that, $\quad\quad\quad H_g = H_o\left[a + b\left(\dfrac{L_a}{L_m}\right)\right]$ ...[Eqn. (2.22)]

where, $H_o = \dfrac{24}{\pi} I_{sc}\left[\left\{1 + 0.033\cos\left(\dfrac{360}{365}n\right)\right\}(\omega_s \sin\phi \sin\delta + \cos\phi \cos\delta \sin\omega_s\right]$ kJ/m². day

...[Eqn. (2.25)]

Calculations for δ, ω_s and L_m:

$$\delta = 23.45\sin\left[\dfrac{360}{365}(284 + n)\right]$$ ... [Eqn. (2.3)]

Here, $\quad\quad\quad n$ = number of days = 148 for May 28.

$\quad\quad\quad\quad\quad\quad [n = 31 + 28 + 31 + 30 + 28 = 148]$

∴ $\quad\quad\quad \delta = 23.45\sin\left[\dfrac{360}{365}(284 + 148)\right] = 24.44°$

Now, sunshine hour angle, $\omega_s = \cos^{-1}(-\tan\phi \cdot \tan\delta)$ ... [Eqn. (2.16)]

Hence, $\quad\quad\quad \omega_s = \cos^{-1}(-\tan 22° \cdot \tan 24.44°) = 100.58°\ (= 1.755\ \text{rad.})$

Now, $\quad\quad\quad L_m = \dfrac{2}{15}\omega_s = \dfrac{2}{15} \times 100.58 = 13.41$ hours

Substituting the various values in the eqn. of H_0, we obtain:

$$H_o = \dfrac{24}{\pi} \times 4870.8\left[\left\{1 + 0.033\cos\left(\dfrac{360}{365} \times 148\right)\right\}\right.$$

$$(1.755 \sin 22° \sin 24.44° + \cos 22° \cos 24.44° \sin 100.58°$$

$$= 37210\ (0.9726)\ (0.2720 + 0.8297) = 39871\ \text{kJ/m}^2.\text{day}.$$

Substituting the values in eqn. of H_g (above), we get:

$$H_g = 39871\left[0.28 + 0.48\left(\dfrac{10.5}{13.41}\right)\right]$$

$$= 26149\ \text{kJ/m}^2.\text{day}\ \left(26149 \times \dfrac{1353}{4870.8}\right)$$

$$= 7263.6\ \text{W/m}^2.\ \textbf{day (Ans.)}$$

Example 2.10. *Calculate the average value of global solar radiation on a horizontal surface for March 21, at the latitude of 12°N if the constants are given as equal to 0.28 and 0.50 respectively. The ratio of average length of solar day and length of the longest solar day is 0.68.*

Solution. *Given:* $\quad\quad \phi = 12°;\ a = 0.28;\ b = 0.50;\ \dfrac{L_a}{L_m} = 0.68\ ;\ Day\ of\ year = March\ 21.$

Average value of global solar radiation; H_g:

$$H_g = H_o\left[a + b\left(\dfrac{L_a}{L_m}\right)\right]$$ ...[Eqn. (2.22)]

Declination, $\quad\quad \delta = 23.45\sin\left[\dfrac{360}{365}(284 + n)\right]$ ...[Eqn. (2.3)]

Here, $\quad\quad\quad n = 31 + 28 + 21 = 80$ for March 21.

∴ $\quad\quad\quad \delta = 23.45\sin\left[\dfrac{360}{365}(284 + 80)\right] = -0.404°$

Sunshine hour angle, $\omega_s = \cos^{-1}(-\tan\phi\tan\delta)$ ... [Eqn. (2.16)]

$$= \cos^{-1}[-\tan 12° \tan(-0.404°)] = 89.9°$$

The empirical relation for average insolation at *the top of the atmosphere* is given by:

Now, $H_o = \dfrac{24}{\pi} I_{sc}\left[\left\{1 + 0.033\cos\left(\dfrac{360}{365}n\right)\right\}\left(\cos\phi\cos\delta\sin\omega_s + \dfrac{2\pi\omega_s}{360}\sin\phi\sin\delta\right)\right]$ ...[Eq. 2.26]

$$= \dfrac{24}{\pi} \times 4870.8\left[\left\{1 + 0.033\cos\left(\dfrac{360}{365}\times 80\right)\right\}\{\cos 12° \cos(-0.404)\sin 89.9°\right.$$

$$\left. + \dfrac{2\pi\times 89.9}{360}\times\sin 12°\times\sin(-0.404)\}\right]$$

$$= 37210\,(1.006)\,(0.9781 - 0.0023) = 36527 \text{ kJ/m}^2.\text{ day}$$

Hence, $H_g = 36527\,(0.28 + 0.50 \times 0.68) = \mathbf{22647}$ **kJ/m².day (Ans.)**

2.6. SOLAR RADIATION ON AN INCLINED SURFACE

The following three types of solar radiation constitute the total solar radiation on a surface:

   (i) Beams solar radiation (I_b);

   (ii) Diffuse solar radiation (I_d);

  (iii) Solar radiation reflected from the ground and the surroundings.

Usually, I_b and I_d on a horizontal surface are recorded. In case of *non-availability of data* for beam and diffuse radiation, the following *expression* for beam and diffuse radiation on the horizontal surface may be used:

$$I_b = I_N \cos\theta_z \qquad\qquad ...(2.27)$$

and, $$I_d = \dfrac{1}{3}(I_{ext.} - I_N)\cos\theta_z \qquad\qquad ...(2.28)$$

Li and Jordon (1962) suggested the following *formula* to evaluate *total radiation on a surface of "arbitrary orientation"*:

$$I_T = I_b R_b + I_d R_d + \rho R_r (I_b + I_d) \qquad\qquad ...(2.29)$$

where, R_b, R_d and R_r = "*Conversion factors*" for *beam, diffuse* and *reflected* components respectively;

          ρ = The reflection coefficient of the ground

             = 0.2 and 0.7 for ordinary and snow covered ground respectively.

Expressions for Conversion factors:

Refer to Fig. 2.12.

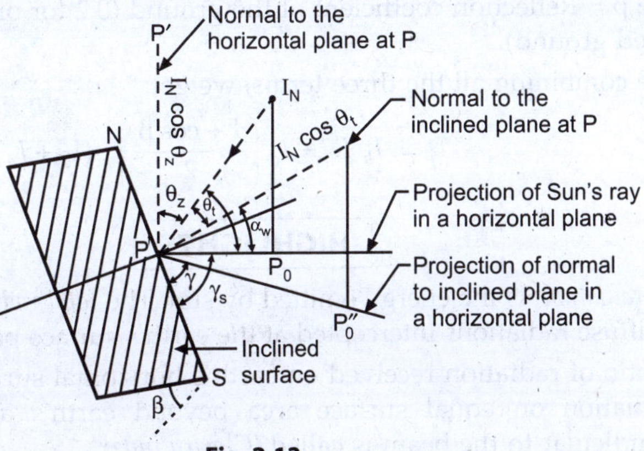

Fig. 2.12

(i) **R_b:** It is defined as the ratio of flux of beam radiation (I_b) incident on an inclined surface (I_{bt})to that on a horizontal surface. It is called *tilt factor* for beam radiation.

$$I_b = I_N \cos \phi_z \qquad \text{...on horizontal surface}$$

and, $$I_{bt} = I_N \cos \phi_t \qquad \text{...on tilted/inclined surface}$$

Mathematically, $$R_b = \frac{I_b}{I_{bt}} = \frac{I_N \cos \theta_t}{I_N \cos z} = \frac{\cos \theta_t}{\cos z} \qquad ...(2.30)$$

where, I_N = Intensity of beam radiation,

θ_t = Angle of incidence on the inclined surface (it depends on several variables associated with the location and orientation of the surface and the direction of sun rays), and

θ_z = Angle of incidence on the horizontal surface.

For beam radiation, in most cases, the *tilted surface* faces due south *i.e.*,

$\gamma = 0$, for this case,

$$\cos \theta = \cos \theta_t = \sin \delta \sin (\phi - \beta) + \cos \delta \cos \omega \cos (\phi - \beta)$$

For *horizontal force* ($\theta = \theta_z$),

$$\cos \theta = \cos \theta_z = \sin \phi \sin \delta + \cos \phi \cos \delta \cos \omega$$

Hence, $$R_b = \frac{\cos \theta_t}{\cos \theta_z} = \frac{\sin \delta \sin (\phi - \beta) + \cos \delta \cos \omega \cos (\phi - \beta)}{\sin \phi \sin \delta + \cos \phi \cos \delta \cos \omega}$$

$$...[2.30(a)]$$

(ii) **R_d:** It is the ratio of the flux of diffuse radiation falling on the tilted surface to that on the horizontal surface.

If one "*radiation shape factor*" for the tilted surface with respect to sky, is $\frac{1 + \cos \beta}{2}$, then

$$R_d = \frac{1 + \cos \beta}{2} \qquad ...(2.31)$$

(iii) **R_r:** The reflected component comes *mainly from the ground and other surrounding objects.* If the considered reflected radiation is *diffuse and isotropic*, then the situation is opposite to that in the above case, and

$$R_r = (1 - \cos \beta)\frac{\rho}{2}$$

where ρ = Reflection coefficient of the ground (0.2 for ordinary and 0.7 for snow covered ground).

Hence combining all the three terms, we get:

$$I_T = I_b R_b + I_d \left(\frac{1 + \cos \beta}{2} \right) + (I_b + I_d) \left(\frac{1 - \cos \beta}{2} \right) \rho \qquad ...(2.32)$$

HIGHLIGHTS

1. '*Solar radiation*' is the energy emitted by sun. The *total radiation* is the sum of beam and diffuse radiations intercepted at the earth's surface per unit area of location.

2. The ratio of radiation received on earth's horizontal surface over a given period to radiation on equal surface area beyond earth's atmosphere in direction prependicular to the beam is called '*Clarity index*'.

3. The declination (δ) in degrees for any day may be calculated from the approximate equation of Cooper:

$$\delta = 234.5 \sin\left[\frac{360}{365}(284 + n)\right]$$

   where, n is the day of the year.

4. Hour angle (ω) may be calculated from the follow relation:

$$\omega = 15\,(12 - LST)$$

   where, LST means local solar time.

5. *General equation* for $\cos\theta$ is given

$$\cos\theta = [\sin\phi\,(\sin\delta\cos\beta + \cos\delta\cos\gamma\cos\omega\sin\beta)$$
$$+ \cos\phi\,(\cos\delta\cos\omega\cos\delta - \sin\delta\cos\gamma\sin\beta)$$
$$+ \cos\delta\sin\gamma\sin\omega\sin\beta]$$

(i) *For surface facing south,* $\gamma = 0$:

$$\cos\theta = \sin\phi\,(\sin\delta\cos\beta + \cos\delta\cos\omega\sin\beta\,)$$
$$+ \cos\phi\,(\cos\delta\cos\omega\cos\beta - \sin\delta\sin\beta)$$

   or, $\cos\theta = \sin\delta\sin(\phi - \beta) + \cos\delta\cos\omega\cos(\phi - \beta)$

(ii) *For horizontal surface,* $\beta = 0°;\ \theta = \theta_z$:

$$\cos\theta\,(= \cos z) = \sin\phi\sin\delta + \cos\phi\cos\delta\cos\omega$$

   Also, $\cos\theta = \cos\theta_z = \sin\alpha$ $\left(\because \theta_z = \dfrac{\pi}{2} - \alpha\right)$

(iii) *For vertical surface;* $B = 90°$:

$$\cos\theta = \sin\phi\cos\delta\cos\gamma\cos\omega - \cos\phi\sin\delta\cos\gamma$$
$$+ \cos\delta\sin\gamma\sin\omega$$

   For vertical surface facing south, $B = 90°,\ \gamma = 0$:

$$\cos\theta = \sin\phi\cos\delta\cos\omega - \cos\phi\sin\delta$$

   Also, $\cos\gamma_s = \dfrac{\cos\theta_z\sin\phi - \sin\delta}{\sin\theta_z\cos\phi}$

   where: θ = Incident angle; α = Altitude angle; ϕ = Latitude angle;
                 δ = Declination; ω = Hour angle; θ_z = Zenith angle;
                 γ_s = Solar azimuth angle;
                 γ = Surface azimuth angle; β = slope or tilt angle.

6. The value of ω_s (hour angle for sunrise and sunset) can be calculated from the following relation:

$$\omega_s = \cos^{-1}(-\tan\phi\tan\delta)$$

7. Total day-length (t_{day}) is given as:

$$t_{day} = \frac{2}{15}\left[\cos^{-1}(-\tan\phi\tan\delta)\right]\text{hours}$$

8. In India standard time is based on *"82.5° E longitude"*.

9. Solar radiation may be measured by following devices:

   (i) Pyranometer;

   (ii) Pyrheliometers;

   (iii) Sunshine recorder.

10. Angstrom's equation for the estimation of average solar radiation is given by:

$$\frac{H_g}{H_o} = a + b\left(\frac{L_a}{L_m}\right)$$

Here,

$$H_o = \frac{24}{\pi} I_{sc}\left[\left\{1 + 0.033 \cos\left(\frac{360}{365}n\right)\right\}\right.$$

$$\left.\left(\cos\phi \cos\delta \sin\omega_s + \frac{2\pi\omega_s}{360}\sin\phi \sin\delta\right)\right]$$

where, H_g = Monthly average of daily global radiation on a horizontal surface at the location (kJ/m^2. day)

H_o = Monthly average of the daily extraterrestrial radiation, which would fall on a horizontal surface at the location under consideration,

I_{sc} = Solar constant, 4870.8 kJ/m^2 hour,

a, b = Constants determined from various cities in the world by measurements,

L_a = *Average length of solar day* for a particular month calculated/observed (hours), and

L_m = Length of the *longest solar day* in the month (hours).

11. Liu and Jorden (1962) suggested the following formula to calculate total radiation on a surface of "arbitrary orientation":

$$I_T = I_b R_b + I_d\left(\frac{1 + \cos\beta}{2}\right) + (I_b + I_d)\left(\frac{1 - \cos\beta}{2}\right)\rho$$

where, I_T = Total radiation; I_b = Beam radiation; R_b = Tilt factor for beam rodiation; I_d = Diffuse radiation; β = Tilt angle; ρ = Reflection coefficient of the ground (0.2 for ordinary and 0.7 for snow covered ground).

THEORETICAL QUESTIONS

1. List the renewable energy sources which find their origin in sun.
2. State the advantages, disadvantages and applications of solar energy.
3. Define the following terms as applied to solar energy:
   (*i*) Solar radian;   (*ii*) Extraterrestrial radiation;
   (*iii*) Beam radiation;   (*iv*) Diffuse radiation.
4. State the reasons for variation in solar radiation reaching the earth than received at the outside of the atmosphere.
5. What is solar constant? Explain briefly.
6. What is air mass?
7. Define the terms: (*i*) Clarity index; (*ii*) Concentration ratio.
8. Define the following angles:
   (*i*) Latitude angle;   (*ii*) Hour angle;
   (*iii*) Zenith angle;   (*iv*) Surface azimuth angle.

9. Give the solar altitude angle expression for angle between incident beam and normal to a plane surface.

10. Explain briefly the following:
    (i) Day-length; (ii) Local apparent time (LAT)

11. Explain briefly the following devices used for the measurement of solar radiation:
    (i) Pyranometer; (ii) Pyrheliometer

12. Write a short note on 'Sun recorder'.

13. Define the term 'solar insolation'. State the parameters of solar insolation for a given flat horizontal surface.

14. Discuss briefly the "Angstrom's equation" used for the estimation of average solar radiation.

UNSOLVED EXAMPLES

1. Calculate the number of day-light hours at Delhi on June 21 in a leap year.
   (Ans. 13.82 hours)

2. Calculate the sunset hour angle and day-length at a location latitude of 35°N, on Feb. 14. **(Ans. 80.23°; 10.7 hours)**

3. Calculate the local apparent time (LAT) and declination at Ahmedabad (longitude 72°41'; E, latitude 23°00'N) corresponding to 1430, LST on December 15. (Given E = 5'13"). **(Ans. 13 45' 27"; –23° 20' 6.79")**

4. Calculate the zenith and solar azimuth angles at 9 A.M. solar time on September 1 at latitude 23°N. **(Ans. 45.87°; 78.01°)**

5. Calculate the angle made by beam radiation with the normal to a flat plate collector, pointing due south location in New Delhi (28° 38'N, 77° 17'E) at 9:00 hour solar time on December 1. The collector is tilted at an angle of 36° with the horizontal. Also calculate the day-length. **(Ans. 45.7°; 10.28 hours)**

6. Calculate the daily global radiation on a horizontal surface at Baroda (22° 13' 13N, 73°13'E) during the month of March if constants a and b are given equal to 0.28 and 0.48 respectively and average sunshine hours for day are 9.5.
   (Ans. 22485 kJ/m². day)

7. Calculate monthly average of daily global solar radiation on a horizontal surface located in Ahmedabad, (22° 00'N, 73° 10'E) for the month of April. Average solar day hours are 10 hours. Angstrom's constant for Ahmedabad are $a = 0.28$, $b = 0.48$. **(Ans. 28270 kJ/m². day)**

8. Estimate the average value of solar radiation on a horizontal surface for June 22, at the latitude of 10° N, if constants a and b are given as equal to 0.31 and 0.51 respectively. The ratio of average length of solar day and length of longest solar day is 0.55. **(Ans. 21182. 4 kJ/m². day)**

3

SOLAR THERMAL ENERGY COLLECTORS

3.1 Solar thermal energy – General aspects – Collectors in various ranges and applications – Principles (physical) of conversion of solar energy into heat – Greenhouse effect – Collector systems – Characteristic features of a collector system – Factors adversely affecting collector system's efficiency; 3.2 Types of collectors; 3.3. Flat-plate collector (FPC) – Description – Selective absorber coatings/surfaces – Advantages, disadvantages and applications of flat-plate collectors – Evacuated collectors – Performance analysis of a flat-plate collector; 3.4 Concentrating (or focusing) collectors – Need of orientation in concentrating collectors – Types of concentrating collectors – Advantages and disadvantages of concentrating collectors – Parabolic trough collector – Mirror-strip collector – Fresnel lens collector – Flat-plate collector with adjustable mirrors – Compound parabolic concentrator (CPC) – Paraboloidal disk collector – Comparison between flat-plate and concentrating collectors – Performance analysis of a concentrating collector; 3.5 Solar-thermodynamic conversion. *Highlights – Theoretical Questions – Unsolved Examples.*

3.1 SOLAR THERMAL ENERGY

3.1.1. General Aspects

The solar thermal energy is a *clean, cheap* and *abundantly available renewable energy* which has been used since ancient times. The sun is a sustainable source of providing solar energy in the form of radiations, *visible light* and *infrared radiation*. This solar energy is captured naturally by different surfaces to produce thermal effect or to produce electricity by means of photovoltaic or day lighting of the buildings. Solar energy can be converted into *'thermal energy' by using solar collector*. It can be converted into *'electricity'* by *using photovoltaic cell*.

'Solar collector' surface is designed for *high absorption and low emission*.

Advantages:

1. Solar energy is *easily and abundantly* available.

2. It is *re-usable* source of energy.

3. It is *eco-friendly* (*i.e.* pollution free).

4. It reduces Green-house gas emissions.

Disadvantages:

1. Availability is *limited* to sun hours.

2. *Need of storage.*

3. Large area entails *high capital cost.*

4. Owing to change in the position of sun, *tracking is required.*

Applications:

1. Solar energy is used in solar water heating.

2. It is used for solar pumping.

3. It is employed in solar distillation.

4. It finds use in solar cooking.

5. It is used in the generation of electric power.

- In the solar energy utilisation, the first step is the *collection* of this energy. This is done through *"collectors'* whose surfaces are designed for *high absorptivity and low emissivity.*

Solar energy conversion can be achieved by the following two completely different routes:

    (i) Solar thermodynamic;   *(ii) Solar-photovoltaic.*

When an object receives radiant energy, a proportion, depending upon the angle of incidence and nature of surface, is *reflected,* a part is *absorbed* and some of it *transmitted* through the object. With a few important exceptions (*e.g.,* photovoltaic cells), energy of the absorbed radiation is *rapidly degraded to heat.*

The temperature attained is determined by a *balance between the input of absorbed energy, the rate of heat removal and the heat loss to the environment.* The heat loss *increases* with temperature and *limits the ultimate temperature attained by a 'collector system'.* It also *reduces* the proportion of useful heat extractable from the system. The *highest temperature* and *maximum output of useful power* are therefore obtained when a *highly absorbent, well-insulated body* is exposed to a *high intensity of solar radiation.*

- Solar collectors, based on their geometry, can be divided into a number of generic types. These *vary in efficiency* and, consequently, *useful heat output, depending on demand temperature,* as shown in Fig. 3.1.

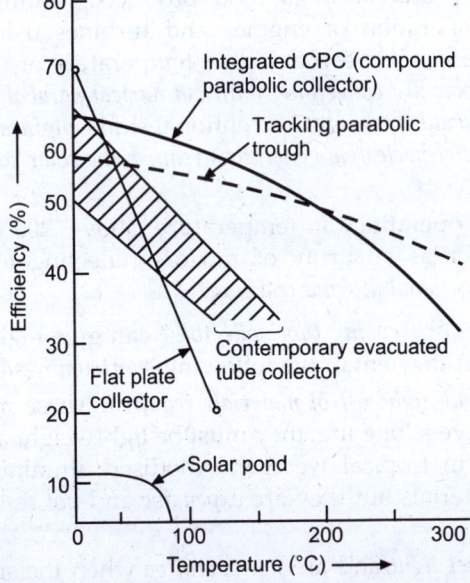

Fig. 3.1. Efficiency for solar collectors (Halcrow/ITP, 1983b).

3.1.2. Collectors in Various Ranges and Applications

The following list gives the *thermal applications of solar energy and possible temperature ranges*:

1. **Low temperature:**
   ($t = 100°C$)

   (*i*) Water heating

   (*ii*) Space heating $\Big\}$ ...*Flat plate*

   (*iii*) Space cooling

   (*iv*) Drying.

2. **Medium temperature:**
   (t: 100 to 200°C)

   (*i*) Vapour engines and turbines

   (*ii*) Process heating $\Big\}$ ...*Cylindrical Parabola*

   (*iii*) Refrigeration

   (*iv*) Cooking.

3. **High temperature:**
   ($t > 200°C$)

   (*i*) Steam engines and turbines

   (*ii*) Stirling engine $\Big\}$ ...*Parabolloid Mirror arrays*

   (*iii*) Thermo-electric generators.

The above classification of low, medium and high temperature ranges is *somewhat arbitrary*.

— *Heating water* for domestic applications, space heating and cooling and drying of agricultural products (and industrial products) is generally at temperature *below 100°C, achieved using "flat plate collectors" with one or two glass plate covers.*

— *Refrigeration* for preservation of food products, heating for certain industrial processes, and operation of engines and turbines using low boiling organic vapours is possible at somewhat higher temperature of 100 to 200°C and may be *achieved using "focusing collectors" with cylindrical-parabola reflectors requiring only one directional diurnal tracking.* Conventional *steam engines and turbines, stirling hot air engines,* and *thermoelectric generators require the solar collectors to operate at high temperatures.*

— Solar collectors operating at temperature above 200°C generally consist of *parabolloid reflector* as an array of mirrors reflecting to a central target, and requiring *two directional diurnal tracking.*

☞ ● The *"concentrators or focusing type collectors"* can give high temperatures than *flat plate collectors,* but they entail the following *shortcomings/limitations.*

1. *Non-availability and high cost of materials required.* These materials must be easily shapeable, yet have a long life; they must be lightweight and capable of retaining their brightness in tropical weather. Anodised aluminium and stainless steel are two such materials but they are *expensive* and *not readily available in sufficient quantities.*

2. They require *direct light* and are *not operative* when the sun is *even partly covered with clouds.*

3. They *need tracking systems* and reflecting surfaces undergo deterioration with the passage of time.

4. These devices are also subject to *similar vibration and movement problems as radar antenna dishes.*

3.1.3. Principles (physical) of Conversion of Solar Energy into Heat— Green-house Effect

When solar radiation from the sun, in the form of light (a *shortwave radiation*), reaches earth, visible sunlight is absorbed on the ground and converted into heat energy but *non-visible light* is re-radiated by earth (a *longwave radiation*). CO_2 in atmosphere *absorbs this light and radiates back a part of it to the earth*, which results in the *increase in temperature*. This whole process is called **Green-house effect**. Hence, the Greenhouse effect brings about an accumulation of energy of the ground.

- The name *'Green-house effect'* related to its first use in green houses, in which it is possible to *grow exotic plants in cold climes through better utilisation of the available light.*

3.1.4. Collection Systems

- **Solar thermal collection system:**

  A solar thermal collection system works in the following manner:

  (*i*) It gathers the heat from the solar radiation and gives it to *the heat transport fluid* (also called *primary coolant*).

  (*ii*) The fluid delivers the heat to the *thermal storage tank* (*viz.* boiler steam generator, heat exchanger etc.).

  (*iii*) The storage system *stores heat for a few hours*. The heat is released during cloudy hours and at night.

- **Thermal-electric conversion system:**

  This system receives thermal energy and drives steam turbine generator or gas turbine generator. The electrical energy is supplied to the electrical load or to the grid.

- **Co-generation plants:**

  In *co-generation plants heat in the form of hot water or steam may also be supplied to the consumer in addition to the electrical energy.* In this case, hot water/steam from the reservoir may be pumped through outlet pipes to the load side.

3.1.5. Characteristic Features of a Collector System

The characteristic features of a collector system include the following:

1. The type of collector – *Focussing* or *non-focussing*.

2. The temperature working fluid attained – *Low* temperature, *medium* temperature, *high* temperature.

3. *Non-tracking* type or *tracking* in one plane or tracking in two planes.

4. *Distributed* receiver collectors or *central* receiver collectors.

5. *Layout* and *configuration* of collectors in the solar field.

6. *Simple* and *low cost* or *complex* and *costly*.

- *'Solar collector cost'* is a significant *component of installation cost*. Hence it is important to keep *unit cost of collectors low* and *total surface area of collectors as small as possible.*

- 'Flat plate collectors' are used for *low temperature applications only*. They are *not* economical for high temperature applications. They are *not* suitable for high temperature applications and solar electric power plants.

3.1.6. Factors Adversely Affecting Collector System's Efficiency

The following *factors which adversely affect the efficiency of a collector system* are: *Shadow, Cosine loss, Dust etc.*

1. Shadow factor:

When the angle of elevation of the sun is *less than 15°* (*i.e.* around sunrise and sunset), the *shadows* of some of the neighbouring collector panels fall on the collector's surface. The shadow effect is reduced with the increase of sun's elevation angle.

The shadow factor is given as:

$$Shadow\ factor = \frac{Collector's\ surface\ receiving\ light}{Total\ collector's\ surface}$$

Its value is less than *0.1* when the angle of elevation of sun is less than 15° and *1* during noon when angle of sun's elevation angle is nearly 90°.

2. Cosine loss factor:

When the collector's surface receives the sun rays *perpendicularly, maximum power collection* is realised. If the angle between the perpendicular to collector's surface and the direction of sun ray is θ, the area of sun beam intercepted by the collector's surface is *proportional to* cos θ. Hence solar power collected in proportional to cos θ (Fig. 3.2).

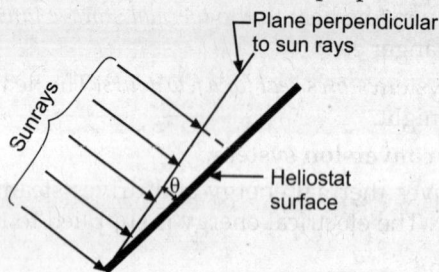

Fig. 3.2. Exhibiting cos θ loss.

- In case of *fixed type collector panels* cosine loss *varies* due to the daily variation and seasonal variation of the direction of sun rays.

3. Reflective loss factor:

The glass surface of the *collector* and the surface of the *reflector* collect dust, dirt and moisture. As a result, the reflector surface gets rusted, deformed and looses the shine. Hence, with the passage of time, the collector's efficiency is *reduced* significantly. Thus, to prevent the loss, daily maintenance, seasonal maintenance and yearly overhaul (change of seals, cleaning after dismantling) should be undertaken.

3.2 TYPES OF COLLECTORS

A. Solar collectors are broadly *classified* into the following types:

1. *"Non-concentrating"* or *"Flat-plate type solar collector"*.

In such collectors, the area of a collector to grasp the solar radiation is *equal to the absorber plate and has concentration ratio of 1.*

2. *"Concentrating"* or *"Focusing type solar collector"*.

In these collectors, the area of collector is kept *less than the aperture* through which the radiation passes, to concentrate the solar flux and has *high concentration ratio*.

B. Solar collectors may be *categorised* as follows:

   1. Flat-plate collectors

   2. Evacuated collectors

   3. Solar ponds

   4. Stationary concentrators

   5. Linear-focus collectors

   6. Point-focus collectors

   7. Central receivers.

● One of the *disadvantages of concentrating solar collectors is the need to align the collector's aperture with the sun's direct beam. This not only consumes power but also increases costs and the risk of failure. A single axis, tracking, time-focus, solar collector may use a number of "tracking mechanisms"*.

3.3 FLAT-PLATE COLLECTORS (FPC)

3.3.1. Description

Fig. 3.3 shows a Flat Plate Collector which consists of *four essential components*:

1. **An absorber plate.** It *intercepts* and *absorbs* solar radiation. This plate is usually metallic (copper, aluminium or steel), although plastics have been used in some low temperature applications. In most cases it is *coated with a material to enhance the absorption of solar radiation. The coating may also be tailored to minimise the amount of infrared radiation emitted*.

A *heat transport fluid* (usually air or water) is used to extract the energy collected and passes over, under or through passages which form an integral part of the plate.

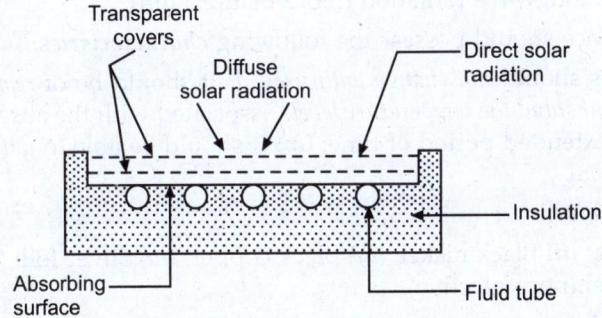

Fig. 3.3. Flat-plate solar collector.

2. **Transparent covers.** These are one or more sheets of solar radiation transmitting materials and are placed above the absorber plate. They allow solar energy to reach the absorber plate while reducing convection, conduction and re-radiation heat losses.

3. **Insulation beneath the absorber plate.** It *minimises and protects* the absorbing surface from heat losses.

4. **Box-like structure.** It contains the above components and keeps them in position.

- Various types of flat-plate collectors have been designed and studied. These include *tube in plate, corrugated type, spiral wound type etc.* Other criteria is *single exposure, double exposure or exposure and reflector type.* The collector utilizes sheets of any of the *highly conducting material viz.* copper, aluminium, or galvanized iron. The *sheets are painted dead black for increasing the absorptivity.* The sheets are provided with one or more glass or plastic covers with *air gap in between to reduce the heat transfer losses.* The sides which are not exposed to solar radiation are *well insulated.* The whole assembly is fixed in airtight wooden box which is mounted on *simple device* to give *the desired angle of inclination.* The dimensions of collectors should be such as to make their handling easy. The collector will absorb the sun energy (*direct as well as diffused*) and transfer it to the fluid (air, water or oil) flowing within the collector.

  Basically, a flat-plate collector is *effective* most of time, *reliable* for good many years and also *inexpensive.*

- Use of *flat mirrors in the flat-plate collectors improves the output, permitting higher temperatures of operation. Side mirrors are used either at north and south edges or at east and west edges of the collector or a combination of both.* The mirrors may be of reversible or non-reversible type.

Materials for flat-plate collectors:

1. *Absorber plate*: Copper, Aluminium, Steel, Brass, Silver etc.
2. *Insulation*: Crown white wool, Glass wool, Expanded polystrene, foam etc.
3. *Cover plate*: Glass, Teflon, Tedlar, Marlex etc.

3.3.2. Selective Absorber Coatings/Surfaces

In order to *reduce thermal losses* from the absorber plate of a solar heating panel, an efficient way is to *use selective absorber coatings.* An ideal selective coating is a *perfect absorber of solar radiation* as well as a *perfect reflector of thermal radiation.* A selective coating, thus, increases the temperature of an absorbing surface.

A *"selective surface"* has a *high absorptance* for shortwave radiation (less than 2.5 μm) and *low emittance* of longwave radiation (more than 2.5 μm).

A selective surface should possess the following *characteristics:*

(*i*) Its properties should *not change with use;* (*ii*) It should be of *reasonable cost;* (*iii*) It should be able to *withstand the temperature levels* associated with the absorber plate surface of a collector over extended period of time; (*iv*) It should be able to *withstand atmospheric corrosion and oxidation.*

Some selective coatings are:

(*i*) Black chrome; (*ii*) Black nickel; (*iii*) Black copper; (*iv*) Silver foil; (*v*) Enersorb (non-selective); (*vi*) Nextel (non-selective).

3.3.3. Advantages, Disadvantages and Applications of Flat-plate Collectors

Advantages:

1. *Both beam and diffuse solar radiations* are used.
2. Require *little maintenance.*
3. The orientation of the sun is *not required* (*i.e.* no tracking device needed)
4. Mechanically *simpler* than the focusing collectors.

Disadvantages:

1. *Low temperature* is achieved.
2. *Heavy* in weight.
3. Large heat losses by conduction due to large area.

Applications:

1. Used in *solar water heating*.
2. Used in *solar heating and cooling*.
3. Used in *low temperature power generation*.

3.3.4. Evacuated Collectors

Planar solar collectors of *evacuated type often achieve efficiencies with an output temperature of above 80°C. In these devices a vacuum occupies the space between the absorber and the aperture cover.* The absorber may consist of a heat pipe that is thermally bonded to collecting this, possibly in an evacuated glass tube.

Efficiencies in excess of 40% or an output temperature of 200°C can be reached (Collins and Duff, 1983).

3.3.5. Performance Analysis of Flat-plate Collector

Analysis:

Consider an object exposed to sun radiations of intensity I, per unit area at the surface of the body. These radiations will *partly be absorbed* by the body, while the remaining will be *partly transmitted* and *rest reflected*. If we take the incident radiations equal to unity, then the absorbed, reflected, and transmitted parts of energy will add up to unity. These parts are called *absorption* coefficient, *reflection* coefficient and *transmission* coefficient and represented by the symbols α, ρ and τ respectively.

Using the above symbols we can write

$$\alpha + \rho + \tau = 1 \qquad \qquad ...(3.1)$$

The absorbed part of the solar radiations, which is equal to α, *is responsible for increasing the temperature of the body.* However, the body also loses energy by conduction, convection and radiation. The *equilibrium temperature of the body will be that at which the heat loses from the body are equal to the absorbed radiations.*

For analysis purposes, if we represent the body by a flat a plate and assume that the *convection and conduction losses are negligible to begin with*, then at equilibrium temperature the absorbed solar radiations should be equal to the radiation losses from the flat plate. The *radiation losses are equal to* $\varepsilon\sigma T^4$, where ε and T are the emission coefficient and absolute temperature respectively of a flat plate and σ is the Boltzman's constant.

Therefore, at equilibrium

$$\alpha I = \varepsilon\sigma T^4 \qquad \qquad ...(3.2)$$

or,

$$\frac{\alpha I}{\varepsilon} = \sigma T^4 \qquad \qquad ...(3.3)$$

From equation (3.3), it is evident that comparatively higher equilibrium temperature will be obtained where the quantity $\dfrac{\alpha}{\varepsilon}$ *i.e.*, the ratio of absorption coefficient to emission coefficient of the flat plate is *more*. However, this has been demonstrated by an equation obtained under *idealised condition*. In the realistic conditions too, its nature will remain the same, but it will get *modified* by other influencing factors.

The collectors for which ratio is equal to unity are called 'Neutral collectors' and those for which the ratio is greater than unity are called 'Selective collectors'.

The amount of energy collected, however, does not depend on $\dfrac{\alpha}{\varepsilon}$ ratio. It primarily depends on higher value of α. So to obtain higher energy collection, one should use such flat plate where *absorption coefficient is as high as possible.*

A flat plate painted black is placed on a well insulated base. If it is exposed to solar radiations where $I = 800 \ W/m^2$, a typical summer value for a tropical region, we obtain from equation (3.3) the equilibrium temperature as 70°C. In spite of the simplifications here, it is a fair estimate of the temperature reached by a black plate left for a time in the tropical sun.

- This method can be refined by including the *convection losses and the energy gain as a result of absorption of diffused radiations by the flat plate.*

If I' is the intensity of the diffused radiations and α' the absorption coefficient, then equation (3.2) becomes

$$\alpha I + \alpha' \ I' \ = \ h_c(T - T_a) + \varepsilon \sigma T^4 \qquad \qquad ...(3.4)$$

This is valid, where the base is insulated, hence conduction losses are neglected. Here T_a is the atmospheric temperature and h_c is the convection heat transfer coefficient.

Transmissivity-absorptivity product ($\tau.\alpha$):

The effective part radiation absorbed is given by:

$$(\tau.\alpha)_e \ = \ \frac{\tau.\alpha}{1-(\tau-\alpha)\rho_d} \qquad \qquad ...(3.5)$$

The value of ρ_d for an incident angle of 60° is about 0.16, 0.24 and 0.2 for *one, two* and *three glass covers* respectively.

Performance:

The *"performance of a flat-plate collector"* is described by *an energy balance that indicates the distribution of incident solar energy into useful energy gain and various losses.*

Under steady conditions:

Useful heat delivered by a solar collector

= Energy absorbed in the metal surface — Heat losses from the surface directly and indirectly to the surroundings.

Useful heat output of a flat-plate collector is given by:

$$Q_c \ = \ A_{cs}\Big[I_{cs}(\tau\alpha)_e - U_{oc}(t_{fi}-t_a)\Big] \ \text{watts} \qquad \qquad ...(3.6)$$

Q_c = *Useful heat output* of flat-plate collector (*W*),

A_{cs} = Collector *surface area* (m²),

I_{cs} = *Intensity* of solar radiation incident on the collector surface (W/m²),

τ = *Transmission coefficient* (*i.e.,* fraction of incoming solar radiation that reaches the absorbing surface)

α = *Absorption coefficient* (*i.e.* fraction of the solar radiation reaching the surface that is absorbed)

$(\tau\alpha)_e$ = *Effective* product of transmittivity of the transparent cover and absorptivity of the absorber,

U_{oc} = *Overall total heat loss coefficient of the collector* (W/m°C),

t_{fi} = collector fluid inlet temperature (°C), and

t_a = Ambient air temperature (°C).

Introducing heat *"removal factor F_R"* in (Eqn. (3.6), we get,

$$Q_c = A_{cs} [I_{cs} F_R (\tau\alpha)_e - F_R U_{oc} (t_{fi} - t_a)] \text{ watts} \qquad ...(3.7)$$

The "**efficiency** *of a solar collector* (η_c)" is defined as the ratio of *the useful heat output of the collector to the solar energy flux incident on the collector.*

Mathematically,

$$\eta_c = \frac{Q_c}{A_{cs} I_{cs}} \qquad ...(3.8)$$

Inserting the value of Q_c from Eqn. (3.7) in Eqn. (3.8), we get:

$$\eta_c = F_R (\tau\alpha)_e - F_R U_{oc} \left(\frac{t_{fi} - t_a}{I_{cs}} \right) \qquad ...(3.9)$$

Eqn. (3.9) indicates that if the efficiency is plotted against $\left(\dfrac{f_i - t_a}{I_{cs}} \right)$, a straight line will result, with a slope of $F_R U_{oc}$ and Y-intercept of $F_R (\tau\alpha)_{e}$,

If,

$$\frac{t_{fi} - t_a}{I_{cs}} = 0 \text{ } i.e. \text{ } t_{fi} = t_a, \text{ then}$$

$$\eta_c = F_R (\tau\alpha)_e \qquad (...3.10)$$

This is the effective *optical efficiency.*

- The *energy balance equation* on the *whole collector* can be written as:

$$A_{cs} [I_{cs} F_R (\rho.\alpha)_b + I_{cs} F_R (\tau.\alpha)_d] = Q_u + Q_l + Q_s \qquad ...(3.11)$$

where,

Q_u = Rate of *useful heat transfer* to a working fluid in the solar heat exchanger,

Q_l = Rate of *energy losses* from the collector to the surroundings by re-radiation, convection and by conduction through supports for the absorber plate and so on. The losses due to reflection from the covers are included in the ($\tau.\alpha$) terms, and

Q_s = Rate of *energy storage in the collector.*

Suffices, *b* and *d* stand for beam and diffuse radiations respectively.

The outlet temperature of collector heat transfer fluid, t_{fo} (0°C)

The outlet temperature of collector heat transfer fluid t_{fo} is given by:

$$t_{fo} = t_{fi} + \frac{Q_c}{\dot{m}.c_p} \qquad ...(3.12)$$

where,

t_{fi} = Collector fluid inlet temperature (°C),

Q_c = Useful heat output of collector (W),

$\dot{m}$ = Mass flow rate of collector fluid (kg/s), and

c_p = Specific heat of collector fluid (J/kg K).

The *stagnant temperature* (t_s) of the collector is *defined as the temperature of the absorber which is achieved when there is no flow of heat transfer fluid in the collector* and therefore, its useful heat output and efficiency both are equal to *zero*. Hence, for Eqn. 3.7, we get:

$$O = A_{cs} [I_{cs}F_R (\tau\alpha)_e - F_R U_{oc} (t_{fi} - t_a)]$$

or, $$I_{cs} F_R (\tau\alpha)_e = F_R U_{oc} (t_{fi} - t_a)$$

or, $$t_s = t_{fi} = t_a + I_{cs} \frac{F_R(\tau\alpha)_e}{F_R U_{oc}} \qquad ...(3.13)$$

It is evident from Eqn. (3.13) that t_s will be high, if I_{cs} and $(\tau\alpha)_e$ are *high* and $F_R U_{oc}$ is *low*.

Factors affecting the performance of a flat-plate collector:

The following *factors* affect the performance of a flat-plate collector:

1. Incident solar radiation.   5. Selective surface.
2. Number of cover plates.   6. Fluid inlet temperature.
3. Spacing between absorber plate and glass cover.   7. Dust on cover plate.
4. Tilt of the collector.

1. **Incident solar radiation.** The collector's efficiency is *directly* related to solar radiation falling on it and *increases* with rise in temperature.

2. **Number of cover plates.** The *increase* in number of cover plates *reduces* the internal connective heat losses but also *prevents* the transmission of radiation inside the collector.

3. **Spacing between absorber plate and glass cover.** The *more the space* between the absorber and the cover plate, the *less is the internal heat loss*.

4. **Tilt of the collector.** In order to achieve better performance, flat-plate collector should be *tilted at an angle of latitude of the location*.

    — The collector is placed with south facing at northern hemisphere to receive maximum radiation throughout the day.

5. **Selective surface.** The selective surface should be able to *withstand high temperature*, should *not oxidise* and should be *corrosion resistant*.

6. **Fluid inlet temperature.** With the increase in the inlet temperature of the fluid, there is an increase in operating temperature of the collector and this leads to *decrease in efficiency*.

7. **Dust on cover plate.** The collector's efficiency *decreases* as dust particles *increase* on the cover plate. Thus, *frequent cleaning* is required to get the maximum efficiency of the collector.

Example 3.1. *The following data relate to an evacuated tube collector:*

The intensity of solar radiation on the collector's surface	= 800 W/m²;
The inlet temperature of the fluid	= 38 °C;
The ambient air temperature	= 25 °C
Effective optical efficiency	= 0.76
Effective heat loss coefficient	= 1.65 W/m²K
Mass flow rate of water	= 0.019 kg/s/m²
Specific heat of water at constant pressure	= 4187 J/kg K

Calculate the following:

(i) *Useful heat output, per m² of the surface area.*

(ii) *Outlet temperature of the fluid.*

(iii) *Stagnation temperature.*

Solution. *Given:* $I_{cs} = 800$ W/m²; $A_{cs} = 1$m²; $t_{fi} = 38$ °C; $t_a = 25$ °C; $F_R(\tau\alpha)_e = 0.76$; $F_R U_{oc} = 1.65$ W/m²K; $\dot{m} = 0.019$ kg/s/m², $c_p = 4187$ J/kg K.

(i) **Useful heat output; Q_c:**

$$Q_c = A_{cs}[I_{cs} F_R(\tau\alpha)_e - F_R U_{oc}(t_{fi} - t_a)] \quad \text{watts} \qquad \text{...[Eqn. (3.7)]}$$

$$= 1 \times [800 \times 0.76 - 1.65(38 - 25)] = \textbf{586.5 W (Ans.)}$$

(ii) **Outlet temperature of the fluid, t_{fo}:**

$$t_{fo} = t_{fi} + \frac{Q_c}{\dot{m} c_p} \qquad \text{...[Eqn. (3.12)]}$$

$$= 38 + \frac{586.5}{0.019 \times 4187} = \textbf{45.4 °C (Ans.)}$$

Example 3.2. *The following data relate to a flate plate collector used for heating the building:*

Location and latitude = Baroda, 22°N; Day and time: January 22,11:30 – 12:30 (IST); Annual average intensity of solar radiation = 340 W/m² hr; Tilt of the collector = latitude + 14°; Number of glass covers = 2; Heat removal factor for collector = 0.82; Transmittance of glass = 0.87; Aborptance of glass = 0.89; Top loss coefficient for collector = 7.9 W/m² hr °C; Collector fluid inlet temperature = 48 °C; Ambient temperature = 16 °C.

Calculate the following:

(i) *Solar altitude angle;*   (ii) *Incident angle;*

(iii) *Efficiency of the collector.*

Solution. *Given:* $\phi = 22°$; Day and time: Jan. 22, 11:30 – 12:30 (IST);

$H_b = 340$ W/m² hr ; $\beta = \phi + 14° = 22° + 14° = 36°$;

ρ_d (diffuse reflectance for *two* glass covers) = 0.24; $F_R = 0.82$; τ (for glass) = 0.87; α (glass) = 0.89; Top loss coefficient for collector, $U_{oc} = 7.9$ W/m²hr °C; $t_{fi} = 48$°C; $t_a = 16$ °C.

(i) **Solar altitude angle, α:**

Solar declination, $\delta = 23.45 \sin\left[\dfrac{360}{365}(284+n)\right]$ ...[Eqn. (2.3)]

∴ $\delta = 23.45 \sin\left[\dfrac{360}{365}(284+22)\right] = -19.93°$

Solar hour angle $\omega = 0$, (at mean of 11:30 and 12:30).

Solar altitude angle α is given by:

$$\sin\alpha = \sin\phi \sin\delta + \cos\phi \cos\delta \cos\omega \qquad \text{...[Eqn. (2.10)]}$$

$$= \sin 22° \sin(-19.93°) + \cos 22° \cos(-19.93°) \cos 0°$$

$$= -0.1277 + 0.8716 = 0.7439$$

∴ $\alpha = \textbf{48.6° (Ans.)}$

(ii) **Incident angle, θ:**

$$\theta = \frac{\pi}{2} - \alpha = 90° - 48.06° = \textbf{41.94° (Ans.)}$$

(iii) **Efficiency of the collector, η_c:**

Tilf for the beam radiation (R_b) is given by:

$$R_b = \frac{\sin\delta\,\sin(\phi-\beta)+\cos\delta\,\cos\omega\,\cos(\phi-\beta)}{\sin\phi\,\sin\delta+\cos\phi\,\cos\delta\,\cos\omega} \qquad ...[\text{Eqn. (2.29}(a)]$$

$$= \frac{\sin(-19.93°)\,\sin(22°-36°)+\cos(-19.93°)\cos0°\,\cos(23°-36°)}{\sin22°\,\sin(-19.93°)+\cos22°\,\cos(-19.93°)\,\cos0°}$$

$$= \frac{-0.0825+0.9121}{-0.1277+0.8716} = 1.115$$

Effective transmittance absorptance product is given by;

$$(\tau.\alpha)_e = \frac{\tau.\alpha}{1-(1-\alpha)\rho_d} \qquad ...[\text{Eqn. (3.5)}]$$

(where, ρ_d = diffuse reflectance for *two* glass covers = 0.24)

$$= \frac{0.87\times0.89}{1-(1-0.89)\times0.24} = 0.795$$

Beam solar radiation intensity, $H_b = 340$ W/m²hr   ...(Given)

Now, solar radiation, $H = H_b R_b\,(\tau.\alpha)_e$

$$= 340 \times 1.115 \times 0.795 = 301.38 \text{ W/m}^2\text{hr}$$

Useful gain, $Q_c = F_R[H - U_{oc}(t_{fi} - t_a)]$

$$= 0.82\,[301.38 - 7.9\,(48 - 16)] = 39.83 \text{ W/m}^2\text{hr}$$

∴ Collecton efficiency, $\eta_c = \dfrac{Q_c}{H_b R_b} = \dfrac{39.83}{340\times1.115} = \textbf{0.105 or 10.5\% (Ans.)}$

3.4. CONCENTRATING (OR FOCUSING) COLLECTORS

Concentrating collector *is a device to collect solar energy with high intensity of solar radiation on the absorbing surface by the help of reflector or refractor.*

☞ 3.4.1. Need of Orientation in Concentrating Collectors

Such collectors generally use optical system in the form of reflectors or refractors. A concentrating collector is a special form of flat-plate collector modified by introducing a reflecting (or refracting) surface (*concentrator*) between the solar radiations and the absorber. *These types of collectors can have radiation increase from low value of 1.52 to high values of the order of 10,000.* In these collectors radiation falling on a relatively large area is *focused on to a receiver* (or absorber) of *considerably smaller area.* As a result of the *energy concentration,* fluids can be heated to temperatures of *500°C or more.*

Orientation of sun from earth *changes from time to time.* So to harness maximum solar rays it is *necessary to keep our collector facing to sun rays direction.* This is the *reason why orientation in concentrating collector is necessary.* This is *achieved* by the use of "*Tracking device*".

3.4.2. Types of Concentrating Collectors

The different types of focusing/concentrating type collectors are:

1. Parabolic trough collector.
2. Mirror strip collector.
3. Fresnel lens collector.
4. Flat-plate collector with adjustable mirrors.
5. Compound parabolic concentrator (CPC).
6. Parabolic dish collector.

3.4.3. Advantages and Disadvantages of Concentrating Collectors

Advantages:

1. *High* concentration ratio.
2. *High* fluid temperature can be achieved.
3. *Less* thermal heat losses.
4. System's efficiency *increases at high temperatures.*
5. *Inexpensive* process.

Disadvantages:

1. *Non-uniform flux* on absorber.
2. Collect *only beam radiation components* because diffuse radiation components *cannot be reflected*, hence these are *lost*.
3. Need *costly tracking device*.
4. *High initial cost.*
5. *Need maintenance* to retain the quality of reflecting surface against dirt and oxidation.

3.4.4. Parabolic Trough Collector

Fig 3.4. shows the *principle of the parabolic trough collector* which is *often used in focusing collectors*. Solar radiation coming from the particular direction is *collected over*

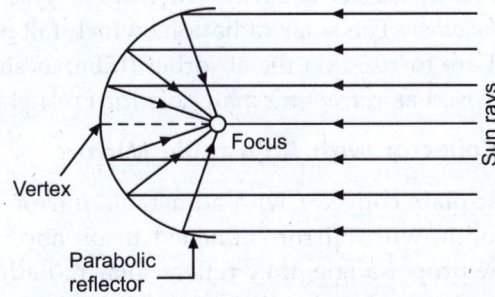

Fig. 3.4. Cross-section of parabolic trough collector.

the area of reflecting surface and is concentrated at the focus of the parabola, if the reflector is in the form of a trough with parabolic cross-section, the solar radiation is focused along a line. Mostly *cylindrical parabolic concentrators* are used in which *absorber is placed along focus axis* [Fig. 3.5].

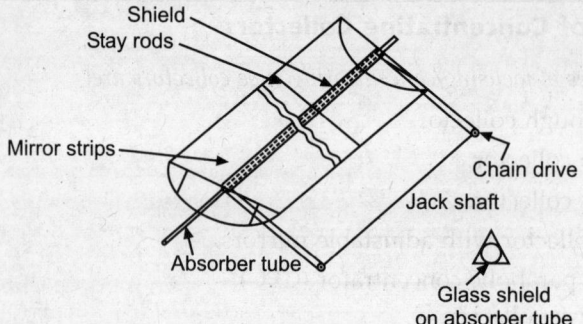

Fig. 3.5. Cylindrical parabolic system.

3.4.5. Mirror Strip Collector

Refer to Fig. 3.6. A mirror strip collector has a number of planes or slightly curved or concave mirror strips which are mounted on a base. These individual mirrors are placed at such angles that the reflected solar radiations fall on the same focal line where the pipe is placed. In this system, *collector pipe is rotated so that the reflected rays on the absorber remain focused with respect to changes in sun's elevation.*

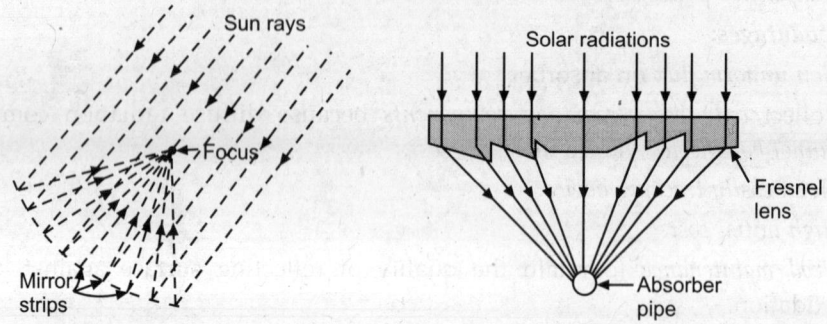

Fig. 3.6. Mirror strip collector.   **Fig. 3.7.** Fresnel lens collector.

3.4.6. Fresnel Lens Collector

In this collector a *Fresnel lens* is used in which *linear grooves are present on one side* and *flat surface on the other*. The solar radiations which fall normal to the lens are *refracted* by the lens and are focused on the absorber (tube) as shown in Fig. 3.7. Both glass and plastic can be used as refracting materials for Fresnel lenses.

3.4.7. Flat-plate Collector with Adjustable Mirrors

Fig. 3.8. shows a flat-plate collector with adjustable mirrors. It consists of a flat-plate collector facing south, with mirrors attached to its north and south edges. If the mirrors are set at the proper angle, they reflect solar radiation on to the absorber plate. Thus, the latter receives *reflected radiation* in *addition* to that normally falling on it. In order to make the mirrors *effective*, the *angles should be adjusted continously* as the sun's altitude changes. Since the mirrors can provide only a relatively small increase in the solar radiation falling on the absorber, flat-plate collectors with mirrors are not widely used.

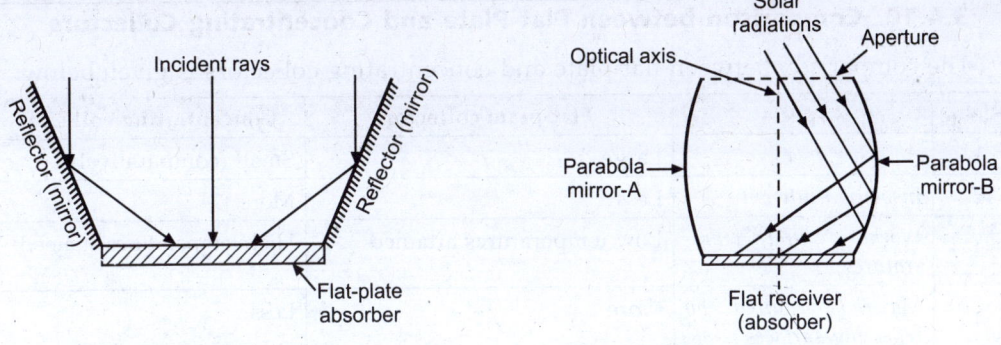

Fig. 3.8. Flat-plate collector absorber with adjustable mirrors.

Fig. 3.9. Compound parabolic concentrator (CPC).

3.4.8. Compound Parabolic Concentrator (CPC)

Fig. 3.9 shows the compound parabolic concentrator. It was designed by Winston (and Baranov). It consists of two parabolic segments, oriented such that focus of one is located at the bottom end point of the other and vice versa. The receiver is a flat surface *parallel to the aperture joining of two foci of the reflecting surfaces.*

For thermal and economic reasons the *fin and the tubular type of absorbers are preferable.* It is claimed that Winston collectors are capable of competitive performance at high temperatures of about 300°C required for power generation, if they are *used with selectively coated, vacuum enclosed receivers.*

The maximum concentration ratio available with paraboloidal system is of the order of 10,000.

Advantages:

1. *High concentration ratio.*
2. *No need of tracking.*
3. *Efficiency* for accepting diffuse radiation is *much larger* that conventional concentrators.

3.4.9. Paraboloidal Dish Collector

Refer to Fig. 3.10. In this type of collector all the radiations from the sun are focussed at a point. This collector can generate temperature up to 300°C and contraction ratio from 10 to few thousands. Its diameter is of the range between 6 to 7 m and can be commercially manufactured.

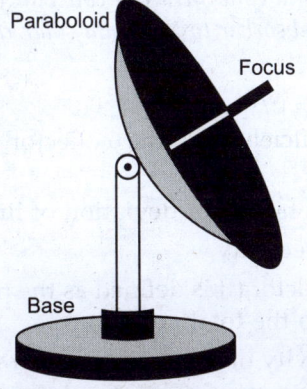

Fig. 3.10. Paraboloidal dish collector.

3.4.10. Comparison between Flat-Plate and Concentrating Collectors

The comparison between flat-plate and concentrating collectors is given below:

S.No.	Aspects	Flat-plate collector	Concentrating collector
1.	*Absorber area*	Large	Small (comparatively)
2.	*Insolation intensity*	Less	More
3.	*Working fluid temp-erature*	Low temperatures attained	High temperatures attained
4.	*Material required by reflecting surfaces*	More	Less
5.	*Use for power gen-eration*	Cannot be used	Can be used
6.	*Need of tracking system*	No	Yes
7.	*Flux received on the absorber*	Uniform	Non-uniform
8.	*Collection of beam and diffuse solar radiation components*	*Beam as well as diffuse* solar radiation components collected.	Only beam component collected (because diffuse component cannot be reflected and is thus lost).

3.4.11. Performance Analysis of a Concentrating Collector

The *useful heat output* of a concentrating collector (Q_C) is given by:

$$Q_C = F_R \, A_{ua} \, [I_{bc} \, \eta_{opt} - (U_{oc}/C)(t_{in} - t_a)] \qquad \qquad ...(3.14)$$

where, F_R = Heat removal factor of the collector,

A_{ua} = Unshaded aperture area (m^2),

I_{bc} = Intensity of beam solar radiation incident on the concentrator aperture (W/m^2),

η_{opt} = Optical efficiency of the collector,

U_{oc} = Overall/total heat loss coefficient (W/m^2°C),

C = Correction factor,

t_{in} = The inlet temperature of the collector (°C), and

t_a = Ambient temperature.

The *"optical efficiency"* of a concentrating collector (η_{opt}) is defined as the ratio of the solar radiation absorbed by the absorber to the beam solar radiation on the concentrator. It is given by:

$$\eta_{opt} = \eta_{opt \, 0°} \, C_{opt} = \rho \gamma \tau \alpha_a \cdot C_{opt} \qquad \qquad ...(3.15)$$

where, $\eta_{opt 0°}$ = Optical efficiency of the collector at 0°-incident angle of beam radiation,

C_{opt} = Correction factor for deviation of incidence angle from 0°;

ρ = Mirror reflectivity,

γ = Intercept factor (It is defined as the ratio of radiation intercepted by absorber to the total radiation),

τ = Transmittivity of transparent cover of the absorber, and

α_a = Absorptivity of the absorber.

$\therefore$ The *"efficiency"* of a concentrating collector (η_c) is given by:

$$\eta_c = \frac{Q_c}{A_a I_{bc}}$$

or, $\qquad \eta_c = F_R \eta_{opt} - \frac{F_R U_{oc}}{C.I_{bc}} (t_{in} - t_a)$ ...(3.16)

The outlet enthalpy (h_{out}) of heat transfer fluid in a concentrating collector with a phase change (vapourisation) of fluid is given by:

$$h_{out} = h_{in} + \frac{Q_c}{\dot{m}} \ \ (\text{J/kg})$$ ...(3.17)

where, $\quad h_{in}$ = Inlet enthalpy of heat transfer fluid (J/kg),

$\qquad Q_c$ = Useful heat output of collector (W), and

$\qquad \dot{m}$ = Mass flow rate of heat transfer fluid (kg/s).

3.5. SOLAR-THERMODYNAMIC CONVERSION

3.5.1. Introduction

Since several years, this has been scientists's endeavour to *convert heat into mechanical energy*, which in turn, may be *used to generate electricity*. Several ambitious solar power systems were constructed during 19th and 20th centuries.

Carnot discovered the following formula for the theoretical maximum efficiency for *converting heat energy into mechanical energy*:

Carnot efficiency, $\quad \eta_{carnot} = \dfrac{T_h - T_c}{T_h}$ ...(3.18)

where, $\qquad\qquad T_h$ = Absolute temperature of the *heat source*, and

$\qquad\qquad\qquad T_c$ = Absolute temperature of the *heat sink*.

In practice, however, it is *not possible* even with a good design to obtain an efficiency *nearly as high as the theoretical one*. In most cases, the true thermal efficiency for converting heat to work will be 30 to 60 per cent of η_{carnot}. Whereas thermodynamics dictate that to achieve maximum conversion efficiency, the difference in operating temperature should be *as large as possible* (Fig. 3.11), the *efficiency of a solar collector decreases with increasing temperature*.

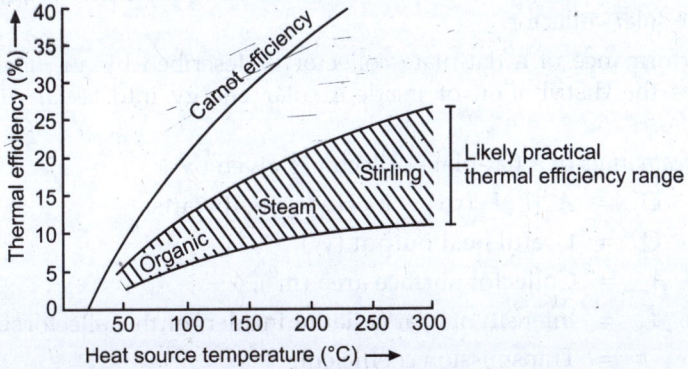

Fig. 3.11. Comparison of theoretical efficiency with those obtained in practice at a sink temperature of 25°C (Halcrow/ITDG 1981 b)

- The type of solar collector required is *governed by the choice of heat engine*. The *"medium temperature steam engines and high-temperature engines"* need linear focusing, parabolic dish or heliostat devices that concentrate the sun's direct beam.
— In case of *"gas-cycle engines"* which demand high temperature, the solar collector should be *parabolic dish or heliostat (power tower) devices*.

Practically, so far, only the *Rankine and Stirling cycles* have been used for small scale applications.

3.5.2. Rankine-cycle Engines

The Rankine and similar vapour-cycle engines can be of two types: (*i*) *Low-temperature ORC engines* that use *organic fluids with low boiling points*; (*ii*) Medium-temperature vapour cycles, which generally use water as the working fluid. *"Rankine cycle engines"* are the most well-developed of the heat engines used in solar-thermodynamic systems.

— The components of this engine *can be powered by flat-plate* collectors for tasks such as *operating a "water pump"*. The working fluid is evaporated by solar heat in a boiler before passing to the expander, from which mechanical work is extracted.

- Following *"working fluids"* are suitable for operations at *"lower temperatures"*:

Water, Freons (F-11 widely used); Ammonia, Ethylene; Ethane, Propylene; Sulphur dioxide:

3.5.3. High-temperature Gas-Cycle Engines

High-temperature engines (including using the Brayon, Stirling and Ericsson cycles) have high Carnot efficiencies and their technology is being developed mostly for large scale-systems in the range of 7kW to 10 MW. *Rankine-cycle steam turbines* which were developed for conventional power stations, have been *adopted for high-temperature solar applications*. However, *ORC engines are now frequently used, with toluene as the working fluid*.

HIGHLIGHTS

1. The *factors* which adversely affect the collector's efficiency are; *Shadow, Cosine loss, Dust* etc.

2. Solar collectors are mainly of two types:

   (*i*) Non-concentrating or flat-plate solar collector, (*ii*) Concentrating or focusing type solar collector.

3. The performance of a flat-plate collector is described by an *energy balance* that indicates the distribution of incident solar energy into *useful energy gain* and *various losses*.

   Useful heatoutput of a flat-plate collector is given by:

   $$Q_c = A_{cs}[I_{cs} F_R(\tau\alpha)_e - F_R U_{oc}(t_{fi} - t_a)] \text{ watts}$$

   where, Q_c = Useful heat output (W),

            A_{cs} = Collector surface area (m^2),

            I_{cs}' = Intensity of solar radiation incident on the collector surface (W/m^2),

            τ = Transmission coefficient,

            α = Absorption coefficient,

$(\tau\alpha)_e$ = Effective product of transmissivity of the transparent cover and absorptivity of the absorber,

U_{oc} = *Overall heat loss coefficient* of the collector ($W/m^2°C$),

t_{fi} = Collector fluid inlet temperature (°C),

t_a = Ambient air temperature (°C), and

F_R = Heat removal factor.

Stagnation temperature (t_s) of the collector is gives by:

$$t_s = t_{in} = t_a + I_{cs} \cdot \frac{F_R(\tau\alpha)_e}{F_R U_{oc}}$$

4. *Concentrating collector* is a device to collect solar energy with high intensity of solar radiation on the absorbing surface by the help of reflector or refractor.

5. The useful heat output of a concentrating collector (Q_c) is given by:

$$Q_c = F_R A_{ua} [I_{bc} \eta_{opt} - (U_{oc}/C)(t_{in} - t_a)]$$

where, F_R = Heat removal factor,

A_{ua} = Unshaded aperture area (m^2),

I_{bc} = Intensity of beam solar radiation incident on the concentrator aperture (W/m^2),

η_{opt} = Optical efficiency of the collector,

U_{oc} = Overall/total heat loss coefficient ($W/m^2°C$),

C = correction factor,

t_{in} = The inlet temperature of the collector (°C), and

t_a = Ambient temperature.

Efficiency of a concentrating collector (η_c) is given by:

$$\eta_c = F_R \eta_{opt} - \frac{F_R U_{oc}}{C.I_{bc}}(t_{in} - t_a)$$

Here, $\eta_{opt} = \eta_{opt\,0°} C_{opt}.\rho\gamma\tau\alpha_a C_{opt}.$

where, $\eta_{opt\,0°}$ = Optical efficiency of the collector at 0° incident angle of beam radiation, C_{opt} = Correction factor for derivation of incidence angle from 0°, ρ = Mirror reflectivity, γ = Intercept factor, τ = Transmissivity of transparent cover of the absorber, and α_a = Absorptivity of the absorber.

THEORETICAL QUESTIONS

1. List the advantages, disadvantages and applications of solar thermal energy.
2. What is Green-house effect? Explain briefly.
3. What are the characteristic features of a collector system?
4. Explain briefly the factors which adversely affect the efficiency of a collector system.
5. How are solar collectors classified?
6. Give brief description of a flat-plate collector.
7. Discuss briefly selective coatings/surfaces.
8. State the advantages, disadvantages and applications of flat-plate collectors.
9. Explain briefly "Evacuated collectors"
10. What do you mean by performance of a flat-plate collector? Explain briefly
11. Explain briefly the factors which affect the performance of a flat-plate collector.
12. What is a concentrating collector?

13. Why is there a need of orientation in concentrating collectors?

14. Explain briefly, with neat sketches, any two of the following concentrating collectors:
    (i) Parabolic trough collector   (ii) Fresnel lens collector
    (iii) Flat-plate collector with adjustable mirrors   (iv) Paraboloidal dish collector.

15. Give the comparison between flat-plate collectors and concentrating collectors.

UNSOLVED EXAMPLES

1. An evacuated tube collector is working under the following conditions:

   The intensity of solar radiation on the collector surface = 760 W/m²; the collector fluid inlet temperature = 43°C; the ambient air temperature = 26°C; effective optical efficiency $F_R(\tau\alpha)_e = 0.77$; effective heat loss coefficient = $F_R U_C = 1.65$ W/m²°C; mass flow rate of water = 0.017 kg/s/m²; specific heat of water at constant pressure, $c_p = 4187$ J/kg°C.

   Calculate the Following:
   (i) Useful heat output.
   (ii) Outlet temperature of water.
   (iii) Stagnation temperature.

   [Ans. (i) 557.15 W/m² (ii) 50.83°C (iii) 380.66°C]

2. Following data relate to a flat-plate collector used for heating the building:

   Location and latitude = Baroda, 22°N; day and time = January 1, 11:30 to 12:30 (IST); annual average intensity of solar radiation = 350 W/m² hr; collector tilt = latitude + 15°; No. of glass covers = 2; heat removal factor for collector = 0.81; transmittance of the glass = 0.88; absorption of the glass = 0.90; top loss coefficient for collector = 7.88 W/m²°C; collector fluid temperature = 60°C; ambient temperature = 15°C:

   Calculate:
   (i) Solar altitude angle.
   (ii) Incident angle.
   (iii) Collector efficiency.

   [Ans. (i) 44.7°;, (ii) 45.3°; (iii) 6%]

3. The following data relate to a flat-plate collector used for heating:

   Location and latitude = Coimbatore, 11°00 N°; day and time = March 22, 2.30 – 3.30 (LST); average intensity of solar radiation = 560 W/m²; collector tilt = 26°; number of glass covers = 2; heat removal factor for collector = 0.82; transmittance of glass = 0.88; absorptance of the plate = 0.93; top loss coefficient for collector = 7.95 W/m²°C; collector fluid inlet temperature = 75°C; ambient temperature = 25°C

   Calculate:
   (i) Solar latitude angle.
   (ii) Incident angle.
   (iii) Collector efficiency.

   [Ans. (i) 43.9°; (ii) 46.1°; (iii) 8.87%]

4

SOLAR ENERGY STORAGE
AND APPLICATIONS

4.1 Energy storage system; 4.2 Classification of solar energy storage systems; 4.3 Solar thermal energy storage – Sensible heat storage—latent heat storage; 4.4 Thermo-chemical energy storage; 4.5 Electrical storage – Battery storage – Capacitor storage – Inductor storage; 4.6 Electromagnetic storage; 4.7 Mechanical energy storage – Pumped hydro-electric storage – Compressed air – Flywheel storage; 4.8 Solar pond; 4.9 Solar pond electric power plant; 4.10 Introduction to solar energy applications; 4.11 Solar water heating – Natural circulation solar water heater – Forced circulation solar water heater; 4.12 Space heating; 4.13 Space cooling – Introduction – Vapour absorption cooling system; 4.14 Solar distillation; 4.15 Solar pumping; 4.16 Solar air heaters and drying – Solar air heaters – Solar drying; 4.17 Solar cooking—Box-type solar cooker – Dish solar cooker – Community solar cooker for indoor cooking; 4.18 Solar furnace; 4.19 Solar green-houses and global warming–Solar green-houses–Global warming; 4.20 Solar power plants – Low temperature solar power plant – Medium temperature systems using focusing collectors – High temperature systems – Solar farm and solar power plant; 4.21 Solar photovoltaic (SPV) systems – Semiconductors – Photovoltaic effect – Conversion efficiency and power output of solar cell – Solar photovoltaic cells – Classification of solar cells – Silicon cell modules – Photovoltaic (PV) systems – Water pumping systems – SPV lighting systems – PV hybrid system 4.22 Solar production of hydrogen – Advantages of hydrogen as a substitute for fossil fuels – Methods of producing hydrogen from solar energy – Solar hydrogen energy cycle. *Highlights – Theoretical Questions.*

A. SOLAR ENERGY STORAGE

4.1. ENERGY STORAGE SYSTEM

In order to take care of intermittency of solar energy availability and to fill up the gap between power demand and supply of the power plant, there is *need for energy storage system.*

An **energy storage system** *stores the collected amount of energy in excess of requirement of the demand and supplies this energy when the demand exceeds the supply energy.*

- Following are the *factors* on which the *optimum capacity of an energy storage system depends:*

(*i*) The expected time dependence of solar radiation availability; (*ii*) The nature of loads to be expected on the process; (*iii*) The degree of reliability required for the process; (*iv*) The size of the solar thermal power system or solar electric generator; (*v*) The cost per kWh of the stored energy; (*vi*) Environmental and safety considerations.

4.2. CLASSIFICATION OF SOLAR ENERGY STORAGE SYSTEMS

The solar energy storage systems may be *classified* as follows:

1. **Thermal energy storage**
   (*i*) Sensible heat storage
      — Solids
      — Liquids
      — Solar ponds
      — Absorbents
   (*ii*) Latent heat storage.
2. **Thermo-chemical storage**
3. **Electrical storage**
   (*i*) Battery storage
   (*ii*) Capacitor storage
   (*iii*) Inductor storage.
4. **Electromagnetic energy storage**
5. **Mechanical energy storage**
   (*i*) Pumped hydro electric storage
   (*ii*) Compressed air storage
   (*iii*) Flywheel storage.

4.3. SOLAR THERMAL ENERGY STORAGE

Thermal energy storage is *the storage of energy by heating, melting or vaporisation of material and the energy becomes available as heat.*

Thermal energy storage is of *two* types:

1. Sensible heat storage.
2. Latent heat storage.

4.3.1. Sensible Heat Storage

The storage by causing a material to rise in temperature is *called* **sensible heat storage**. It involves a material that undergoes *no change in phase* over the temperature domain encountered in the storage process. The quality of heat stored is *proportional* to the temperature rise of the material.

In a sensible heat storage system, energy is stored by heating a *liquid* or a *solid*. In such a system the following materials are used:

Water; Inorganic molten salts and solids like *rock, gravel, refractories, iron shot, concrete* etc.

The choice of the material used depends on the *temperature level of its utilisation* (*e.g.* water is used for temperature below 100°C whereas refractory bricks can be used for temperatures up to 1100°C).

The **basic equation**(s) for an energy storage unit, operating over a finite temperature difference is:

$$Q_s = m \int_{T_1}^{T_2} dT \qquad \qquad ...(4.1)$$

where,
Q_s = The quantity of heat stored,
m = Mass of the storage material,
T_1, T_2 = The initial and` final tempertures, respectively, and
c_p = The specific heat of the storage material,

If V, ρ, c_p are the volume, density and specific heat of the storage material, the energy storage (Q_s) is given by:

$$Q_s = V\rho \int_{T_1}^{T_2} c_p \, dT \qquad \qquad ...(4.2)$$

Storage with water. The easiest way to store thermal energy is by storing the **water** directly in a well insulated tank. For flat-plate collector system, the optimum tank size is usually about 70 kg/m².

Following are the *characteristics of water for storage medium:*

(*i*) Inexpensive and readily available; (*ii*) High thermal storage capacity; (*iii*) Small pumping cost.

Storage with packed bed exchange. In this type of storage **air** is used as the energy transport mechanism; the storage materials used include *rock, gravel* or *crushed stone in bin.* Such materials provide a *large and cheap heat transfer surface.*

- **Water** is *superior* because of its *lower cost* and *lower volume required per unit of energy stored.*

4.3.2. Latent (or phase change) Heat Storage

In this system of heat storage, *heat is stored in a material when it melts and extracted from the material when it freezes.* Heat can also be stored when a liquid changes to gaseous state but as the volume change is large, such a system is not *economical.*

The following materials which melt on heating are suitable for solar energy applications:

(*i*) **Organic materials:**
- Paraffin wax– It has a high heat of fusion (209 kJ/kg) and is known to freeze without inter cooling;
- Fatty acids
(*ii*) **Hydrated salts:**
- Calcium chloride hexo hydrate ($CaCl_2.6H_2O$);
- Sodium sulphate deca hydrate ($Na_2SO_4.10H_2O$) – *Glauber's salt.*
- It melts at 32°C, with heat of fusion of 241 kJ/kg; the reaction is:

$$Na_2SO_4.10H_2O \underset{Energy\ release}{\overset{Energy\ storage}{\rightleftharpoons}} Na_2SO_4 + 10H_2O \qquad ...(4.3)$$

It has been proposed mainly for *storing domestic heat.*

(*iii*) **Inorganic materials:**

- I_{ce} (H_2O);
- Sodium nitrate ($NaNO_3$);
- Sodium hydroxide (NaOH)

4.4. THERMO-CHEMICAL ENERGY STORAGE

In this process, the *heat of a chemical reaction is used to store thermal energy.* Actually, the thermo-chemical thermal energy is the *binding energy of reversible chemical reactions.*

Fig. 4.1. shows the schematic representation of thermo-chemical energy storage reaction

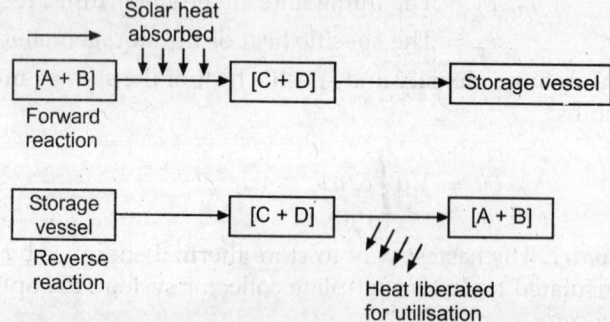

Fig. 4.1. Representation of thermo-chemical energy storage reaction.

Table 4.1. illustrates some chemical energy storage reactions.

Table 4.1. Some chemical energy storage reactions

S. No.	Reaction	Temperature of forward reaction (°C)	Temperature of reverse reaction (°C)	Energy stored per unit volume of storage material (kJ/m³)
1.	$CH_4 + H_2O \rightleftharpoons CO + 3H_2$	780	610	209.4×10^3
2.	*$SO_3 \rightleftharpoons SO_2 + \frac{1}{2} O_2$	1025	590	460.6×10^3
3.	$Mg(OH)_2 \rightleftharpoons MgO + H_2O$	199	335	3098.3×10^3

*This storage system has been suggested for use in 100 MW central tower solar power plant operating on Brayton cycle with helium as the working fluid.

- *Chemical energy storage system*, like latent heat storage system, has the *advantage of releasing heat at "constant temperature".*

4.5. ELECTRICAL STORAGE

4.5.1. Battery Storage

A rechargeable storage battery (called *secondary battery* – a group of interconnected cells) *receives electrical energy as direct current* which is *stored in the form of chemical energy* by a *reversible electro-chemical reaction.*

The capacity of a battery is given in terms of *ampere-hours on discharge.* This is determined by the following factors:

(i) Final limiting voltage of the cells;

(ii) Discharge rate;

(iii) Number, design and dimensions of the plates;

(iv) Design of separators;

(v) Quantity of electrolyte;

(vi) Density of electrolyte;

(vii) Temperature etc.

The **efficiency** *of a battery* is defined as *"the ratio of the output of battery to the input required to restore the initial state of charge under specified conditions of temperature, current rate and final voltage.*

Examples *of secondry batteries are:*

(i) Lead acid; (ii) Nickel-cadmium; (iii) Nickel-hydrogen, (iv) Zinc–air; (v) Sodium-sulphur etc.

- The number of times the battery can be charged and discharged under specified conditions is called *"Cycle life"*. The cycle life of a battery may vary greatly with the depth of the discharge, deep discharge tending to result in short cycle life.

4.5.2. Capacitor Storage

In this storage process, the "**capacitors**" are used to store a large amount of electrical energy for fairly long periods. The total energy stored is given by:

$$\text{(Total energy stored)}_{cap.} = \frac{1}{2}V\varepsilon E^2 \qquad \qquad ...(4.4)$$

where, V = Volume of dielectric,

ε = Permittivity, a constant, and

E = Electric field strength.

- *The capacitors store electrical energy at high voltage* and *low current*:

4.5.3. Inductor Storage

The energy stored by an *inductor* is given as:

$$\text{(Total energy stored)}_{Ind.} = \frac{1}{2}V\mu B^2 \qquad \qquad ...(4.5)$$

where, μ = Permeability of the material, and

B = Magnetic flux density.

The total energy stored will be high when both μ and B have large values; this will need high magnetic fields. This will create large mechanical forces which should be supported by *strong structures*. The reverse operation of discharging the stored energy entails the problem of *opening of circuit carrying large currents*.

- *The inductors store energy at low voltage* and *high current*

4.6. ELECTROMAGNETIC ENERGY STORAGE

In this process, the *energy is stored in the magnetic field of a superconducting coil carrying direct current*. By attaching the coil to a load, the stored energy could be *recovered as electrical energy (direct current)*. This is basis of the Superconducting Magnetic Energy Storage (**SMES**) system. Useful superconducting materials available commercially are:

(*i*) A niobium titanium (Nb – Ti) alloy at temperatures below –263°C, and (*ii*) A compound of niobium and tin (Nb$_3$S$_n$) below –255°C; these materials (metals and alloys) suddenly lose essentially all resistance to the flow of electricity when cooled below these very low temperatures.

Following *problems* are associated with superconducting storage:

1. The operation and maintenance of a *cryogenic plant* for producing the liquid helium required for the very low superconductivity temperatures.

2. Special structures are required to withstand the strong magnetic field of an SMES unit.

4.7. MECHANICAL ENERGY STORAGE

4.7.1. Pumped Hydro-electro Storage

Refer to Fig. 4.2.

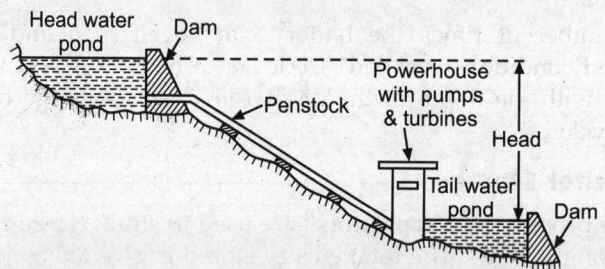

Fig. 4.2. Pumped storage plant.

Pumped storage plants are employed at the places where the quantity of water available for power generation is *inadequate*. Here the water passing through the turbines is stored in '*tail race pond*'. During low load periods this water is pumped back to the head reservoir using the extra energy available. This water can again be used for generating power during peak load periods. Pumping of water may be done seasonally or daily depending upon the conditions of the site and the nature of the load on the plant.

Such plants are *usually interconnected* with steam or diesel engine plants so that off peak capacity of interconnecting stations is used in pumping water and the same is used during peak load periods. Of course, the energy available from the quantity of water pumped by the plant is *less* than the energy input during pumped operation. Again while using pumped water the *power available is reduced* on account of losses occurring in prime movers.

4.7.2. Compressed Air

Compressed air energy storage follows the similar principles as pumped storage system. The excess electric energy generated by turbine during low loads *can be stored by compressing the air in large vessels at high pressures*. This energy *can be utilised during peak hours of power requirement*.

4.7.3. Flywheel Storage

When an electric motor, during its *off peak hours*, drives a flywheel, its speed is *increased* due to storage of mechanical (rotational) energy.

The basic idea of flywheel energy storage (sometimes known as '*super flywheel*'), is to accelerate a suitably designed physical rotor to a very high speed in a vacuum, as via electric motor, at which state *high energy storage densities are achieved*. The energy is stored as kinetic energy (*K.E.*), most of which can be *electrically regained when the flywheel is run as a generator*.

4.8. SOLAR POND

Solar pond, also called solar '*salt pond*', is *an artificially designed pond, filled with salty water, maintaining a definite concentration gradient*. It combines solar energy radiation and sensible heat storage, and as such, it is *utilised for collecting and storing solar energy*.

A solar pond reduces the convective and evaporative heat losses by *reversing* the temperature gradient with the help of *non-uniform vertical concentration of salts*.

Fig. 4.3 illustrates the *principle* of solar pond.

The vertical configuration of "*salt gradient solar pond*" normally consist of the following *three zones*:

1. "*Surface (homogeneous) convective zone (SCZ)*". It is adjacent to the surface and serves as a buffer zone between environmental fluctuations at the surface and conductive heat transport from the layer below. It is about 10 to 20 cm thick with a low uniform concentration at nearly the ambient air temperature.

2. "*Lower connective zone (LCG)*". It is at the bottom of the pond, and this is the layer with *highest salt concentration*, where *high temperatures are built up*.

3. "*Concentration/Intermediate gradient zone (CGZ)*". This zone keeps the two convective zones (*SCG* and *LCG*) apart and gives the solar pond its unique thermal performance. It provides *excellent insulation for the storage layer*, while transmitting the solar radiation. To maintain a solar pond in this non-equilibrium stationary state it is necessary to *replace* the amount of salt that is transported by *molecular diffusion* from the *LCG* to *SCZ*. This means that salt must be added to the *LCG*, and fresh water to the *SCG* whilst brine is removed. The brine can be recycled, divided into water and salt (by solar distillation) and returned to the pond.

The major heat loss occurs from the surface of the solar pond. This heat loss can be *prevented* by spreading a plastic grid over the pond's surface to prevent disturbance by the wind. Disturbed water tends to lose heat transfer faster than when calm.

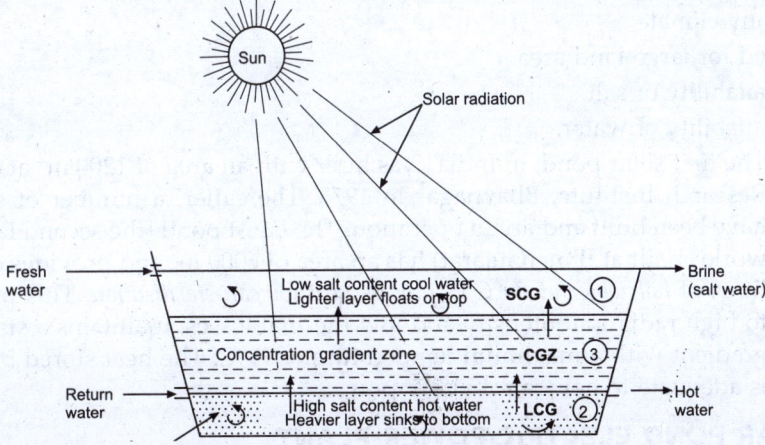

Fig. 4.3. Principle of solar pond.

Due to the excessively high salt concentration of the LCZ, a *plastic liner* or *impermeable soil* must be used to *prevent infiltration* into the nearby ground water or soil. The *liner* is a factor that increases the *cost* of a solar pond. A site where the soil is naturally impermeable, such as the base of a natural pond or lake, or can be made impermeable by compaction or other means, will allow considerably lower power costs.

The *optical transmission properties and related collection efficiency vary greatly* and *depend* on the following *factors*:

(*i*)   Salt concentration.

(*ii*) The quantity of suspended dust or other particles.

(*iii*) Surface impurities like leaves or debris, biological material like bacteria and algae.

(*iv*) The type of salt.

It becomes obvious that much higher efficiencies and storage can be achieved *through the utilization of refined or pure salt* whenever possible, as this *maximizes optical transmission*.

☞ *The solar pond is an effective collector of diffuse, as well as direct radiation, and will gather useful heat even on cloudy or overcast days.* Under ideal conditions, the pond's absorption efficiency can reach 50% of incoming solar radiation, although actual efficiencies average about 20% due to heat losses. *Once the lower layer of the pond reaches over 60°C the heat generated can be drawn off through a heat exchanger and used to drive a low temperature organic Rankine cycle* (ORC) *turbine.* This harnesses the pressure differentials created when a low boiling point organic fluid (or gas) is boiled by heat from the pond *via* a heat exchanger and cooled by a condenser to drive a turbine to generate electricity. The conversion efficiency of an organic Rankine cycle turbine driving an electric generator is 5—8% (which mean 1—3% from insolation to electricity output).

Applications of solar ponds:

1. Power generation.
2. Space heating and cooling.
3. Crop drying.
4. Desalination.
5. Process heat.

Limitations:

1. Sunny climate.
2. Need for large land area.
3. Availability of salt.
4. Availability of water.

   ● The first solar pond, in India, was built with an area of 1200 m^2 at Central Salt Research Institute, Bhavnagar in 1973. Thereafter, a number of solar ponds have been built and are in operation. The latest pond (the second largest in the world) built at Bhuj (Gujarat) has an area of 6000 m^2 and provides daily 90,000 litres of *hot water at 80°C as process heat for can-sterilisation.* This pond, owing to high radiation intensity and low thermal losses, maintains a stable salinity gradient with a maximum temperature of 99°C. The heat stored by this pond is adequate to generate 150 kW power.

4.9. SOLAR POND ELECTRIC POWER PLANT

A low temperature thermal electric power production scheme using solar pond is shown schematically in Fig. 4.4. The energy obtained from a solar pond is used to drive a

Rankine cycle heat engine. Hot water from the bottom level of the pond is pumped to the evaporator where the *organic working fluid is vapourized*. The vapour then flows under high pressure to the *turbine* where it expands and work thus obtained *runs an electric generator producing electricity*. The exhaust vapour is then *condensed* in a condenser and the *liquid is pumped back to the evaporator* and the cycle is repeated.

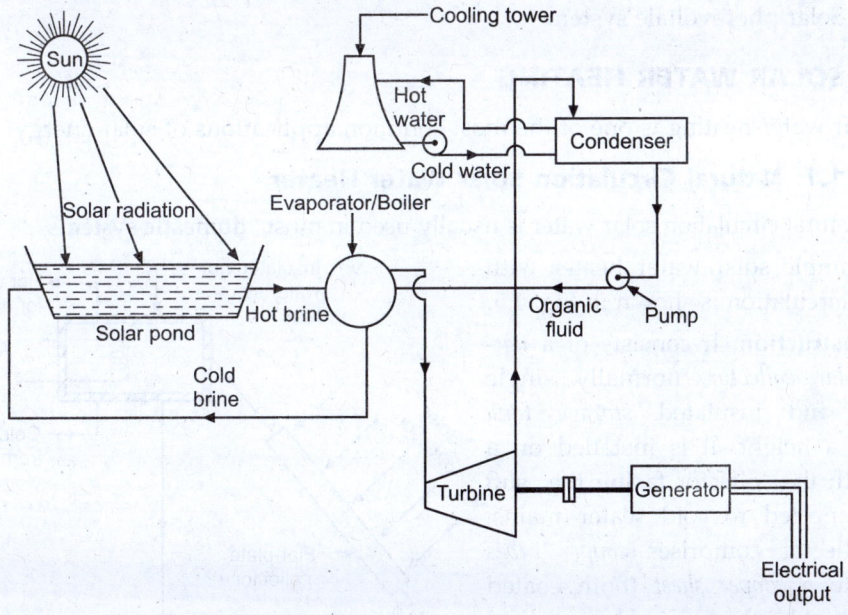

Fig. 4.4. Solar pond electric power plant.

B. SOLAR ENERGY APPLICATIONS

4.10. INTRODUCTION TO SOLAR ENERGY APPLICATIONS

The solar energy applications may be considered under the following *three* general categories:

1. *Direct thermal applications*

2. *Solar electric applications*

    — Solar thermal electric conversion

    — Photovoltaic conversion

    — Thermoelectric conversion

    — Ocean thermal energy conversion

3. *Biomass energy applications.*

Based on the above classification, some of the commonly used direct solar energy applications are enumerated and discussed henceforth:

1. Solar water heating.

2. Solar heating and cooling.

3. Solar distillation.

4. Solar pumping.

5. Solar drying.

6. Solar cooking.

7. Solar Green-house.

8. Solar power plant.

9. Solar photovoltaic system.

4.11. SOLAR WATER HEATING

Solar water heating is one of the most common applications of solar energy.

4.11.1. Natural Circulation Solar Water Heater

A natural circulation solar water is usually used in most "**domestic systems**".

A simple solar water heater with natural circulation is shown in Fig. 4.5.

Construction. It consists of a *flat-plate solar collector*, normally single glazed, and insulated *storage tank* kept at a height. It is installed on a roof with the collector facing the sun and connected to cold water mains. The collector comprises *copper tubes welded* to a *copper sheet* (both coated with a highly absorbing *black coating*) with a *toughened glass sheet* on top and insulating material on the rear.

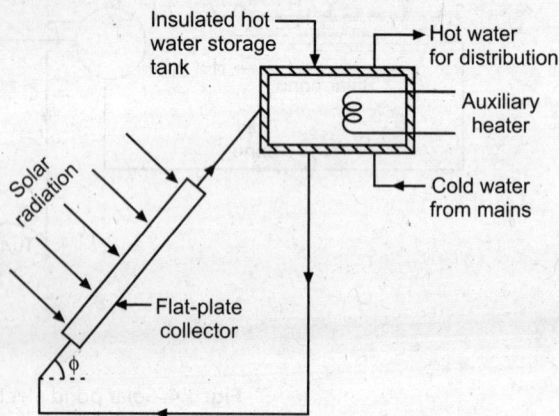

Fig. 4.5. Natural circulation solar water heater.

Working. The collector collects the sun radiation due to which water flowing through the tubes absorbs solar heat and is stored in a tank. The water circulation is entirely *based on the density difference between the solar-heated water in the collector and the cold water in the storage tank*. Hot water for use is taken out from the top of the tank.

To provide heat during long, cloudy periods, an *electrical immersion heater (i.e. 'auxiliary heater')* can be used as a back-up for the solar system. A non-freezing fluid may be used in the collector circuit.

● *Most domestic systems are of capacity ranging from 100 to 150 litres of hot water per day.*

4.11.2. Forced-Circulation Solar Water Heater

Solar water heaters of this type which are suited to supply large quantity of hot water, are suitable for *industries, hospitals, hostels* and *offices*. Fig. 4.6. shows a forced-circulation solar water heater with a *pump*. Water is pumped through flat-plate solar collector(s) where it is heated and flows back into the storage tank. Whenever hot water is withdrawn for use, cold water takes its place. When the difference in water temperature at the collector(s) outlet and that at the storage tank exceeds 7°C, the pump motor is activated by a differential thermostate. A non-return/check valve is needed to prevent reverse circulation and resultant night time thermal losses from the collector.

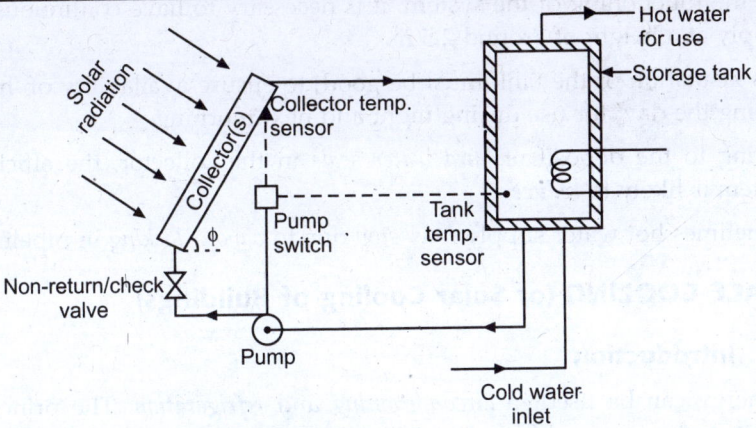

Fig. 4.6. Forced-circulation water heater.

4.12. SPACE HEATING (or Solar Heating of Buildings)

All solar heating systems may be divided into the following *two* categories:

1. **Active systems.** These systems generally consist of the following *components*:

   (i) *Separate solar collectors*, which may heat either *water* or *air*,

   (ii) *Storage devices* which can accumulate the collected energy for use at nights and during *inclement days*, and

   (iii) A *back-up system* to provide heat for protracted periods of bad weather.

2. **Passive systems.** In such systems solar radiation is collected by some element of structure itself, or admitted directly into building through large, south facing windows.

Fig. 4.7. shows a **basic space heating system**.

The *collector array* collect the solar radiation and heat up the water. This heated water is stored in the *tank* and the energy is transferred to the air circulating in the building *by water to air heat exchanger*. Tow *pumps* are provided for *forced circulation* between the collectors and the tank, and between the tank and heat exchanger.

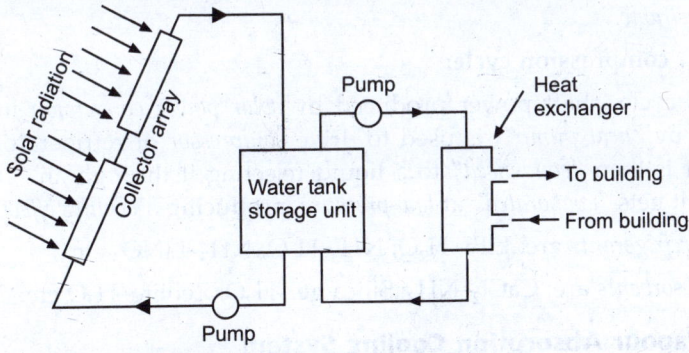

Fig. 4.7. Basic space heating system.

Limitations of uses of solar water heating system:

1. The energy received from the sun in cloudy days is almost nil and so the output of solar collector(s) will be almost zero and hence the system cannot provide the necessary service required.

2. For the functioning of the system, it is necessary to have continuous cold water supply at a height of around 2.5 m.

3. The insulation of the tank must be good, to ensure availability of 'heated water during the day' for use during night and next morning.

4. Owing to the deposit of *hard water scale* in the collector, the efficiency of the system is likely to *reduce*.

5. Sometimes hot water supply may *stop* due to *vapour locking* in pipeline.

4.13. SPACE COOLING (or Solar Cooling of Buildings)

4.13.1. Introduction

Solar energy can be used in *airconditioning* and *refrigeration*. The principles of the following two cycles are used for refrigeration using solar energy.

1. Absorption cycle   2. Vapour compression cycle

1. **Absorption cycle:**

In this cycle, two working fluids are used:

(*i*) A *refrigerant*, and   (*ii*) *Absorbent-refrigerant solution.*

The **absorbent cooling** is *based on the principle that the refrigerant can be bound by a liquid or solid solvent*, known as **absorbent** *to release heat during operation, while it absorbs heat during evaporation*, thus producing a *"cooling effect."*

Common refrigerant-absorbent combinations are:

(*i*) Lithium bromide (Li) – Water (LiBr–H_2O).

(*ii*) Ammonia-water (NH_3–H_2O).

— *"LiBr–H_2O system"* is simple and requires lower generator temperature of the order of 85°C to 95°C which are *achievable* by a flat-plate collector. Also it possesses *higher C.O.P.* than the NH_3–H_2O system

— Absorption cooling with solar energy, which is regarded as *more practical*, is possible with current technology although *improvement in design would be desirable.*

2. **Vapour compression cycle:**

In this cycle, shaft power produced by *solar power conversion* into mechanical power by *"heat pump"* is used to drive *compressor* of refrigerator. Compressed vapour is then *"condensed"* to a liquid rejecting it through an *"expansion valve"* where it gets *"evaporated"* at *low pressure*, producing a *cooling effect.*

Liquid refrigerants are: LiBr–H_2O; NH_3–H_2O; NH_3–$LiNO_3$ etc.

Solid absorbents are: $CaCl_2$–NH_3; Silica gel–H_2O; Zeolite–H_2O etc.

4.13.2. Vapour Absorption Cooling System

Refer to Fig. 4.8. A vapour absorption cooling system consists of the following *components*:

(*i*) *Generator (or Heat Exchanger)*; (*ii*) *Condenser*; (*iii*) *Evaporator*; (*iv*) *Absorber.*

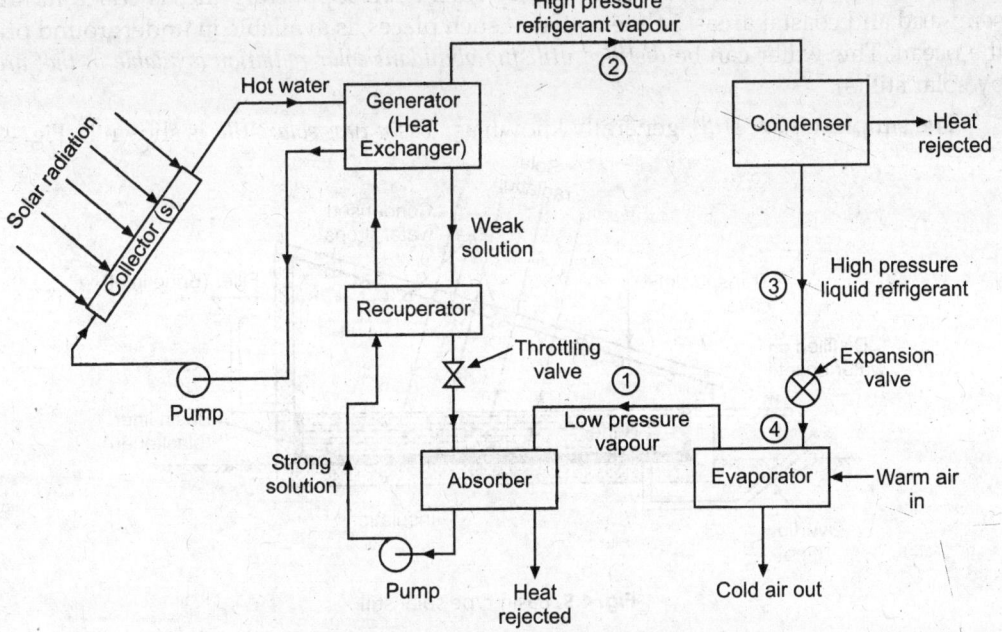

Fig. 4.8. Vapour absorption cooling system.

Working of the system:

— The low pressure refrigerant vapour leaving the **evaporator** at '1' is readily absorbed in the low temperature hot solution in the **absorber**, releasing the latent heat of condensation. The temperature of the solution tends to rise, while the absorber is cooled by the circulating water, absorbing the heat of solution, and maintaining a constant temperature.

— *Strong solution*, rich in refrigerant, is pumped (pump increases the pressure of the solution) to the **generator** where heat is supplied by the collector array. The high pressure refrigerant vapour is given off, and the weak solution returns to the **absorber** through throttling (or pressure reducing) valve.

— The high pressure refrigerant vapour from the generator at '2' is condensed in the **condenser** to a high pressure liquid at '3' refrigerant.

— The liquid refrigerant is throttled by the **expansion valve** at '4', and then evaporates, absorbing the heat of evaporation from the warm air or surroundings.

Advantages:

This system claims the following *advantages over vapour compression cycle*:

1. *No moving part* in the system except the pump-motor.

2. *Quiet in operation*, very little wear, and low maintenance cost.

3. *More compact and less bulky.*

4. The space requirements and automatic control requirements favour the absorption system more and more as the desired evaporator temperature drops.

4.14. SOLAR DISTILLATION (SOLAR STILL)

Solar still *is a device which is used to convert saline water into pure water by using solar energy.*

Soft drinking water, an essential requirement for supporting life, is *scarce* in arid, semi-arid and coastal areas. Saline water, at such places, is available in underground or in the ocean. This water can be *distilled utilising abundant solar radiation available in that area*, by solar still(s).

The simplest *'solar still'*, generally known as, *'basin type solar still'* is shown in Fig. 4.9.

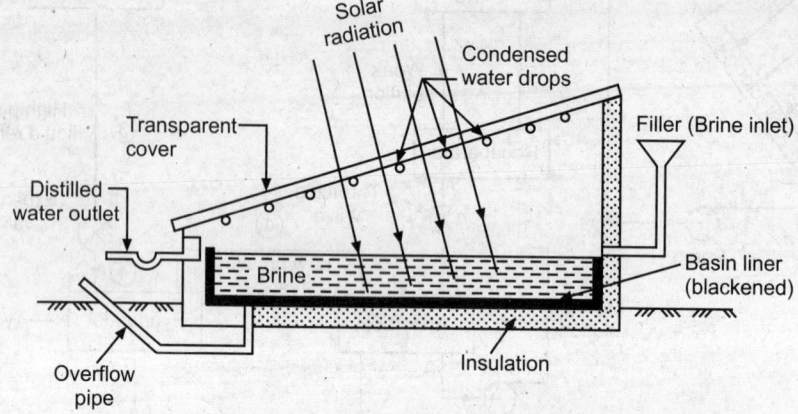

Fig. 4.9. Basin type solar still.

Construction. It is a shallow basin having blackened surface called *basin liner*. The filler supplies the saline water to the basin and an *overflow pipe* allows the excess water to flow out from the basin. The top of the basin is covered with a sloping airtight *transparent cover* that encloses the space above the basin. This cover is made of glass or plastic and slope is provided towards a collection trough

Working. Solar radiation passes through the transparent cover and is absorbed and converted into heat by the black surface of the still. The saline water is then heated and the water vapours condense over the cool interior surface of the transparent cover. The condensate flows down the sloping roof and gets collected in troughs installed at the outer frame of the solar still. The distilled water is then transferred into a storage tank.

"Desalination output" *increases* with the *rise* in ambient temperature and is independent of the salt content in raw feed water.

The *"solar still performance"* is expressed as the *quantity of water produced by each unit of basin area per day.*

Solar still installations may provide about 15 to 50 litres/day/10 m^2.

Advantages of distilling process:

1. *Low* energy consumption.
2. *Less* skilled labour required.
3. *Simple* technique required.
4. *Low* maintenance cost.

4.15. SOLAR PUMPING

Solar pumping *utilises the power generated by solar energy for water pumping, useful for irrigation.* Water pumps can be driven directly by solar heated water or fluid which operates either a heat engine or turbine. For *low heads*, a flat-plate collector is used to heat a low-boiling point liquid, the vapour of which drives the pump. For *larger heads*, a *parabolic trough concentrator or parabolic bowl concentrator* is installed to drive a steam turbine, which in turn drives the pump.

Fig. 4.10 shows a typical *solar-powered water pumping system.*

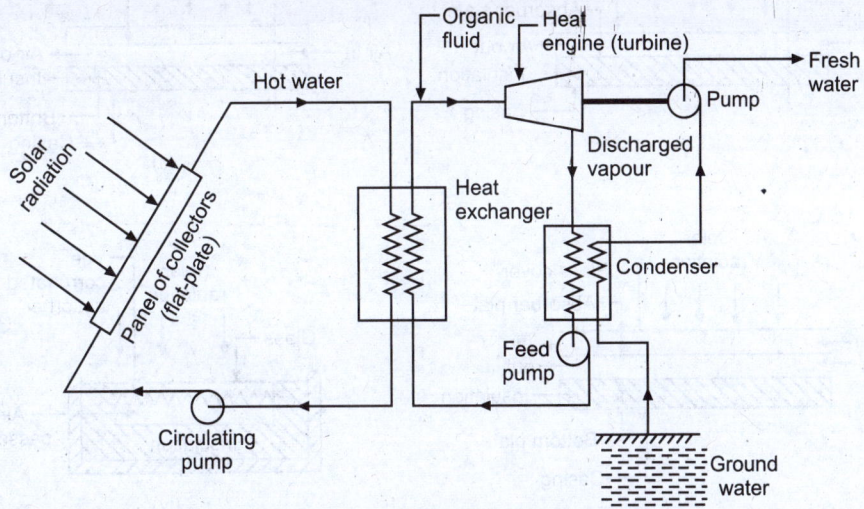

Fig. 4.10. Solar-powered water pumping system.

Working:

The panel of solar flat-plate *'collectors'*, on receiving solar radiation, heat water or an organic fluid (R-115 is an acceptable working fluid as it has *low cost* and *high cycle efficiency*).

— The hot fluid then flows to a mixing/storage tank and then to a *heat exchanger* to convert the working fluid of the heat engine from liquid to vapour.

— Water or hot transport fluid is fed again into the collector circuit by a *'circulating pump'*

— The discharged vapour, from the turbine, flows into *'condenser'* where the vapour gets condensed. The working fluid is then fed into the heat exchanger by a 'feed pump' to complete the cycle. The water, which is pumped, is used as a coolant in condenser of the turbine.

● A *high engine efficiency is achieved if the temperature in heat exchanger or boiler is high.*

4.16. SOLAR AIR HEATERS AND DRYING

4.16.1. Solar Air Heaters

A **conventional solar air heater** *is essentially a flate-plate collector with an 'absorber plate', a 'transparent cover system' at the top and 'insulation' at the bottom and on the sides.* The whole assembly is encased in a sheet metal container. The working fluid is *"air"*, though the passage for its flow varies according to the type of air heater.

The 'materials' used for construction of air heaters are similar to those of liquid flat-plate collectors. 'Selective coating' on the absorber plate can be used to improve the collection efficiency but cost effectiveness criterion should be kept in mind.

Figs. 4.11 [(*i*) to (*iv*)] shows the *basic features of non-porous absorber type air heaters.*

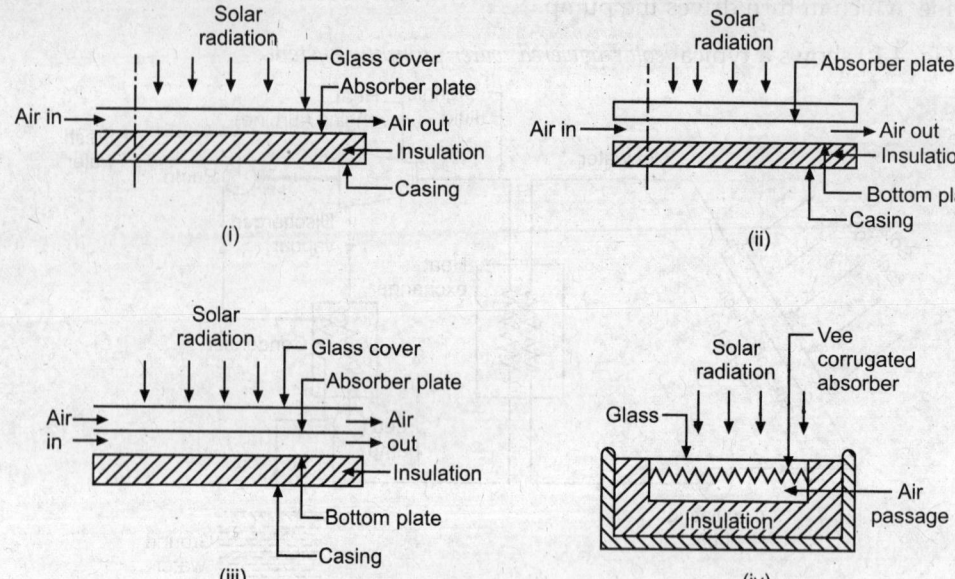

Fig. 4.11. Basic features of non-porous absorber type air heaters.

Advantages:

The solar air heaters claim the following *advantages* over other solar heat collectors:

1. The system is compact and less complicated.
2. The *pressure* inside the collector *does not become very high.*
3. Leakage of air from the duct does not pose any major problem.
4. Corrosion is completely eliminated.
5. Freezing of working fluid does not exist.

Disadvantages:

1. Poor heat transfer properties.
2. Need for handling large volumes of air.
3. The thermal capacity of air being low, it cannot be used as a storage fluid.
4. In the absence of proper design, the cost of air heater can be very high.

Important areas of applications:

The important areas of applications of non-porous absorber type air heaters are:

1. Heating and cooling buildings (such heaters are used only in *actively* heated or cooled buildings).
2. Air heaters are also used as desiccant beds for solar air-conditioning.
3. Heating of green-house.
4. Industrial processes such as drying agricultural crops and timber.

4.16.2. Solar Drying

Drying has been the oldest and most widely used application of solar energy in the developing countries. The methods have been based on open air drying.

Fig. 4.12. shows the schematic diagram of an *indirect crop dryer* (Headly and Springer design).

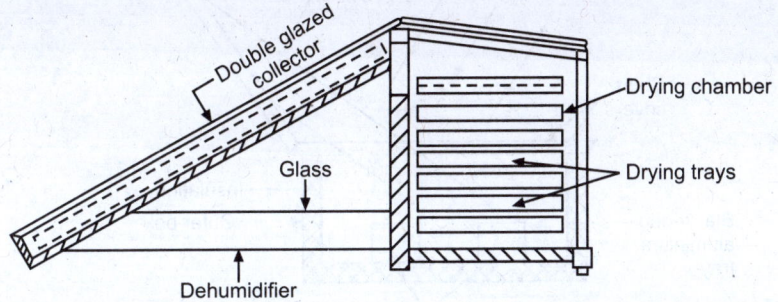

Fig. 4.12. Indirect crop dryer

In this type of dryer, ambient air is heated and is passed into drying chamber by *natural convection*. The hot air removes moisture from the crops, becomes cooler and falls to the bottom of drying chamber. By this system, yam, sweet potato, sorrel and grasses can be dried.

- *For large scale drying, i.e. seasoning of timber, corn drying, tea processing, tobacco curing, fish and fruit drying,* **solar kins** *are used.*

Advantages and disadvantages of solar drying:

Advantages:

1. Dried products improve family nutrition (Dried fruit and vegetables contain high quantities of *vitamins*, *minerals* and *fibre*.

2. Dried fruit can be used in stews, soups, making ice-cream and baked products.

3. Improves the bargaining position of farmers.

Disadvantages:

1. The dried product is often of poor quality as a result of grit and dirt.

2. The product is often unhygienic as a result of microorganism and insects.

4.17. SOLAR COOKING

In India, **cooking** *is the common application of solar energy*. Several varieties of solar cookers are available to suit different requirements.

Following types of cookers will be briefly discussed:

1. Box-type solar cooker.

2. Dish solar cooker.

3. Community solar cooker for indoor cooking.

4.17.1. Box-type Solar cooker

Fig. 4.13 shows a box-type solar cooker.

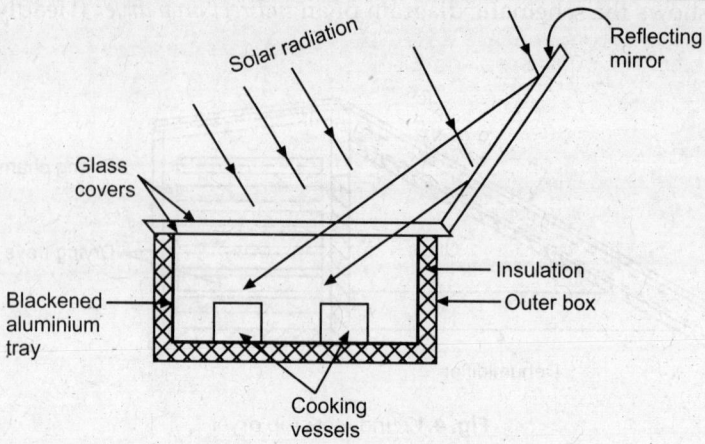

Fig. 4.13. Box-type solar cooker.

It consists of an *'outer box'* made of either fibre glass or aluminium sheet, a *'blackened aluminium tray'*, a *'double glass lid'*, a *'reflector'*, *'insulation'* and *'cooking pots'*. The blackened aluminium tray is fixed inside the box, and the sides are covered with an insulation cover to *prevent heat loss*. The *reflecting mirror* provided on the box cover *increases the solar energy input*. Metal pots are painted black on the outside.

Food to be cooked is placed in cooking pots and cooker is kept facing the sun to cook the food. An electric heater may also be installed to serve as a back during non-shine hours.

Technical drawbacks of the design:

The Government of India is trying to introduce box-type solar cookers with financial subsidies but with poor success due to the following technical drawbacks of the design of the box-type solar cooler:

1. The temperature obtained is *less* than 100°C and is *insufficient* for chapati making and frying which are very important cooking processes.

2. There is no provision for storage of heat and cooking gas to be carried out while the sun is shining, which is an odd time for preparing breakfast and dinner.

3. Cooking has to be carried out in open without privacy and in the sun which is very inconvenient to housewife.

4.17.2. Dish Solar Cooker

In this type of cooker, a *'parabolic dish'* is used to concentrate the incident solar radiation. A typical dish solar cooker has an *aperture* of diameter *1.4 m* with *focal length* of *0.8 m*. The *'reflecting material'* is an *anodized aluminium sheet* having reflectivity of over 80%.

The cooker *needs to track the sun* (which requires power of about 0.6 kW). The temperature at the bottom of the vessel may reach up to 400°C which is sufficient for boiling, roasting and frying.

Such a cooker can meet the requirements of cooking for *fifteen* persons.

4.17.3. Community Solar Cooker for Indoor Cooking

A community solar cooker is a *'parabolic reflector cooker'*. It has a *'large reflector'* ranging from 7 to 12 m² of aperture area. The reflector is placed *outside the kitchen* so *as to reflect solar rays into the kitchen.* A *'secondary reflector'* further concentrates the rays on to the bottom of the cooking pot painted black. Temperature can reach up to 400°C and food can be cooked quickly for *fifty* persons.

4.18. SOLAR FURNACE

A **solar furnace** *is an optical equipment to get high temperatures by concentrating solar radiations on to a specimen.*

The *"primary components"* of a solar furnace are:

1. **Concentrator.** In solar furnaces, either a paraboloidal reflection concentrator or a spherical reflector concentrator is used; the *former* is considered *superior* due to acceptable spherical aberration in a spherical reflector.

2. **Heliostat.** In a solar furnace, the function of heliostats is to *orient* solar radiation *parallel* to the optical axis of the concentrator. As a guide, the size of the heliostat should be 1.4 D × 1.4 D where D represents the size of aperture of concentrator.

3. **Sun tracking.** For a solar furnace to function with optimum output, it is imperative that heliostats should follow the sun from morning till evening.

In a solar furnace the temperatures may reach up to 3500°C.

Solar furnace with multiple heliostat:

The world's first 1000 kW solar furnace was operated at Odeillo, France, in 1973. Solar intensity was 1 kW/m², with bright sunshine for about 1200 hours a year. It consisted of 63 heliostats installed at 8 elevations which reflected sun rays to the concentrator parallel to its optical axis as shown in Fig. 4.14. The *receiver* diameter was changed to obtain different temperatures.

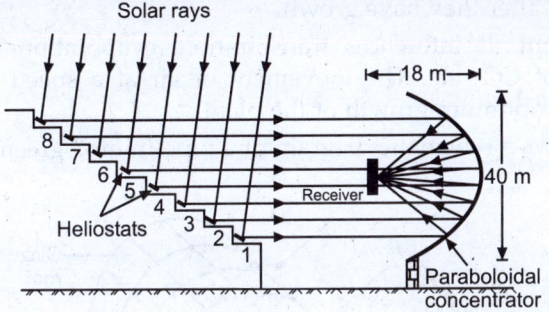

Effective mirror area = 1920 m²; Aperture ratio = 2.8 (nearly)

Solar energy input = 1800 kW

Fig. 4.14. 1000 kW solar furnace with heliostats.

Advantages and Uses of a Solar Furnace:

Advantages:

1. Simple working.
2. High heat flux is obtainable.
3. Heating without contamination.
4. Easy control of temperature.

Uses:

1. For phase and vaporisation studies.

2. Melting behaviour analysis.

3. Purification of ceramic and refractory materials.

4.19. SOLAR GREEN-HOUSES AND GLOBAL WARMING

4.19.1. Solar Green-houses

A **green-house** *is an enclosed space which provides the required environment for growth and production of plants under adverse climatic conditions.* Its design depends upon the local climatic conditions and the environment needed for the growth.

Plants *manufacture their food by 'photosynthesis process'* which maintains a balance with respiration.

The various parameters *for a plant growth are*:

1. **Light intersity.** For plant growth a minimum intersity of 25,000 lux is adequate. A green-house structure, with two glaziings can have maximum light intensity up to 50,000 lux on a clear day.

2. **Temperature.** Temperature is an important environmental factor for plant growth. For *winter crops*, the ideal temperature range is from 5° C to 15° C with a variation up to 3° C. For *summer crops*, the range is from 20° C to 30° C with a variation of 5° C.

   The soil temperature of 20° C to 25° C, for most plants, has been reported to be *optimum*.

3. **Humidity.** For plant growth, relative humidity (RH) of air between 30 and 70% is good. Saplings and germinating seeds need high humidity 100% but its value (RH) *reduces* after they have grown.

4. **Air movement**. It influences transpiration, evaporation of water from soil, availability of CO_2 etc. The movement of air at a speed of 0.8 to 2 cm/s is adequate for optimum growth of the plant.

   Fig. 4.15 shows a schematic diagram of a pipe-framed green-house.

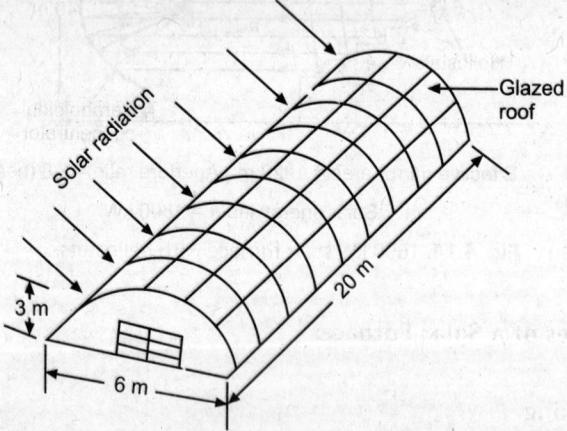

Fig. 4.15. Schematic diagram of pipe-framed greenhouse

Advantage of green-houses:

This type of structure is less expensive to build than a fully insulated structure. It provides the following *advantages* :

1. Inexpensive, good quality food can be grown.

2. An additional heat source (temperature control) is available for the house attached to it.

3. A source of moderator for the humidity (humidity control) in the house.

4.19.2. Global Warming

Global warming *is the term which indicates the increase in the average temperature of the atmosphere.* The increased volume of CO_2 and other green-house gases released by the burning of fossil fuels and other human activities contribute to the warming of the earth. The *amount of heat trapped in the atmosphere depends mostly on the concentrations of green-house gases.*

The major green-house gases are: *Carbon dioxide, ozone, methane, chlorofluorocarbons* (CFCs) and *water vapours.* Due to anthropogenic activities, there is an increase in the concentration of the green-house gases in the air, which results in the increase in average surface temperature.

Deterimental effects:

1. If the emission of green-house gases continues, the global temperature will increase.

2. With the increase in global temperature sea water will expand, which will result in rise in sea level. Heating will melt the polar ice caps resulting in further rise in sea level.

3. Results in floods.

4. Global warming will lead to changes in the rainfall pattern.

Control measures:

1. Use of renewable energy sources.

2. Afforestation.

3. Stabilization of population growth.

4. To cut down the current rate of use of CFCs and fossil fuels.

4.20. SOLAR POWER PLANTS

The solar thermal power generation involves the collection of solar heat which is utilised to increase the temperature of a fluid in a turbine operating on a cycle such as Rankine or Brayton. In another method, hot fluid is allowed to pass through a heat exchanger to evaporate a working fluid that operates a turbine coupled with a generator. These may be *classified* as : Low temperature, medium temperature and high temperature systems.

4.20.1. Low Temperature Solar Power Plant

Fig. 4.16. shows a schematic diagram of a *low temperature solar power plant*

In this system an array of flat-plate collectors is used to heat water to about 70°C and then this heat is used to boil *butane* in a heat exchanger. The high pressure butane vapour thus obtained runs a butane *turbine* which in turn operates a *hydraulic pump*. The pump pumps the water from a well which is used for *irrigation purposes*. The exhaust butane vapour (from butane turbine) is condensed with the help of water which is pumped by the pump and the condensate is returned to the heat exchanger (or boiler).

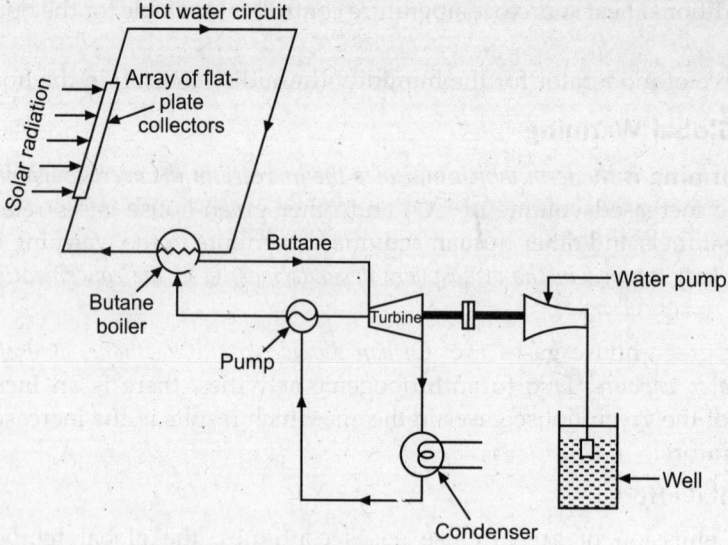

Fig. 4.16. Low temperature solar power plant.

4.20.2. Medium Temperature Systems Using Focusing Collectors

A circular or rectangular parabolic mirror can collect the radiation and focus it on to a small area, a mechanism for moving the collector to follow the sun being necessary. Such devices are used for *metallurgical research* where *high purity and high temperatures are essential*, an example being a 55 m diameter collector giving about 1 MW (th) at Mont Louis in Pyrenees. Smaller units having 20 m diameter reflector can give temperatures of about 300°C over an area of about 50 m². The collector efficiency is about 50%. On a small scale, units about 1 m diameter giving temperatures of about 300°C have been used for cooking purposes.

Fig. 4.17 shows a concave solar energy collector focusing sun's rays on boiler at a focal point. Generation of steam at 250°C could give turbine efficiencies up to 20–25 per cent.

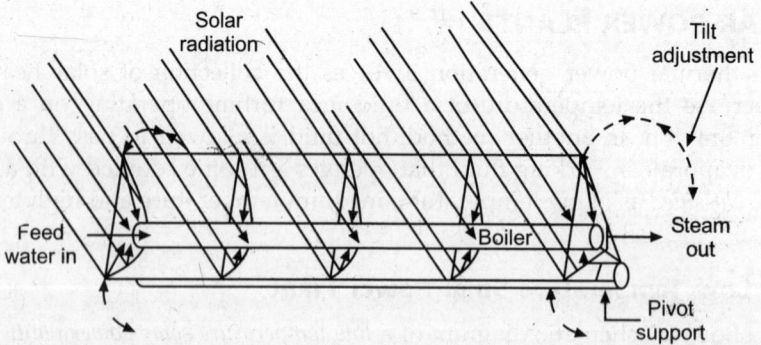

Fig. 4.17. Concave solar energy collector focuses sun's rays on boiler at focal point.

4.20.3. High Temperature Systems—Solar Farm and Solar Power Plant

For a large scale production of process-heat the following *two concepts are available:*

1. **The solar farm.** It consists of a whole field covered with parabolic trough concentrators.

2. **The solar tower.** It consists of a central receiver on a tower and a whole field of tracking.

In case of a *'solar farm'* temperature at the point of focus can reach several hundred degrees celsius. Fig. 4.18 shows a *solar tower system*.

Fig. 4.18. Solar tower system.

In case of central receiver "solar tower" concentrators, temperature can reach thousands of degrees celsius, since a field of reflectors (heliostats) are arranged separately on sun-tracking frames to reflect the sun on to a boiler mounted on a central tower (Figs. 4.19, 4.20)

With both systems ('solar farm' and 'solar tower'), a heat transfer fluid or gas is passed through the point or line of insolation concentration to collect the heat and transfer it to the point of use. Such heat can be used either directly in industrial or commercial processes or indirectly in electricity production via. steam and a turbine.

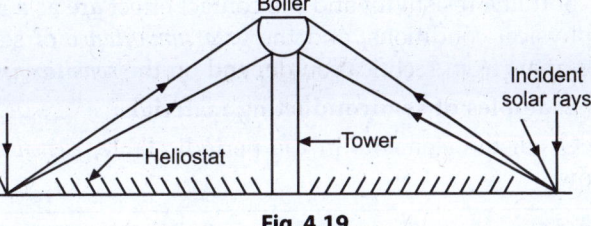

Fig. 4.19

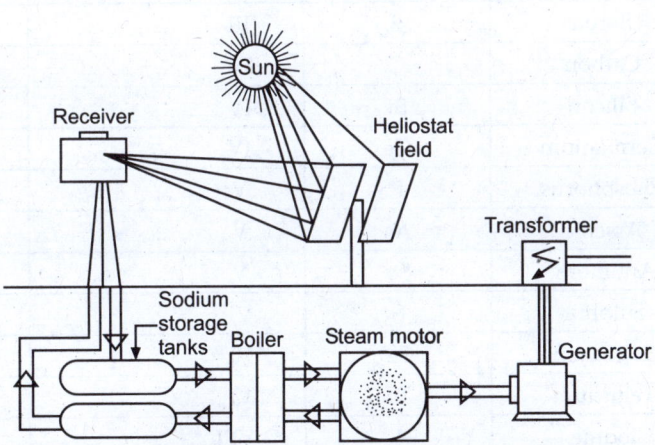

Fig. 4.20. Solar tower power plant.

The solar technologies such as the above two systems that produce very hot water or steam are currently still under development and, in general, these technologies are not cost competitive with conventional power sources such as oil or gas.

4.21. SOLAR PHOTOVOLTAIC (SPV) SYSTEMS

4.21.1. Semiconductors

"Semiconductors" are solid materials, either non-metallic elements or compounds, which allow electrons to pass through them so that they conduct electricity in much the same way as a metal.

4.21.1.1 Characteristics of Semiconductors

Semiconductors possess the following *characteristics*:

1. The resistivity is usually high.
2. The temperature coefficient of resistance is *always negative*.
3. The contact between semiconductor and a metal forms a layer which has a higher resistance in one direction than the other.
4. When some suitable metallic impurity (*e.g.,* Arsenic, Gallium etc.) is added to a semiconductor, its *conducting properties change appreciably*.
5. They exhibit a rise in conductivity in the increasing temperature, with the decreasing temperatures their conductivity falls off, and at low temperatures semiconductors become dielectrics.
6. They are usually metallic in appearance but (unlike metals) are generally hard and brittle.

Both the resistivity and the contact effect are as a rule very sensitive to small changes in physical conditions, and the *great importance* of semiconductors for a wide range of uses apart from rectification depend on the *sensitiveness*.

Examples of semiconducting materials:

Of all the elements in the periodic table, *eleven are semiconductors* which are listed below:

S. No.	Element	Symbol	Group in the periodic table		Atomic No.
1.	Boron	B	III	... ...	15
2.	Carbon	C	IV	... ...	6
3.	Silicon	Si	IV	... ...	14
4.	Germanium	Ge	IV	... ...	32
5.	Phosphorus	P	V	... ...	15
6.	Arsenic	As	V	... ...	33
7.	Antimony	Sb	V	... ...	51
8.	Sulphur	S	VI	... ...	
9.	Seleinium	Se	VI	... ...	
10.	Tellurium	Te	VI	... ...	
11.	Iodine	I	VIII	... ...	

Examples of *"semiconducting compounds"* are given below:

(*i*) Alloys : Mg_3Sb_2, ZnSb, Mg_2Sn, CdSb, AlSb, InSb, GeSb.

(*ii*) Oxide : ZnO, Fe_3O_4, Fe_2O_3, Cu_2O, CuO, BaO, CaO, NiO, Al_2O_3, TiO_2, UO_2, Cr_2O_3, WO_2, MoO_3.

(iii) Sulphides: Cu_2S, Ag_2S, PbS, ZnS, CdS, HgS, MoS_2.

(iv) Halides: AgI, CuI.

(v) Selenides and Tellurides.

PbS is *used in photo-conductive devices, BaO in oxide coated cathodes, caesium antimonide in photomultipliers etc.*

4.21.1.2 Differences between Semiconductors and Conductors

Semiconductors	Conductors
1. Their resistivity is usually high and temperature coefficient of resistance is always *negative*.	1. The resistivity is very low and the temperature coefficient of resistant is *not constant*.
2. Both resistivity and contact effects are very sensitive to small changes in physical conditions.	2. Both resistivity and contact effects are very sensitive to small changes in physical conditions.
3. These are materials having filled energy bands and small forbidden zones.	3. These are materials with unfilled or overlapping energy bands.
4. Moving carriers of electric current are originated due to absorption of thermal, radiant or electric energy from an external source.	4. Current carriers here are free electrons which exist whether external energy is applied or not.
5. At low temperatures, they become *dielectrics*.	5. At very low temperatures, *i.e.*, at absolute zero temperatures they become *superconductors*.
6. They show a rise in conductivity with increase in temperature due to increase in current carriers and *vice versa*.	6. Their conductivity increases with decreasing temperature up to near absolute zero stage.

4.21.1.3 Differences between Semiconductors and Insulators

Semiconductors	Insulators
1. These are the materials whose valence electrons are bound *somewhat loosely* to their atom.	1. These are materials in which valence electrons used are bound *very tightly* to their parent atoms.
2. They require an energy *less* than insulators and more than good conductors to remove an electron from the parent atom.	2. They require *very large* electric field to remove them from the attraction of nuclei.
3. They have: (i) *an empty conduction band* (ii) *almost filled valence band* (iii) *very narrow energy gap (of 1eV) separating the two bands.*	3. They have: (i) *a full valence band* (ii) *an empty conduction band* (iii) *a large energy gap.* *(or several eV) between them.*

4.21.1.4. Differences between Conductors and Insulators

Conductors	Insulators
1. These are the materials in which an electromotive force causes appreciable drift of electrons.	1. The ionically and covalently bounded materials are known as insulators or poor conductors.

2. All metals are good conductors of electricity.	2. Materials like glass, plastics, ceramics, diamond, wood, asbestos, bakelite, mica, PVC, rubber and porcelain having high resistance are known as insulators.
3. The conductors do not break **down** under high electrical voltage.	3. These materials generally break down under high electrical voltage.
4. Metals like silver, copper and aluminium having low resistance are good conductors.	4. The energy gap in insulator is very high.

4.21.1.5. Atomic Structure

To understand how semiconductors work, it is necessary to study briefly the structure of matter. All atoms are made of electrons, protons and neutrons. Most solid materials are classed, from the stand-point of electrical conductivity, as conductors, semiconductors or insulators. *To be conductor, the substance must contain some mobile electrons—one that can move freely between the atoms.* These free electrons come only from the valence (outer) orbit of the atom. Physical force associated with the valence electrons bind adjacent atoms together. The inner electrons below the valence level do not normally enter into the conduction process.

Conductivity depends on the number of electrons in the valence orbit. Electron diagrams for three typical elements, aluminium, phosphorus and germanium are shown in Figs. 4.21, 4.22, 4.23.

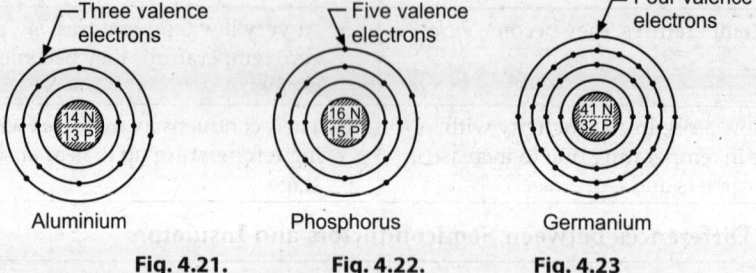

Aluminium	Phosphorus	Germanium
Fig. 4.21.	**Fig. 4.22.**	**Fig. 4.23**

These elements can all be used in semiconductor manufacture. The *degree of conductivity* is determined as follows:

1. Atoms with *fewer than four valence* electrons are *good conductors*.
2. Atoms with *more than four valence* electrons are *poor conductors*.
3. Atoms with *four valence electrons are semiconductors*.

Fig. 4.21 shows aluminium which has *three valence electrons*. When there are less than four valence electrons they are loosely held so that at least one electron per atom is normally free; hence aluminium is a good conductor. This ready availability of free electrons is also true of copper and most other metals.

Fig. 4.22 shows Phosphorus with *five valence electrons*. When there are more than four valence electrons, they are lightly held in orbit so that normally *none are free*. Hence phosphorus and similar elements are poor conductors (insulators).

Germanium (Fig. 4.23) has four *valence electrons*. This makes it neither a good conductor nor a good insulator, hence its name "semiconductor". Silicon also has four valence electrons and is a semiconductor.

Note. The energy level of an electron increases as its distance from the nucleus increases. Thus an electron in the second orbit possesses more energy than the electron in the first orbit; electrons in the third orbit have higher energy than in the second orbit and so on. It follows, therefore, that electrons in the last orbit will possess very high energy. These high energy electrons are less bound to the nucleus and hence they are more mobile. It is the mobility of last orbit electrons that they acquire the property of combining with other atoms. Further it is due to this combining power of last orbit electrons of an atom that they are called *valence electrons*.

4.21.1.6 Intrinsic Semiconductor

A *pure semiconductor is called "intrinsic semiconductor"*. Here no free electrons are available since all the co-valent bonds are complete. A *pure semiconductor, therefore, behaves as an insulator*. It exhibits a peculiar behaviour even at room temperature or with rise in temperature. The *resistance of a semiconductor decreases with increase in temperature*.

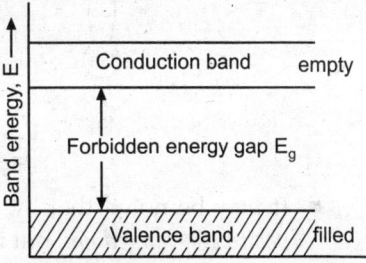

Fig. 4.24. Energy diagram for intrinsic (pure) semiconductor at absolute zero.

When an electric field is applied to an intrinsic semiconductor at a temperature greater than 0°K, conduction electrons move to the anode and the holes (when an electron is liberated into the conduction band a positively charged hole is created in valence band) move to cathode. Hence semiconductor current consists of movement of electrons in opposite direction.

4.21.1.7. Extrinsic Semiconductor

In a pure semiconductor, which behaves like an insulator *under ordinary conditions*, if small amount of certain *metallic impurity* is added, it attains *current conducting properties*. The impure semiconductor is then called *"impurity semiconductor"* or *"extrinsic semiconductor"*. *The process of adding impurity (extremely in small amounts about 1 part in 10^8) to a semiconductor to make it extrinsic (impurity) semiconductor is called* **Doping.**

Generally following doping agents are used:

(i) *Pentavalent atom* having five valence electrons (arsenic, antimony, phosphorus) calle *donor atoms.*

(ii) *Trivalent atoms* having three valence electrons (gallium aluminium, boron) called *acceptor atoms.*

With the addition of suitable impurities to semiconductor, two type of semiconductors are:

(i) N-type semiconductor.

(ii) P-type semiconductor.

N-type semiconductor:

The presence of *even a minute quantity of impurity* can produce N-type semiconductor. If the impurity atom has *one valence electron more* than the semiconductor atom which it has substituted, this *extra electron* will be loosely bound to the atom. For example, an atom of *Germanium* possesses *four valence electrons*; when it is replaced in the crystal lattice of the substance by an impurity atom of antimony (Sb) which has *five valence electrons*, the fifth valence electron (free electron) produces extrinsic N-type conductivity *even at room temperature*. Such an impurity into a semiconductor is called *"donor impurity"* (or donor). The conducting properties of germanium will depend upon the *amount of antimony* (i.e., impurity) *added*. This means that controlled conductivity can be obtained by proper addition of impurity. Fig. 4.25 (*a*) shows the loosely bound excess electron controlled by the donor atom.

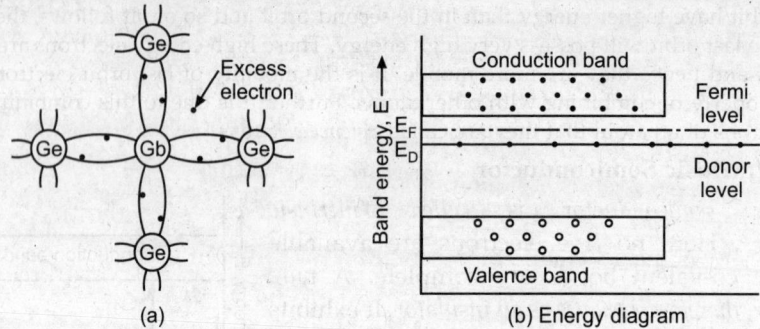

Fig. 4.25. N-type semiconductor.

- It may be noted that by giving away its one electron, the *donor atom* becomes a *positively charged ion*. But it cannot take part in conduction because it is firmly fixed or tied into the crystal lattice. In addition to the electrons and holes *intrinsically available in germanium*, the addition of antimony greatly increases the number of conduction electrons. Hence, *concentration of electrons in the conduction band is increased and exceeds the concentration of holes in the valence band*. Consequently, *Fermi level shifts upwards towards the bottom of the conduction band* as shown in Fig. 4.25 (*b*). [Since the number of electrons as compared to the number of holes increases with temperature, the *position of Fermi level also changes considerably with temperature*].

- It is worth noting that even though N-type semiconductor has excess of electrons, still it is *ekectrically neutral*. It is so because by addition of donor impurity, number of electrons available for conduction purposes becomes more than the number of holes available intrinsically. But the *total charge of the semiconductor does not change* because the donor impurity brings in as much negative charge (by way of electrons) as positive charge (by way of protons).

Note: In terms of energy levels, the fifth antimony electron has as energy level (called donor level) just below the conduction band. Usually, the donor level is 0.01 eV below conduction band for *germanium* and 0.054 eV for *silicon*.

P-type semiconductor:

- P-type extrinsic semiconductor can be produced if the impurity atom has *one valence electrons less* than the semiconductor atom that it has replaced in the crystal lattice. This impurity atom cannot fill all the *interatomic* bonds, and the free bond can accept an electron from the neighbouring bond; leaving behind a vacancy of *hole*. Such an impurity is called an *"acceptor impurity"* (or *acceptor*). Fig. 4.26 (*a*) shows structure of P-type semiconductor (Germanium and Boron).

- In this type of semiconductor, conduction is by means of holes in the valence band. Accordingly, *holes form the majority carriers whereas electrons constitute minority carriers*. The process of conduction is called '*deficit conduction*'.

- Since the concentration of holes in the valence band is more than the concentration of electrons in the conduction band, Fermi level shifts nearer to the valence band [Fig. 4.26 (*b*)]. The acceptor level lies immediately above the Fermi level. *Conduction is by means of hole movement at the top of valence band, the acceptor level readily accepting electrons from the valence band.*

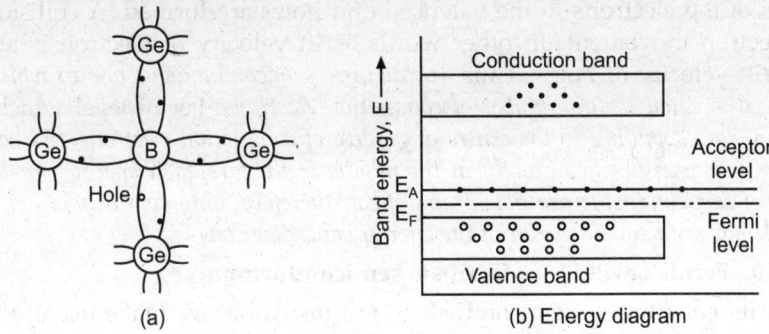

Fig. 4.26. P-type semiconductor.

It may be noted again that even though P-type semiconductor has excess of holes for conduction purposes, as a whole it is electrically neutral for the same reasons as discussed earlier.

4.21.1.8. Atomic Binding in Semiconductors

The atoms of semiconductors are arranged in an ordered array called *crystal lattice* because they have a crystalline structure *e.g.*, *germanium* and *silicon*. Since both these materials are tetravalent having four valence electrons in their outermot shell, therefore they form *covalent bonds* with the neighbouring atoms. In order to achieve inert gas structure having 8 electrons in the outermost orbit, they *share four electrons with each other*. In case of germanium atom only four electrons out of 32 electrons take part in its electrical characteristics because the remaining 28 electrons are *tightly bound* to the nucleus. The four electrons revolving in the outermost shell are called *valence electrons*. Therefore, each atom of the semiconductor surrounded symmetrically by four other atoms forms a *tetrahedral crystal*. Hence a stable structure is formed when each shares a valence electron with each of its four neighbours.

In case of pure (intrinsic) germanium, to allow the electrons for conduction of current the *covalent bonds should be broken*. For setting the electron free, there are different ways of *rupturing the covalent bond*. This can be done by *increasing the crystal temperature above 0° K*. *There are mainly two properties, hardness and brittleness, of covalent crystals. The hardness is characterised by the great strength of the covalent bond itself. The brittleness is characterised by the fact that adjacent atom must remain in accurate alignment because the bond is strongly directional and formed along a line forming the atoms.*

4.21.1.9. Formation of Holes in Semiconductors

Refer to Fig. 4.27. In the semiconductor, the *hole formed is a positive charge carrier*. When a covalent bond is broken at its edge, electrons move through the crystal lattice leaving behind a hole in the bond. An electron from the side lattice jumps into the first vacant hole. Later on, an electron from another point N will jump into the hole at M and so on. Thus a hole would appear at a point (R) in the lattice opposite to the first hole (L) by a succession of electron movements. A negative charge would move from R to L. In other words, a positive charge is said to have moved from L to R. Therefore, a hole in this case is regarded as a *positive charge carrier* or an electron with a negative charge. Hence due to

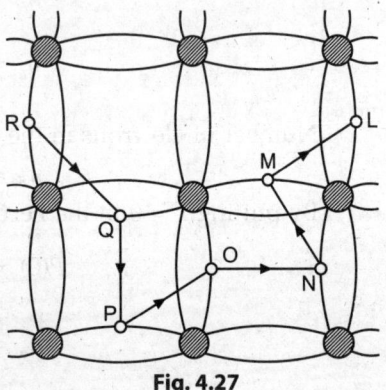

Fig. 4.27

movements of the electrons in the valence bond holes are formed. A collision is caused by each electron movement. In other words, drift velocity of electron is always more than the drift velocity of holes. Thus an *intrinsic semiconductor is one in which number of holes produced is equal to the number of conduction electrons*. Fermi level which is defined as the *energy corresponding to the centre of gravity of conduction electrons and holes weighted according to their energies lies exactly in the middle of the forbidden energy gap. As the lower filled bonds of semiconductor are not of any effect, therefore, only two bands, i.e., valence and conduction bands are usually shown in the energy band diagrams.*

4.21.1.10. Fermi Level in an Intrinsic semiconductor

To classify conductors, semiconductors and insulators we make use of the reference energy level, called *Fermi level*. In case of an *intrinsic semiconductor, Fermi level (E_F) lies in the middle of energy gap or mid-way between the conduction and valence bands.*

Let (at any temperature T°K),

$$n_v = \text{Number of electrons in the valence band.}$$
$$n_c = \text{Number of electrons in the conduction band.}$$
$$N = \text{Number of electons in both bands.}$$
$$= n_v + n_c$$

Assumptions:

(*i*) In valence band energy of all levels is zero.

(*ii*) In conduction band, energy of all levels is equal to E_g (energy at gap).

(*iii*) As compared to forbidden energy gap between two bands, the widths of energy bands are small.

(*iv*) All levels in a band consist of same energy due to small width of band.

Let the zero energy reference level is arbitrarily taken at the top of valence band.

∴　No. of electrons in the conduction band, $n_c = N. \, P \, (E_g)$ where $P(E_g)$ =probability of an electron having energy E_g.

Fermi-Dirac probability distribution function gives its value given below:

$$P(E) = \frac{1}{1 + e^{(E - E_F)/kT}} \qquad \qquad ...(4.1)$$

where $P(E)$ = Fermi-Dirac distribution function or probability of finding an electron having any particular value of energy E and E_f in the Fermi level.

∴ $$P(E_g) = \frac{1}{1 + e^{(E_g - E_F)/kT}}$$

∴ $$n_c = \frac{1}{1 + e^{(E_g - E_F)/kT}} \qquad \qquad ...(4.2)$$

Number of electrons in the vlaence band is

$$n_v = NP(0)$$

By putting $E = 0$ in the Fermi-Dirac probability distribution function,

$$P(0) = \frac{1}{1 + e^{(0 - E_F)/kT}} = \frac{1}{1 + e^{-E_F/kT}}$$

∴ $$n_v = \frac{N}{1 + e^{-E_F/kT}}$$

Now
$$N = n_v + n_c = \frac{N}{1 + e^{-E_F/kT}} + \frac{N}{1 + e^{(E_g - E_F)/kT}}$$

After simplification, we get

$$E_F = \frac{E_g}{2} \qquad \qquad ...(4.3)$$

This *shows that in an 'Intrinsic semiconductor', the Fermi level lies mid-way between the conduction and valence bands.*

4.21.2. Photovoltaic Effect

When a solar cell (*p-n* junction) is illuminated, electron-hole pairs are generated and the electric current I is obtained. *I* is the difference between the solar light generated current I_L and the diode current I_j

Mathematically, $I = I_L - I_j$ ...(4.4)

or, $I = I_L - I_0\left[\exp\left(\frac{eV}{kT}\right) - 1\right]$ ...(4.5)

where, I_0 = Saturation current,

    e = Electron charge,

    V = Voltage across the junction,

    k = Boltzmann's constant, and

    T = Absolute temperature.

This *phenomenon is known as the* **Photovoltaic effect**.

4.21.3. Conversion Efficiency and Power Output of a Solar Cell

A solar cell uses a *p-n* junction its physical configuration. The current and voltage relationship is given by:

$$I_j = I_0\left[\exp\left(\frac{eV}{kT}\right) - 1\right] \qquad\qquad ...(4.6)$$

where, I_j = Junction current,

    I_0 = Saturation current (also called *dark current*),

    e = Electron charge,

    V = Voltage across the junction, and

    T = Absolute temperature.

Open circuit voltage V_{oc} for the *ideal cell* is given by:

$$V_{oc} = \left(\frac{kT}{e}\right)\ln\left(\frac{I_L}{I_0} + 1\right) \qquad\qquad ...(4.7)$$

$\because$ $I_L \gg I_0$, the 1 in the equation can be neglected. Then open circuit voltage (V_{oc}) becomes:

$$V_{oc} = \frac{kT}{e}\ln\left(\frac{I_L}{I_0}\right) \qquad\qquad ... (4.8)$$

In practice the *'open circuit voltage of the cell' decreases with increasing temperature.*

The voltage, current and power delivered by the solar cell are influenced by the following *factors:*

(*i*) Conditions of sunlight, intensity, wavelength and angle of incidence etc.;

(*ii*) Conditions of the junction, temperature, termination, etc.

(*iii*) External resistance(*R*).

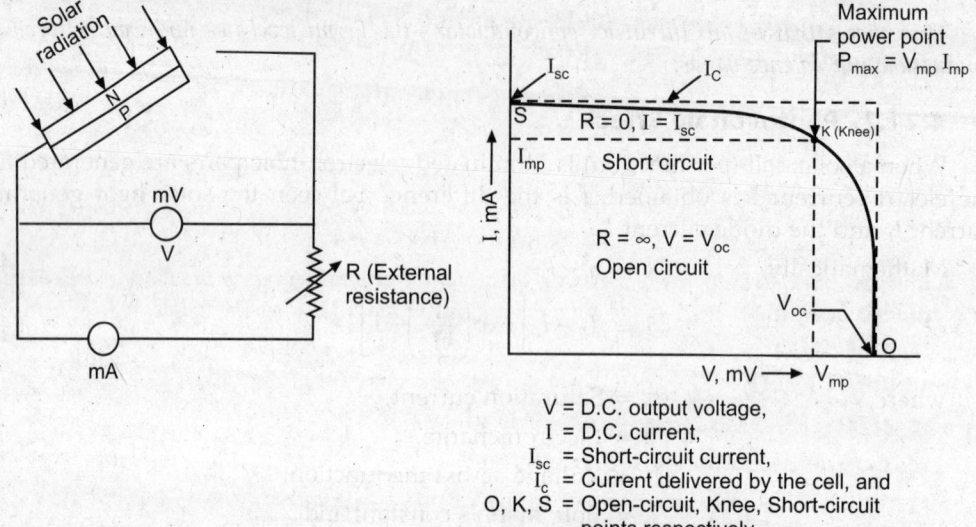

V = D.C. output voltage,
I = D.C. current,
I_{sc} = Short-circuit current,
I_c = Current delivered by the cell, and
O, K, S = Open-circuit, Knee, Short-circuit points respectively.

Fig. 4.28. Test condition.   **Fig. 4.29.** V-I characteristics of a solar cell.

The ratings of a solar cell are specified for particular reference conditions and with the help of *V-I* characteristics.

Fig. 4.28 represents a test condition. Fig. 4.29 gives *V–I characteristics* of a typical commercially available solar cell.

— When external resistance *R* is **high** (mega-ohms range or infinity) the condition is called '*Open-circuit*'. The open-circuit voltage V_{oc} of a solar cell is about 0.5 V.D.C. It is the *maximum voltage* across a PV (photovoltaic) cell. Open-cirucit current is zero.

— If *R* is *reduced gradually* and the readings of the terminal voltage *V* and load current *I* are taken, we get V-I, characteristics of the PV cell (Fig. 4.29).

— As *R* is *reduced* from high value to low value, the terminal *voltage* of the cell *falls* and *current increases*. A *steep* characteristic OK is obtained.

— At knee point '*K*', the characteristic undergoes a smooth change and becomes *flat* for the portion Ks.

— When the external resistance is completely shorted, the short-circuit current I_{sc} is obtained. The *terminal voltage* for the short-circuit conditions is *zero* and *maximum current delivered by the cell is I_{sc}.*

The *maximum power* (P_{max}) that can be derived from the device is given by:

$$P_{max} = V_{mp} I_{mp} \qquad \qquad ...(4.9)$$

where, V_{mp} and I_{mp} are the voltage and current at maximum power point as shown in Fig. 4.29.

Maximum efficiency (η_{max}) *of a solar cell is defined as the ratio of maximum electric power output to incident solar radiation.*

Mathematically, $\qquad \eta_{max} = \dfrac{V_{mp} I_{mp}}{I_s A_c}$ ...(4.10)

where, $\qquad\qquad I_s$ = Incident solar flux, and

$\qquad\qquad\qquad A_c$ = Cell's area

Fill factor (FF). The *fill factor* for a solar cell is defined as *the ratio of two areas shown* (*Fig. 4.29*). Mathematically,

$$FF = \dfrac{I_{mp} V_{mp}}{I_L V_{oc}}$$ ...(4.10)

Maximum power can be defined in terms of I_L and V_{oc} and is given by

$$P_{max} = I_L \times V_{oc} \times FF$$ ...(4.11)

Solar cell designers strive to *increase the FF values*, to *minimise internal losses*.

FF for a good silicon cell is about 0.8.

Voltage factor. The voltage factor $\dfrac{eV_{oc}}{E_g}$ is determined by the basic properties of the materials in the cell and is typically *about 0.5 for a 'silicon cell'* (E_g = Forbidden energy gap).

- *"Leakage"* across the cell increases with temperature which reduces voltage and maximum power.

Causes of low efficiency of a solar cell:

The efficiency of a photovoltaic cell is *15% only*. The major losses which lead to the low efficiency of the cell are:

1. As the *temperature* of the cell *rises* due to solar radiation, *leakage* across the cell *increases*. Consequently, power output, relative to solar energy input, decreases. For *silicon*, the *output decreases by 0.5% per °C*.

2. The excess energy of active photons given to the electrons beyond the required amount to cross the band gap *cannot be recovered* as useful electric power. It appears as heat, about 33 per cent, and is lost.

3. The electric current (generated) flows out of the top surface by a mesh of *metal contacts* provided to *reduce* series resistance losses. These *contacts cover a definite area which reduces the active surface and proves an obstacle to incident solar radiation*.

☞ • To *achieve maximum efficiency the semiconductor with optimum band gap should be used.*

Power output of solar panel, array and module:

Let, $\qquad\qquad n$ = Number of solar cells in a module,

$\qquad\qquad\qquad m$ = Number of modules in an array or a panel, and

$\qquad\qquad\qquad P_c$ = Power per solar cell, watts

Then, *Power per module,* $P_{mod} = nP_c$, watts ...(4.12)

and, *Power per array or panel* $P_p = m \times nP_c$ watts ...(4.13)

Voltage across panel, $V_p = \dfrac{P_p}{I_p}$

Current delivered by the panel, $I_p = \sqrt{\dfrac{P_p}{R}}$ ...(4.14)

4.21.4. Solar Photovoltaic Cells

The *"photovoltaic or solar cell"* is a *samiconductor device*. The 'photovoltaic effect' was first observed in *1839* by *Becquerel* who found that, *when light was directed on to one side of an electrochemical cell, a voltage was created*. The development of *selenium and cuprous oxide photovoltaic cells* led to many applications, including photographic exposure meters. In the late 1950s, silicon solar cells were made with a conversion efficiency high enough for power generators.

4.21.4.1. Photovoltaic materials

The solar cells are made of various materials and with different structures in order to *reduce the cost and optimize efficiency*. Various types of solar cell materials available in the market are:

- The *single crystal, polycrystalline and amorphous silicon, compound thin material and also semiconductor absorbing layer which gives highly efficient cells for specialised applications.*

- **Thin film solar cells** are manufactured from $CuInSe_2$, CdS, $CdTe$, Cu_2S, InP.

- The *amorphous silicon thin solar cells* are *less expensive while crystalline silicon cells are expensive and more popular*. The amorphous silicon layer is used with both hydrogen and fluerine incorporated in the structure.

- *Higher efficiency of photovoltaic generator can be achieved by combination of different bond gap materials in the tandem configurations.*

4.21.4.2. Silicon photovoltaic cell (Single crystal solar cell)

The main feature of a silicon photovoltaic cell is a *thin wafer of high purity silicon crystal, doped with a minute quantity of boron* (Fig. 4.30).

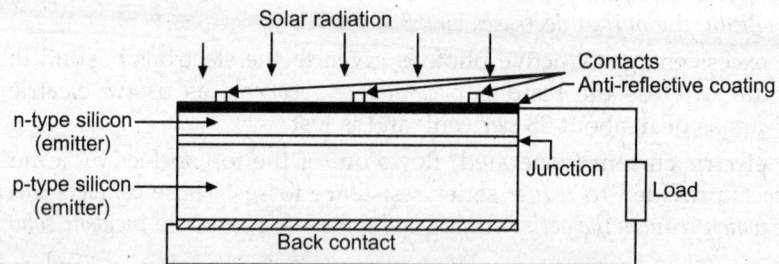

Fig. 4.30. Construction of silicon photo voltaic cell.

Phosphorous is diffused into the active surface of the slice by means of a high temperature process. The top electrical contact is made by a *metallic grid*, and the *back metal contact* covers the whole surface. The top surface usually has an anti-reflective coating (ARC).

Working theory :

— The *'phosphorous'* in the silicon causes an *excess of conduction-band electrons*, and the *'boron'* causes an *excess of valence electron vacancies, or holes*, which act like *positive charges*. At the "junction" between the two types of silicon, conduction electrons from the negative (*n*) region diffuse into the positive (*p*) region and combine with holes, thus *cancelling* their charges. The opposite action also occurs, with holes from the *p*-region crossing into the *n*-region and combining electrons. The area around the junction is thus 'depleted' by disappearance of electrons and nearby holes. Layers of charged impurity atoms, positive in the *n*-region and negative in the *p*-region, are formed on either side of the junction, which sets up a '*reverse*' electric field.

— When light falls on the active-surface, photons with energy exceeding a certain critical level *known* as *band* or *energy gap* (1.1 electron volt in the case of silicon) interact with the valence electrons and elevate them to the conduction band. This activity also leaves holes, so that the photons are said to generate '*electron-hole pairs*'. These electron-hole pairs are produced throughout the thickness of the silicon in concentrations that depend on the intensity and spectral distribution of the light. The electrons move throughout the crystal, and the less mobile holes also move by valence-electron substitution from atom to atom. Some *recombine, neutralising their charges and their energy is converted into heat*. Others reach the junction and are separated by the reverse field, at which point the electrons are accelerated towards the negative contact and the holes towards the positive. A *potential difference is established across the cell, and this will drive a current through an external load*.

- Silicon voltaic cells *require the use of very pure silicon*. The best source of silicon is silica (silicon dioxide), which occurs abundantly in nature as quartz rock and sand. Quartz rocks are reduced in an arc-furnace with the help of carbon-based agents to produce 'metallurgical grade' silicon.

4.21.4.3. Polycrystalline silicon cells

The cost of production of single crystal silicon cell is *quite high* compared to the polycrystalline silicon cell. Polysilicon can be obtained in thin ribbons drawn from molten silicon bath and cooled very slowly to obtain large size crystallites. Cells are made with care so that the *grain boundaries cause no major interference with the flow of electrons and grains are larger in size than the thickness of the cell*.

Following are the *three designs* in which the polycrystalline silicon solar cell can be fabricated:

1. **p-n junction cells.** In such a cell, a polycrystalline silicon film is deposited by chemical vapour deposition on substrates like glass, graphite, metallurgical grade silicon and metal.

2. **Metal insulator semiconductor (MIS) cells.** This type of cell can be developed by inserting a thin insulating layer of SO_2 between the metal and the semiconductor.

3. **Conducting oxide-insulator semiconductor cells.**

4.21.5. Classification of Solar Cells

Solar cells can be *classified* on the basis of:

(*i*) Cell size; (*ii*) Thickness of active material; (*iii*) Type of junction structure; (*iv*) Type of active material.

1. **Cell size.** The size of the silicon solar cells can be divided into *four groups*; (*i*) Round single crystalline having 100 mm diameter; (*ii*) Square single crystalline having area of 100 cm^2, (*iii*) 1000 mm × 1000 mm square multicrystalline, and (*iv*) 125 mm × 125 mm square multicrystalline.

- *Larger size solar cells* are used in *terrestrial applications*. Due to brittleness property of the silicon, *area* of silicon solar cells is *limited*.

   2. **Thickness of active material.** Such solar cells are of *two* types: (*i*) Bulk material cell, and (*ii*) Thin film cell.

   — *Bulk material* single crystal and multicrystalline cells are most successful for terrestrial applications.

   — *Thin film cells* are *not* commercially successful.

   3. **Type of junction structure.** These cells are *classified* as:

   (*i*) *p-n* homojunction cell, (*ii*) *p-n* hetrojunction cell, (*iii*) *p-n* multijunction cell, and (*iv*) Metal semiconductor Schottky junction.

   4. **Type of active material.** Such cells are *classified* as: (*i*) Single crystal silicon cell, (*ii*) Multicrystalline silicon cell, (*iii*) Amorphous silicon cell, (*iv*) Gallium arsenide cell, (*v*) Copper indium diselenide cell, (*vi*) Cadmium telluride cell, and (*vii*) Organic P-V cell.

4.21.6. Silicon Cell Modules

- *'Solar cells'* are *electrically connected in series and parallel to give suitable voltages and currents for a particular application.* A number of cells are generally encapsulated into a *module*, which is the *building block of a photovoltaic system.*

   — A *typical module* measures about 1000 mm × 300 mm × 50 mm and contains 36 cells, which produce 30-35 W at 12 V in bright sunshine.

Several modules combine to form a *'photovoltaic array'*. Galvanized steel, aluminium or chemically treated wood have been used to support the modules, although the latter is not recommended in developing country environments. Arrays are usually fixed in position on concrete foundations.

   — *Arrays that can track the apparent motion of the sun and intercept more energy are naturally more expensive and complex.*

- *"Photovoltaic modules"* are *rated in peak watts* (W_{pk}), which refer to the maximum power output from the module when operating at a cell temperature of 28°C (or sometimes 25°C) under a solar irradiance of 1000 W/m^2. It is a higher output than that usually achieved in the field.

   — If the cell efficiency is 10%, a 1-kW_{pk} array would contain a cell area of 10 m^2. Typically, the *packing factor* (the ratio of cell area to array area) is about 75%, giving a gross area of 13.3 m^2 for a 1-kW_{pk} array.

   — The *efficiency* and, consequently, the power output of the cell depends on the *electrical load* because of the relationship between current and voltage for the cells (Fig. 4.31).

   — The *efficiency of solar cells falls off about 0.5 percent /°C as the operating temperature rises.*

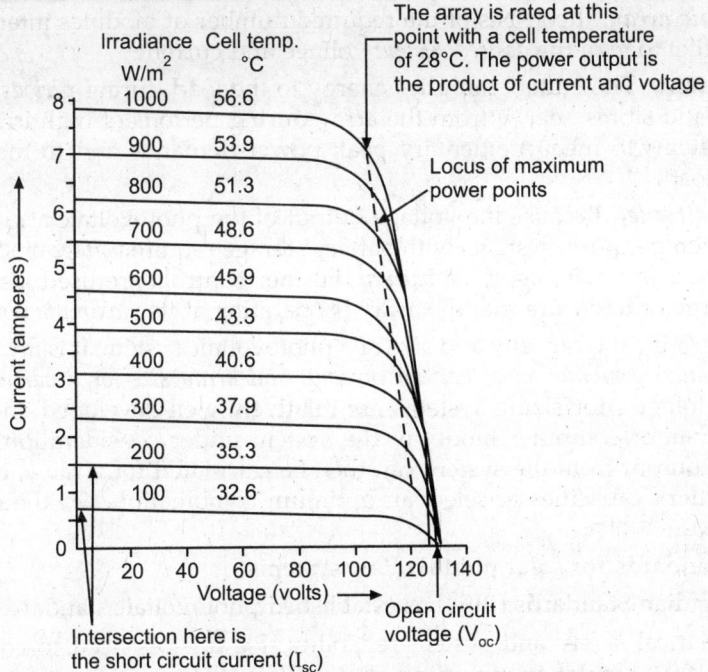

Fig. 4.31. A typical relationship between current and voltage for a photovoltaic array (ITP/Halcrow, 1983).

4.21.7. Photovoltaic (PV) Systems

Solar photovoltaic systems refer to a wide variety of solar electricity systems. Such a system *uses solar array made of silicon to convert sunlight into electricity.* Components other than PV array are collectively known as *'balance of system (BOS)' which includes storage batteries, an electronic charge controller and an inverter.*

These systems are of the following *two* types:

1. **Stand-alone power systems.** In such a system, the *photovoltaic array is the principal or only source of energy.* Energy is stored, often in batteries, for periods when there is insufficient solar radiation. There may also be a back-up power supply such as an engine-generator set.

2. **Grid connected power systems.** In this type of system, *load is connected to both a photovoltaic power system and an electricity grid.* In periods when there is sufficient solar radiation, the array powers the load, otherwise grid is used. In some cases, any surplus electricity produced by the array (i.e. when the load output exceeds the load) is fed back into the grid. This type includes *large MW-sized systems.*

4.21.7.1 Stand-alone power systems

A photovoltaic system can be designed to meet any electrical load. The principal components of stand-alone photovoltaic systems are shown in Fig. 4.32.

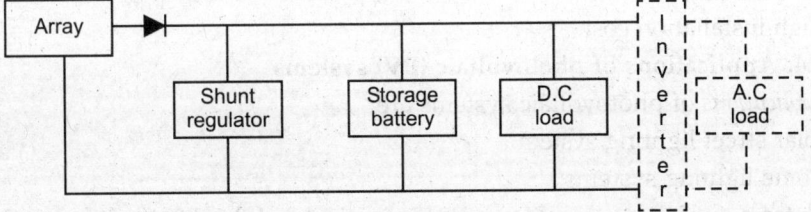

Fig. 4.32. Principal components of stand-alone photovoltaic systems.

Photovoltaic array. It consists of the required number of modules interconnected in series and parallel to give the *desired system* voltage and current.

Storage battery. The battery supplies energy to the load during *periods of little or no solar irradiance* and stores energy from the array during periods of high irradiance. This enables the systems to meet momentary peak power demands and to maintain stable voltage to the load.

Power conditioner. Because the voltage output of the photovoltaic array varies with insolation and temperature, systems with battery storage require *voltage or shunt regulator to prevent excessive overcharging of the battery.* Further controls are used, as required, to prevent discharge or to ensure that the array is operating at its *maximum power point.*

- To determine the capacity and size of a photovoltaic system, it is *necessary to select an optimum combination of battery capacity and array size for a particular location.* Methodologies for sizing systems are relatively well-developed and employ an hour-by-hour computer model of the system under consideration. The annual energy output from the system can then be calculated for a range of array sizes and battery capacities to select an optimum combination, *i.e.,* the one with the *lowest cost.*

4.21.7.2. Standards for solar photovoltaic systems

Bureau of Indian Standards (BIS) has established photovoltaic standards in India.

- For electrical safety and system reliability, PV devices need to conform to IS-12839 (1989) regulation regarding photovoltaic parts.
- Measurement of current and voltage is covered by IS-12762 (1989) and IS-12763 (1989) which deal with electrical characteristics of crystalline silicon cells.

4.21.7.3. Advantages and limitations of photovoltaic systems

Advantages and limitations of photovoltaic systems are as follows:

Advantages:

1. Systems are durable.
2. No operational cost.
3. Low maintenance.
4. More flexibility available.
5. Systems are eco-friendly.
6. Highly reliable.
7. Long effective life.
8. Absence of moving parts.
9. Can function unattended for long periods.
10. High power to weight ratio.

Limitations:

1. Weather dependent.
2. Low efficiency
3. High installation cost.

4.21.7.4. Applications of photovoltaic (PV) systems

The *applications* of photovoltaic systems are:

1. Solar street lighting system.
2. Home lighting systems.
3. Water pumping systems (for micro irrigation and drinking water supply)

4. Solar vehicles.

5. Radio beacons for ship navigation at ports.

6. Community radio and television sets.

7. Cathodic protection of oil pipelines.

8. Railway signalling equipment.

9. Weather monitoring.

10. Battery charging

4.21.8. Water Pumping Systems

The photovoltaic *water pumping systems* (*major application of PV systems*) essentially consist of:

(*i*) A photovoltaic (PV) array,

(*ii*) Storage battery,

(*iii*) Power control equipment,

(*iv*) Motor pump sets, and

(*v*) Water storage tank.

Solar pump configurations:

Because the batteries have several disadvantages (*e.g.* power loss, increased risk of failure, shorter operational life than the rest of the solar pump and regular maintenance requirement), at present, most solar pumping systems do not include batteries, although where water storage is needed, a viable alternative may be provided.

The *four* principal combinations of motor and pump that are suitable for solar pump are:

1. Submerged motor pump set;

2. Submerged pump with surface motor;

3. Floating motor-pump set;

4. Surface motor with surface mounted pump.

These combinations are shown in Fig. 4.33.

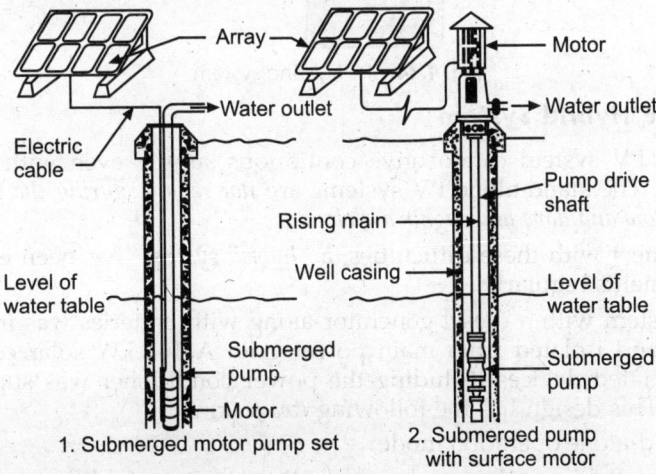

Fig. 4.33 (a). Solar pump configurations (ITP/Halcrow, 1983)

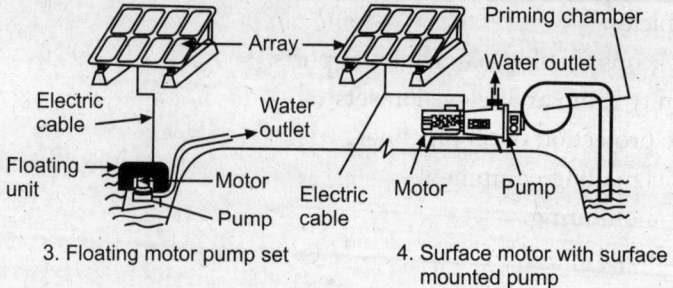

3. Floating motor pump set   4. Surface motor with surface
                                          mounted pump

Fig. 4.33 (b). Solar pump configurations (ITP/Halcrow, 1983).

4.21.9. SPV Lighting System

Solar street light, as shown in Fig. 4.33, describes a stand-alone PV power generating device.

SPV lighting system has the following *advantages*:

1. Compact in size.
2. Highly durable.
3. Highly efficient.
4. Low maintenance.

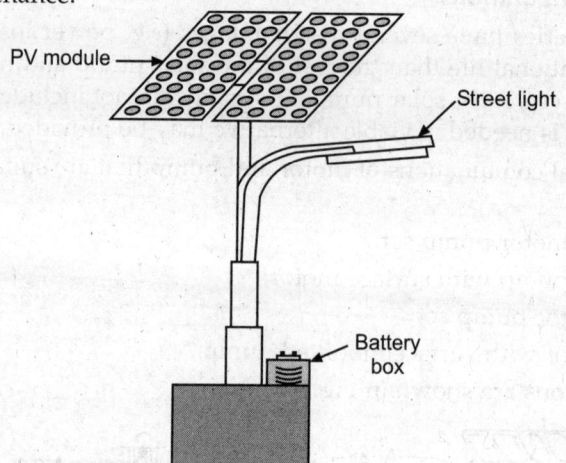

Fig. 4.34. SPV lighting system.

4.21.10. PV Hybrid System

A dedicated PV system cannot give continuous supply even with the use of the storage batteries. The stand-alone PV systems are *not reliable during the periods when the solar incidence is low and days and nights are cloudy*.

In order to meet with these difficulties, a *"hybrid system"* has been evolved to meet the load requirements regularly.

- A PV system with a diesel generator along with batteries was installed in 1987 in an island isolated from main power grid. A 100 kW solar cell module and the associated devices including the power conditioner was supplied by M/s. Hitachi. This design has the following *two features*:

  (*i*) Stand-alone operation mode;

  (*ii*) A parallel operation mode with a diesel unit; an additional diesel generator can also be added in the system.

4.22. SOLAR PRODUCTION OF HYDROGEN

4.22.1. Advantages of Hydrogen as a Substitute for Fossil Fuels

Hydrogen offers the following **advantages** *as a substitute for fossil fuels*:

1. It can *replace natural gas* in the natural gas pipeline system and could replace petroleum products for use in automobiles etc.
2. It can *provide fuel for devices like the fuel cell* which could decentralise electricity production by generating electricity at the site where it is needed.
3. It can be used in *home* as a natural gas substitute, for cooking, cooling and heating, and in the industry for cooling and heating with only minor adjustments or redesign of burners.
4. It can also be used *as a raw material* in many industries including fertilizer, food stuffs, petrochemical and metallurgical industries.

The technology for storing and handling hydrogen in its concentrated form as a liquid is already existing.

4.22.2. Methods of Producing Hydrogen from Solar Energy

The various methods of producing hydrogen from solar energy are enumerated and discussed as follows:

1. Direct thermal method.
2. Thermo-chemical method.
3. Electrolytic method.
4. Photolytic method.

1. **Direct thermal method.** This method is based on the fact that if *water (steam) is heated to about 3000K or above, it decomposes into hydrogen and oxygen*. The energy for heating can be obtained from *solar radiation* by using an optical system which collects solar radiation and concentrates it into small area.

Advantages:

1. High thermal efficiency.
2. Intermediate chemicals are not required.
3. Less environmental impacts.

However, for commercial realization of this method extensive research is needed because of high temperature requirement.

2. **Thermo-chemical method.** In this method *thermo-chemical reactions are used to decompose water*. Water and one or more chemical elements (or compounds) are made to react while heat is being added, resulting in a combination of hydrogen and/or oxygen of water with compounds and *either hydrogen or oxygen being liberated*. Then in one and more chemical reactions the new chemical compounds of first reaction are *reduced* to their original composition with the help of other intermediary chemicals and/or heat releasing oxygen or hydrogen. *The intermediary chemicals can be used again and again and the only inputs are heat, water and work, the outputs being hydrogen, oxygen and low grade heat.*

3. **Electrolytic method.** The electrolytic method uses an *electrolytic cell to produce hydrogen and oxygen from water*. The cell consists of two electrodes immersed in an electrolyte (water plus some conducting material). When electrodes are connected to D.C. supply of sufficient voltage, a current flows, *oxygen is liberated at 'anode' and hydrogen at the 'cathode'*.

— Solar energy is first converted to D.C. electric power which is then used for electrolysis. An overall conversion efficiency of about 25% is possible.

● This method is *especially suited for coupling with ocean, thermal, wind, hydro and photovoltaic forms of solar energy since in these cases solar energy is converted to electricity.*

4. **Photolytic method.** Sun's photons, under certain circumstances, can be absorbed by water molecules and when the energy absorbed reaches a certain level (1.1622 Wh per mole of water) *hydrogen can be released.* Photons in the ultra-violet region of the radiation spectrum possess the energies needed for direct photolysis of water. The intensity of solar radiation reaching the earth is very low in ultra-violet region of the spectrum but high in the visible region. On the other hand, water is almost transparent to the visible light. Hence a *photo-catalyst* is needed for photolysis of water on earth. *The function of hydro-catalyst is to absorbs solar radiation and then impart this energy to water in order to decompose.*

The conversion efficiency of this process is low.

4.22.3. Solar-Hydrogen Energy Cycle

Fig. 4.35 shows a solar-hydrogen energy cycle.

— A portion of solar energy is *converted to electrical energy* for local use. The remaining solar energy is used to *produce hydrogen* from water.

— A part of the hydrogen produced is used to meet the local needs. The remaining hydrogen is transmitted by pipeline (if it is in gaseous form) or by surface tankers (if it is in liquid form) to energy consumption centres for use as fuel for homes, automobile industries etc.

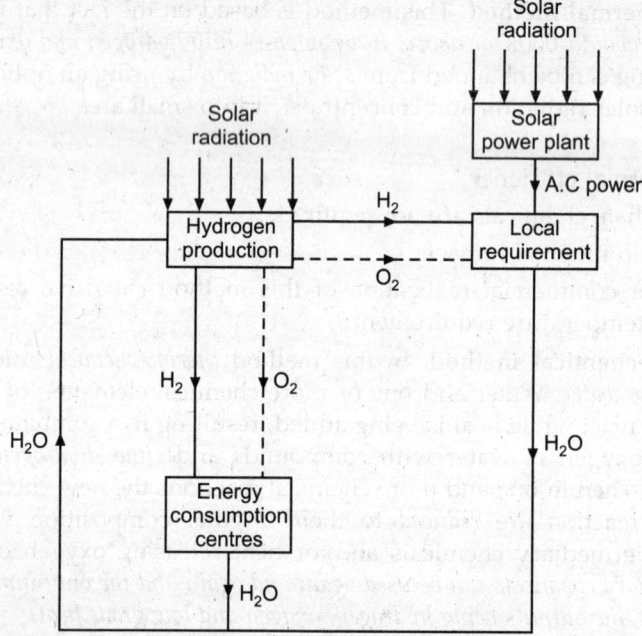

Fig. 4.35. Solar-hydrogen energy cycle.

— Some oxygen, which is always obtained as a *by-product* (during manufacture of hydrogen), is also used locally and in energy consumption centres for industrial and other purposes.

— After use, hydrogen becomes water vapour or water which is used again for production of hydrogen thus *completing the cycle*.

HIGHLIGHTS

1. An *energy system* stores the collected amount of energy in excess of requirement of the demand and supplies this energy when the demand exceeds the supply energy.

2. *Solar pond* is an artificially designed pond, filled with salty water, maintaining a definite concentration gradient. It is utilised for collecting and storing solar energy.

3. A *solar furnace* is an optical equipment to get high temperatures by concentrating solar radiations on to a specimen.

4. A *green-house* is an enclosed space which provides the required environment for growth and production of plants under adverse climatic conditions.

5. *Global warming* is the term which indicates the increase in the average temperature of the atmosphere.

6. *Semiconductors* are solid materials, either non-metallic elements or compounds, which allow electrons to pass through them so that they conduct electricity in much the same way as a metal.

7. Maximum efficiency (η_{max}) of a solar cell is defined as the ratio of maximum electric power output to incident solar radiation.

i.e.
$$\eta_{max} = \frac{V_{mp} \, I_{mp}}{I_s \, A_c}$$

where, V_{mp}, I_{mp} = Voltage and current at maximum power point,

I_s = Incident solar flux, and

A_c = Cell's area

Leakage across the cell increases with temperature which reduces voltage and maximum power.

THEORETICAL QUESTIONS

1. What is the function of an 'energy storage system'?
2. How are solar energy systems classified?
3. Discuss briefly "Solar thermal energy storage".
4. Explain briefly the following:
   (i) Thermal-chemical storage, and (ii) Electrical storage.
5. What is "Solar pond"? Explain its working principle, with a neat diagram.
6. List the applications and limitations of solar ponds.
7. Explain briefly, with a neat diagram, the working of a "Solar pond electric power plant".
8. How does a 'forced-circulation solar water heater' differ from 'natural circulation solar water heater'?
9. Explain briefly the construction and working of a 'Forced-circulation solar water heater'.
10. State the difference between the active and passive space heating systems.
11. Describe a 'basic space heating system'.
12. What are the limitations of uses of solar water heating system?
13. Explain briefly, with a neat diagram, the construction and working of a 'vapour absorption cooling system'.

14. Describe briefly a basic type solar still.
15. What are the advantages of distilling process?
16. Discuss briefly a solar-powered water pumping system.
17. What is a conventional solar air heater?
18. Explain the basic features of non-porous absorber type air heaters.
19. List the advantages and disadvantages of solar air heaters?
20. What are the important areas of applications of solar air heaters.
21. Explain briefly an 'Indirect crop dryer'.
22. State the advantages and disadvantages of solar drying.
23. Describe with a neat sketch the construction and working of a box-type solar cooker.
24. What is a 'Solar furnace'? What are its primary components?
25. Explain a 'Solar furnace'. State its advantages and uses also.
26. What is a solar green-house?
27. Explain a 'pipe-framed green-house'. State the advantages of a green-house.
28. What is global warming? What are its detrimental effects?
29. How can 'global warming' be controlled?
30. Describe with a neat sketch a 'Low temperature solar power plant.'
31. What are semiconductors? What is the difference between intrinsic and extrinsic semiconductors?
32. What is 'Photovoltaic effect'? Explain briefly.
33. Draw *V-I* characteristics of a solar cell and explain briefly.
34. What is 'fill factor' for a cell?
35. What is the effect of 'leakage' across a cell?
36. What are the causes of low efficiency of a solar cell?
37. What is a 'Photovoltaic cell'? Explain briefly with a neat diagram a 'silicon photovoltaic cell'.
38. How are solar cells classified?
39. Write a short note on 'Silicon cell modules'.
40. Describe a 'Stand-alone photovoltaic system'.
41. What are the advantages and limitations of photovoltaic systems?
42. List the applications of photovoltaic (PV) systems.
43. Write a short note on PV hybrid system.
44. Explain briefly solar production of hydrogen.

5

WIND ENERGY

5.1 Introduction; 5.2 Utilisation aspects of wind energy; 5.3 Characteristics of wind; 5.4 Advantages and disadvantages of wind energy; 5.5 Environment aspects of wind energy; 5.6 Sources/Origins of wind; 5.7 Wind availability and measurement; 5.8 Principle of wind energy conservation and wind power; 5.9 Wind energy pattern factor (EPF); 5.10 Basic Components of Wind energy conversion systems (WECS); 5.11 Advantages and disadvantages of wind energy conversion system (WECS); 5.12 Considerations for selection of site for wind energy conversion systems (WECS); 5.13 Terms and Definitions; 5.14 Lift and drag – The basis for wind energy converstion; 5.15 Extraction of wind energy; 5.16 Classification and description of wind mills/machines – Classification of wind mills/machines – description of wind mills/machines – Blade construction methods – Comparison between horizontal and vertical axis wind machines – Parameters to be considered while selecting a wind mill – Design considerations for wind turbine – Performance of wind mills; 5.17 Variation of power output with wind speed; 5.18 Analysis of aerodynamic forces on blade; 5.19 Design of wind turbine rotor – Thrust on turbine rotor – Torque on turbine rotor – Solidity; 5.20 Wind electric generating power plants; 5.21 Generating systems – Constant speed-constant frequency (CSCF) system – variable speed-constant frequency (VSCF) system – variable speed-variable frequency (VSVF) system; 5.22 Wind powered battery chargers; 5.23 Wind elasticity is small independent grids; 5.24 Economic size of WTG; 5.25 Wind electricity economics; 5.26 Problems in operating large wind power generators; 5.27 Selection of site for wind turbine generating system (WTGS). *Highlights – Theoretical Questions.*

5.1. INTRODUCTION

Wind *is air set in motion by small amount of insolation reaching the upper atmosphere of* earth.

It contains *kinetic energy (K.E.)* which can easily be *converted to electrical energy*. Nature generates about 1.67×10^5 kWh of wind energy annually over land area of earth and 10 times this figure over the entire globe.

- This *wind energy*, which is an *indirect source of energy*, can be used to run a wind mill which in turn *drives a generator to produce electricity*.

- Although wind mills have been used for more than a dozen centuries for grinding grain and pumping water, interest in large scale power generation has developed over the past 50 years. A largest wind generator built in the past was 800 kW unit operated in France from 1958-60. The flexible 3 blades propeller was about 35 m in diameter and produced the rated power in a 60 km/hour wind with a rotation speed of 47 r.p.m.

- In India the interest in the wind mills was shown in the last fifties and early sixties. Apart from importing a few from outside, new designs were also developed, but these were not sustained. It is only in last 20-25 years that development work is going on in many institutions. An important reason for this lack of interest in wind energy must be that wind, in India *is relatively low and vary appreciably with seasons. These low and seasonal winds imply a high cost of exploitation of wind energy. In our country high wind speeds are however available in coastal areas of Sourashtra, Western Rajasthan and some parts of central India. In these areas there could be a possibility of using medium and large sized wind mills for generation of electricity.*

5.2. UTILISATION ASPECTS OF WIND ENERGY

Utilisation aspects of wind energy fall into the following *three* broad categories:

1. Isolated continuous duty systems which need suitable energy storage and reconversion systems.
2. Fuel-supplement systems in conjunction with power grid or isolated conventional generating units.
   - This utilisation aspect of wind energy is the *most predominant in use as it saves fuel and is fast growing particularly in energy deficient grids.*
3. Small rural systems which can use energy when wind is available.
   - This category has application in *developing countries with large isolated rural areas.*

5.3. CHARACTERISTICS OF WIND

The main characteristics of wind are:

- *Wind speed increases roughly as $\frac{1}{7}$ th power of height. Typical* tower heights are about 20–30 m.
- *Energy-pattern factor. It is the ratio of the actual energy in varying wind to energy calculated from the cube of mean wind speed.* This factor is always *greater than unity* which means the energy estimates based on mean (hourly) speed are *pessimistic.*

5.4. ADVANTAGES AND DISADVANTAGES OF WIND ENERGY

Following are the *advantages* and *disadvantages* of wind energy:

Advantages:

1. It is a renewable energy source.
2. Wind power systems being non-polluting have no adverse effect on the environment.
3. Fuel provision and transport are not required in wind energy conversion systems.
4. Economically competitive.
5. Ideal choice for rural and remote areas and areas which lack other energy sources.

Disadvantages:

1. Owing to its irregularity, the wind energy needs storage.

2. Availability of energy is fluctuating in nature.

3. The overall weight of a wind power system is relatively high.

4. Wind energy conversion systems are noisy in operation.

5. Large areas are required for installation/operation of wind energy systems.

6. Present systems are neither maintenance free, nor practically reliable.

7. Low energy density.

8. Favourable winds are available only in a few geographical locations, away from cities, forests.

9. Wind turbine design, manufacture and installation have proved to be most complex due to several variables and extreme stresses.

10. Requires energy storage batteries and/or stand by diesel generators for supply of continuous power to load.

11. Wind farms require flat, vacant land free from forests.

12. Only in kW and a few MW range; it does not meet the energy needs of large cities and industry.

☞ 5.5. ENVIRONMENT IMPACTS OF WIND ENERGY

The *possible environment impacts of wind energy* are:

1. Wind energy creates *noise pollution* because of mechanical (gear box) aerodynamic noise.

2. The wind turbine produces *electromagnetic interference* when placed between radio, television etc. stations, as it reflects some electromagnetic radiations.

3. It produces *visual shining* because of reflection and refraction which depends upon turbine size, number of turbines in wind farm, design etc.

4. *Safety consideration* for life because of *accidental braking of blade*.

5. *Fatal collisions of birds* caused by rotating turbine blides.

5.6. SOURCES/ORIGINS OF WIND

Following are the two sources/origins of wind (a natural phenomenon):

1. Local winds.

2. Planetary winds.

1. **Local winds.** These winds are caused by *unequal heating and cooling of ground surfaces and ocean/lake surfaces during day and night*. During the day warmer air over land rises upwards and colder air from lakes, ocean, forest areas, and shadow areas flows towards warmer zones.

2. **Planetary winds.** These winds are caused by *daily rotation of earth around its polar axis and unequal temperature between polar regions and equatorial regions*. The strength and direction of these planetary winds change with the seasons as the solar input varies.

- Despite the wind's intermittent nature, *wind patterns at any particular site remain remarkably constant year by year*.

- Average wind speeds are greater in hilly and coastal areas than they are available in land. The winds also tend to blow more consistently and with greater strength over the surface of the water where there is a less surface drag.

- *Wind speeds increases with height.* They have traditionally been measured at a standard height of 10 meters where they are found to be 20-25 percent greater than close to the surface. At a height of 60 m they may be 30-60 percent higher because of the reduction in the drag effect of the surface of the earth.

5.7. WIND AVAILABILITY AND MEASUREMENT

Wind energy can only be economical in areas of good wind availability. Wind energy differs with region and season and also, possibly to an even greater degree with local terrain and vegetation. Although wind speeds generally increases with height, varying speeds are found over different kinds of terrain. Observations of wind speed are carried out at meteorological stations, airports and lighthouses and are recorded regularly with ten minute mean values being taken every three hours at a height of 10 m. But airports, sometimes are in valleys and many wind speed meters are situated low and combinations of various other factors mean that reading can be misleading. It is difficult, therefore, to determine the real wind speed of a certain place without actual in-situ measurements.

The World Meteorological Organisation (WMO) has accepted the following *four methods of wind recording*:

(*i*) Human observation and log book.

(*ii*) Mechanical cup-counter anemometers.

(*iii*) Data logger.

(*iv*) Continuous record of velocity and direction.

1. **Human observation and log book.** This involves using the Beaufort Scale of wind strengths which defines visible "symptoms" attributable to different wind speeds. The method is *cheap* and *easily implemented* but is *often unreliable*. The best that can be said of such records is that they are better than nothing.

2. **Mechanical cup-counter anemometers.** The majority of meteorological stations use mechanical cup-counter anemometers. By taking the readings twice or three times a day, it is possible to estimate the mean wind speed. This is a *low cost method*, but is *only relatively reliable*. The instrument has to be in good working order; it has to be correctly sited and should be reliably read atleast daily.

3. **Data logger.** The equipment summarizes velocity frequency and direction. It is *more expensive and prone to technical failures but gives accurate data*. The method is tailored to the production of readily interpretable data of relevance to wind energy assessment. It does not keep a time series record but *presents the data in processed form*.

4. **Continuous record of velocity and direction.** This is how data is recorded at major airports of permanently manned meteorological stations. The *equipment is expensive and technically complex*, but it retains a detailed times-series record (second-by-second) of wind direction and wind speed. Results are given in copious quantities of data which require lengthy and expensive analysis.

Variation of wind speed with elevation:

The speed of wind *increases* with height above ground. *Increase in wind speed with height above ground level is called* **wind shear**. The wind speed at the ground is *zero* due to the *friction* between the ground surface and the air. *Increase* in wind speed with height is due to *temperature gradient* and it depends on the type of terrain over which the wind has travelled and atmospheric stability.

The change in wind speed with height, based on the data from several locations, for sites of low ground roughness, can be expressed as:

$$\frac{U_{w_2}}{U_{w_1}} = \left(\frac{H_2}{H_1}\right)^{\alpha} \qquad \qquad ...(5.1)$$

where U_{w1} and U_{w2} are wind speeds at levels H_1 and H_2 respectively. This is known as *power law index "α"* which depends on the roughness of terrain. Its value is taken as $\frac{1}{7}$ for *open land* and **0.10** for calm *sea area*.

The value of power index for a particular site is obtained from the measured wind at two heights by the following relation:

$$\alpha = \frac{\log U_{w2} - \log U_{w1}}{\log H_2 - \log H_1} \qquad \qquad ...(5.2)$$

- The ideal wind energy have a *low value* of α.
- Normally, wind measurements are carried out an elevation of 10 m. However, modern wind turbines are installed at a winds height of 25 to 50 m.
- Wind speed at the required height is calculated from Eqn. (5.1) with $\alpha = \frac{1}{7}$.

5.8. PRINCIPLE OF WIND ENERGY CONVERSION AND WIND POWER

The wind power can be computed by using concept of *kinetics. A wind mill works on the principle of 'converting kinetic energy of the wind to mechanical energy'.*

Let, U_w = Velocity of wind, km/h,

                    ρ = Density of air (1.225 kg/m^3 at sea level), and

(*Air density* is a function of *altitude, temperature and barometric pressure*)

                    A = Area, through which the air flows.

Then, the amount of air passing in unit time (*m*) through area A, with velocity U_w is given by:

Mass, $m = \rho A U_w$

∴   Kinetic energy (K.E.) $= \frac{1}{2} m U_w^2$

$$= \frac{1}{2} \times \rho A U_w \times U_w^2 = \frac{1}{2} \rho A U_w^3 \text{ watts}$$

i.e.   Total power (P_{total}) $= \frac{1}{2} \rho A U_w^3$ $\qquad \qquad ...(5.3)$

From Eqn. (5.3) it is obvious that power output of a wind mill *varies as cube of the wind velocity (i.e.* Power output $\propto U_w^3$)

Further, P_{total} can be expressed as:

$$P_{\text{total}} = \frac{1}{2} \rho \times \frac{\pi}{4} D^2 \times U_w^3 = \frac{1}{8} \rho \pi D^2 U_w^3 \qquad \qquad ...(5.4)$$

where, D = diameter (in meters) in *horizontal axis* aeroturbines.

All this power *cannot be extracted* because, for this, wind velocity *would have to be reduced to zero* which means that the wind mill would accumulate static air around it which would *prevent the wind mill operation*.

Theoretically, a fraction $\frac{16}{27} \approx 0.593$ (59.3%) of the power in the wind is 'recoverable'. This is called **"Gilbert's limit"** or **"Betz coefficient"**. Aerodynamically, efficiency for converting wind energy to mechanical energy *can be reasonably assumed to be 70%*. So the mechanical energy *available at the rotating shaft is limited to 40% or at the most 45% of wind energy*.

Eqn. (5.4) indicates that maximum power available from the wind *varies according to square of the diameter* of the intercept area (or square of the root diameter) normally taken to be swept area of the aeroturbine. The combined effects of wind speed and rotor diameter variations are shown in Fig. 5.1. Thus, *wind machines intended for generating substantial amounts of power should have large rotors and be located in areas of high wind speeds*.

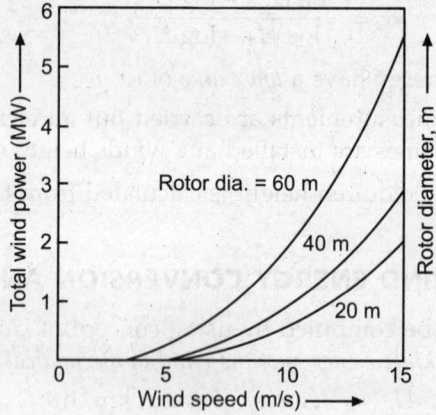

Fig. 5.1. The combined effects of variations of wind speed and diameter.

State-wise wind power potential and wind power addition capacity (as on 31-12-2004) are given in table 5.1 and table 5.2 respectively.

- The best medium–sized and large modern wind turbines (20 kW and above) normally achieve efficiencies of slightly better than 40% (*i.e.*, almost two-thirds of the theoretical maximum) when they convert the kinetic power of the wind to shaft power by passing it through the turbine disc. The more sophisticated smaller units in the 2-20 kW range can achieve 30-40 percent rotor efficiency, while even quite crude designs and small machines of 1 kW or less commonly reach 25 to 30%, which is about half the theoretical value.

Table 5.1 Wind power potential

State	Gross potential (MW) (a)	Technical potential (MW) (b)
Andhra Pradesh	8275	1750
Gujarat	9675	1780
Karnataka	6620	1120
Kerala	875	605
Madhya Pradesh	5500	825
Maharashtra	3650	3020
Orissa	1700	680
Rajasthan	5400	895
Tamil Nadu	3050	1750
West Bengal	450	450
Total	**45195**	**12875**

State	Demonstration projects (MW) (a)	Private sector projects (MW) (b)	MW (Total) capacity (MW) (a) + (b)
Andhra Pradesh	5.4	95.9	101.3
Gujarat	17.3	202.6	219.9
Karnataka	7.1	268.9	276.0
Kerala	2.0	0.0	2.0
Madhya Pradesh	0.6	27.0	27.6
Maharashtra	8.4	402.8	411.2
Rajasthan	6.4	256.8	263.2
Tamil Nadu	19.4	1658.0	1677.4
West Bengal	1.1	0.0	1.1
Others	0.5	0.0	0.5
Total	**68.2**	**2912.0**	**2980.2**

Characteristics of a good wind power site:

A good wind power site should have the following *characteristics*:

1. High annual wind speed.
2. An open plain or an open shore line.
3. A mountain gap.
4. The top of a smooth, well rounded hill with gentle slopes lying on a flat plain or located on an island in a lake or sea.
5. There should be no full obstructions within a radius of 3 km.

5.9. WIND ENERGY PATTERN FACTOR (EPF)

The **energy pattern factor (EPF)** *is the ratio of power from speed distribution to the power from coverage speed of the turbine blades.*

i.e.
$$EPF = \frac{\text{Power from speed distribution}}{\text{Power from average speed}}$$

Generally, *EPF* lies between 2 to 5.

5.10. BASIC COMPONENTS OF WIND ENERGY CONVERSION SYSTEM (WECS)

Fig. 5.2 shows the block diagram of basic components of a wind energy conversion systems.

— **Wind turbines** (Aeroturbines) *convert* the energy of moving air into *rotary mechanical energy*. These turbines require pitch and yaw controls for proper operation.

— A **mechanical interface** consisting of a *step up gear* and a suitable *coupling* transmits the rotary mechanical energy to an **electrical generator**. The output of this generator is connected to the load or power grid as the application demands.

— A **controller** serves purposes of sensing: (*i*) Wind speed, (*ii*) Wind direction, shafts speed and torques at one or more points, (*iii*) Output power and generator temperature as necessary, (*iv*) Appropriate control signals for matching the electrical output to the wind energy input, and (*v*) Protect the system from extreme conditions brought about by strong winds, electrical faults etc.

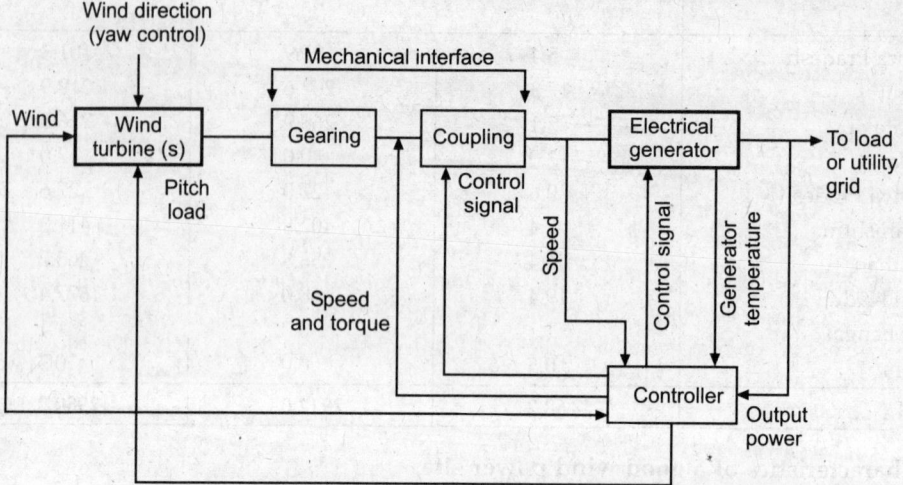

Fig. 5.2. Basic components of a wind energy conversion system (WECS).

5.11. ADVANTAGES AND DISADVANTAGES OF WIND ENERGY CONVERSION SYSTEMS (WECS)

The *advantages and disadvantages* of wind energy conversion systems as follows:

Advantages:

1. Wind energy, a renewable energy source, can be *tapped free of fuel cost*.
2. The wind turbine generation (WTG) produces electricity which is *environmentally friendly*.
3. Wind power generation is *cost effective*.
4. It is *economically competitive* with other modes of power generation.
5. Quite *reliable*.
6. Electric power can be supplied to *remote inaccessible areas*.

Disadvantages:

1. As the wind speed is variable, wind energy is *irregular*, *unsteady* and *erratic*.
2. Wind turbine design is *complex*.
3. Wind energy systems require storage batteries which contribute to *environmental pollution*.
4. Wind energy systems are *capital intensive* and *need government support*.
5. Wind energy has *low energy density* and normally available at only selected geographical locations away from cities and load centers.
6. For wind farms (which are located in open areas away from load centres), the *connection to state grid is necessary*.
7. 'Large units' have *less cost* per kWh, but *require capital intensive technology*. In contrast 'small units' are *more reliable* but have *higher capital cost per kWh*.

5.12. CONSIDERATIONS FOR SELECTION OF SITE FOR WIND ENERGY CONVERSION SYSTEMS (WECS)

Following *factors* should be given due considerations while selecting the site for WECS:

1. Availability of anemometry data.
2. High annual average wind speed.
3. Availability of wind curve at the proposed site.
4. Wind structure at the proposed site.
5. Altitude of the proposed site.
6. Terrain and its aerodynamic.
7. Local ecology.
8. Distance to roads or railways.
9. Nearness of site to local centre/users.
10. Favourable land cost.
11. Nature of ground.

5.13. TERMS AND DEFINITIONS

1. **Aerodynamics.** It is the branch of science which *deals with air and gases in motion and their mechanical effects.*
2. **Airfoil or aerofoil.** A *streamlined air surface designed for air* to flow around it in order to produce *low drag* and *high lift forces.*
3. **Angle of attack.** It is the angle between the relative air flow and the closed of the air foil (Fig 5.3).

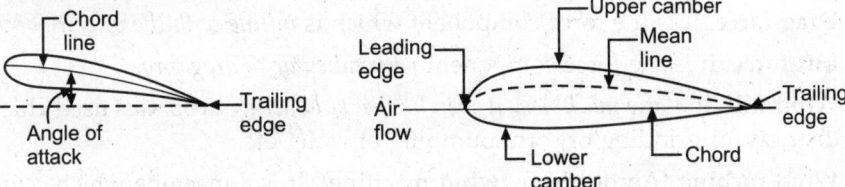

Fig 5.3. Angle of attack   **Fig 5.4.** Air foil showing edges, camber and chord.

4. **Blade.** An important part of a wind turbine that *extracts wind energy.*
5. **Leading edge.** It is the front edge of the blade that *faces towards the direction of flow* (Fig. 5.4).
6. **Trailing edge.** It is the rear edge of the blade that *faces away from the direction of wind flow* (Fig. 5.4).
7. **Mean line.** A line that is *equidistant* from the upper and lower surfaces of the air foil (Fig. 5.4).
8. **Camber.** It is the *maximum distance between the mean line and the chord line, which measures the curvature of the airfoil.*
9. **Rotor.** It is the *primary part* of the wind turbine that *extracts energy* from the wind. It *constitutes the blade-and-hub assembly.*
10. **Hubs.** Blades are fixed to a hubs which is a *central solid part of the turbine.*
11. **Pitch angle.** It is the *angle between the direction of wind and the direction perpendicular to the planes of blades.*

12. **Pitch control.** It is the *control of pitch angle by turning the blades or blade tips* [Fig. 5.5 (*a*)].

13. **Yaw control.** It is the *control for orienting (steering) the axis of wind turbine in the direction of wind* [Fig. 5.5 (*b*)].

14. **Teethering.** It is see-saw like swinging motion with hesitation between two alternatives. The *plane of wind turbine wheel is swung in inclined position at higher wind speeds by teethering control* [Fig. 5.5 (*b*)].

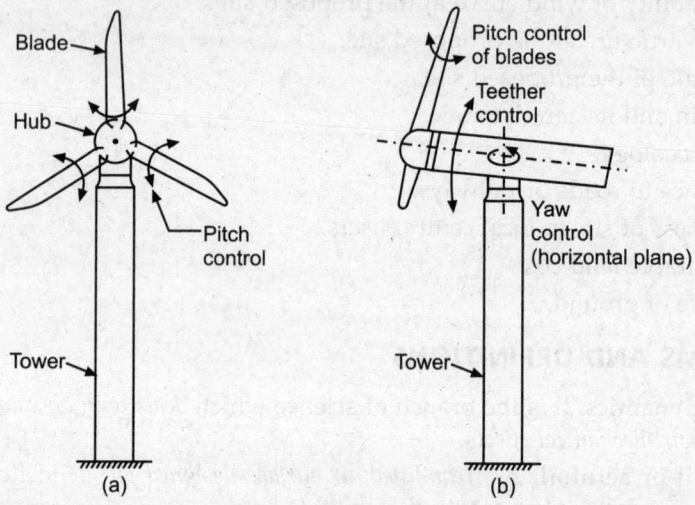

Fig. 5.5. Controls in wind–turbines: Pitch control; Yaw control; Teether control.

15. **Solidity.** *It is ratio of blade area to the swept area* (area covered by the rotating rotor).

16. **Drag force.** It is the force component which is *in line with the velocity of wind.*

17. **Lift force.** It is the force component *perpendicular to drag force.*

18. **Windmill.** It is the *machinery driven by the wind acting upon sails* used chiefly in flat districts for grinding of corn, pumping of water etc.

    Wind turbine (Aeroturbine, wind machine). It is a machine which *converts wind power into rotary mechanical power.* A wind turbine has aerofoil blades mounted on the rotor. The wind drives the rotor and produces rotary mechanical energy.

    Wind turbine generator unit. It is an *assemblage* of a wind turbine, gear chain, electrical generator, associated civil works and auxiliaries.

    Wind farm (wind energy park). It is a *zone comprising several turbine-generator units, electrical and mechanical auxiliaries, substation, control room* etc.

    Wind farms are located in areas having *continuous favourable wind.* Such locations are on-shore or off-shore away from cities and forests.

    Nacelle. It is an *assemblage* comprising of the wind turbine, gears, generator, bearings, control gear etc. *mounted in a housing.*

19. **Wind speeds for turbines:**

    (*i*) *Cut-in-speed.* It is the *wind speed at which wind–turbine starts delivering shaft power.* For a typical horizontal shaft propeller turbine it may be *around 7 m/s.*

(*ii*) *Mean wind speed.* $U_{wm} = \dfrac{U_{w_1} + U_{w_2} + ... U_{w_n}}{n}$

(*iii*) *Rated wind speed.* It is the velocity at which the wind–turbine generator delivers rated power.

(*iv*) *Cut-out wind velocity (furling wind velocity).* It is the *speed at which power conversion is cut out.*

5.14. LIFT AND DRAG–THE BASIS FOR WIND ENERGY CONVERSION

The extraction of power, and hence energy, from the wind depends on creating certain forces and applying them to rotate (or to translate) a mechanism.

Following are the *two primary mechanisms for producing* forces from the wind: Refer to Fig. 5.6

(*i*) Lift force (F_L),   (*ii*) Drag force (F_D).

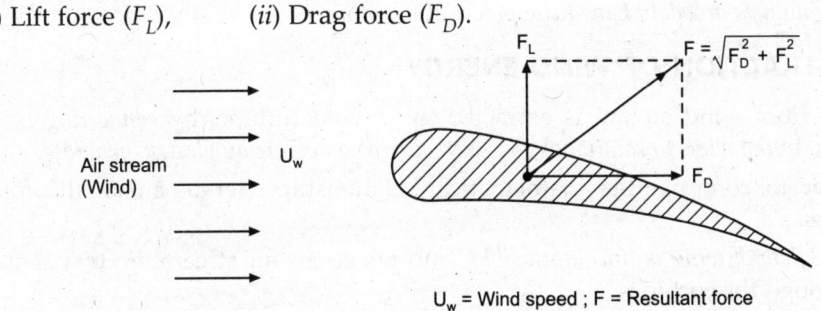

U_w = Wind speed ; F = Resultant force

Fig. 5.6. Lift and drag on an airfoil.

Lift force. The component of force at *right angles* to the direction of air stream on the airfoil is called the *lift force* (F_L) .

Drag force. The component of force *in the direction of stream* is called *drag force* (F_D).

— When air stream approaches the airfoil along the axis of symmetry, the force acting on the body is only the drag force, in the direction of flow and there is *no lift force. The production of lift force requires asymmetry of flow while drag force exists always. It is possible to create drag without lift but impossible to create lift without drag.* 'Lift forces' are produced by changing the velocity of air stream flowing over either side of the lifting surface–speeding up the air flow causes the pressure to *drop,* while slowing the air stream down leads to *increase* in pressure. In other words, *any change in velocity generates a pressure difference across the lifting surface.* The pressure difference *produces a force* that begins to act on the high pressure side and moves towards the low pressure side of the lifting surface which is called an *airfoil.*

— A *good airfoil has high* **lift/drag ratio** (LDR); in some cases it can generate lift forces perpendicular to air stream direction, 30 times as great as the drag force parallel to the flow.

● The *lift increases as the angle formed at the junction of the airfoil and the air stream (the angle of attack) becomes less and less acute, upto the point where the angle of the air flow on low pressure side becomes excessive.* When this happens, the air flow breaks away from the low pressure side, a *lot of the turbulence ensues, the lift decreases* and the *drag increases quite substantially; this phenomenon is known as* **stalling.**

For *'efficient operation'* a wind turbine blade needs to function with as *much lift* and *as little drag as possible* because the *drag dissipates energy.*

The design of each wind turbine specifies the *angle* at which the airfoils should be set to achieve the *maximum LDR.*

- Besides air foils, following are other two *mechanisms for creating lift:*

(*i*) **Magnus effect.** This effect is caused by spinning a cylinder in air stream at a high speed of rotation. The spinning *slows down* the air speed on the side where the cylinder is moving into wind and *increases* it on the other side; the result is similar to an airfoil. This principle has been put to practical use in one or two cases but is *generally not applied.*

(*ii*) **Blowing air through narrow slots in a cylinder.** In this mechanism, air is blown through narrow slots in cylinder, so that it energies *tangentially*; this is known as "*Thwaites slot*". This also creates a rotation (or circulation) of the airflow, which in turn generates lift. *Because the LDR of airfoils is generally much better than those of rotating and slotted cylinders the latter techniques probably have little practical potential.*

5.15. EXTRACTION OF WIND ENERGY

Energy from wind stream is extracted by a wind turbine, by *converting the kinetic energy (K.E.) of the wind to rotational motion required to operate an electric generate.*

In order to compute the mathematical relationships, let us make the following assumptions:

1. The *flow of wind is 'incompressible'*, and hence the air stream *diverges* as it passes through the turbines

2. The *mass flow rate of wind is 'constant'* at *far upstream, at the rotor and at far down stream.*

Let,
p = Atmospheric wind pressure,

p_{us} = Pressure on *upstream* of wind turbine,

p_{ds} = Pressure on *downstream* of wind turbine,

U_w = Atmospheric wind velocity,

$(U_w)_{us}$ = Velocity of wind upstream of wind turbine,

$(U_w)_{bl}$ = Velocity of wind at blades,

$(U_w)_{ds}$ = Velocity of wind downstream of wind turbine before the wind front reforms and regains the atmospheric level,

A_{bl} = Area of blades,

$\dot{m}$ = Mass flow rate of wind, and

ρ = Density of air.

The kinetic energy of wind stream passing through the turbine rotor is given by:

$$\text{K.E.} = \frac{1}{2}\dot{m}(U_w)^2_{bl}$$

And,
$$\dot{m} = \rho\, A_{bl}(U_w)_{bl} \qquad \qquad ...(5.5)$$

$\therefore$
$$\text{K.E.} = \frac{1}{2}\rho A_{bl}(U_w)_{bl} \times (U_w)^2_{bl} = \frac{1}{2}\rho A_{bl}(U_w)^3_{bl}$$

The force on the rotor disc, F is given as:

$$F = (p_{us} - p_{ds})A_{bl} \qquad \qquad ...(5.6)$$

Also, $\qquad F = \dot{m}[(U_w)_{us} - (U_w)_{ds}] \qquad$ ...(5.7)

[momentum per unit time from upstream to downstream winds]

Applying Bernoulli's equation to upstream and downstream sides, we get:

$$p + \frac{1}{2}\rho(U_w)_{us}^2 = p_{us} + \frac{1}{2}\rho(U_w)_{bl}^2 \qquad ...(5.8)$$

and, $\qquad p_{ds} + \frac{1}{2}\rho(U_w)_{bl}^2 = p + \frac{1}{2}\rho(U_w)_{ds}^2 \qquad$ ...(5.9)

Solving the above equations, we obtain:

$$p_{us} - p_{ds} = \frac{1}{2}\rho\left[(U_w)_{us}^2 - (U_w)_{ds}^2\right] \qquad ...(5.10)$$

Equating Eqns. (5.6 and 5.7), we get:

$$(p_{us} - p_{ds})\, A_{bl} = \dot{m}[(U_w)_{us} - (U_w)_{ds}] = \rho A_{bl}\,(U_w)_{bl}\,[(U_w)_{us} - (U_w)_{ds}] \quad ...(5.11)$$

Solving Eqns. (5.10 and 5.11), we get:

$$\frac{1}{2}\rho\left[(U_w)_{us}^2 - (U_w)_{ds}^2\right] = \rho(U_w)_{bl}\,[(U_w)_{us} - (U_w)_{ds}]$$

or, $\qquad (U_w)_{bl} = \dfrac{[(U_w)_{us} + (U_w)_{ds}]}{2} \qquad$ ...(5.12)

In a wind turbine system "*Speed flow work*", W, is equal to the difference in kinetic energy between upstream and downstream of turbine for unit mass flow, $\dot{m} = 1$. Therefore,

$$W = (K.E.)_{us} - (K.E.)_{ds}$$

or, $\qquad W = \dfrac{1}{2}\left[(U_w)_{us}^2 - (U_w)_{ds}^2\right] \qquad$ ...(5.13)

The *power output 'P'* of wind turbine (rate of doing work) is given as:

$$P = \frac{1}{2}\dot{m}\left[(U_w)_{us}^2 - (U_w)_{ds}^2\right] = \dot{m}\left[\frac{(U_w)_{us}^2 - (U_w)_{ds}^2}{2}\right]$$

$$= \rho A_{bl}\left[\frac{(U_w)_{us} + (U_w)_{ds}}{2}\right]\left[\frac{(U_w)_{us}^2 - (U_w)_{ds}^2}{2}\right]$$

$$\left[\because \dot{m} = \rho A_{bl}\,(U_w)_{bl} = \rho A_{bl}\left\{\frac{(U_w)_{us} + (U_w)_{ds}}{2}\right\}\right]$$
$$...\text{using Eqn. (5.12)}$$

or, $\qquad P = \dfrac{1}{4}\rho A_{bl}\,[(U_w)_{us} + (U_w)_{ds}]\left[(U_w)_{us}^2 - (U_w)_{ds}^2\right] \qquad$ ...(5.14)

To get P_{max} (maximum turbine output), differentiating Eqn. (5.14) w.r.t. $(U_w)_{ds}$ and equating to zero, we get:

$$\frac{dP}{d(U_w)_{ds}} = 3(U_w)_{ds}^2 + 2(U_w)_{us}\,(U_w)_{ds} - (U_w)_{us}^2 = 0$$

The above quadratic equation has the following *two solutions*.

$$(U_w)_{ds} = \frac{1}{3}(U_w)_{us} \quad \text{and} \quad (U_w)_{ds} = (U_w)_{us}$$

For power generation $(U_w)_{ds} < (U_w)_{us}$, so we can have only

$$(U_w)_{ds} = \frac{1}{3}(U_w)_{us} \qquad ...(5.14\,(a))$$

Substituting Eqn. (5.14 (a)) in Eqn. (5.14), we get:

$$P_{max} = \frac{1}{4}\rho A_{bl}\left[(U_w)_{us} + \frac{1}{3}(U_w)_{us}\right]\left[(U_w)^2_{us} - \left\{\frac{1}{3}(U_w)_{us}\right\}^2\right]$$

$$= \frac{1}{4}PA_{bl}\left[\frac{4}{3}(U_w)_{us}\right]\left[\frac{8}{9}(U_w)^2_{us}\right]$$

or,
$$P_{max} = \frac{8}{27}\rho A_{bl}\,(U_w)^3_{us} \qquad\qquad ...(5.15)$$

$$= \frac{16}{27}\left[\frac{1}{2}\rho A_{bl}\,(U_w)^3_{us}\right]$$

$$= 0.593\left[\frac{1}{2}\rho A_{bl}\,(U_w)^3_{us}\right]$$

Total power in the wind, stream, $P_{total} = \frac{1}{2}\rho A_{bl}\,(U_w)^3_{us}$ ...(5.15 (a))

$$\therefore \qquad\qquad P_{max} = 0.593\,P_{total}$$

Now, *"coefficient of power"*, $C_P = \dfrac{P_{max}}{P} = 0.593$ ...(5.16)

The factor **0.593** is known as **Blitz limit**.

5.16. CLASSIFICATION AND DESCRIPTION OF WIND MILLS/MACHINES

5.16.1. Classification of Wind Mills/Machines

The wind mills machines are *classified* as follows:

1. **Based on the type of rotor:**
    (*i*) Propeller type (horizontal axis)
    (*ii*) Multiblade type (horizontal axis)
    (*iii*) Savonius type (vertical axis)
    (*iv*) Darrieus type (vertical axis).
2. **Based on orientation of the axis of rotor:**
    (*i*) Horizontal axis
    (*ii*) Vertical axis.

5.16.2. Description of Wind Mills/Machines

1. **Propeller type wind mill:** Refer to Fig. 5.7

These are *most commonly used wind mills*. Such a wind mill has *two or three* blades for *economical reasons*. Though the two blade design is *most efficient*, yet it faces the difficulty of *vibrations* during orientation to wind direction called *'Yaw control'*. These machines are rated from 1 to 3 MW.

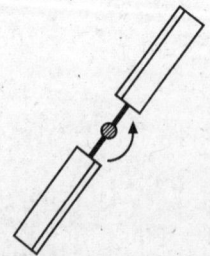 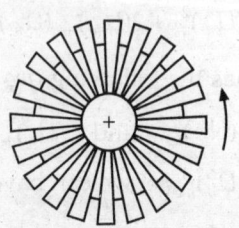 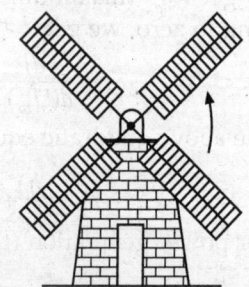

Fig. 5.7. Propeller type (two blade design).   **Fig. 5.8.** Multiblade type.   **Fig. 5.9.** Four-blade dutch wind mill.

2. Multiblade type wind mill: Refer to Figs. 5.8 and 5.9.

The multiblade wind turbines are *high solidity* turbines used for *pumping the water* because of *high starting torque characteristics.*

The multiblade rotors are *less efficient* because of interference of blades in each other but they are *less noisy.*

3. Savonius type wind mill: Refer to Fig. 5.10.

This type of wind mill has hallow circular cylinder sliced in half and the halves are mounted on vertical shaft with a gap in between. Torque is produced by pressure *difference* between the two sides of the half facing the wind.

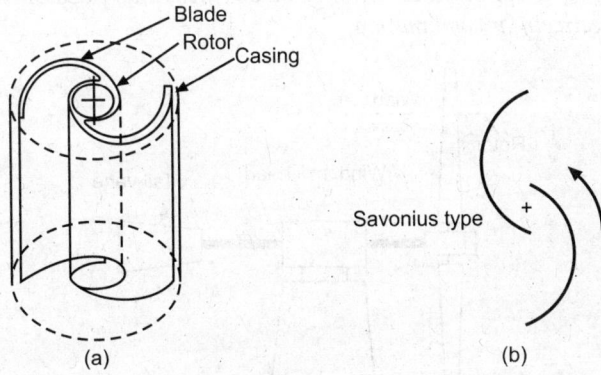

Fig. 5.10. Savonius type wind mill.

This is *quite efficience but needs a large surface area.*

Advantages:

1. Low cost.
2. Operation at low wind velocity.
3. No need of yaw and pitch control.
4. Generator can be mounted at the ground level.

Applications. It is useful for *grinding grains, pumping water* etc.

4. **Darrieus type wind mill :** Refer to Fig. 5.11.

- This wind mill *needs much less surface area.*
- It is shaped like an egg beater and has two or three blades shaped like airfoils.

Characteristics of Darrieus rotor:

(i) Not self starting, needs auxiliary starter.
(ii) High speed.
(iii) High efficiency
(iv) Potentially low capital cost.

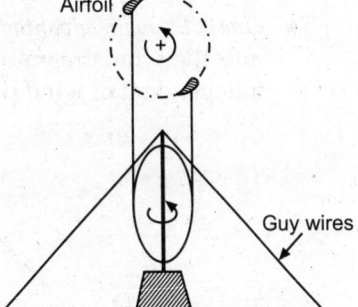

Fig. 5.11. Darrieus type wind mill.

Advantages:

(i) The generator, gear box etc. are placed on the ground.
(ii) No need of yaw mechanism to turn the motor against the wind.

It may be noted that:

- Both the Savonius and Darrieus types are mounted on a *vertical axis* and hence they *can run independently of the direction of wind.*

- The horizontal axis mills *have to face the direction of the wind* in order to generate power.

5. Horizontal axis wind machines :

Fig. 5.12 shows a schematic arrangement of a horizontal axis machine.

- Although the common wind turbine with a horizontal axis is *simple in principle*, yet the design of a complete system, especially a large one that would produce electric power economically, is *complex*.
- It is of paramount importance that the components like rotor, transmission, generator and tower should not only be as *efficient* as possible but they must also function *effectively in combination*.

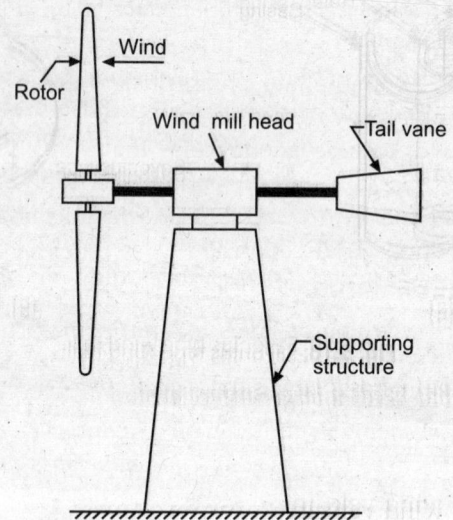

Fig. 5.12. Horizontal axis wind machine.

6. Vertical axis wind machines :

Fig. 5.13 shows vertical axis type wind machine.

- *One of the main advantages* of vertical axis rotors is that they do *not* have to be turned into the windstream as the wind direction changes, because their operation is independent of wind direction. These vertical axis machines are called *panemones*.

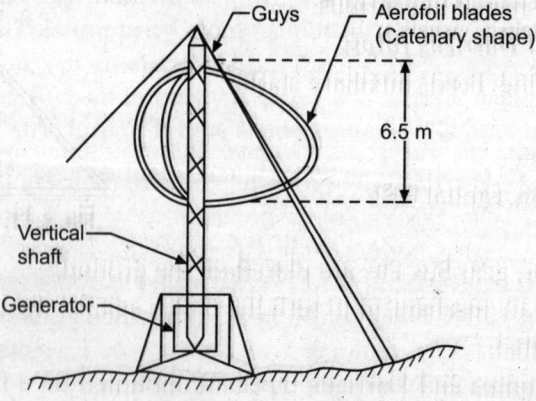

Fig. 5.13. Vertical axis wind machine.

Advantages of vertical axis wind machines:

1. The rotor is *not* subjected to continuous cyclic gravity loads since the blades do not turn end over end (Fatigue induced by such action is a major consideration in the design of large horizontal axis machines).

2. Since these machines would react to wind from any direction, therefore, they do *not* need yawing equipment to turn the rotor into the wind.

3. As heavy components (*e.g.* gear box, generator) can be located at ground level these machines may need *less* structural support.

4. The installation and maintenance are *easy* in this type of configuration.

Fig. 5.14 shows horizontal axis and vertical axis wind machines.

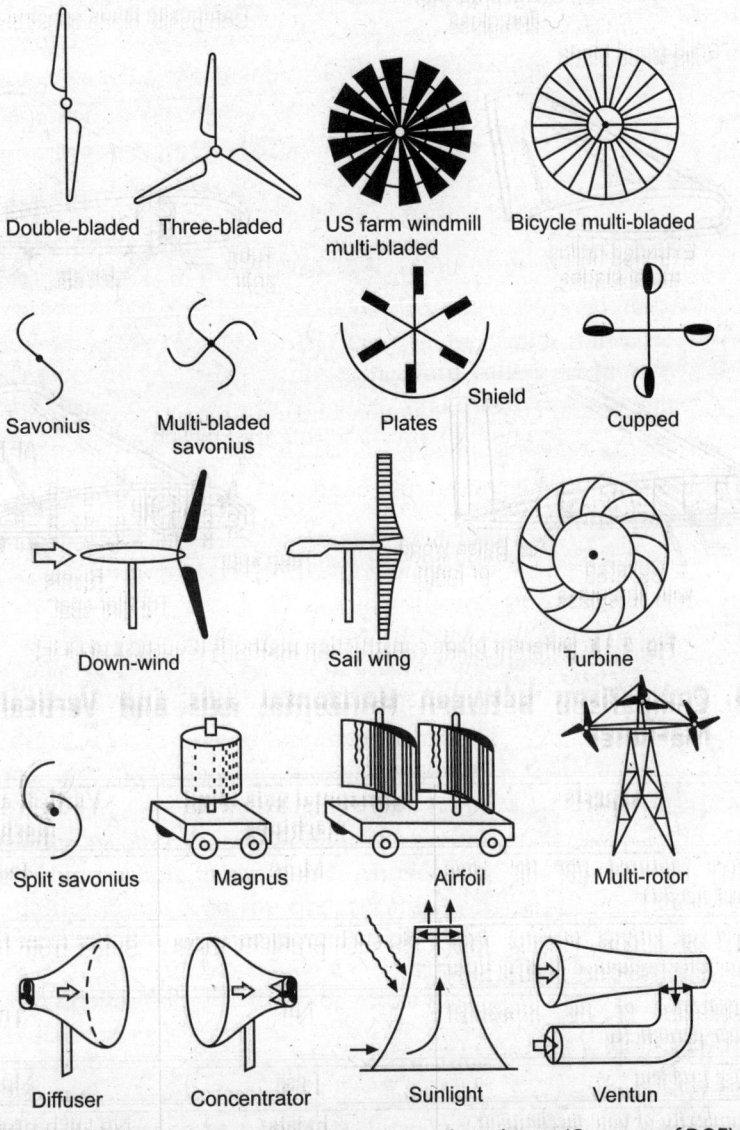

Fig. 5.14. Horizontal axis and Vertical axis wind machines. (Courtesy of DOE)

5.16.3. Blade Construction Methods

The major construction variations one finds while selecting a 'Wind energy conversion system (WECS)' generally will involve the blade. One popular blade material is *wood*, either laminated or solid, with or without fibreglass coatings, as shown in Fig. 5.15.

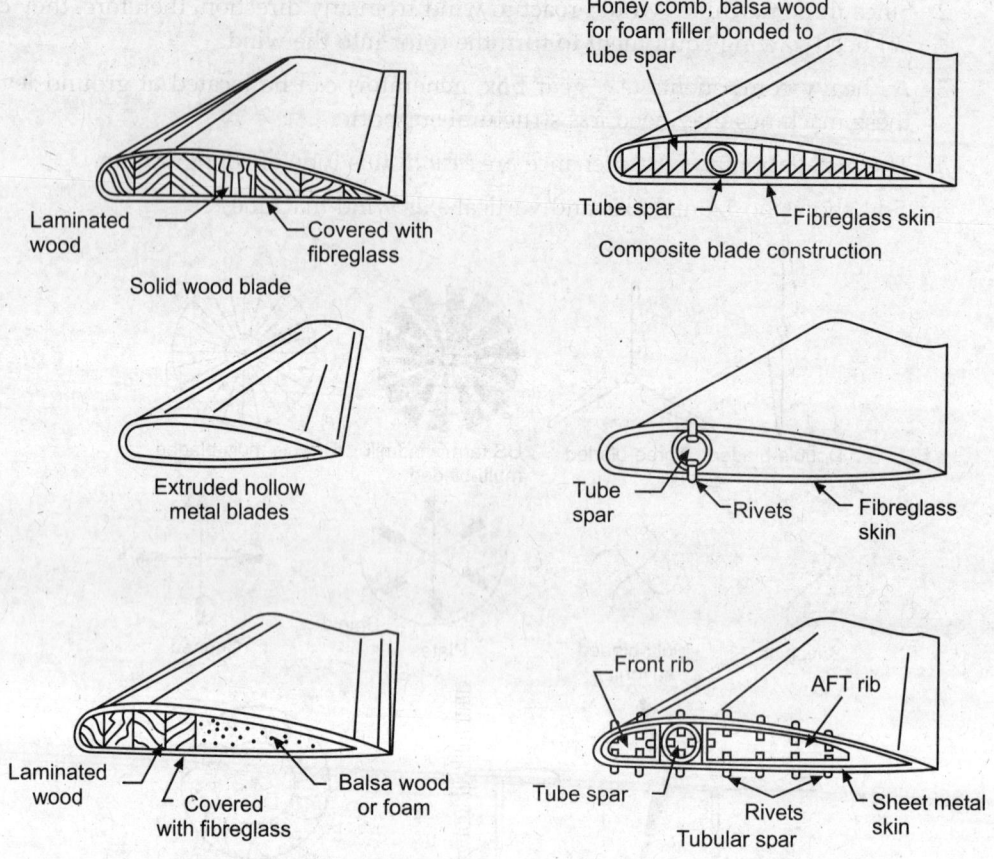

Fig. 5.15. Different blade construction methods (Courtesy of DOE).

5.16.4. Comparison between Horizontal axis and Vertical axis Wind Machines

S. No.	Aspects	Horizontal axis wind machines	Vertical axis wind machines
1.	*Power captured (for the same tower height)*	More	Less
2.	*Effect of fatigue (arising from numerous resonance in structure)*	No such problem arises	Suffer from fatigue effect
3.	*Appearance of the unwanted power periodicity*	Nil	Yes
4.	*Noise problem*	Less	More
5.	*Complexity of yaw mechanism*	Exists	No such problem arises
6.	*Complexity of design*	More	Less

5.16.5. Parameters to be Considered While Selecting a Wind Mill

The following *parameters* should be considered while selecting a wind mill/wind generator.

1. Low land cost.
2. The area should be open and away from cities.
3. Flat open area should be selected, as the wind velocities are high in flat open area.
4. The proposed altitude is to selected by taking average wind speed data.
5. Minimum wind speed should be available throughout the year.
6. Ground surface should be stable and have high soil strength.
7. It should be atleast 5 km away from the cities to reduce the effect of sound pollution.
8. The wind power should be near the customers, so that the transmission losses are minimised.
9. Approach road should be available upto site.

5.16.6. Design Considerations for Wind Turbine

The wind turbine must be able to meet the following *design considerations/criteria.*

1. It should be *small in size* and *suitable for roof mounting in urban area.*
2. *No riks* for its neighbourhood.
3. The efficiency should be good.
4. *Insensitive* to *turbulence.*
5. Suitable for *mass production for low price.*

5.16.7. Performance of Wind Mills

The performance of a wind mill is defined as *'Co-efficient of performance'* (C_p)

$$C_p = \frac{\text{Power delivered by the rotor}}{\text{Maximum power available in the wind}}$$

or

$$C_p = \frac{P}{P_{\max}} = \frac{P}{\frac{1}{2}\rho A U_w^3} \qquad \qquad ...(5.17)$$

where,

p = Density of air,

A = Swept area, and

U_w = Velocity of wind.

Fig. 5.16 shows a plot between C_p and tip speed ratio ($TSR = U_{bt}/U_w$) where, U_{bt} = Speed of blade tip.

It can be seen that C_p is the *lowest of Savonius and Dutch types* whereas the *propeller types have the highest value.*

In the designing of wind mills, it is upper most to keep the *power to weight ratio at the lowest possible level.*

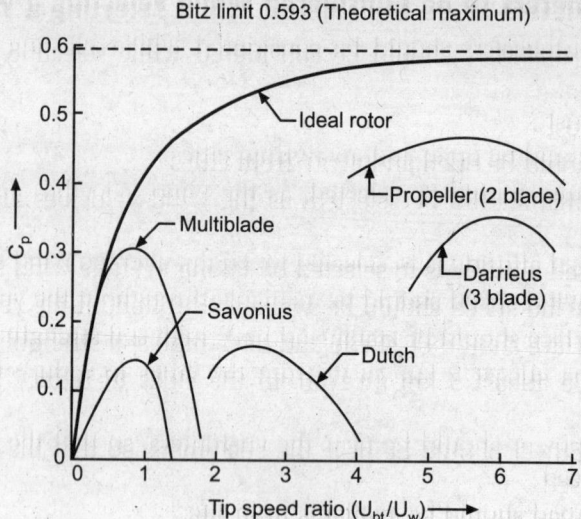

Fig. 5.16. C_p of wind mills.

- *The maximum theoretical coefficient of performance (C_p) is **0.593**.*

5.17. VARIATION OF POWER OUTPUT WITH WIND SPEED

The *actual power output* of a *wind turbine generator* (WTG) *system* depends on the following *wind speeds*:

(*i*)   Actual wind speed,

(*ii*) Rated wind speed,

(*iii*) Cut-in speed (*i.e. speed at which system losses 'equal' the extracted wind power*), and

(*iv*) Cut-out speed (*i.e., speed at which the wind mill has to be 'shut down' for safety reasons*)

The power can be approximated by the relation:

Actual power output, $P_a = 0$ for $0 \leq U_w \leq (U_w)_{ci}$

$\qquad\qquad\qquad\qquad = (A + BU_w + CU_w^2)\, P_r$ for $(U_w)_{ci} \leq U_w \leq (U_w)_r$ ... (5.18)

$\qquad\qquad\qquad\qquad = P_r$ for $(U_w)_r \leq U_w \leq (U_w)_{co}$

$\qquad\qquad\qquad\qquad = 0$ for $U_w \geq (U_w)_{co}$

where, U_w = Actual wind speed,

$\qquad\qquad\qquad\qquad P_r$ = Rated power output,

$\qquad\qquad\qquad (U_w)_{ci}$ = Cut-in speed,

$\qquad\qquad\qquad (U_w)_r$ = Rated wind speed, and

$\qquad\qquad\qquad (U_w)_{co}$ = Cut-out wind speed.

A, B, C are functions of $(U_w)_{ci}$ and $(U_w)_r$.

5.18. ANALYSIS OF AERODYNAMIC FORCES ON A BLADE

Fig. 5.17 shows the vector diagram of forces on an elemental blade section of an *aeroturbine*, the blade rotating to the *left*. The *blade is a typical cross-sectional element of a two-bladed aeroturbine*. The element shown is at some radius '*r*' from the axis of rotation, and is moving to the *left*.

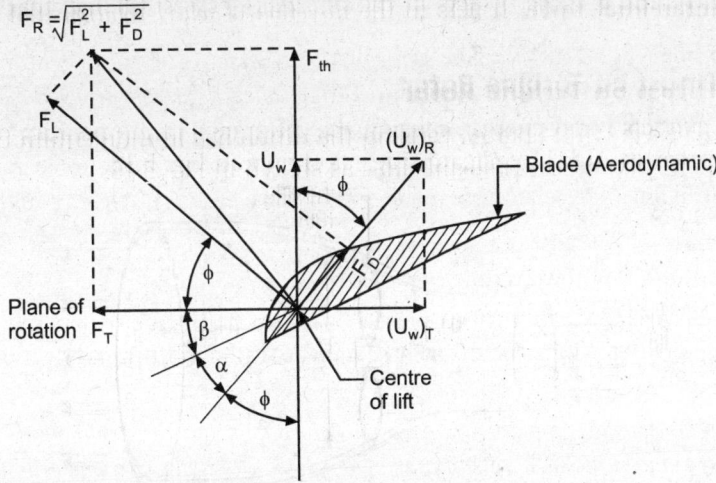

Fig. 5.17. Forces on an elemental blade section of an aeroturbine, the blade rotating to the left.

Refer to Fig. 5.11.

$$U_w = \text{Wind velocity that inpinges on the blade,}$$

$$(U_w)_T = \text{Wind velocity in plane of rotation due to blade burning,}$$

$$(U_w)_R = \text{Resultant wind velocity seen by Aeroturbine blade}$$

$$F_L = \text{Lift force [Normal to } (U_w)_R],$$

$$F_D = \text{Drag force [Parallel to } (U_w)_R],$$

$$F_R = \text{Resultant force on the blade,}$$

$$F_T = \text{Torque producing component of } F_R \text{ making}$$
$$\text{aeroturbine rotate,}$$

$$F_{th} = \text{Thrust force component of } F_R,$$

$$\alpha = \text{Angle of attack of the blade, and}$$

$$\beta = \text{Blade pitch angle.}$$

— $(U_w)_t$ when added vectorially to U_w gives $(U_w)_R$, seen by the rotating blade element.

— F_L caused by the acrodynamic shape of the blade is *perpendicular* to $(U_w)_R$. F_D is *parallel* to $(U_w)_R$.

— The vector sum of F_L and F_D is F_R which has a torque producing component F_T and and a thrust producing component F_{th}. The component F_T *drives the aeroturbine rotationally*, while the component F_{th} tends to *flex* the blade and also *overturn the aerogenerator.*

— The angle of attack 'α' of the acrodynamic element *determines lift and drag forces and hence speed and torque output of the aeroturbine.* These quantities can be altered by changing the blade pitch angle 'β', and this is the *basic torque control method used on large variable pitch 'wind-electric generators'.*

5.19. DESIGN OF WIND TURBINE ROTOR

In a propeller type wind turbine, the following *two forces operate on the blades*:

1. **Axial thrust.** It acts in the *same direction* as that of the *flowing wind stream.*

2. **Circumferential force**. It acts in the *direction of wheel rotation* that provides the *torque*.

5.19.1. Thrust on Turbine Rotor

A turbine extracts wind energy, causing the difference in momentum of air streams between the upstream and downsteam sides as shown in Fig. 5.18.

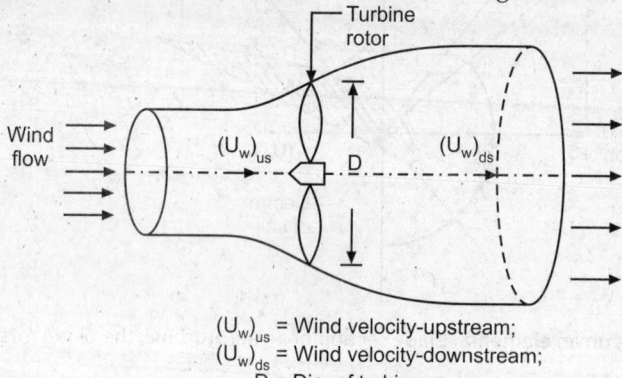

$(U_w)_{us}$ = Wind velocity-upstream;
$(U_w)_{ds}$ = Wind velocity-downstream;
D = Dia. of turbine.

Fig. 5.18. Wind flow across turbine rotor.

Actual force,
$$F_x = \rho A_{bl} (U_w)_{bl} [(U_w)_{us} - (U_w)_{ds}]$$

$$= \rho A_{bl} \frac{[(U_w)_{us} + (U_w)_{ds}]}{2} [(U_w)_{us} - (U_w)_{ds}]$$

or
$$F_x = \frac{1}{2}\rho \frac{\pi}{4}D^4 \left[(U_w)_{us}^2 - (U_w)_{ds}^2\right]$$

$$\left[\because (U_w)_{bl} = \frac{[(U_w)_{us} + (U_w)_{ds}]}{2} \quad ...\text{Eqn. (5.12)}\right]$$

or,
$$F_x = \frac{\pi}{8}\rho D^2 \left[(U_w)_{us}^2 - (U_w)_{ds}^2\right] \qquad ...(5.19)$$

For maximum output, $(U_w)_{ds} = \frac{1}{3}(U_w)_{us}$ $\qquad\qquad ...[\text{Eqn. 5.14 }(a)]$

$\therefore$
$$F_{x(max)} = \frac{\pi}{8}\rho D^2 \left[(U_w)_{us}^2 - \frac{1}{9}(U_w)_{ds}^2\right]$$

i.e.,
$$F_{x(max)} = \frac{\pi}{9}\rho D^2 (U_w)_{us}^2 \qquad\qquad ...(5.20)$$

Eqn. (5.20) indicates that for designing a wind energy generator (WEG) a *large axial force can be obtained by using large diameter turbines*. The upper limit of the diameter needs to be *optimized* by matching structural design with economy.

5.19.2. Torque on Turbine Rotor

The torque on a turbine rotor would be maximum when maximum thrust can be applied at the *blade tip farthest from the axis*, Maximum torque (T_{max}) in a propeller turbine of radius 'R' is given as:

$$T_{max} = F_{max}. R$$

From Eqn. (5.19), F_x becomes maximum when $(U_w)_{ds} = 0$,

i.e.,
$$T_{max} = \frac{1}{2}\rho A_{bl} (U_w)_{us}^2 .R \qquad\qquad ...(5.21)$$

(where, A_{bl} = area of blades)

For wind turbine producing a shaft torque T, the *torque coefficient* C_T is defined by:

$$T = C_T.T_{max}$$

The "**tip speed ratio (TSR), λ** " is the *ratio of the blade's outer tip speed* (U_{bt}) *to the* upstream (free) wind speed, U_w.

i.e.
$$\lambda = \frac{U_{bt}}{(U_w)_{us}} = \frac{\omega R}{(U_w)_{us}} \qquad ...(5.23)$$

(where, ω = angular velocity of the rotor)

Inserting the value of R in eqn. (5.21), we get,

$$T_{max} = \frac{1}{2}\rho A_{bl}(U_w)^2_{us}\left[\frac{\lambda(U_w)_{us}}{\omega}\right]$$

i.e.,
$$T_{max} = P_{total}\times\frac{\lambda}{\omega} \qquad ...(5.24)$$

$$\left[\because P_{total} \text{ (wind power in upstream)} = \frac{1}{2}\rho A_{bl}(U_w)^3_{us}\right]$$

The *maximum power* (P_{max}) obtained from the turbine is given by:

$$P_{max} = T.\omega = C_T..T_{max}\,\omega \qquad ...(5.25)$$

Equating Eqns. (5.16), (5.24) and (5.25), we get:

$$C_p. P_{total} = C_T.T_{max}.\,\omega = C_T.P_{total}\frac{\lambda}{\omega}.\omega$$

i.e.
$$C_p = \lambda\,C_T \qquad ...(5.26)$$

$\therefore$ Ideally, $\qquad (C_T)_{max} = \dfrac{0.593}{\lambda} \qquad ...(5.27)$

[C_p = 0.593 as per Eqn. (5.16), the Blitz limit]

5.19.3. Solidity

Solidity (S) is *defined as the ratio of the blade area to the rotor circumference*. It determines the quantity of blade material required to intercept a certain wind area.

Mathematically, *solidity*, $S = \dfrac{Nb}{\pi D}$ $\qquad\qquad ...(5.28)$

where, N = No. of blades,

b = Blade width and

D = Diameter of the circle described by a blade.

"*Solidity*" represents the *fraction of the swept area of the rotor which is covered with metal.*

Variation of solidity (S) with tip speed ratio (TSR):

The variation of 'S' with 'TSR' is shown in Fig. 5.19.

● The *faster a rotor runs with respect to the wind speed less 'solidity' is required to intercept the entire stream-tube of wind passing through the rotor disc. In other words, a rotor with *high TSR needs less material* and can have *relatively slender blades*. However, because it rotates at higher speeds and its twist angles require only slight angles of attack at the tip, a high LDR (Lift/drag ratio) is essential to produce a driving torque. Therefore, *high speed low-solidity wind rotors need good* *quality airfoils* (similar to those used for aircraft). They *must also be made with good surface finish and the structural integrity necessary to withstand rotational speeds.*

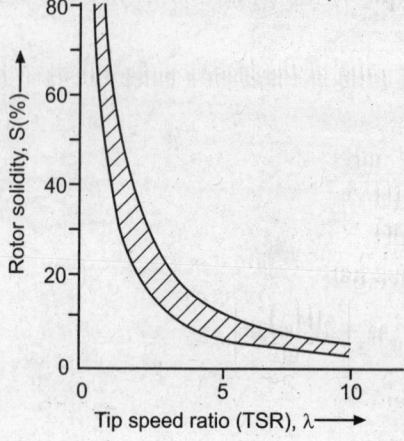

Blade number Vs. TSR	
Tip-speed ratio	*Number of blades*
1	6-20
2	4-12
3	3-8
4	3-5
5-8	2-4
8-15	1-2

Fig. 5.19. (*a*) Range of appropriate rotor solidity (*S*) as a function of tip speed ratio (TSR); (*b*) Number of blades as a function of tip speed ratio (Park, 1981)

— Conversely, *low TSRs* require rotor discs with *relatively solid blades or soils,* so that wind energy will not be lost through the gaps between them. These blades need courser pitch setting. Because lift is stronger in the plane of rotation and drag is weaker, the LDR is less critical. For this reason, mindmills with low TSRs tend to have many blades, and the lifting surfaces can be much cruder without causing any serious loss of efficiency.

— As TSR *increases*, the number of blades *decreases.*

Effect of tip speed ratio (TSR) on torque and solidity:

● Fig. 5.20 shows how the relative torque of various Wind energy conversion systems (WECS) decreases *with increasing TSR. High torque requires a high solidity,* and that type of WECS works *best at low tip speed ratios.*

● Fig. 5.21 shows how the best operating *tip speed ratio changes with solidity.*

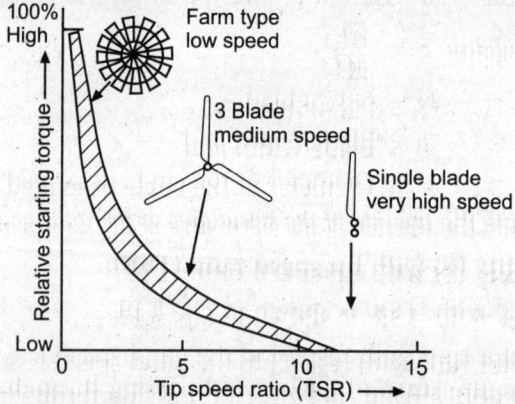

Fig. 5.20. Plot between relative starting torque and tip speed ratio (Courtesy of DOE)

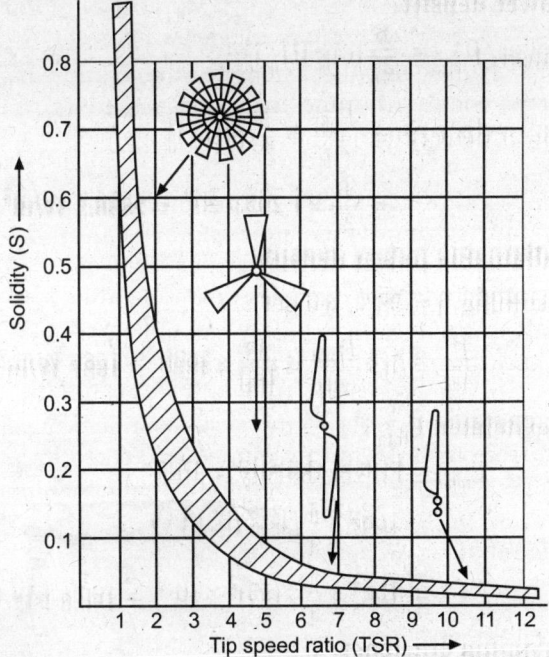

Fig. 5.21. Solidity of several wind machines (courtesy of DOE)

Examples 5.1. *The following data relate to a propeller turbine:*

Velocity of wind at 20°C = 20 m/s.

(at atmospheric pressure)

        Turbine diameter = 12 m, and

Operating speed of the turbine = 45 r.p.m. at maximum efficiency.

Calculate:

(i) *Total power density in the wind stream.*

(ii) *Maximum obtainable power density.*

(iii) *Reasonably obtainable power density.*

(iv) *Total power generated.*

(v) *Maximum torque and maximum axial thrust.*

Solution. *Given:* U_w = 20 m/s; D = 12 m; N = 45 r.p.m. at η_{max}

(i) **Total power density in the wind stream:**

Air density, $\rho = \dfrac{p}{RT} = \dfrac{1.0132 \times 10^5}{287\,(20 + 273)} = 1.205 \text{ kg/m}^3$

($\because$ Gas constant, R = 287 J/kgK and p_{atm} = 1.0132 × 10⁵ Pa)

Total power, $P_{total} = \dfrac{1}{2}\rho A_{bl}\,(U_w)^3_{us}$ ...[Eqn. (5.3)]

$\therefore$ Total power density $= \dfrac{P_{total}}{A_{bl}} = \dfrac{1}{2}\rho\,(U_w)^3_{us}$

                            $= \dfrac{1}{2} \times 1.205 \times (20)^3 = \textbf{4820 W/m}^2 \textbf{ (Ans.)}$

(*us* stands for up-stream)

(*ii*) **Maximum power density:**

Maximum power, $P_{max} = \dfrac{8}{27}\rho A_{bl}\,(U_w)_{us}^3$ ...[Eqn. (5.15)]

$\therefore$ Maximum power density $= \dfrac{P_{max}}{A_{bl}} = \dfrac{8}{27}\rho(U_w)_{us}^3$

$$= \dfrac{8}{27}\times 1.205 \times 20^3 = \textbf{2856.3 W/m}^2\ \textbf{(Ans.)}$$

(*iii*) **Reasonably attainable power density:**

Assuming $\eta = 35\%$, we get

$$\dfrac{P}{A_{bl}} = \eta \times \dfrac{P_{total}}{A_{bl}} = \dfrac{35}{100}\times 4820 = \textbf{1687 W/m}^2\ \textbf{(Ans.)}$$

(*iv*) **Total power generated P_{total}:**

P_{total} = Power density × area

$$= 1687 \times \dfrac{\pi}{4}D^2 \times 10^{-3}\ kW$$

$$= 1687 \times \dfrac{\pi}{4}\times (12)^2 \times 10^{-3} = \textbf{190.8 kW (Ans.)}$$

(*v*) **Torque at maximum efficiency, T_{max}:**

We know that, $P_{max} = T_{max}\cdot \omega$

where, ω (angular velocity) $= \dfrac{2\pi N}{60}\ rad/s$

$\therefore$ $T_{max} = \dfrac{P_{max}}{\omega} = \dfrac{60}{2\pi N}\left[\dfrac{8}{27}\rho A_{bl}\,(U_w)_{us}^3\right]$

$$= \dfrac{60}{2\pi \times 45}\left[\dfrac{8}{27}\times 1.205 \times \dfrac{\pi}{4}\times 12^2 \times (20)^3\right] = 68551\ Nm$$

i.e. $T_{max} = \textbf{68551 Nm (Ans.)}$

Maximum axial thrust, $F_{x(max)}$:

$F_{x(max)} = \dfrac{\pi}{9}\rho D^2\,(U_w)_{us}^2$ ...[Eqn. (5.20)]

$$= \dfrac{\pi}{9}\times 1.205 \times (12)^2 \times (20)^2 = \textbf{24228 N (Ans.)}$$

Example 5.2. *The following data relate to a multiblade wind turbine:*

Wind speed = 35 km/h; Rate of pumping water = 7 m³/h with a lift of 6.5 m; Water pump efficiency = 45 percent; Efficiency of rotor to pump = 78 percent; Co-efficient of performance = 0.3; Tip speed ratio (TSR) = 1.0; Air density = 1.2 kg/m³; Water density = 1000 kg/m³; g = 9.8 m/s².
Determine:

(*i*) *Rotor radius;*

(*ii*) *Angular velocity of the rotor.*

Solution. *Given:* $U_w = 35$ km/h; Rate of pumping water = 7 m³/h; Lift, $H = 6.5$ m; $\eta_{pump} = 45\%$, $\eta_{rotor} = 78\%$ $C_p = 0.3$; λ(TSR) = 1.0; $\rho_a = 1.2$ kg/m³, $\rho_w = 1000$ kg/m³; $g = 9.8$ m/s².

(i) **Rotor radius R:**

Power required to pump water,

$$P_{pump} = \frac{WH}{3600}$$

$$= \frac{mgH}{3600} = \frac{(\rho_w V_w)gH}{3600} = \frac{(1000 \times 7) \times 9.8 \times 6.5}{3600} = 123.86 \text{ W}$$

Power required at the rotor,

$$P_{rotor} (= P_{max}) = \frac{P_{pump}}{\eta_{pump} \times \eta_{rotor}} = \frac{123.86}{0.45 \times 0.78} \simeq 352.9 \text{ W}$$

Also, $\qquad C_p \times P_{total} = P_{max}$ $\qquad\qquad\qquad$ ...[Eqn (5.16)]

or, $\quad C_p \left[\frac{1}{2} \rho A_{bl}(U_w)^3 \right] = P_{max}$

or, $\quad 0.3 \left[\frac{1}{2} \times 1.2 \times \pi R^2 \left(\frac{35 \times 1000}{60 \times 60} \right)^3 \right] = 352.9$

or, $\qquad\qquad 519.66 \, R^2 = 352.9$

$\therefore \qquad\qquad R = \left(\frac{352.9}{519.66} \right)^{0.5} = \textbf{0.824 m (Ans.)}$

(As $\lambda = 1$, the number of blades in the multiblade turbine varies from 6 to 20).

(ii) **Angular velocity of the rotor, ω:**

Tip speed ratio, $\qquad \lambda = \frac{\omega R}{(U_w)_{us}}$ $\qquad\qquad\qquad$ ...[Eqn. (5.23)]

$\therefore \qquad\qquad \omega = \frac{\lambda (U_w)_{us}}{R}$

or, $\qquad\qquad \omega = \frac{1.0 \times \left(\frac{35 \times 1000}{60 \times 60} \right)}{0.824} = 11.8 \text{ rad/s}$

Also, $\qquad\qquad \omega = \frac{2\pi N}{60}$

$\therefore \qquad\qquad N = \frac{60\omega}{2\pi} = \frac{60 \times 11.8}{2 \times \pi} = \textbf{112.7 r.p.m. (Ans.)}$

Example 5.3. *A wind energy generator generates 1600 W at rated speed of 7 m/s at atmospheric pressure and temperature of 20° C. Calculate the power generated and the change in output if the wind generator is operated at an attitude of 1750 m, temperature 11°C, wind speed 8.5 m/s and air pressure 0.9 atmosphere.*

Solution. *Given:* $P = 1600$ W; $U_w = 7$ m/s; $P_{atm} = 1.0132 \times 10^5$ P_a; $T_{alt} = 11 + 273 = 284$;

$\qquad\qquad P_{air(alt.)} = 0.9 \times 1.0132 \times 10^5 = 0.912 \times 10^5$ Pa; $(U_w)_{alt.} = 8.5$ m/s

Power generated and change in output:

Power generated at an altitude of 1750 m, P_{alt}.

Air density at 1750 m, $\quad \rho_{air(alt.)} = \frac{P_{air(alt.)}}{RT}$

$$= \frac{0.912 \times 10^5}{287 \times 284} = 1.12 \text{ kg/m}^3$$

(where R = gas constant = 287 J/kg K)

Now,
$$P = \frac{1}{2}\rho A_{bl} U_w^3$$

or,
$$1600 = \frac{1}{2} \times 1.205 \times A_{bl} \times (7)^3$$

or,
$$A_{bl} = \frac{1600 \times 2}{1.205 \times (7)^3} = 7.74 \ \text{m}^2$$

$$\left(\because \rho_{atm.} = \frac{p}{RT} = \frac{1.0132 \times 10^5}{287 \times 293} = 1.205 \ \text{kg/m}^3 \right)$$

$$\therefore \quad P_{alt.} = \frac{1}{2}\rho_{alt.} \times A_{bl} \times (U_w)_{alt.}^3$$

i.e.
$$P_{alt.} = \frac{1}{2} \times 1.12 \times 7.74 \times (8.5)^3 = 2661.9 \ \text{W}$$

Hence,   *Change in output* $= P_{alt.} - P$
$$= 2661.9 - 1600 = \textbf{1061.9 W (increase) (Ans.)}$$

5.20. WIND-ELECTRIC GENERATING POWER PLANT

Fig. 5.22 shows the various parts of a wind-electric generating power plant. These are:

1. Wind turbine or rotor.
2. Wind mill head–it houses speed increaser, drive shaft, clutch, coupling etc.
3. Electrical generator.
4. Supporting structure.

- The most important component is the **rotor**. For an effective utilisation, all components should be properly designed and matched with the rest of the components.
- The *wind mill head* performs the following *functions*:

(*i*) It supports the rotor housing and the rotor bearings.

(*ii*) It also houses any control mechanism incorporated like changing the pitch of the blades for safety devices and tail vane to orient the rotor to face the wind, the latter is facilitated by mounting it on the top of the supporting structure on suitable bearings.

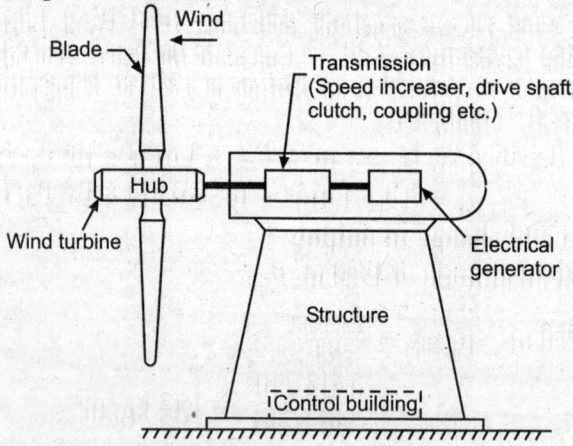

Fig. 5.22. Wind-electric generating power plant.

- The *wind turbine* may be located either *unwind* or *downwind* of the power. In the unwind location the wind encounters the turbine before reaching the tower. *Downwind rotors are generally preferred especially for the large aerogenerators.*

- The **supporting structure** is designed to withstand the wind load during gusts. Its type and height is related to cost and transmission system incorporated. Horizontal axis wind turbines are mounted on towers so as to be above the level of turbulence and other ground related effects.

Applications of wind plants:

Following are the main *applications* of wind plants:

1. Electrical generation.
2. Pumping.
3. Drainage.
4. Grinding grains.
5. Saw milling.

5.21. GENERATING SYSTEMS

The **wind turbine-generator unit** comprising *wind turbine, gears* and *generator, converts wind power into electrical power.* Several *identical units are installed in* a **"Wind farm"**. The total electrical power produced by the wind farm is *fed into the distribution network* or *stand-alone electrical load.*

The choice of electrical system for an aeroturbine is guided by *three* factors:

(*i*) *Type of electrical output:*
   — D.C.
   — Variable-frequency A.C.
   — Constant-frequency A.C.

(*ii*) *Aeroturbine rotational speed :*
   — Constant speed with variable blade pitch.
   — Nearly constant speed with simpler pitch-changing mechanism.
   — Variable speed with fixed pitch blades.

(*iii*) *Utilisation of electrical energy output :*
   — In conjunction with battery or other form of storage.
   — Interconnection with power grid.

5.21.1. Constant Speed-Constant Frequency (CSCF) System

Large scale electrical energy generated from wind is expected to be fed to the power grid to displace fuel generated kWh. For this application present economics and technological developments are heavily weighted in favour of CSCF system with alternator as the generating unit. It must be reminded here that to obtain high efficiencies the blade pitch varying mechanism and controls have to be installed.

   — Wind turbines of electrical rating of *100 kW and above* are of constant-speed type and are coupled to synchronous generators (conventional type). The turbine rated at *less than* 100 kW is coupled to fairly constant speed induction generators connected to grid and so operating at constant frequency having their excitation VARs from the grid or capacitor compensators.

5.21.2. Variable Speed-Constant Frequency (VSCF) System

Variable-speed drive is *typical for small wind generators* used in autonomous applications, generally producing variable frequency and variable voltage output.

The variable speed operation of wind electric system *yields higher outputs* for both low and high wind speeds. This results in higher annual energy yields per rated installed kW capacity. Both horizontal axis and vertical axis turbines will exhibit this gain under variable speed operation.

The following schemes are used *to obtain constant frequency output*:

(*i*) A.C.—D.C.—A.C. link

(*ii*) Double output induction generator.

(*iii*) A.C. commutation generator.

With the advent of power switching technology (*viz* high power diodes and thyristors) and chip-based associated control circuitry, it has now become possible to use VSCF systems. VSCF and wind electrical systems and its associated power conditioning system operate as shown in Fig. 5.23.

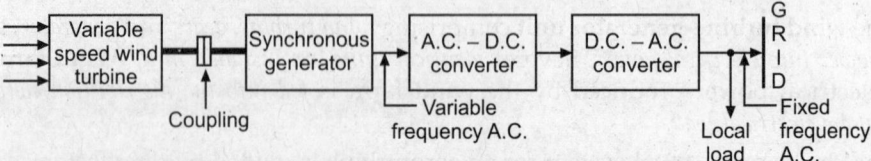

Fig. 5.23. Block schematic of VSCF wind electrical system.

VSCF wind electrical systems claim the following *advantages:*

1. Significant reduction in aerodynamic stresses, which are associated with constant-speed operation.

2. It is possible to extract extra energy in the high wind region of the speed-duration curve.

3. Complex pitch changing mechanism is not required.

4. Wind turbine/Aeroturbine always operates at maximum efficiency point (constant tip-speed ratio.)

5.21.3. Variable Speed-Variable Frequency (VSVF) System

The generator output is affected by the variable speed. The frequency of the induced voltage depends on the impedance of the load and speed of the prime mover. The variable voltage can be converted to constant D.C. using choppers or rectifier and then to constant A.C. by the inverters.

5.22. WIND-POWERED BATTERY CHARGERS

One application of wind energy system which is of considerable potential importance (to developing countries) is the use of small wind generators to charge batteries for powering lighting, radio communication and hospital equipment. Wind generators have been in use in Europe and North America since the 1920s, although their use declined considerably.

A battery charging system has to include the following:

(*i*) A wind powered generator.

(ii) A converter.

(iii) A container for the batteries.

Fig. 5.24 shows a set-up of wind powered battery charging system. It is worth noting that 12 volt batteries, which are rechargeable using wind generators, can be used to power fluorescent tube lighting which is six times more efficient than tungsten filament lamps. Such lighting opens up a number of important development opportunities in areas which normally have no lighting.

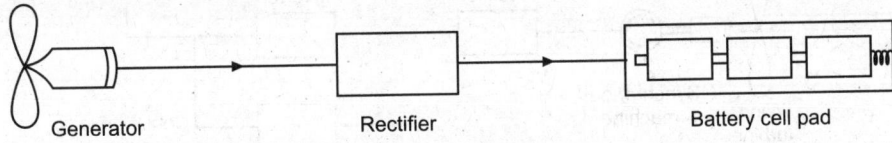

Fig. 5.24. Wind powered battery charging system.

For small wind generators the total system efficiency is made up as follows:

Wind regime matching efficiency	60% (approx.)
Rotor efficiency	35% (approx.)
Generator and wiring efficiency	70% (approx.)
Battery charge/discharge efficiency	70% (approx.)

Cumulatively, a total energy capture efficiency of about 10% is generally obtained from small wind generators utilized for battery charging.

Battery charging wind generators are produced in several countries, notably Australia, France, Sweden, Switzerland, the U.K., the U.S.A. and West Germany. In developing countries production is underway in China and has started in India.

5.23. WIND ELECTRICITY IN SMALL INDEPENDENT GRIDS

Refer to Fig. 5.25. *In such systems electricity consumption fluctuates constantly as does the availability of wind energy.* The degree of coincidence of supply and demand can be calculated by statistical means and it has been found that electricity supply with an acceptable degrees of reliability *cannot be based solely on wind energy.* If an extensive grid does not exist, electricity storages (batteries) or a back-up system (diesel) is required. Loads for remote systems of upto 6 kWh/day equivalent to an average power consumption of 250 W with a duty cycle of 24 hours, can be provided with battery storage.

Combined wind/diesel power generation:

A wind turbine generator (WTG) is very suitable for an *isolated and remote area.* To feed such an area from a power grid requires long transmission line and huge investment. Such areas are mostly served by diesel plants. If strong winds persist for sufficiently long periods, *WTG system can be installed to work in combination with diesel plant.* For such an operation the following *factors* need consideration:

1. Gross output of WTG and power requirements of auxiliaries of WTG.

2. Change in diesel plant efficiency due to decrease in load.

3. Change in diesel plant efficiency due to the throttle activity.

    The following *operations need special attention*:

(i)   Start up and synchronisation of WTG with the diesel plant system, and

(ii) Normal shut down and cut-out of WTG.

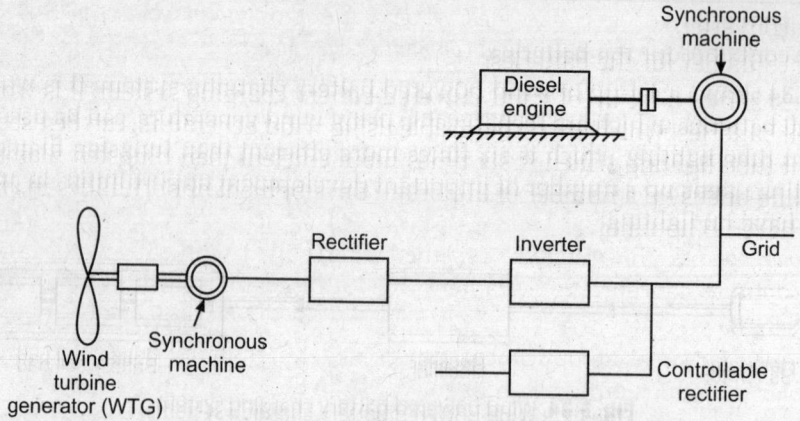

Fig. 5.25. Principle of combined wind/diesel power generation.

Following are *two ways*, out of many ways, in which *WTG can be used*:

1. **Isolated:**
   - Wind electric supply
   - Hybrid system
     — Wind-Diesel
     — Wind-Photovoltaic-Batteries
     — Wind-Photovoltaic-Diesel-Batteries
2. **Grid connected.**

5.24. ECONOMIC SIZE OF WTG

After several studies having been taken in different countries of the world to determine the economic size of individual wind turbine generators, it has been found that each generator *can have a rated output of only upto '200 kW'*.

The small size of WTG requires the installation of large number of WTGs for a wind power plant of significant size. A 10 MW plant needs atleast 50 turbine WTGs generators. The installation of such a large number of generators requires a 'Wind farm' of large size and a *huge investment*.

5.25. WIND ELECTRICITY ECONOMICS

Wind generator power costs are heavily linked to the characteristics of a wind resource in a specific location. The cost of supplied power declines as wind speeds increase, and the power supplied increases in proportion to the *cube of the wind speed*.

Matching available energy and load requirements is also important in wind energy economics. The correct size of wind generator must be chosen together with some kind of storage or cogeneration with an engine or a grid to obtain the best economy. The ideal application is a task that can utilise a variable power supply, *e.g.*, ice making or water purification.

Regarding the *economics*, the choice of interest rate obviously has a major effect on the overall energy cost. With low interest rates, capital intensive power sources such as solar and wind are favoured. Other factors bearing a strong influence on the economics of wind electricity are the standard of maintenance and service facilities and the cost of alternative energy supplies in the particular area.

5.26. PROBLEMS IN OPERATING LARGE WIND POWER GENERATORS

The operation of large wind power generators entails the following *problems:*

1. **Location of site.** The most important factor is locating a site *big enough* which has a reasonable average high wind velocity.

   Sourashtra and Coastal Regions in India are promising areas.

2. **Constant angular velocity.** A constant angular velocity is a *must* for generating A.C. (alternating current) power and this means very sensitive governing.

3. **Variation in wind velocity.** The wind velocity varies with time and varies in direction and also varies from the bottom to top of *a large rotor* (some rotors are as long as 50 meters). This *causes fatigue in blades.*

4. **Need of a storage system.** At zero velocity conditions, the power generated will be zero and this means some *storage system* will have to be *incorporated* along with the wind mill.

5. **Strong supporting structure.** Since the wind mill generator will have to be located at a height, the supporting structure will have to be designed to *withstand* high wind velocity and impacts. This will *add* to the initial costs of the wind mill.

6. **Occupation of large areas of land.** Large areas of land will become unavailable due to wind mill gardens (places where many wind mills are located). The whole area will have to be protected to avoid accidents.

   Inspite of all these difficulties, interest to develop wind mills is there since this is a *clean source of energy.*

5.27. SELECTION OF SITE FOR WIND TURBINE GENERATING SYSTEM (WTGS)

Following are the main *factors* which govern the selection of site for a proposed WTGS:

1. **Availability of wind.** It has been estimated that an annual mean wind speed of about 18 km/h (or 5 m/s) or more is needed for the installation of an economically viable WTGS.

2. **Availability of land.** Large wind farms are required for the installation of WTGS. An installation of *1 MW requires about 25 acres.* The actual land used directly for WTGS is only about 10% of this value. The remaining land may be used for cultivation of low height vegetation. However, it is necessary that this vegetation should not disturb 'wind resource'.

3. **Access to land.** The land should be accessible for transportation of machinery and construction material.

4. **Grid stability.** Since WTGS will feed the power grid, therefore, the grid should be *free from voltage and frequency variations as far as possible.*

1. A wind mill works on the principle of *'converting kinetic energy of the wind to mechanical energy'*.

$$\text{Total power } (P_{total}) = \frac{1}{2}\rho A U_w^3$$

2. The *energy pattern factor* (EPF) is the ratio of power from speed distribution to the power from average speed of the turbine blades. EPF generally lies between 2 to 5.

3. *Coefficient of performance,*

$$C_p = \frac{\text{Power delivered by the rotor}}{\text{Maximum power available in the wind}}$$

$$= \frac{P}{P_{max}} = \frac{P}{\frac{1}{2}PAU_w^3}$$

where, ρ = Density of air; A = Swept area, and U_w = Velocity of wind.

The maximum theoretical coefficient of performance (C_p) is **0.593**.

4. *Solidity* (S) is defined as the ratio of the blade area to the rotor circumference.

$$S = \frac{Nb}{\pi D}$$

where, N = No. of blades, b = Blade width, and D = Diameter of the circle described by a blade.

1. Explain briefly wind and wind energy.
2. Discuss the utilisation aspects of wind energy.
3. What are the characteristics of wind?
4. State the advantages and disadvantages of wind energy.
5. Discuss environment impacts of wind energy.
6. What are the sources/origins of wind? Explain briefly.
7. Enumerate and explain four methods of wind recording, accepted by the World Metero-logical Organisation (WMO).
8. What is the principle of wind energy conversion?
9. What is "Betz coefficient"? Explain briefly.
10. List the characteristics of a good wind power site.
11. Define wind 'energy pattern factor (EDF)'.
12. What are the basic components of wind energy conversion systems (WECS)? Explain them with a neat block diagram.
13. Give the advantages and disadvantages of WECS.
14. Enumerate the factors which should be given due consideration while selecting the site for WECS.
15. Define the following terms:
    (*i*) Angle of attack; (*ii*) Pitch angle; (*iii*) Nacelle; (*iv*) Cut-in speed
16. What is the basis of wind energy conversion? Explain.
17. Give the classification of wind mills/machine.

18. Explain briefly any *two* of the following:
    (*i*) Propeller type wind mill;   (*ii*) Savonius type wind mill;
    (*iii*) Darrieus type wind mill;   (*iv*) Horizontal axis wind machine

19. Give the comparison between horizontal and vertical axis wind machines.

20. List the parameters which should be considered while selecting a wind mill.

21. What should be design consideration for wind turbine?

22. What is 'Coefficient of performance'? Explain briefly.

23. Give the analysis of aerodynamic forces on a blade.

24. Define and explain the following terms:
    (*i*) Tip speed ratio;   (*ii*) Solidity.

25. Discuss briefly the effect of tip speed ratio (TSR) on torque and solidity.

26. Explain with a neat sketch the construction and working of wind-electric generating plant.

27. List the applications of wind plants.

28. Explain briefly the following:
    (*i*) Constant speed-constant frequency (CSCF) system;
    (*ii*) Variable speed-constant frequency (VSCF) system.

29. Explain the principle of combined wind/diesel power generation.

30. What are the problems in operating large wind power generators? Explain.

31. What are the main factors which govern the selection of site for a proposed wind turbine generating system (WTGS)?

UNSOLVED EXAMPLES

1. The following data relate to a wind turbine:
   Velocity of wind at 15° C = 10 m/s;
   Turbine diameter = 10 m
   Operating speed of the machine = 35 rpm at maximum efficiency of 40%.
   Calculate:
   (*i*) The total power density in wind stream;
   (*ii*) The maximum power density;
   (*iii*) The actual power density;
   (*iv*) The power output of the turbine, and
   (*v*) The axial thrust on the turbine structure

   (**Ans.** (*i*) 613 W/m^2; (*ii*) 363 W/m^2; (*iii*) 245.2 W/m^2;
   (*iv*) 19.33 kW; (*v*) 4277.4 N)

2. Determine the rotor radius for a multiblade wind turbine that operates in a wind of 10 m/s to pump water at a rate of 6 m^3/h with a lift of 6 m. Also calculate the angular velocity of rotor:
   Given: Density of water = 1000 kg/m^3; Water pump efficiency = 50 percent; Efficiency of rotor to pump = 80 per cent; Coefficient of performance = 0.3; Density of air = 1.2 kg/m^3; g = 9.8 m/s^2; speed ratio = 1.0.   (**Ans.** 0.66m; 144 r.p.m.)

3. A wind energy generator generates 1500 W at rated speed of 24 km/h at atmospheric pressure and temperature of 20°C. Calculate the change in output if the wind generator is operated at an altitude of 1800 m, temperature 10°C, wind speed 30 km/h, and air pressure 0.88 atmosphere.   (**Ans.** 1186 W)

BIOMASS ENERGY

6.1. INTRODUCTION TO BIOMASS

Biomass is *an organic matter from plants, animals and micro-organisms grown on land and water and their derivatives.* The *energy obtained from biomass is called* **biomass energy.**

"Biomass" is considered as a *renewable source of energy because the organic matter is generated everyday.*

— Coal, petroleum, oil and natural gas *do not* come in the category of 'biomass', because they are produced from dead, buried biomass under pressure and temperature during millions of years.

● *"Biomass"* can *also be considered a form of solar energy as* the latter *is used indirectly* to grow these plants by *photosynthesis.*

- *"Biomass fuel"* is used over 85 percent of rural households and in about 15 percent urban dwellings. Agriculture products rich in starch and sugar like wheat, maize, sugarcane can be fermented to produce *ethanol* (C_2H_5OH). *Methanol* (CH_3OH) is also produced by distillation of biomass that contains cellulose like wood and begasse. *Both these alcohols can be used to fuel vehicles and can be mixed with 'diesel' to make* **biodiesel**.

6.2. BIOMASS RESOURCES

In our country, there is a great potential for *application of biomass as an 'alternate source of energy'*. We have plenty of agricultural and forest resources for reproduction of biomass.

The following are the *biomass resources* :

1. **Concentrated wastes :**

   (*i*) Municipal solid   (*ii*) Sewage wood products

   (*iii*) Industrial waste   (*iv*) Manure at large lots.

2. **Dispersed waste residue :**

   (*i*) Crop residue   (*ii*) Logging residue

   (*iii*) Disposed manure.

3. **Harvested biomass :**

   (*i*) Standing biomass   (*ii*) Biomass energy plantations.

The biomass sources are highly dispersed and bulky and contain large amount of water (50 to 90%). Thus, it is not economical to transport them over long distances, and as such *conversion into usable energy must take place close to the source, which is limited to particular regions. However, biomass can be converted to liquid or gaseous fuels, thereby increasing its energy density and making transportation feasible over long distances.*

6.2.1. Availability of Biomass

The total terrestrial crop alone is about 2×10^{12} metric tonnes. This includes : (i) Sugar crops such as sugarcane and sweet sorghum; (ii) Herbaceous crops, which are non-woody plants that are easily converted into liquid or gaseous fuels; and (iii) Sihiriculture (forestry) plants such as cultured hybrid poplar , sycamore, sweatgum, alder, eucalyptus, and other hardwoods.

- The terrestrial crops have energy potential of 3×10^{22} joules. The efficiency of solar energy utilisation in natural photosynthesis is only 0.1 to 2 percent. At present only 1 percent of world biomass is used for energy conversion.

  Current *research* focuses on the screening and identification of species that are suitable for short-rotation growing and on optimum techniques for planting, fertilization, harvesting, and conversion. Fast growing trees, sugar, starch and oil containing plants can be cultivated which have about 5 percent efficiency of solar energy utilisation.

- The estimated production of agricultural residue in India is 200 million tonnes per year and that of wood is 130 million tonnes.

- Aquatic crops are grown in fresh sea and backish waters, both submerged and emergent plants. These include seaweeds, marine algae etc.

- *Animal and human wastes* are an *indirect terrestrial crop* from which *methane* for combustion, and *ethylene* can be produced while *retaining the fertilizer value of the manure.* The daily produce of cowdung is about 13.5 kg per cattle which can be used to produce 0.46 m³ of *biogas* in 'Gobar gas plant'. This gas is sufficient to produce 1 kWh of electricity in a biogas engine.

The *human waste* can also be used for production of biogas. *Community latrines* can be planned in the villages for collection of night soil for feeding to biogas plants. Wastes of 200 persons can be used to produce about 5 m³ of gas per day to extract 12 kWh of equivalent energy by running a biogas engine.

6.2.2. Limitations of Utilising Biomas

Following are the *limitations* of utilising biomass :

1. Relatively *expensive energy conversion.*
2. *Low conversion efficiency (i.e.* small percentage of sun light is converted to biomass by plants).
3. Relatively low concentration of biomass per unit area of land and water.

☞ 6.3. ENVIRONMENTAL EFFECTS OF BIOMASS/BIOFUELS

Following are the *effects/benefits* of biomass/biofuels :

1. Biomass can pollute air when it is burned but less than those of fossil fuels.
2. Biomass, when burned, does not release green-house gases (or CO_2).
3. Burning biofuels do not produce pollutants like sulphur, that results in acid rain.
4. When biomass crops are grown, nearly equivalent amount of CO_2 is captured through photosynthesis.

6.4. PHOTOSYNTHESIS

The preparation of food by the leaves of green plant and micro-organism in presence of sunlight, chlorophyll, water and "CO_2" is called **photosynthesis.** In this process, the CO_2 from the atmosphere combines with water and light energy to produce *carbohydrates* (i.e., sugars, starches etc.) and *oxygen.*

The *photosynthesis process* can be represented by the following *reaction* :

$$6CO_2 + 6H_2O + \text{light energy} \rightarrow C_6H_{12}O_6 + 6CO_2$$

Biomass does *not add* CO_2 to the atmosphere as it *absorbs* the same amount of carbon in growing the plants as it is released when consumed as fuel. It is a *superior fuel* as the energy produced by biomass is *'carbon cycle neutral'.*

The **conditions** *necessary for photosynthesis* are :

(*i*) **Light :** The intensity of solar radiation of 400-700°A wavelength is one of the important inputs for biomass production; this range of light is called *'Photosynthetically active radiation (PAR)'.* The upper limit of photosynthesis efficiency is about 5 per cent.

(*ii*) **CO_2 concentration :** CO_2 *is the primary raw material for photosynthesis. The main sources of CO_2 are :*

— Animal respiration;
— Combustion of fuel;
— Decay of organic matter by bacteria;
— Ocean (respiration of marine plants and animals releases CO_2 into the water).

(*iii*) **Temperature :** The process of photosynthesis is *restricted* to temperature range of 0°C to 60°C which can by *tolerated by proteins.*

The energy stored in the plants by way of carbon fixation in the form of chemical bond energy when expressed as a fraction of total insolation falling on the plant, is called as **photosynthetic efficiency.**

6.5. BIOMASS CONVERSION PROCESSES

The following *processes* are used for the **biomass conversion to energy or to biofuels:**

1. Densification.
2. Combustion and incineration.
3. Thermo-chemical conversion.
4. Biochemical conversion.

6.5.1. Densification

In this process bulky biomass is *reduced to a better volume-to-weight ratio by compressing in a die at a high temperature and pressure.* The biomass pressed into briquettes or pellets (easier to transport and store) can be *used* as clean fuel in *domestic chulhas, bakeries and hotels.*

6.5.2. Combustion and Incineration

Combustion :

Combustion is the process of burning in presence of oxygen to produce heat (utilised for cooking, space heating, industrial purposes and for electricity generation), *light* and *by-products.*

The combustion of biomass is *more difficult* than other fuels since it contains relatively *higher moisture content. Biomass is free from toxic metals and its ash.*

This method is *very inefficient* with heat losses to 30 to 90% of the original energy contained in the biomass.

- The technology of *"fluidised bed combustion"* may be used for the efficient combustion of forestry and agricultural waste material such as sawdust, wood chips, hog fuel, rice husks, straws, nutshells and chips.

In **fluidised bed combustion** of *biomass, the biomass is fed into a bed of hot inert particles,* such as sand kept in fluidised state with air at sufficient velocity from below. The operating temperature is normally controlled within the range 750-950°C; ideally it is kept as high as possible in order to maximise the rate of combustion and heat transfer but low enough to avoid the problem of sintering of the bed particles. *The rapid mixing and turbulence within the fluidised bed enables efficient combustion to be achieved with high heat releases, as well as effective transfer,* than in a conventional boiler. This can *result in more compact boiler with less number of tubes.*

Incineration :

It is the process of burning completely the solid masses to 'ashes' by high temperature oxidation.

Although the terms 'combustion' and 'incineration' are synonymous, yet the *"combustion process"* is *applicable to all fuels* (i.e., solid, liquids and gases); *"incineration"* is a special process which is used for incinerating municipal solid waste to *reduce* the volume of solid refuse (90 per cent) and to *produce heat steam and electricity.*

Pyrolysis. Wood, dung, vegetable waste can be *dried and burnt to provide heat or converted into low calorific value by **pyrolysis***. In the pyrolysis process, the organic material is converted to gases, solids and liquids by heating to 500 to 900°C in the *absence of oxygen*.

6.5.3. Thermo-chemical Conversion

It is a process to decompose biomass with various combinations of temperatures and pressures.

Thermo-chemical conversion takes the following *two* forms :

(*i*) Gasification;

(*ii*) Liquification.

(*i*) Gasification:

It is the process of heating the biomass with limited oxygen to produce 'low heating value' or by reacting it with steam and oxygen at high pressure and temperature to produce 'medium heating value gas'.

The *output gas is known as **"producer gas"***, a mixture of H_2 (15-20%), CO (10 to 20%), CH_4 (1 to 5%), CO_2 (9 to 12%) and N_2 (45 to 55%). As compared to solid mass the gas is more *versatile*, it can be burnt to *produce heat and steam, or used in I.C. engines or gas turbines to generate electricity*.

(*ii*) Liquification:

- Biomass can be *liquified* through *fast* or *flash pyrolysis*, called *"pyrolytic oil"* which is a dark brown liquid of low viscosity and a mixture of hydrocarbons.

- Biomass can also be liquified by *"methanol synthesis"*. Gasification of biomass produces synthetic gas containing a mixture of H_2 and CO. The gas is *purified* by adjusting the composition of H_2 and CO. Finally, the purified gas is subjected to liquefication process, *converted to methanol* over a zinc, chromium catalyst.

— *Methanol* can be used as *liquid fuel*.

6.5.4. Biochemical Conversion

In biochemical conversion there are *two principal conversion processes*:

1. Anaerobic digestion;

2. Fermentation.

1. Anaerobic digestion:

This process involves *'microbial digestion'* of biomass and is done in the *'absence of oxygen'*.

The process and *end products* depend upon the *micro-organisms cultivated under culture conditions*. (An *anaerobe* is a microscopic organism that can live and grow *without external oxygen or air*; it extracts oxygen by decomposing the biomass at low temperatures upto 65°C, in presence of moisture).

This process generates mostly *methane* (CH_4) and CO_2 *gas* with small impurities such as hydrogen sulphide.

The *output gas* obtained from anaerobic digestion can be *directly burnt*, or *upgraded to superior fuel gas (methane)* by removal of CO_2 and other impurities. The residue may consist of protein-rich sludge and liquid effluents which can be used as animal feed or for soil treatment after certain processing.

- **Aerobic decomposition** is done in the *presence of oxygen* and it produces CO_2, NH_3 and some other gases in small quantities and *large quantity of heat*. The final by-product of this process can be used as *fertilizer*.

2. Fermentation:

Fermentation is the process of decomposition of organic matter by micro-organisms especially bacteria and yeasts.

- It is a well established and widely used technology for the conversion of grains and sugar crops into *ethanol* (ethyl alcohol):
- Ethanol can be blended with gasoline (petrol) to produce gasohol (90% petrol and 10% ethanol). Processes have been developed to produce various fuels from various types of fermentations.

$$\underbrace{C_{12}H_{22}O_{11}}_{\text{Sucrose}} + H_2O \rightarrow \underbrace{2C_6H_{12}O_6}_{\text{Glucose}}$$

$$C_6H_{12}O_6 \xrightarrow{\text{Fermentation}} \underbrace{2C_2H_5OH}_{\text{Ethanol}} + 2CO_2$$

Fig. 6.1, shows the biomass and conversion technologies and products.

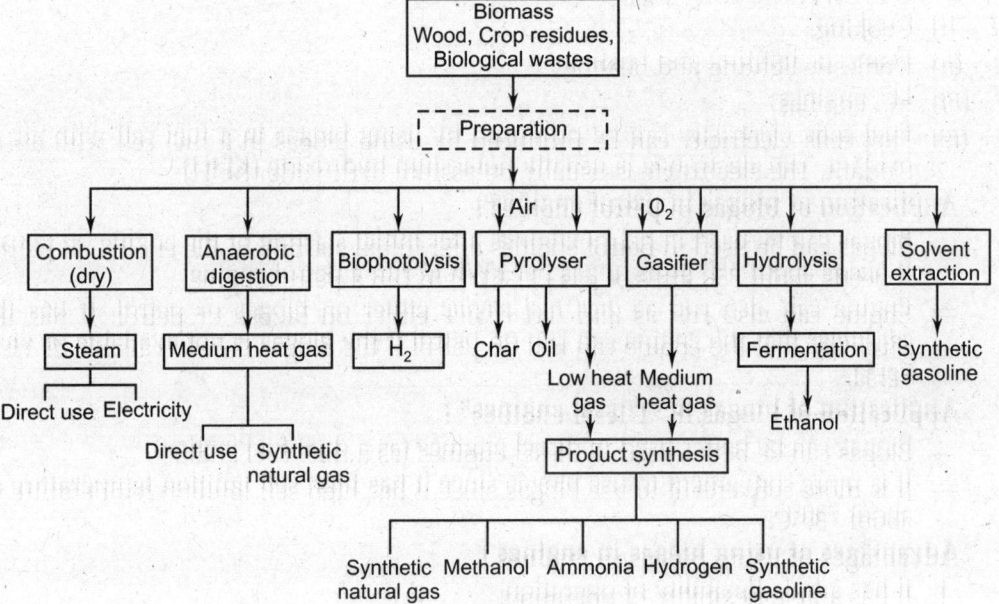

Fig. 6.1. Biomass conversion technologies and products .

6.6. BIOGAS

6.6.1. Introduction

The *main source for biogas is* **wet cattle dung**. Some of the *other sources are*:

(*i*) Sewage	(*ii*) Crop residue
(*iii*) Vegetable wastes	(*iv*) Water hyacinth
(*v*) Alga	(*vi*) Poultry droppings
(*vii*) Pig-manure	(*viii*) Ocean kelp

- Biomas, a mixture containing *55-65% methane, 30-40% carbon dioxide* and the rest being the impurities (hydrogen, hydrogen sulphide and some nitrogen), can be produced from the decomposition of animal, plant and human waste.
- It is a clean but slow-burning gas and usually has a heating value about 18 kJ/m³.
- It can be used directly in cooking, reducing the demand for firewood.

- The material from which the biogas is produced retains its value as *fertilizer* and can be returned to soil.
- 'Biogas' *is produced by digestion, pyrolysis or hydrogasification.* **Digestion** is a biological process that occurs in the absence of oxygen and in the presence of anaerobic organisms at ambient pressures and temperatures of 35-70°C. The container in which this digestion takes place is known as the *digester*.

6.6.2. Biogas Applications

Biogas is a flammable fuel gas usually with 60% CH_4 and rest CO_2. The gas can be upgraded by removal of CO_2 with water scrubbing and the gas with high heating value can be used in I.C. engine.

The main *applications of biogas* are :

(*i*) Cooking.

(*ii*) Domestic lighting and heating.

(*iii*) I.C. engines.

(*iv*) Fuel cells–electricity can be produced by using biogas in a fuel cell with air as oxidant. The electrolyte is usually potassium hydroxide (KOH).

Application of biogas in petrol engines :

— Biogas can be used in petrol engines after initial starting of the engine on petrol. It needs about 550 litres of gas per kWh to run a petrol engine.

— Engine can also run as *duel fuel engine* either on biogas or petrol. It has the *advantage* that the engine can run on petrol if the biogas is not available or vice-versa.

Application of biogas in "Diesel engines" :

— Biogas can be better used in diesel engines (as a duel fuel engine).

— It is more convenient to use biogas since it has high self ignition temperature of about 730°C.

Advantages of using biogas in engines :

1. It has ample flexibility of operation.

2. A uniform gas-air mixture is available in multicylinder engines.

3. Clean combustion reduces the wear of engine parts.

4. Lubricating oil consumption is reduced.

5. Emissions of CO are greatly reduced.

6. NO_x emissions are also reduced.

6.6.3. Anaerobic Digestion System (Biogas Technology)

Anaerobic digestion system consists of the following *three* stages :

1. Hydrolysis (Stage–1)

2. Acidification (Stage–2)

3. Methane formation (Stage–3).

These stages are shown in the Fig. 6.2 shown below:

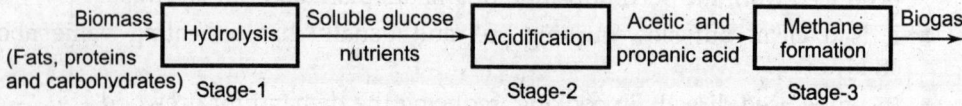

Fig. 6.2. Anaerobic digestion system.

1. **Hydrolysis (Stage–1) :**

    In this stage, the complex compounds, such as *fats, proteins* and *carbohydrates* are *broken into small size compounds* through the effluence of *water* and *enzymes* called *"Hydrolysis"*.

    Reaction of hydrolysis:

    $$(C_6H_{10}O_5)n + nH_2O \xrightarrow{\text{Enzymes}} n(C_6H_{12}O_6)$$

    $\underbrace{\textit{Cellulose + Nutrients}} \qquad \underbrace{\textit{Soluble glucose + Nutrients}}$

2. **Acidification (Stage 2).**

    In this stage, the soluble glucose and nutrients are converted into *simpler volatile fatty acid and acetic acid as byproduct*, that accounts for 70 per cent of *methane* by-product with the help of *acid forming bacteria.*

    Reaction of acidification:

    $$C_6H_{12}O_6 \xrightarrow[\text{bacteria}]{\text{Acid producing}} 2CH_3CH_2OH \text{ (Ethylalcohol)} + 2CO_2$$
    $$+ \text{ Fatty acid} + H_2 + \text{other products}$$

3. **Methane formation (Stage-3) :**

    During this stage, the *methane producing bacteria converts the organic acid into* **biogas** having its *main constituent as "methane".*

    Reactions of methane formation:

    $$2CH_3CH_2OH + CO_2 \xrightarrow[\text{bacteria}]{\text{Methane producing}} 2CH_3COOH + CH_4$$
    $$CH_3COOH \xrightarrow{\hspace{2cm}} \underset{\text{Methane}}{CH_4} + CO_2$$

6.6.4. Advantages of Anaerobic digestion

Anaerobic digestion claims the following *advantages* :

1. The *biogas* produced as a byproduct, has a *calorific value* (which can be used as energy source for producing steam or hot water).
2. Low nutrient requirement.
3. Low odour.
4. Reduction of pathogens.
5. Stable sludge.
6. New sludge production.
7. Sludge acts as a soil conditioner.
8. Low running cost.

6.6.5. Advantages of Biogas Production

1. Gas production is cheap.
2. Less pollution.
3. Waste material can be used as fertilizer.
4. Gas is used for cooking, lighting, as fuel etc.

6.6.6. Factors affecting Generation of Biogas

The generation of biogas is affected by the following *factors :*

1. Temperature,   2. Loading rate,
3. Solid concentration,   4. pH value,

5. Retention period,   6. Nutrients concentration, and

7. Toxic substances.

1. **Temperature.** The anaerobic fermentation process is *temperature dependent*. The process of the digestion and gasification proceeds at the *highest rate* when the temperature lies between 35°C – 38°C. The process becomes *slow* within temperature range of 45°C – 45°C and then *rises to a peak* between 55°C – 60°C. Thus, the *rate of gas production 'increases' with the increase in temperature* but the *percentage of methane 'decreases'*.

2. **Loading rate.** *"Loading rating"* is the *weight of volatile solids fed to a digester per day.* It depends upon the *plant capacity* and also the *retention period.* Thus, for a given capacity of the digester, if the *loading rate* is *increased* the *'retention period'* is correspondingly *decreased.*

3. **Solid concentration.** Normally, 7 to 9 parts of solid in 100 parts of the slurry is considered ideal.

— It is recommended that 4 parts of the *cattle dung* to be mixed with *5 parts* of *water*.

4. **pH value.** pH denotes the acidity and alkalinity of the substrate. The pH *less than* 7 is called *'acidic'* and pH *more* than 7 is called *'alkaline'* and pH solution of 7 is called *'neutral'.*

5. **Retention period.** It is the time period for which fermentable material *resides inside the digester*. This period ranges from 30 days to 50 days depending upon the *climatic conditions.*

● Generally it is observed that *maximum gas production takes place within 'first four weeks'* and it *tapers off gradually.*

6. **Nutrients concentration.** The major nutrients required by the bacteria in the digester are C, H_2, O_2, N_2, P and S. To maintain proper balance of nutrients an extra raw material, rich in P and N_2, should be added along with cattle dung *to obtain maximum gas production.*

7. **Toxic substance.** The presence of ammonia, pesticides, detergents and heavy metals are considered as toxic substances to micro-organisms, since their *presence reduces fermentation rate.*

6.7. BIOGAS PLANTS

6.7.1. Introduction

Biogas plants *converts wet biomass into biogas (methane) by the process of "anaerobic fermentation".*

The bacteria called *"anaerobe"* carrier out digestion of biomass *without oxygen* and produces methane (CH_4) and carbon dioxide (CO_2).

Biogas plants are very popular in India particularly in rural areas.

6.7.2. Raw Materials used in Biogas Plant

1. *Animal wastes :* Cattle dung, urine, fish wastes, piggery wastes etc.

2. *Human wastes :* Waste matter, urine etc.

3. *Agriculture wastes :* Sugarcane trash, tobacco waste, oil cake, vegetable wastes etc.

4. *Industrial wastes :* Sugar factory, paper factory etc.

6.7.3. Main Components of a Biogas Plant

The *main components of a biogas plant* are enumerated and briefly described below:

1. Digester;   2. Gas holder;
3. Inlet;   4. Outlet;
5. Slurry mixing tank;   6. Gas outlet pipe;
7. Stirrer.

1. **Digester.** A digester is also called *"fermentation tank"* and is mostly embedded partly or fully in the ground. It is generally cylindrical in shape and is made of bricks. It holds the slurry for a sufficiently long time to complete the digestion.

2. **Gas holder.** Its function is to *keep the gas for subsequent use.* The gas connection for use is taken from the top of the gas holder. In some designs of biogas plants, it may be separable from the digester whereas in other designs it may be an integral part of the digester.

3. **Inlet.** An inlet is provided to *add* the mixture of dung and water to the digester, and is sloped accordingly.

4. **Outlet.** The provision of an outlet is made to *take out* the digested portion of slurry.

5. **Slurry mixing tank.** This tank carries out *mixing of the dung with water* for induction in the digester, through the inlet.

6. **Gas outlet pipe.** It is used for *taking out gas* from the gas holder and is connected to its *top.* The other end of the pipe is connected with the *device using biogas.*

7. **Stirrer.** The stirrers are *provided in biogas plants of large size* for *stirring the slurry for fermentation* inside the fermentation chamber to *ensure the normal production of gas.*

— The small size biogas plants can function without a stirring device.

● Some biogas plants also have the *arrangement of external heating by solar / electrical energy etc. under colder climates.*

● The biogas plants are built in several sizes, small (0.5 m³/day) to *very large* (2500 m³/day). Accordingly, the configurations are *simpler to complex.*

6.7.4. Classification of Biogas Plants

The biogas plants may be *classified* as follows :

1. **Continuous type biogas plant :**
   (i) *Single stage type;*
   (ii) *Double stage type.*
2. **Batch type biogas plant.**
3. **Floating drum type biogas plant.**
4. **Fixed dome type (Janta model or chinese model).**
5. **Modified fixed dome type biogas plant** – This type of plant has an *additional displacement tank and water seal gas tank.*

Description of these biogas plants is given in the following Articles.

6.7.5. Continuous Type Biogas Plant

In this type of plant the *biomass is fed regularly to the digester and it supplies the gas continuously*.

These are *two* types of continuous biogas plant :

1. Single stage type   2. Two stage type

1. **Single stage continuous biogas plant.** In this type of plant, the entire process of conversion of biomass into biogas are carried in a *single digester*. This chamber is regularly fed with the raw materials while the spent residue keeps moving out.

Advantages :

(*i*) Simple in construction.

(*ii*) It does not need skilled labour.

(*iii*) It is easy to operate and control.

(*iv*) These are preferred for *small and medium sized biogas plants*.

● *Serious problems are encountered with agricultural residues when fermented in a single stage continuous process.*

2. **Two stage continuous type biogas plant.** These plants have *two digesters* for digestion of biomass. In the *'first digester'* the biomass is fed in which the *acid production* is carried out and then only dilute acids are fed into the *'second digester'* where *bio-methanation* takes place and biogas can be collected from the second digester/chamber.

Advantages :

(*i*) It produces *more gas* than the single stage plants.

(*ii*) It requires *lesser period of digestion* as compared to single stage plants.

(*iii*) These plants are preferred for *large size biogas plants*.

Advantages of continuous type biogas plants :

Following are the *advantages* of continuous type biogas plants :

1. *Continuous* gas production.

2. *Less* retention period.

3. *Small digestion chambers* required.

4. *Less problems* as compared to batch type plants.

6.7.6. Batch Type Biogas Plant

In a batch type plant, the biomass feeding is done in *batches with large time interval between two consecutive batches*.

A batch loaded digester is filled to its capacity and given sufficient retention time (35 to 45 days) for digestion of biomass. After the completion of digestion, the residue is emptied and filled again. Gas production is *'uneven' due to slow start of bacterial* digestion and to overcome this difficulty, *several digesters are used which are fed and emptied in sequential manner*. Thus, the regular supply of gas is maintained.

The **"salient features"** *of batch type plant are :*

1. Gas production *uneven/intermittent*, depending upon the cleaning of the digester.

2. *Several digesters required,* to get continuous supply of gas.

3. *High space requirements,* due to several digesters.

4. *High initial cost,* due to large volume of digester.

5. *Needs addition of fermented slurry to start the digestion process.*

6. Operational and maintenance problems.

6.7.7. Floating Drum Type Biogas Plant

Khadi Village Industries Commission (KVIC) standardised a model in 1961 (Fig. 6.3). It consists of an underground *"cylindrical masonary digester"* having an *"inlet pipe"* for feeding animal dung slurry and an *"outlet pipe"* for sludge. There is a *"steel dome"* for gas collection which *floats on slurry*. It moves up and down depending upon accumulation and discharge of gas guided by the domeguide itself. A *"partition wall"* provided in the digester *improves circulation*, necessary for fermentation.

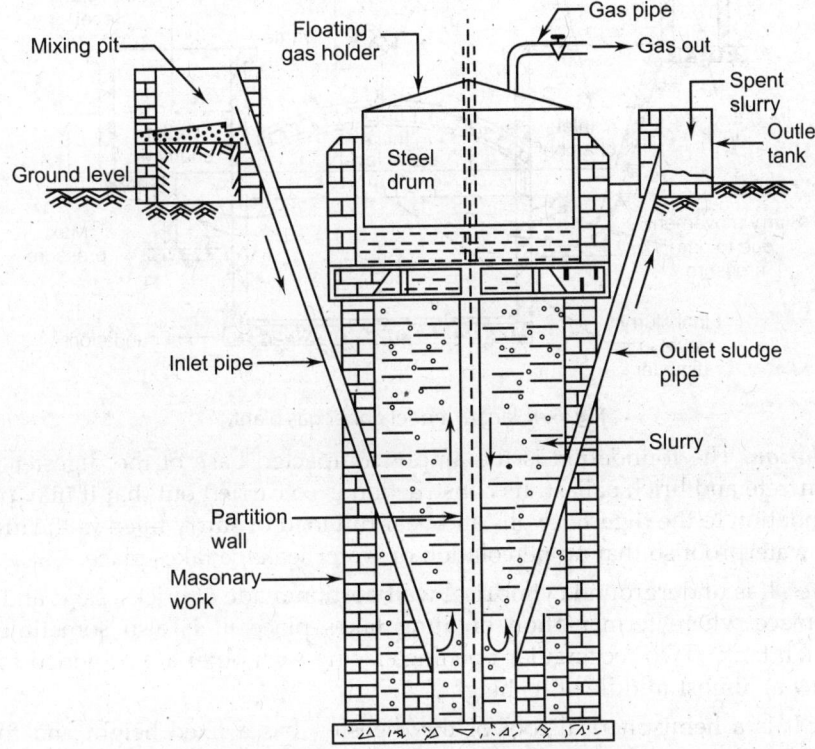

Fig. 6.3. Floating drum biogas plant (KVIC model)

A pressure of about 100 mm of water column is built in the *"floating gas holder"*, which is sufficient to supply gas upto 100 metres. This gas pressure also forces out the *spent slurry* through a *sludge pipe*.

Advantages :

1. *Gas pressure is constant.*

2. *Less scum problem.*

3. *No danger of explosion* since there is no possibility of mixing of biogas and external air.

4. *No gas leakage problem.*

Disadvantages :

1. *High cost.*

2. *High maintenance cost.*

3. There is a *loss of heat* through gas holder.

4. The outlet pipe, which should be flexible, *requires regular attention.*

6.7.8. Janata Model Biogas Plant (Fixed dome)

Constructional features. This plant consists of the following *parts*:

Refer to Fig. 6.4.

1. Foundation
2. Digester
3. Dome
4. Inlet Chamber
5. Outlet Chamber.
6. Mixing Tank
7. Gas outlet pipe.

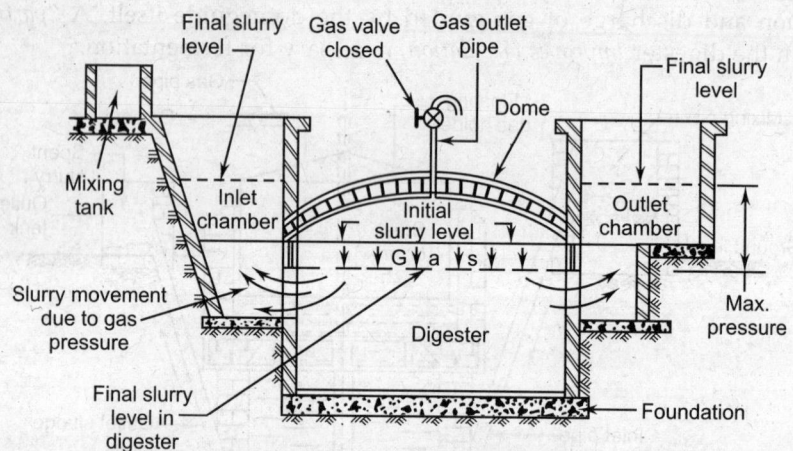

Fig. 6.4. Janata model gober gas plant.

Foundation. The foundation is the amply compacted base of the digester made of cement concrete and brick ballast. Its construction is so carried out that it may provide a stable foundation to the digester walls and bear full load of slurry filled in the digester. It should be waterproof so that no percolation or water leakage takes place.

Digester. It is underground cylindrical wall portion made of bricks, sand and cement. It is this place where fermentation of dung takes place. It is also sometimes called 'fermentation tank'. Two rectangular openings facing each other are provided for inflow and outflow at almost middle of its height.

Dome. It is a hemispherical roof of the digester; has a fixed height and forms the *critical part* in the construction of Janata gobar gas plant. The gas gets collected in the space of the dome and exerts pressure on the slurry in the digester.

Inlet chamber. An inlet chamber has a bell mouth shape and is made of bricks, cement and sand. It has its top opening at the ground level. Its outlet wall is made inclined/slopy to enable the daily cattle dung feed to move easily into the digester.

Outlet chamber. It is that part of the plant through which digested slurry moves out of the digester at a predetermined height. It has a small rectangular cross-section and above this it becomes larger to a defined height. For easy cleaning of the digester two steps are provided in it which enable a man to climb down. Its top opening is also at the ground level. Just near the top opening is provided a small outlet through which the digested/spent slurry flows to a compost pit.

Mixing tank. It is this tank where dung and water are mixed properly in the ratio of 1 : 1 to make slurry which is then poured into the inlet chamber.

Gas outlet pipe. It is a small piece of G.I. Pipe which is fitted at the top of the dome for conveying the gas to the points of use. A valve is fitted at its end to regulate the flow of gas to the gas connections.

Site selection. While selecting the construction of Janata gober gas plant the following *points* should be given due consideration :

(i) The surface should be plane/even and be at a higher elevation so that during rainy season the water level is at least at 3 m depth. The higher level will discourage water logging and ensure easy discharge of spent slurry from the outlet chamber. Preferably this place should be beyond the reach of children.

(ii) The site should be as far as possible near the cattle shed and points of gas utilisation.

(iii) It should be at least 2 metres away from the foundation of the house/building.

(iv) The sun light should be available during whole of the day round the year.

(v) The site should be atleast 10-15 metres away from the any water drinking source.

(vi) There should not be any big tree near the plant whose roots may cause any harm with the passage of time.

(vii) There should be an easy availability of water near the plant.

(viii) To avoid carrying spent slurry to a very far distance there should be some space for making compost pit.

(ix) The earth should have adequate bearing stress to avoid any possibility of caving in or collapse of the plant.

Nature of soil and corresponding precautions :

1. *Soil formed of clay :*

— Related problem   Easy expansion and contraction of soil as the moisture content increases and decreases.

— Precautions   Do not disturb the primitive soil; Or Take measures for excess or less of water; Or Drain the surface water.

2. *Non-uniform soil structure :*

(*e.g.* partly soft soil and partly rocky)

— Related problem   Crack may appear in digester wall due to uneven settlement.

— Precautions   Provide a uniform kind of soil under the digester; Or

— Remove a portion of the soft soil and put lime concrete and rubble in its place; Or

— Remove a part of the rock and add medium and coarse sand, cinders, clay or clay gravel.

3. *Soil with high water table :*

— Related problem   There is an upward force on the digester due to pressure of ground water.

— Precautions   Construct the plant in low water seasons.

— Dig a trench around the digester to collect the water; this water may be occassionally pumped out.

— Divert the ground water, if possible, to other places *e.g.,* well points.

— Increase the strength of the base of the digester by increasing its thickness. It will counter act the buoyancy force of ground water.

Initial loading of the plant. After completion of the plant it is to be charged with cattle dung and water in the ratio of 1 : 1. Since Janata Plant is designed for 50 days retention period, therefore initial loading will comprise of 50 times of gobar needed for daily feed. Naturally to arrange such a large quantity of gobar by farmers who have a few cattle is a problem. This difficulty can be overcome if the gobar is collected regularly during the period of construction of plant and by the time the plant is completed the large quantity of gobar collected can be then fed.

— Before feeding into the digester the dung should be properly mixed with water in the ratio of 1 : 1 in the mixing tank to get the slurry.

— The slurry is then fed to the digester upto the step provided on the Outlet chamber. Thereafter the fermentation starts.

— *Initially allow the gas which comprises of large percentage of a carbondioxide, oxygen and hydrogen sulphide to escape since it will not burn.*

— The gas obtained after this release will have a right combination of methane (60%) and carbondioxide (40%) and burn with odourless blue flame in the gas chulha/burner/stove.

— Now, at this stage start feeding the plant with dung slurry.

— Regular, proper and careful feeding will ensure flawless gas service.

How does the plant function ?

— The cattle dung and water are mixed properly in the ratio 1 : 1 to form slurry which is then filled in the digester upto the height of its cylindrical portion.

— As soon as the fermentation of dung picks up the generated gas starts accumulating in the dome. This gas then *exerts pressure on the slurry and displaces it into the Inlet and Outlet chambers.* Consequently the slurry level in the digester falls and rises in the Outlet chamber.

— This fall in the level of slurry continues (if the outlet pipe valve is closed) till the slurry level reaches the upper ends/edges of the Inlet and Outlet gates.

— After this stage any further accumulation of gas would exert pressure and start escaping through these gates into the atmosphere. This condition is indicated by *bubbling* and froth formation on the surface of the slurry in the inlet and outlet chambers.

— *The quantity of usable gas can be determined by calculating the increase in slurry volume in the inlet and outlet chambers.*

— In the event of using gas the slurry level in the digester rises while that in the inlet and outlet comes down.

— The increase in gas pressure due to increased generation in comparison to its utilisation will push/displace the slurry up in the inlet and outlet chambers while the decrease in pressure is balanced by the return flow of slurry back in the digester.

— During the process *approximately* equivalent quantity of spent/digested slurry (top layer slurry) is discharged from the outlet chamber through the outlet opening; this is so because fresh/undigested slurry which is fed into the inlet chamber is heavier than the digested/spent slurry and settles down in the digester.

Chemistry of gas generation. In this gas plant the feed which comprises mainly of cattle dung is subjected to anaerobic fermentation as a result of which is produced combustible gas and fully matured organic manure which is *superior to green manure* available from dung otherwise. The whole process of feeding slurry of dung and water and extraction of spent slurry in gobar gas plant is *continuous one*.

The cattle dung (and other fermentable materials like night soil, poultry or piggery droppings etc.) when confined in a place where there is *no air* gives rise to mainly two types of bacteria *viz. Acid forming bacteria* and *Gasifying bacteria*.

Acid forming bacteria convert carbohydrates, proteins, fats into volatile acids and carbon dioxide is produced during the process. This phase is also known as *"liquification phase"* and is brought about by a set of *saprophytic bacteria* by means of extracellular enzyme. These bacteria are less sensitive and can exist, develop and *multiply in wide range of conditions*.

Liquification phase is followed by *"gasification phase"* which is actively carried out by *methane bacteria*. They work upon volatile acid produced during the previous phase, with the help of intracellular enzyme and convert it into methane and carbondioxide.

The whole process of gas generation is governed by the factors *viz.* Temperature of substrate, Loading rate, Solid concentration, Detention period, pH value, Nutrients concentration and Toxic substances etc.

The composition of the gas produced varies with the type of fermentable material used. In case of cattle dung on an average the gas produced consists of 55 to 60% methane and 40 to 45% carbondioxide with little quantity of hydrogen, hydrogen sulphide. With night soil the percentage may be : Methane = 65%, carbondioxide 34%, hydrogen sulphide = 0.6% and other gas 0.4%.

Diphasic anaerobic digestion. To tide over the problems generally associated with maintaining optimum fermentation conditions inside the digester, diphasic anaerobic digestion seems to be probable solution. In this system *two separate digesters* are used which separate the hydrogen and acid formation phases from methane formation phase; this separation of phases is brought out by kinetic control. The acid phase is operated at low retention times so that only the fast growing acidogens of first hydrolysis and acid forming stages are retained in the digester and methogens are washed out. The methane phase takes place in a *separate digester* with the acid phase effluent as the feed material.

This system facilitates provision of optimum environment conditions for growth of two physiologically different types of bacteria which in turn ensures *higher efficiency of the process with subsequent reduction in the size of digester and their cost*. In this process system monitoring is also rendered *easier*.

How to accelerate gas generation. The gas production is usually satisfactory during summer season but it falls considerably during coldest months of the year. *The gas production can be enhanced / accelerated by following the **tips** given below :*

(i) Add about *1 litre of cattle urine daily for every 5 kg slurry fed to the plant*. It will augment the fermentation process and eventually increase gas production. The urine can be collected in the sump connected through a drain to the animal shed.

(ii) Feed back about *10% of the fresh spent/digested slurry* to the plant. The micro-organisms available in the spent slurry will *stimulate fermentation*.

(iii) Prepare the dung slurry with *hot water*. Water may be kept exposed to Sun during the day and be used in the evening to produce slurry. This process will prove helpful in increasing the temperature of slurry in the digester and subsequently actify the bacteria to generate more gas.

(iv) *Cover the plant with rice straw or gunny bags* during late evenings and nights; it will check fall of temperature and keep the bacteria *active* in gas generation.

(v) Add *1 kg of powdered leaves for every 50 kg dung* fed to the plant; it will help creating warmer environment suitable to bacteria for actified fermentation.

(vi) Addition of *poultry droppings (5 to 10%) and piggery waste* helps to bring about increase in gas generation.

(vii) Provision of a *compost pit around the plant* assists in keeping the *plant warm* due to which gas generation is improved.

Advantages and disadvantages of "Fixed dome type plants" :

Advantages :

1. No maintenance problems due to absence of moving parts.
2. Low cost.
3. Low operating cost.
4. Longer working life.
5. Due to underground construction, heat insulation is better and therefore, rate of gas production is uniform during night and day.
6. Quantity of gas producd is higher than movable drum type plants.
7. No corrosion problem.
8. Space above the plant can be used for other purposes.

Disadvantages :

1. *Variable gas pressure.*
2. Gas production per cm³ of the digester volume is *less.*
3. Problem of *scum formation.*
4. For construction work *skilled* masons are required.

6.7.9. Deenbandhu Biogas Plant (DBP)

Refer to Fig. 6.5. This is *fixed dome* plant developed by Action for Food Production (AFPRO) in 1984. It is appropriate for *all types of wastes and minimises biogas losses from inlet chamber.*

Construction. It has '*curved bottom*' and '*hemispherical top*' which are joined at their bases with *no* cylindrical portion in between. An '*inlet pipe*' connects '*mixing tank*' with the '*digester*'.

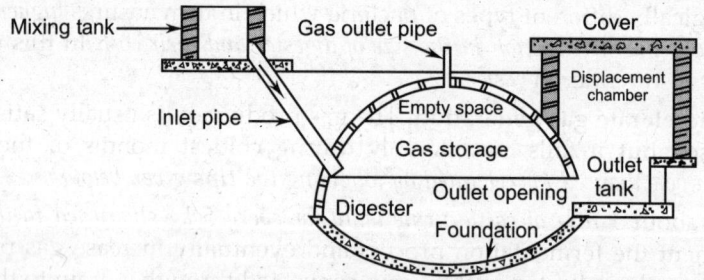

Fig. 6.5. Deenbandhu biogas plant.

Working. Cattle dung slurry prepared in 1 : 1 ratio with water is fed upto the level of second step in the outlet tank. As the gas generates and accumulates in the empty portion of the plant, it presses the slurry of the digester and displaces it into the outlet '*displacement chamber*'. The slurry level in the digester falls whereas in the outlet chamber it starts rising. This fall and rise continues till the level in the digester reaches the upper end of the outlet opening and at this stage the slurry level in the outer tank reaches the height of discharge opening.

Advantages :

1. This plant requires *less space* being mainly underground.
2. Its *cost is reduced* as the surface area is minimised by joining segments of two different diameter spheres at their bases.
3. It is 30 per cent economical as compared to Janata biogas plant.

6.7.10. Comparison Between Fixed Dome Type and Movable Drum Type Biogas Plants

The comparison between fixed dome type and movable drum type biogas plants is given below :

S.No.	Aspects	Fixed dome type plant	Movable drum type plant
1.	*Material used*	Bricks, cement, sand, steel and lime/cement.	Concrete, bricks, sand, lime and steel.
2.	*Cost factor*	Bricks and cement.	Bricks, cement and steel.
3.	*Cost*	Less costlier.	More costly due to steel.
4.	*System used*	Regular filling and irregular discharge.	Regular filling and regular discharge.
5.	*Feed stock*	Agriculture waste, other organic matter, animal and human waste.	Animal and human wastes and chopped agricultural wastes as additive.
6.	*Digestion period*	30-60 days	40-60 days.
7.	*Gas-tightness*	Gas storage dome is required to give special treatment for gas tightness and painted periodically.	No such problems.
8.	*Thermal insulation*	Due to underground construction the temperature and heat insulation is uniform.	Digester heat is lost through gas holder, therefore, less suitable for colder region.

6.7.11. Methods of Maintaining Biogs Production

Following techniques are suggested for maintaining the biogas production :

1. Insulating the gas plant.
2. Compositing.
3. Hot water circulation.
4. Use of chemicals.
5. Solar energy systems.

6.7.12. Community Night-soil Based Biogas Plant

Such plants have been developed to facilitate sanitary treatment of *human waste* at community and institutional level.

Fig. 6.6 shows the cross-section of a community night-soil based biogas plant. It consists of a *'floating metal drum'* with a *'water jacket'*. It is linked with community toilets. It can serve a population of 1000 persons, and can provide fuel for *cooking, operating duel-fuel engines for water supply and generate electric power.*

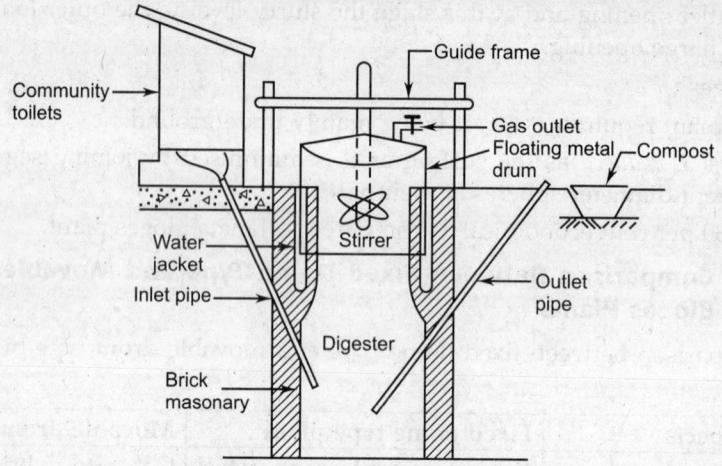

Fig. 6.6. Community night-soil based biogas plant.

6.7.13. Guidelines for fixing Optimum Size of a Biogas Plant

The following *guidelines* may be used to fix optimum size of biogas plant :

1. Type of waste.
2. Daily rate of waste to be digested.
3. Digestion period.
4. Method of stirring, if any.
5. Arrangement for raw waste feeding and discharge of digested slurry.
6. Climatic conditions.
7. Mix of raw waste.
8. Water table and sub-soil conditions.
9. Type of dome.

6.7.14. Site Selection of a Biogas Plant

While selecting a site for a biogas plant, the following *factors* must be considered :

1. **Less distance.** In order to achieve *economy in pumping of gas,* the distance between the plant and site of gas consumption should be *less.* The optimum distance for a plant of 2 *m³ capacity is 10 metres.*

- A *minimum gradient of 1 percent* must be made available for the line to convey the gas.

2. **Open space.** For gas generation at adequate rate, there should an open space sufficient enough for the sunlight to fall on the plant to provide essential temperature between 15 to 30°C.

3. **Space requirements.** To carry out day-to-day operation and maintenance, sufficient space must be available. As a guideline 10 to 12 m² area is required per m³ of the gas.

4. **Availability of water.** Availability of plenty of water must be ensured, to prepare proper cattle dung slurry for gas generation.

5. **Water table.** As the biogas plant is normally constructed underground, care should be taken to prevent seepage of water. In case the water table is less than 3 metres, the plant should *not* be constructed.

6. **Transportation cost of cattle dung/materials.** To reduce the transportation cost of cattle dung/materials for biogas generation, the distance between the materials and site of the biogas plant should be *minimum*.

7. **Distance from wells.** A biogas plant should be constructed at a minimum distance of 15 metres to check the seepage of fermented slurry which *may pollute the water of well*.

8. **Seasonal run-off.** To prevent the interference of run-off water during monsoon, *intercepting ditches or bunds* may be constructed.

6.7.15. Reasons why Biogas Plants are not much Successful in India

Reasons why biogas plants are not much successful in India, even after the subsidies provided by Government of India (or *problems related to biogas plants*) are as follows :

1. Gas forming methanogenic bacteria are very sensitive towards temperature. During *winter* the temperature falls and there is decrease in activity of the bacteria and the *gas production rate reduces*.

2. A lot of problem arises due to *lack of training* to the biogas plant owners for the operation of plant.

3. Collecting dung, mixing with water and draining out the waste is tiresome and time consuming process.

4. Some persons add urea-fertilizer (to augment gas production) in *large quantities* due to which toxicity of ammonia, nitrogen may cause a *decrease in gas production*.

5. The *cleaning* and recharging of whole tank of biogas plant is necessary after 3 or 4 years, which is *not so easy*.

6. *Handling of effluent slurry* is major problem. For domestic plant, 200 litres capacity oil drums can be used to carry this effluent; this will require some human labour.

7. In some areas like Himachal, Jammu and Kashmir the average annual temperature is *low* which is not adequate for anaerobic digestion.

6.7.16. Fuel Properties of Biogas

The important *properties of biogas, generated by anaerobic fermentation of organic wastes* are as follows :

Composition – *Percentage by volume* :

Methane = 50–60; Carbon dioxide = 30–45; Hydrogen = 5–10;

Nitrogen = 0.5–0.7; Hydrogen sulphide and oxygen = Traces.

Calorific value :

60% methane : 22.35 to 24.22 MJ/m^3

without CO_2 : 33.52 to 35.39 MJ/m^3

Octane rating :

Without CO_2 : 130

With CO_2 : 110

Ignition temperature : 650°C

Air-to-methane ratio for complete

Combustion (by volume) : 10 to 1

Explosive limits to air (by volume): 5 to 15

- The main products of the biogas plant are *'fuel gas'* and *'organic manure'*. Biogas is a flammable gas. *'Methane'* is the *only combustible portion in the gas* and hence *around 60% by volume only is usable for combustion*.

6.7.17. Methods to obtain Energy from Biomass

Energy from biomass can be *obtained* by using the following *methods* :

1. Combustion;   2. Anaerobic digestion;
3. Pyrolysis;   4. Hydrolysis and ethanol fermentation;
5. Gasifier.

1. **Combustion.** *'Combustion' is the process*, now in commercial operation, that *uses biomass to produce energy*. Direct combustion requires biomass with a moisture content around 15% or less, so it may require drying prior to combustion for most of the crops.

2. **Anaerobic digestion.** The biogas plants using anaerobic digestion are *simple in construction* with *low capital outlay*. The anaerobic digestion process has the following *advantages* :

(*i*) It utilises biomass with high percentage of water.

(*ii*) Small units are available, which can be operated as individual farms.

(*iii*) The residue has fertilizer value.

The major *"limitation"* with this process is that large quantity of waste water is to be disposed of after digestion.

3. **Pyrolysis :** *'Pyrolysis'* is an *irreversible change* brought about by the action of heat in the *absence of oxygen;* the energy *splits* the chemical bonds and leaves the energy stored in biomass. It may yield either solid, liquid or gaseous fuel.

Advantages :

(*i*) Compactness;

(*ii*) Simple equipment;

(*iii*) Low pressure operation;

(*iv*) High conversion efficiency of the order of 83 percent

(*v*) Negligible waste product.

4. **Hydrolysis and ethanol fermentation.** The process of *hydrolysis* converts cellulose to alcohols through fermentation. Ethyl alcohol can be produced from variety of sugar by fermentation with yeasts.

5. **Gasifier.** *Pyrolysis-gasification* is a promising conversion technology. It appears to be *economically competitive with natural gas*, using *biomass wastes*.

6.8. BIOMASS GASIFICATION

Gasification *implies converting a solid or liquid into a gaseous fuel without leaving any solid carbonaceous residue.* This process is carried out in a *'gasifer'*.

6.8.1. Gasifiers

Gasifiers (*essentially a chemical reactor*) is an *equipment which can gasify a variety* of biomass such as *woodwaste, agricultural waste like stalks*, and *roots of various crops, maize cobs* etc. In a gasifier, the biomass (as it flows) gets dried, heated, pyrolysed, partially oxidised and reduced.

Advantages, following are the *advantages of a gasifier :*

1. Very easy operation.
2. Reliable operation.
3. Easy maintenance.
4. Sturdy construction.

Classification of gasifiers :

Biomass gasifiers may be *classified as follows :*

A. According to the "type of bed" :

1. *Fixed bed gasifiers :*

(*i*) Updraft,

(*ii*) Downdraft, and ⎱ Depending upon the *direction of airflow.*

(*iii*) Crossdraft

B. According to the "output power" :

(i) Small size gasifiers	– Output upto 10 kW.
(ii) Medium size gasifiers	– Output in the range of 10 kW to 50 kW.
(iii) Large size gasifiers	– Output in the range of 50 kW to 300 kW. and
(iv) Very large gasifiers	– Output of 300 kW and above.

6.8.2. Fixed Bed Gasifiers

(*i*) **Updraft (or counter current) gasifier :** Refer to Fig. 6.7

In such a gasifier (where fuel and air move in *counter current manner*) air enters *below* the combustion zone and the *'producer gas'* leaves near the *top* of the gasifier. The gas produced contains tar and water vapour and the *ash content is almost nil.*

- These gasifiers are suitable for *stationary engines* (which use tar free fuels like charcol).

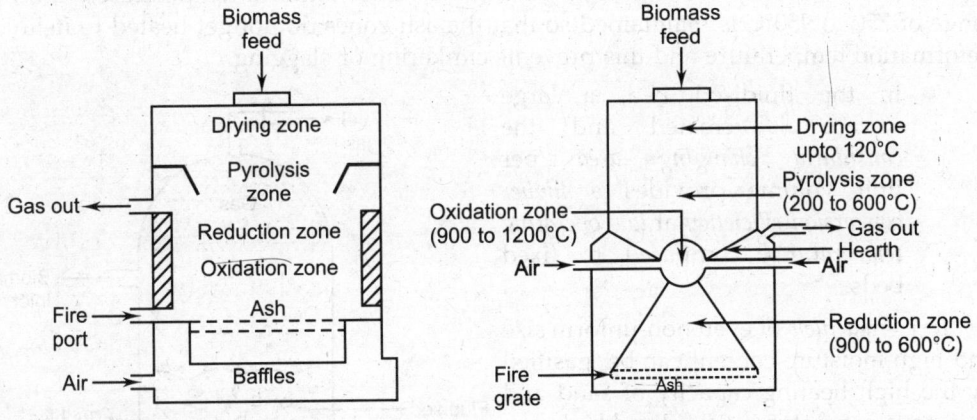

Fig. 6.7. Updraft gasifier.   **Fig. 6.8.** Downdraft gasifier.

2. **Downdraft (or cocurrent) gasifier.** Refer to Fig. 6.8.

In downdraft gasifier (where fuel and air move in a *cocurrent manner,* air *enters at the combustion zone* and the gas produced leaves *near the bottom of the gasifier.*

— Fuel (biomass) is loaded in the reactor from the top. As the fuel moves down it is subjected to *'drying (120°C)* and pyrolysis (200-600°C) where solid char, acetic acid, methanol and water vapour are produced.

— Descending volatiles and char reach the *oxidation zone* (900 to 1200°C) where air is injected to complete the combustion.

— The products moving downwards, enter the *reduction zone (900 to 600°C),* (reaction being *endothermic*) where *'producer gas'* is formed by the action of CO_2 and water vapour on red hot charcoal. The producer gas contains products like CO, H_2 and CH_4; it is *purified* by passing it through coolers, tar is removed by condensation, whereas soot and ash are removed by centrifugal seperation.

The downdraft gasifier is *most commonly used for "engine applications"* because of its ability to produce a *relatively clean gas.*

- *Fixed bed gasifiers* can attain efficiency upto 75 percent for conversion of *solid biomass to gaseous fuel.*

3. **Crossdraft gasifiers.** In this type of gasifier, the gas produced passes upwards in the annular space around the gasifier that is filled with charcoal. The charcoal acts as an insulator and a dust filter.

- These gasifiers are suitable for *power generation upto 50 kW;* however, these do *not* find much application.

6.8.3. Fluidised Bed Gasifiers

A fluidised bed gasifier is *most versatile* and any biomass, including sewage sludge pulping effluents etc., can be gasified by using this type of gasifier.

Refer to Fig. 6.9.

Fluidised Bed Combustion (FBC) constitutes a hot bed of inert solid particles of sand or crushed refractory supported on a fine mesh or grid. An upward air current fludizes the bed material.

The pressurised air starts bubbling through the bed and the particles attain a stage of *high turbulence,* and the *bed exhibits fluid like properties.* A uniform temperature within the range of 750 to 950°C is maintained so that the ash zones do not get heated to its initial deformation temperature and this prevents clinkering or slagging.

- In the fluidised bed, a large surface is created and the *constantly changing area* per unit volume provides a *higher conversion efficiency* at *low operating temperatures,* compared to fixed beds.

Low grade fuels of even non-unform size and high moisture content can be gasified by the high heating capacity of sand and uniform temperature of fluidized bed.

- To put the gasifier in use the bed *material is heated to ignition temperature of the fuel* and *biomass is then injected* causing rapid oxidation and gasification. The

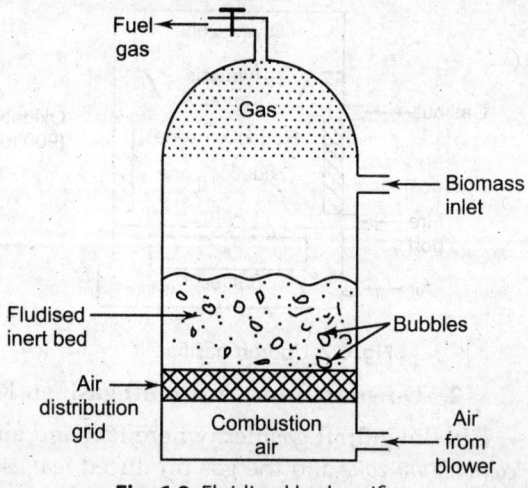

Fig. 6.9. Fluidized bed gasifier.

fuel gas thus obtained is conditioned and cleaned for utilisation as an engine fuel.

Advantages of fluidised bed gasifier :

1. High heat storage capacity.
2. Simple operation.
3. Compact size.
4. Consistent combustion rate.
5. High output rate.
6. Quick startup.
7. Fuel flexibility.
8. High moisture content fuel can be used.
9. Uniform temperature throughout the finance volume.
10. Reduced emission of nitrogen oxides.

6.9. ENERGY RECOVERY FROM URBAN WASTE AND WOOD

6.9.1. Introduction

The *"Urban Waste"* is disposed off suitably by *"Waste-to-Energy"* conversion systems including :

1. Landfill Gas Energy Plants.
2. Waste Incineration Cogeneration Plants.
3. Biochemical Conversion Plants.

- One of the most important biomass conversion technologies is **Incineration (Combustion)**. The *applications* of *"Incineration process"* are given below :

Biomass resource	Conversion technology	Products	Applications
(i) Wood		Heat*	Co-generation plants, heat
(ii) Wood waste	"Incineration (Burning)"	Steam*	steam, electricity for industry
(iii) Forest matter		Electricity	and utility.
(iv) Dry solid biomass			
(v) Urban solid waste (Municipal refuse)	"Incineration"	Heat* Steam* Electricity	— Disposal of urban waste — Energy to urban consumers

* In cold countries heat is supplied by the utilities for heating the houses. Steam is required for process industries.

Following are the other processes of *"waste-to-energy"* and *"wood-to-energy"* :

1. Production of biomass from waste, by methane formation process,
2. Production of wood gas from wood,
3. Production of biogas from landfills, etc.

6.9.2. Urban Solid Waste (Municipal Waste, Municipal Refuse)

The domestic waste (Refuse) in the cities is usually sent to **landfill sites** located far away from the centre of the city. Large cities like Delhi, Bombay have large amount of waste and increasing waste disposal problems. *The emerging solution is to produce useful* *thermal and electrical energy by waste-to-energy plants* (WTE) located in the heart of city. Such energy **plants** are rated in MW (50 to 500 MWe) and serve the following *functions* :

1. **Safe** and economical disposal of urban waste.
2. Supply of electrical and thermal energy to the consumers in the city.
3. Environmental protection from urban waste.

Fig. 6.10. shows the *energy routes of urban waste to energy.*

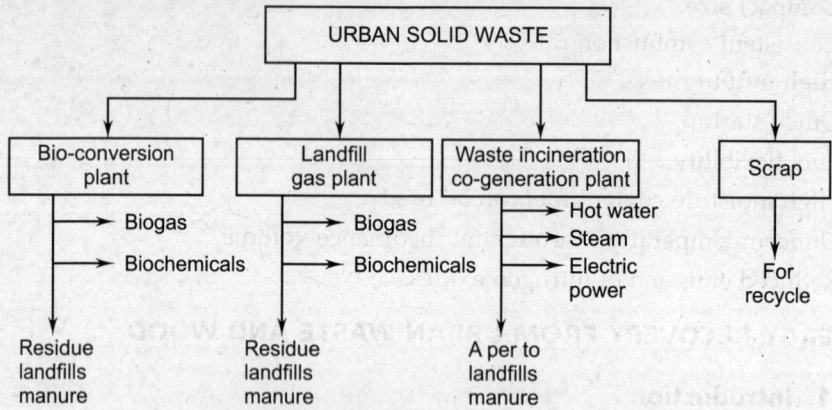

Fig. 6.10. Energy routes of urban waste to energy.

6.9.3. Waste-to-Energy Incineration Process

The *energy route* of the waste-to-electrical energy by *incineration process* is given below:

Biomass energy → **Thermal energy** → **Electrical energy** → **Electrical energy**
  (from *nature*)   (from *incinerator*)   (from *generator*)   (*to users* or
                                                                                  grid users)

The *"inceration process"* accepts a wide variety of *biomass inputs* including :

1. Semi-dried wood, trees, tree residues, wood chips, saw-dust.
2. Semi-dried garbage (urban waste).
3. Semi-dried farm waste (dried cow-dung, straw, sugar, bagasse, etc.).
4. Mixtures of fossil fuels and biomass for higher heat content of the infeed.
5. Steam is supplied to steam-turbine power plant (50 to 150 MW).
6. Heat (hot water) is supplied for district heating in cold countries.
7. Steam is supplied to process industry.

"Waste incineration power plant" is usually *located near the source of waste.* Table below gives the *locations of "Waste-to-Power Plants"* :

S.No.	In Feed	Location of Plant	Output
1.	*Forest produce* (*i*) Trees, tree residue (*ii*) Wood (*iii*) Wood waste	Forest Near furniture industry	Electric power Heat/steam for furniture industries
2.	*Sugar, bagasse*	Near sugar producing plants	(*i*) Electric power (*ii*) Heat, steam for sugar plant
3.	*Urban waste*	In a large city	(*i*) Electric power (*ii*) Heat and steam for urban consumers.

6.9.4. Waste Incineration Energy Plant

The schematic diagram of *'Waste-to-Energy plant' for urban waste incineration* is shown in Fig. 6.11 which is self explanatory.

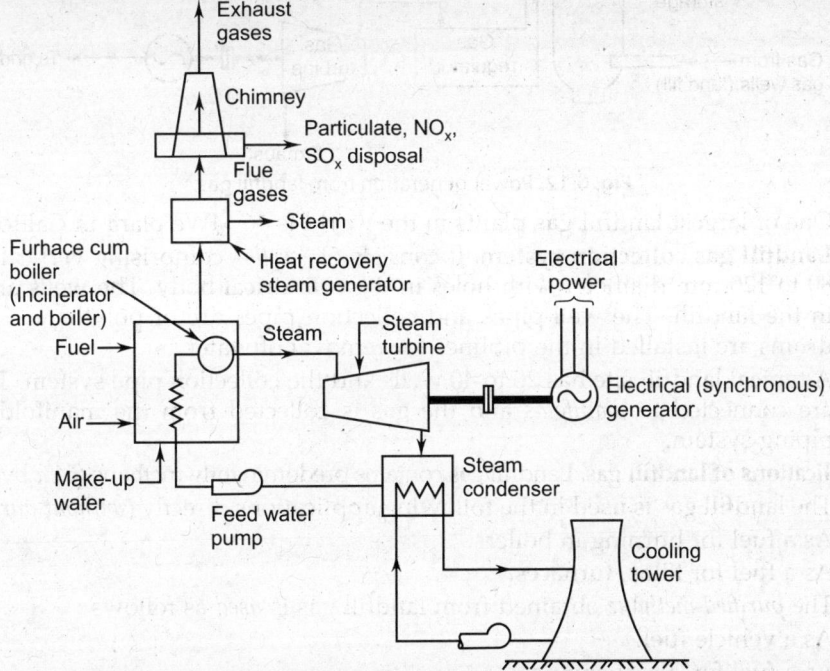

Fig. 6.11. Waste-to-Energy plant for urban waste incineration.

Processing of wood and wood-waste (fuel) for feeding to the incineration plant :

The incineration plant is *usually located in the forest and near saw mill*. This *reduces* the expenditure of transportation of wood and makes it competitive as a fuel for *producing electricity.*

The various *steps involved in the process* are :

   (*i*) Felling of trees in the forest.

  (*ii*) Segregating logs, tree barks, leaves etc.

 (*iii*) Transporting the logs and other residue to central store.

 (*iv*) Storing the logs in a circular store.

  (*v*) Drying of wood in the circular store.

 (*vi*) Collecting dried wood by means of central crane in the circular store and transporting the wood to power plant for incineration.

 (*vii*) Shredding (making smaller pieces).

(*viii*) Feeding to the furnace.

6.10. POWER GENERATION FROM LANDFILL GAS

Refer to Fig. 6.12. In this method of power generation, a large pit at the outskirt is prepared and a pipe system for gas collection is laid down before the waste is filled. Municipal solid waste is then buried for *anaerobic digestion*. Landfill gas is extracted after 2 or 3 months (depending upon the climate) by inserting perforated pipes into landfill. The gas is then stored in a storage tank. The gas after passing through gas regulator, runs gas turbine which in turn runs a generator, producing electrical power.

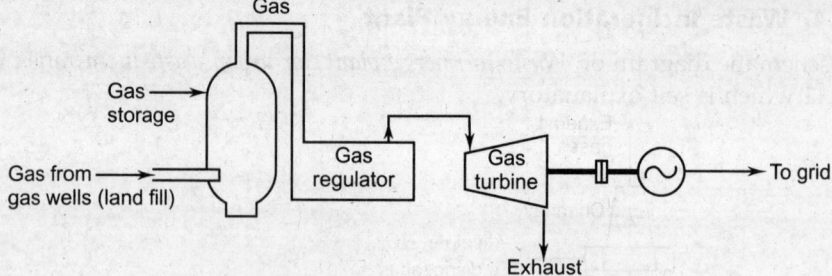

Fig. 6.12. Power generation from landfill gas.

One of largest landfill gas plants in the world is 46 MWe plant in California.

Landfill gas collection system. It consists of *"wells"* comprising vertical pipes of 80 to 120 mm diameter with holes in the cylindrical body. The wells are driven in the landfill. The well-pipes and collection pipes are of polythene. Knockout drums are installed in the piplines for removal of water.

A typical landfill site has 20 to 40 wells and the collection pipe system. The wells are connected to *manifolds* and the gas is collected from the manifolds by the piping system.

Applications of landfill gas. Landfill gas contains predominantly methane (54% by volume).

● The landfill gas is used in the following applications directly *(without purification)*:

(*i*) As a fuel for burning in boilers.

(*ii*) As a fuel for kilns, furnaces.

● The *purified methane* obtained from landfill gas is *used* as follows :

(*i*) As a vehicle fuel.

(*ii*) As a fuel for diesel engines.

(*iii*) As a fuel for diesel engine, to produce electrical energy.

(*iv*) After upgrading, supplied as fuel gas to domestic customers.

Fig. 6.13. shows the applications of Landfill Gas (LFG).

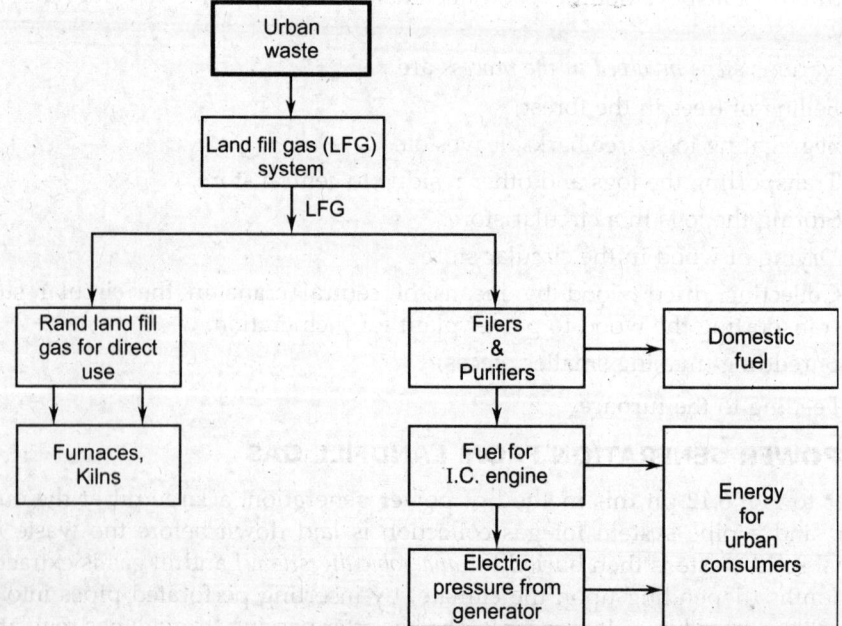

Fig. 6.13. Applications of Landfill Gas (LFG).

6.11. POWER GENERATION FROM LIQUID WASTE

Liquid waste may be of following types :

1. Sewage;
2. Distillary waste;
3. Pulp and paper mill black liquor waste.

1. Sewage:

Anaerobic digestion is used to produce gas for extracting energy from sewage.

Anjana sewage treatment plant (at Surat) has *three sludge digesters* with a total capacity of 82.5 million litres per day; each digester generates about 2500 m³ biogas daily. A scrubber system is employed to clean the gas, to make it suitable for use in a 100 percent biogas engine to generate electricity.

2. Distillary waste:

Fig. 6.14 shows a schematic diagram of power generation from distillary liquid waste (carrying rich raw material for producing gas).

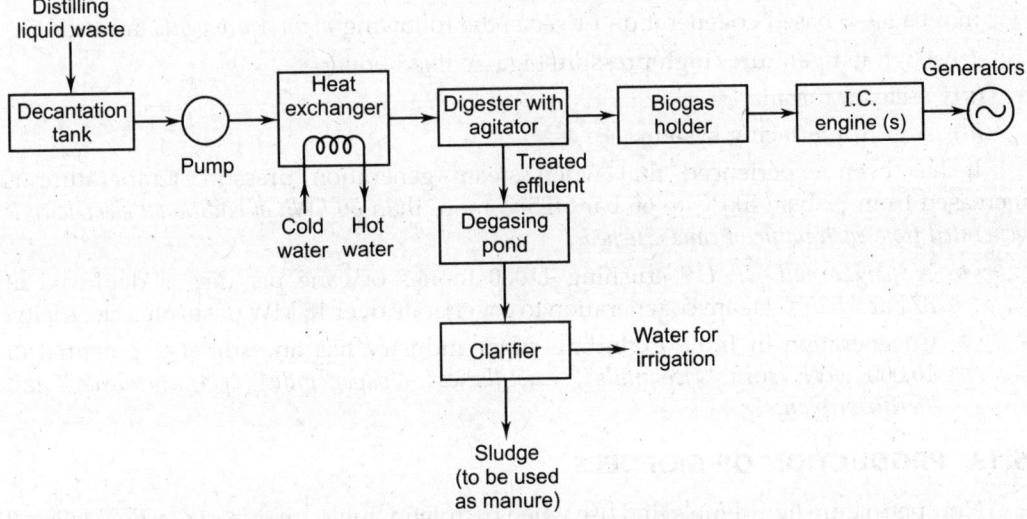

Fig. 6.14. Power generation from distillary liquid waste.

— Distillary liquid waste is collected in a *'decantation tank'* where the suspended solids settle down. Decanted liquid/effluent containing fermented molasses is pumped into a *'digester'* through a *'heat exchanger'*, where it (effluent) is cooled to maintain the digester temperature at 36-38°C.

— Effluent is than allowed to be digested anaerobically for about 12-15 days, during which gas is produced. The biogas accumulates in a biogas holder and is stored under pressure using a pressure control device. The gas is used to run the I.C. engine(s) which in turn generate electrical power through generator(s).

3. Pulp and paper mill black liquor waste:

A large amount of energy and water is consumed by the pulp and paper industry. The waste discharge water contains *compounds of wood and raw material* which are useful for recovery of energy. A plant for *biomethanation of bigasse wash effluent* is installed at Karur (in Tamil Nadu), based on USAB technology. Biochemical Oxygen Demand (BOD) and Chemical Oxygen Demand (COD) removal is 94%

and 89% respectively, with gas production of 0.37 m³/kg. At present, about 1500 m³ of gas is generated per day, which is used in a lime-mud reburning kiln. The gas output from the plant meets 50% heat load of the kiln, equivalent to about 12.5×10^3 litres of furnace oil.

6.12. BIOMASS COGENERATION

Cogeneration *means sequential conversion of energy contained in fuel into two usable forms, typically 'mechanical energy' and 'thermal energy'.*

— *'Mechanical energy'* is used to drive an alternator for *producing electricity.*

— *'Thermal energy'* can be used either for *direct applications* or for *producing steam.*

Cogeneration claims the following *advantages* :

(*i*) Generates power for in-house consumption.

(*ii*) Earns additional revenue from the sale of surplus electricity.

(*iii*) Cleans up the environment.

In India, sugar industry use *bagasse-based cogeneration* for achieving *self-efficiency in steam and electricity.*

In a bagasse-based cogeneration projects, the following *main equipments* are used :

(*i*) High temperature/high pressure bigasse fixed boilers.

(*ii*) A steam turbine.

(*iii*) A grid-interfacing system.

It has been experienced that when steam generation pressure/temperature is increased from 32 bar/400°C to 66 bar/485°C more than *80 kWh of additional electricity is generated from each tonne of cane crushed.*

- A *"sugar mill" in UP* crushing 11000 tonnes of cane per day is deployed at 87 bar / 525°C steam cogeneration to cogenerate over 18 MW of surplus electricity.

- Cogeneration in India excluding sugar industry has an estimated potential of 10,000 MW from *"rice mills", "distillaries", "paper mills", petrochemicals"* and *"fertilizer plants".*

6.13. PRODUCTION OF BIOFUELS

Non-petroleum liquid fuels find use when petroleum fuels are *scarce or costly. Methanol* (CH_3OH), *Ethanol* (CH_5OH) and *Biodiesel* are used as *fuels in I.C. engines.*

- **Ethanol** (or ethyl alcohol) can be produced by *"fermentation of carbohydrates"* which occur naturally and abundantly in some plants like *sugarcane* and *starchy materials* like *corn and* and *potatoes.*

- **Methanol** can be produced from *municipal solid wastes* and *specially grown biomass.*

- **Biodiesel** is produced from *non-edible oil seeds.*

- **Producer gas** is obtained by *partial combustion of wood or any cellulose organic material of plant origin.*

6.13.1. Production of Ethanol

It is one of the *most exotic organic chemicals* used as :

(*i*) Solvent, (*ii*) Germicide, (*iii*) Alcoholic beverage, (*iv*) Antifreeze, (*v*) Fuel, (*vi*) Depressent and as a chemical intermediate for other organic chemicals, and (*vii*) *Fuel for automobiles, used as additive to petrol.*

Ethanol can be *derived by fermentation of material containing sugar:*

Sugars : Cane sugar, sugar beet, fruits, mollases, grapes.

Starches : Grains, potatoes, root crops.

Cellulose : Wood, grasses, crop residue etc.

The *reactions* are as follows :

$$C_6H_{10}O_5 + H_2O \xrightarrow{\text{Enzyme}} C_6H_{12}O_6$$
$$\text{(starch)} \qquad\qquad\qquad \text{(Glucose)}$$

$$C_6H_{12}O_6 \xrightarrow{\text{Yeast}} 2C_2H_5OH + 2CO_2$$
$$\text{(Glucose)} \qquad\qquad \text{(Ethanol)}$$

Fermentation temperature 20° to 30°C process completed in 50 hours, yields 90%. Alcohol content is 10 to 20% depending upon alcohol tolerance of yeast. This is increased by distillation.

Ethanol from sugarcane :

Fig. 6.15. shows the schematic diagram of ethanol production from sugarcane :

— The cane is cut and ground and cane juice is extracted.

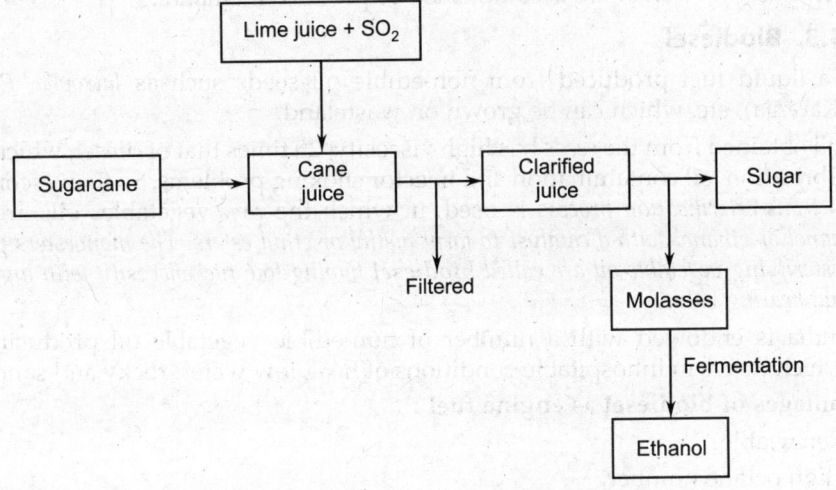

Fig. 6.15. Ethanol production from sugarcane.

— After clarification and concentration through evaporation, the juice is *fermented* with *yeast* to produce *raw ethanol*. A series of distillation steps, including a final extractive distillation with benzene, are used to obtain anhydrous ethanol.

● Ethanol yield is 6 times higher if sugarcane juice is directly fermented instead of molasses. One tonne of sugarcane with sugar content of 13% yields about 70 litres of ethanol through direct fermentation of juice. Sugar content in molasses is only 2%.

● *Ethanol* being a *high octane fuel* raises the octane rating of the mixture (Octane rating is the quality of the fuel to *increase its antiknocking property*).

6.13.2. Production of Methanol

Methanol can be produced from *municipal solid wastes* and *specially grown biomass*.

Fig. 6.16 shows schematic diagram of methanol production from municipal solid wastes:

— The wastes are *shredded* and passed through *magnets to remove iron*.

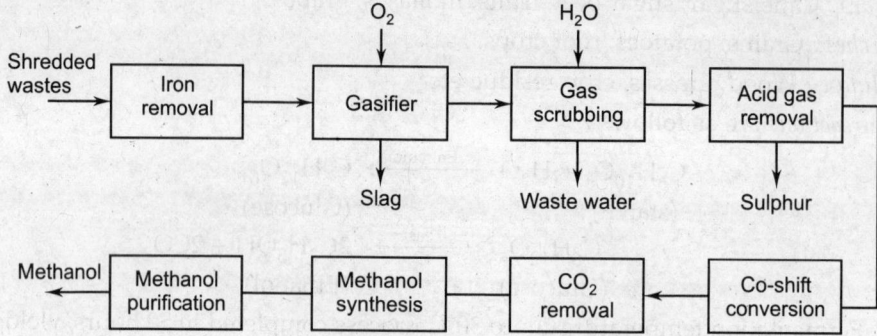

Fig. 6.16. Schematic diagram of methanol production from municipal solid wastes.

— The iron-free wastes are *gasified* with O_2.

— The product synthesis gas is *scrubbed* by water to remove particulates, entrained oil, H_2S and CO_2

— *Co-shift conversion* for $H_2/CO_2/CO$ ratio adjustment, *methanol synthesis* and *methanol purification* are accomplished to produce methanol.

6.13.3. Biodiesel

It is a liquid fuel produced from non-edible oil seeds such as *Jatropha*, *Pongamia pinnata* (Karanja), etc. which can be grown on wasteland.

The oil obtained from the *seeds* has high viscosity, 20 times that of diesel, which causes serious lubrication, oil contamination and injector choking problems. So to overcome such problems *trans-esterification process* is used, in which the raw vegetables oils are *treated with methanol or ethanol with a catalyst to form methy or ethyl esters. The monoesters produced by trans-esterifying vegetable oil are called **biodiesel** having low fuel viscosity with high octane number and heating value.*

— India is endowed with a number of non-edible vegetable oil producing trees which thrive in inhospitable conditions of heat, low water, rocky and sandy soils.

Advantages of biodiesel as engine fuel :

1. Renewable.
2. High octane number.
3. Biodegradable and produces 80% less CO_2 and 100% less SO_2 emissions.
4. Higher flash point (making it safe to transport).
5. Can be used as heat fuel (100% biodiesel) or mixed in any ratio with petrol-diesel.

6.13.4. Producer Gas

● It is *obtained by partial combustion of wood or any cellulose organic material of plant origin.*

● It is a mixture of a few gases (CO_2, CH_4, H_2, CO, N_2), *hydrogen and methane* keep heating value between 4.5MJ/m^3 and 6MJ/m^3, depending upon the volume of its constituents.

● It can be *burnt in a boiler to generate steam.*

● It is used as fuel in '*I.C. engines*' for *irrigation pumps* and '*gas turbines*' for *power generation.*

6.14. SUMMARY OF BIOMASS ENERGY CONVERSION PROCESSES

Summary of biomass energy conversion processes is given in the table 6.1.

Table 6.1 : Summary of biomass energy conversion processes.

S. No.	Process	Input feedstock	Conversion temperature	Conversion pressure	Characteristics of process	Product form	Process yield (% of original mass)
1.	Anaerobic fermentation	Aqueous slurry (30–20% solids)	20% to 50° C	Atmospheric	Fermentation of wastes or algae grown on waste of energy crops.	50 to 70% methane, remainder CO_2 (Biogas)	20 to 26%
2.	Biophotolysis	Aqueous slurry for algae, bacteria and/ or protein-enzyme complexes	20° to 50°C	Atmospheric	Sunlight produces intracellular enzymatic reduction of H_2O.	Hydrogen	
3.	Acid hydrolysis	5% acidified slurry (H_2SO_4 with cellulose)	20° to 50°C	Atmospheric	Glucose fermented to ethyl alcohol. Cellulose hydrolized to glucose.	Ethyl alcohol	
4.	Enzyme hydrolysis	Aqueous slurry (cellulose-rich)	20° to 50°C	Atmospheric	Extracellular enzymatic conversion of cellulose to sugar to alcohol.	Ethyl alcohol	90%
5.	Combustion	Dried feedstock (10% to 25% H_2O)	1200° to 1300°C	Atmospheric	Augments (i.e. 5 to 20%) boiler fuel (i.e., coal, oil or gas).	Heat, steam can be converted to electricity.	
6.	Pyrolysis	Dried feedstock	500° to 1300°C	Atmospheric	All of the gas and 1/3 of the char produced is used to supply heat in typical process. Oxygen-free environment used.	Oil Char gas	40% 20%
7.	Chemical reduction	Aqueous slurry (15% solids)	250° to 400°C	Atmospheric	Uses CO and H_2 3/8 of product oil, used by process.	Oil	23% (2 barrels/ ton)
8.	Hydro-glassification	Animal manure (other wastes can also be used)	550°C		Hydrogen atmosphere produced from manure. Purification and methanation of product gas required.	C_2H_6 (12%) CH_4 (42%) CO_2 (37%) or CH_4/C_2H_6	40%
9.	Catalytic gasification	Dried feedstock, mixed with alkali carbonate catalyst (12%–25% by wgt.)	650° to 750°C		Nickel catalysts used for second conversion step. Inert atmosphere required.	CO_2 CH_4 or CH_4 only	90%

6.15. BIOMASS ENERGY BIOPOWER SCENARIO IN INDIA

Biomass energy (Biopower) scenario in India, Potential vis-a-vis achievement (upto January 2009), is given in Table 6.2.

Table 6.2 : Biopower Scenario in India

S.No.	Resource	Estimated potential (MW)	Achievement upto Jan. 2009 (MW)
1.	*Biopower* (wood biomass)	52,000	683
2.	*Waste-to-energy*		
	(*i*) Grid-interactive power	5,000	34.95
	(*ii*) Distributed power	50,000	11.03
3.	*Biomass gasifiers*	—	87
4.	*Cogeneration bagasse*	5,000	1034
5.	*Family type biogas plants*	120 lakhs	39.8 lakhs

6.16. ENERGY PLANTATION

When land plants are grown purposely for their fuel value, by capturing solar radiation in is called **Energy plantation.** '*Energy plantations*' by design are managed and operated to provide substantial amounts of unusable fuel continuously throughout the year at the costs competitive with other fuels. Annual plants, typical of important farm crops, are unsuitable for providing a year-around supply of fuel.

Biomass energy concepts under study are resulting in the cultivation of large forests in areas not suitable for food production. Energy plantation may yield 10 to 20 tonnes per acre per year. The energy plantation would be perhaps 125-500 m² in land area. The *trees are to be harvested by automated means, then chipped and pulverised for burning in a power plant that would be located in the middle of the energy plantation.*

The properties of plants to be cultivated in India are given in table 6.3.

Table 6.3 : Common species for Energy Plantation.

S.No.	Species	Yield (m³/ha/yr)	Calorific value (MJ/kg)
1.	*Acavia auriculiformis*	10–12	19–20
2.	*Casuarina equisilifolia*	7–10	20
3.	*Eucalyptus camaldulensin*	7–10	19
4.	*Leucaena lencocephala*	30–40	17–19
5.	*Albizia lebbek*	5	22

- Some schemes envision '*acquatic farms*' growing *algae, tropical grasses, floating kelp, water hyacinth and others.* These could *yield several hundred tonnes/acre year in controlled environments.*

- An interesting idea may be *to use hot condenser cooling water from a power plant to grow algae in large quantities or increase the yield of other crops.*

- Fast growing trees, sugar, starch and oil containing plants can also be cultivated for bio-energy.

Advantages of energy plantation :

1. *No storage losses* vis-a-vis harnessing solar energy (Plants, as they grow, serve as their own energy accumulator).

2. *Wide flexibility*, if the crop maturity cycle extends over years.

3. Under optimised conditions, the cost of energy obtained through plantation works out to be *lesser* than present fossil fuel cost on heat value basis as obtained by direct burning of wood.

4. The problem of SO_2 *pollution by combustion of biomass* is practically *nil*.

5. The ash obtained after burning vegetable matter is *rich in plant nutrient-minerals*, unlike ash of fossil fuels, and can be used as a *manure*.

6. Energy plantation will convert large tracks of semi-barren land into green belt and thus *ecological conditions can be* restored.

7. Growth of biomass *consumes CO_2 as much (as produced by consumption of biomass, and as such the atmosphere is not polluted*. Combustion of fossil fuel releases CO_2 in very large amounts and also they consume huge amounts of O_2 from the atmosphere).

THEORETICAL QUESTIONS

1. What is biomass?
2. What are the biomass resources?
3. Discuss briefly about "Availability of biomass".
4. What are the limitations of utilising biomas'?
5. State the environmental effects/benefits of biomass/biofuels.
6. Define the term 'Photosynthesis'. What are the conditions necessary for photosynthesis process?
7. Enumerate the processes which are used for biomas conversion.
8. What is the difference between 'Combustion' and 'Incineration'?
9. Explain briefly the densification process of biomass conversion.
10. What is 'Thermo-chemical conversion'? Explain briefly.
11. Explain briefly any two of the following processes :
    (i) Gasification; (ii) Liquefication; (iii) Pyrolysis.
12. What is biochemical conversion?
13. What is 'Anaerobic digestion'? Explain briefly.
14. What is 'Fermentation'? Explain
15. List the various sources for production of biogas.
16. Discuss the applications of biogas.
17. What are the advantages of using biogas in engines?
18. Explain with a schematic diagram the "Anaerobic digestion system'.
19. Give the advantages of 'anaerobic digestion'.
20. List the advantages of biogas production.
21. What are the factors which affect generation of biogas? Explain briefly.
22. What is the function of a biogas plant?
23. List the raw materials used in biogas plant.
24. Explain briefly the components of a biogas plant.
25. How are biogas plants classified?
26. Give the description of a two-stage continuous biogas plant.
27. States the advantages of a continuous type biogas plant.
28. What are the salient features of a batch type biogas plant?
29. Explain briefly, with a next diagram, the construction and working of floating drum type biogas plant (KVIC model). State also its advantages and disadvantages.

30. Explain with a neat sketch the construction and working of Janta model biogas plant. States also its advantages and disadvantages.

31. Describe with a neat diagram the 'Deenbandhu' biogas plant (DBP)'. State its advantages.

32. Give the comparison bewteen "Fixed dome type and 'Movable drum type' biogas plants.

33. List the methods/techniques by which biogas production can be maintained.

34. Explain with a neat diagram a 'Community night-soil based biogas plant'.

35. Give the guidelines for fixing optimum size of a biogas plant.

36. Which factors should be considered while selecting a site for biogas plant? Explain briefly.

37. What are biogas plants not much successful in India?

38. What are the fuel properties of biogas?

39. Explain briefly the methods used to obtain energy from biomass.

40. What do you understand by the term 'Gasification'?

41. What is a gasifier? What are its advantages?

42. How are gasifiers classified?

43. Explain with a neat diagram the working of a 'downdraft gasifier'.

44. Describe with a neat sketch the 'Fluidised gasifier'.

45. What are the advantages of a fluidised gasifier?

46. Give a schematic diagram of energy routes of urban waste to energy.

47. Explain with a neat diagram the working of waste-to-energy plant for urban waste incineration.

48. Explain briefly the following :
    (i) Power generation from Landfill gas;
    (ii) Power generation from liquid waste.

49. What is "Biomass cogeneration"? Explain briefly.

50. Explain briefly the production of following biofuels?
    (i) Ethanol; (ii) Methanol.

51. Write a short note on 'Biodiesel'?

52. Explain briefly 'Energy plantation'. State also its advantages.

GEOTHERMAL ENERGY

7.1. Introduction; 7.2. Important aspects of geothermal energy; 7.3. Structure of earth's interior; 7.4. Energy of earth – Heat flux; 7.5. Geothermal system – hot spring structure; 7.6. Earthquakes and volcanoes; 7.7. Geothermal gradients; 7.8. Geothermal resources – Hydrothermal resources – Geopressured resources – Petro-thermal systems or hot dry rocks (HDR) resources – Magma resources (Molten–rock–chamber systems) – Hybrid geothermal fossil systems; 7.9. Advantages and disadvantages of geothermal energy over other energy forms; 7.10 Applications of geothermal energy; 7.11. Environmental problems; 7.12. Geothermal energy in India and abroad. *Theoretical Questions.*

7.1 INTRODUCTION

Geothermal energy is *primarily heat energy from earth's own interior.* The word "Geothermal" comes from the Greek words 'geo' meaning earth and 'thermal' meaning heat.

- It is classified as **renewable** *because the earth's interior is and will coutinue in the process of cooling for the indefinite future.* Hence, geothermal energy from the earth's interior is *almost inexhaustible* as solar or wind energy, so long as its sources are actively sought and economically tapped.

- As we travel down earth's surface radially, there exists a temperature gradient of *0.03°C per metre.* Thus a 30°C increase in temperature can be obtained per kilometre depth from the earth crust. There are *many local hot spots just below the surface where the temperatures are much higher than expected.* Ground water, when comes into *contact* with hot spots, *either dry or wet steam is formed.* By *drilling holes* to these locations, hot water and steam can be tapped and these can be *used for power generation or space heating.*

- Geothermal energy is present over the entire earth's surface except that it is *nearer to the surface in the 'volcanic areas'.*

- Heat transfer from the earth's interior is by *three primary means :*
    1. Direct heat conduction,
    2. Rapid injection of *ballastic magma* along natural rifts penetrating deep into earth's mantles.
    3. Bubble-like magma that buoys upwards towards the surface.

7.2 IMPORTANT ASPECTS OF GEOTHERMAL ENERGY

Following are the *important aspects* about geothermal energy:

1. **Form of energy:**
   — *Thermal energy'* in the form of *hot water, steam, geothermal brine, mixture of these fluids.*

2. **Availability:**
   — Generally *available* deep inside the earth at a depth *more than about 80 km.* Hence, generally not possible to extract.
   — In a few locations in the world, *deposits are at depths* of *300 to 3000 m.* Such *locations are called Geothermal fields.*

3. **Method of extraction:**
   — *Deep production wells* are drilled in the geothermal fields. The hot steam/ water/brine is *extracted* from the geothermal deposits by *production wells*, by *'pumping'* or by *'natural pressure'*.

4. **Geothermal fluids:**
   — Hot water;
   — Hot brine;
   — Wet steam;
   — Mixture of above.

5. **Range of geothermal power plants installed capacity:**
   — 5 MW to 400 MW.

6. **Average geothermal gradient:**
   — 30°C per 1000 m length.

7. **Geothermal energy released through earth's crust:**
   — About 0.06 W/m^2.

8. **Total geothermal reserves in the earth:**
   — 4×10^{24} MJ(Estimated).

9. **Renewable energy deposits available for use in upper 3 km zone:**
   — 4×10^{15} MJ (Estimated).

10. **Rate at which the renewable energy can be tapped for production of electricity:**
    — 2×10^{12} to 10×10^{12} MJ/year (Estimated).

11. **Types of geothermal energy deposits:**
    — Hydrothermal: Hot water and steam, hot brine.
    — Petrothermal: Hot dry rock (HDR).

7.3 STRUCTURE OF EARTH'S INTERIOR

The earth consists of the following *parts (concentric shells)*:

1. Crust;   2. Mantle;   3. Core.

1. **Crust.** It is the uppermost shell of earth that extends to variable depths below mountains, continents and oceans. The thickness of crust is believed to be 0.90 km and several substances like limestone, coal, gold, petroleum etc. are found in the crust.

2. **Mantle.** It is the second concentric shell of earth that lies below the crust. The upper *rigid* part of the mantle extends up to 100 km below the separating crust and contains mainly iron and magnesium. The crust and upper mantle form the 'lithosphere'.

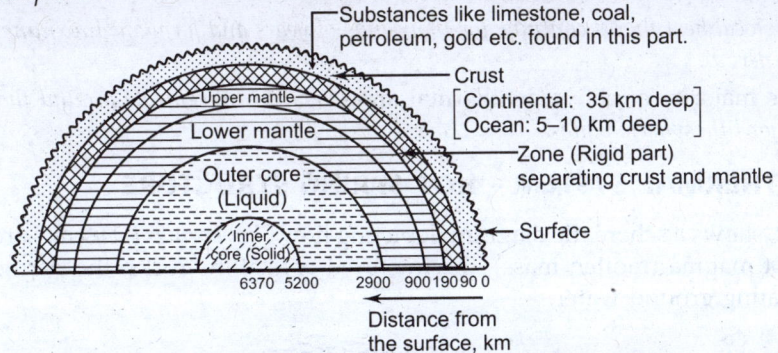

Fig. 7.1. Structure of eath's interior.

The lower mantle extending up to 2900 km below the earth's surface is *less rigid* and is *hotter*. This is known as 'asthenosphere' and is capable of being deformed.

The movement of lithosphere over the asthenosphere results in the 'phenomenon of plate tectonics' i.e. movement of the earth's crust.

3. **Core.** It is the innermost concentric shell of the earth. The core boundary begins at a depth of 2900 km from the surface and extends to the centre of the earth at 6370 km. This layer is further subdivided into *outer core* and *inner core*. The outer core comprises the region from a depth of 2900 km to 5200 km below the earth's surface and behaves mere like a *liquid*. The inner core with a thickness of around 1170 km is believed to be a *solid* metallic body, containing nickel-iron alloy.

- *The hot molten rock of the mantle is called* **Magma.**

7.4 ENERGY OF EARTH – HEAT FLUX

Earth is in a state of *thermiol equilibrium*. The *energy received from sun is lost at night*. The small amount of energy generated by the decay of unstable isotopes of uranium, thorium etc. is *dissipated* from earth's interior to oceans and atmosphere.

The heat generated within earth is around 2700 GW.

- *The heat energy in earth's interior is due to* **radioactivity**. Regions of higher radioactivity have *higher heat flux* and are *potential geothermal sites*.

- Earth's surface consists of about one dozen **tectonic plates** (*e.g.*, American plate, Arabian plate, Indian plate, Philippine plate, Pacific plate etc.). Each of these plates has *thickness around 100 km* and *thousands of kilometres area*. Earth's interior is *unable to lose heat, by conduction, as rapidly as it is generated by 'radioactivity'*. This leads to "*convective instabilities*" which means that these *plates are continuously in motion with respect to each other*. A variety of processes along the margins of the plates lead to *partial melting* at depths between 15 and 200 km. The molten masses penetrate the surrounding rocks and rise towards earth at rates varying from a few cms per day to a few cms per year, thus resulting in **volcanic activity**. The molten masses which *do not reach earth's surface* come to *rest in the middle or upper part of the earth's crust at depths less than 20 km*. These **liquid magmas** may have temperatures around 1000°C.

The crystallisation of these liquid magmas produces *intrusive igneous bodies*. The cooling and crystallisation of igneous bodies give rise to *local heat flux*. This **heat flux** *constitutes the geothermal energy* which may be used for a variety of purposes including generation of electricity.

The local heat fluxes continue for thousands of years and form an 'inexhaustible source of energy'.

- The majority of active geothermal areas *tend to concentrate around the margins of major lithospheric plates.*

7.5 GEOTHERMAL SYSTEM – HOT SPRING STRUCTURE

Fig. 7.2, shows a schematic diagram depicting how *"hot springs (geysers)"* are produced through hot magma (molten mass), the fractured crystalline rocks, the permeable rocks and percolating ground water.

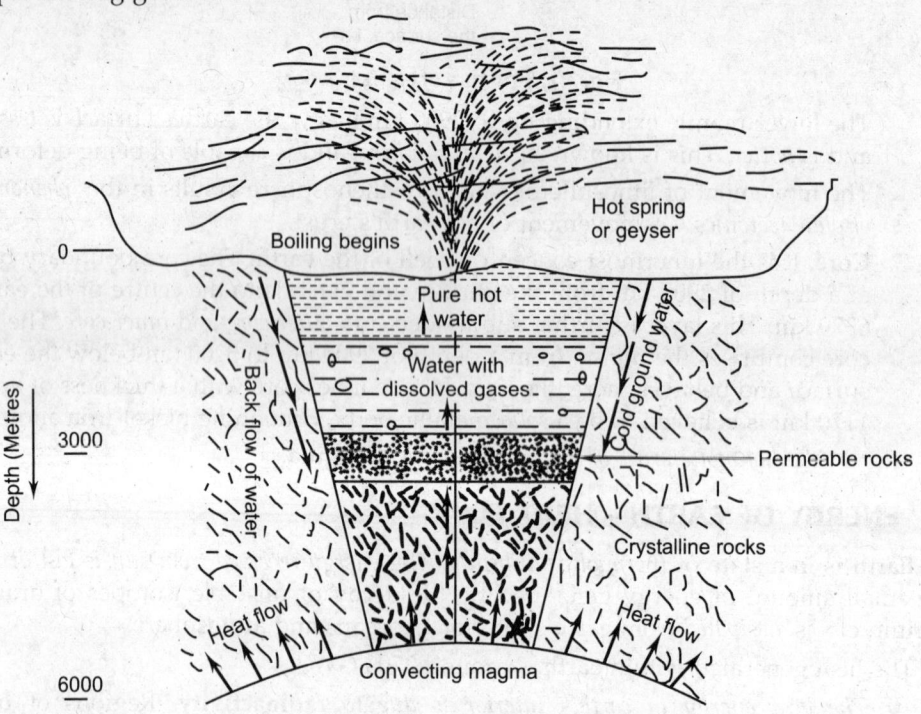

Fig. 7.2. Geothermal system – hot spring system structure.

At a depth of 5000 m or so lies an *impermeable magma*. Above the magma are the *'impermeable rocks'* which are *overlain by localised pockets of 'permeable rocks'*. One such localised pocket is shown in this figure. The localised pockets are bounded by *fracture zones or faults* along which some relative motion of rocks has occurred. *Water circulates along the fault lines.* As it goes down and moves in earth's interior it is *heated by the permeable layer which is in turn heated by conduction of heat from the magma.* The hot water comes out through another fault and forms a hot spring.

7.6 EARTHQUAKES AND VOLCANOES

Geothermal systems/fields need a *combination of* the following *three geological conditions:*

1. A natural underground source of water,
2. An impermeable layer that traps water and allows formation of steam, and
3. A large mass of hot rock in vicinity of water system.

Earthquakes and **volcanoes** *are largely located in the "plate boundaries".*

Most of the world's volcanic activities and geothermal sites are located in the *circum-pacific belt* known as **rim of fire**. It starts from New Zealand, encompasses Philippines, Japan, West Coasts of North America & Mexico; another belt runs from Iceland touching the British Isles, through Azores across the Atlantic to the West Indies, with a branch running through the Mediterranean Sea.

7.7 GEOTHERMAL GRADIENTS

The temperature difference within the earth depends on :

1. The thermal properties of earth's interior and their radial and lateral variation.
2. Movement of fluids or solid rock materials occurring at rates of more than a few millimetres per year.

A potential geothermal source region should have *high thermal gradient*.

Thermal gradient is *defined as the ratio of heat flux* and *thermal conductivity*.

i.e., Thermal gradient $= \dfrac{\text{Heat flux}}{\text{Thermal conductivity}}$

Fig. 7.3 shows the Geothermal gradients. The figures are based on measurements within a few km of earth's surface.

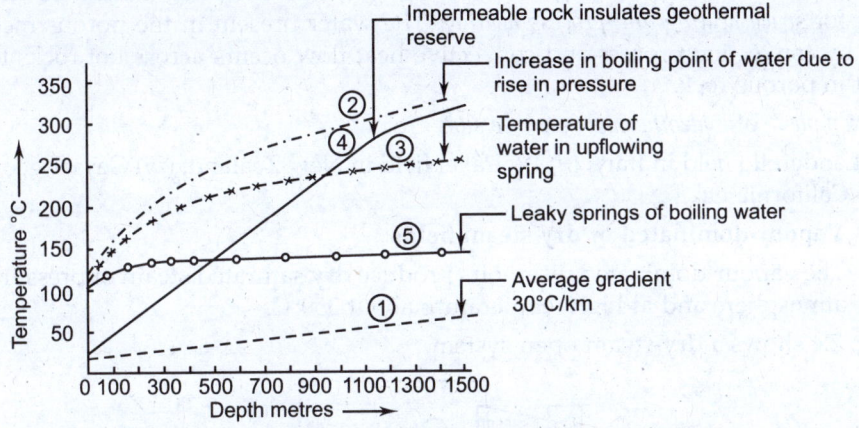

Fig. 7.3. Geothermal gradients.

- **Curve 1:** It represents average uniform gradient.
- **Curve 2:** It represents theoretical *increase in the boiling point* of water at increasing depths *due to higher pressures,* allowing for reduced density due to higher temperatures.
- **Curve 3:** It represents thermal gradients of such regions in which water percolates through upper crust into lower hot region and hot water flows vigorously upwards forming 'hot springs'.
- **Curve 4:** It represents the effect of solid impermeable rock. The rock forms insulating cap on geothermal reserves and does not allow heat flow to upper part.

- **Curve 5:** It represents leaks in the solid impermeable rocks in the form of springs of hot boiling water discharged in large quantities to ground surface. In some locations production of steam occurs at lower depths and the steam is released to the surface in the form of *furmaroles* and *geysers* as shown in curve 5. Such locations are *very few in number.*

7.8 GEOTHERMAL RESOURCES

Geothermal resources are of following *five* types:

1. *Hydrothermal or hydro-geothermal energy resources:*
   (*i*) Vapour-dominated or dry steam fields;
   (*ii*) Liquid-dominated system or wet steam fields;
   (*iii*) Hot-water fields.
2. *Geopressured resources.*
3. *Petro-thermal systems* or hot dry rocks (HDR) resources.
4. *Magma resources (Molten-rock-chamber systems).*
   - The *"hydro-thermal convective systems"* are *best resources* for geothermal energy exploitation at present. 'Hot dry rock' is also considered.

7.8.1 Hydrothermal Resources

Refer to Fig. 7.2. In hydrothermal convective system water is heated by contact with hot rocks. These are wet reservoirs containing steam and hot water or only hot water. If the temperature is *high* enough then *'steam' generates electricity,* otherwise *'hot water'* is used for *space heating and process heating.* The water present in the porous medium is heated by convection process and convective heat flow occurs across hot rocks to water present in porous rock.

"Examples" of *hydrothermal resource sites* are :

(*i*) Landerello field in Italy; (*ii*) Wairakei field in New Zealand; (*iii*) Geyser geothermal field in California etc.

1. **Vapour-dominated or dry steam fields:**

   The vapour-dominated reservoirs produce dry saturated steam of pressure above atmosphere and at high temperature about 350°C.

Fig. 7.4 shows a dry-steam open system.

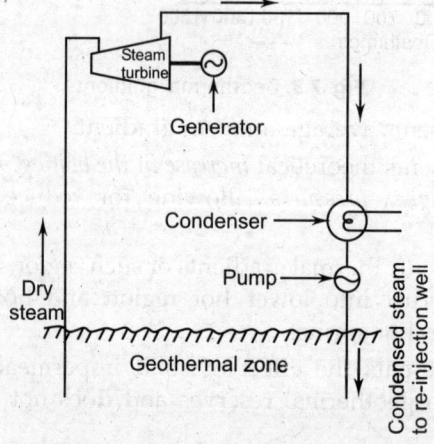

Fig. 7.4. Dry-steam open system.

— Steam extracted from the well is cleaned in *centrifugal separator* which removes solid matters.

— The cleaned steam is then supplied directly into the *'steam turbine'*. The exhaust steam from steam turbine is wet steam, (*i.e.,* mixture of water and steam) which passes through the *condenser*. The condenser condenses wet steam into water (through a cooling tower).

— The *non-condensable gases* present in wet steam are removed by *'steam jet injection method'*.

— The condensed steam is reinjected deep into the ground/well.

  ● This system is used in Landerallo (Italy) and Geyser (USA)

Environmental aspects:

● The steam from hydrothermal resources may contain 0.5 to 5% by weight of non-condensable gas (mainly CO, CH_4 and NH_3) which are *largely harmless* in the quantities present. Gases also contain H_2S (hydrogen sulphide) which is *harmful to plant and animal life.*

● The withdrawal of large amount of steam from the source may result in *surface subsistence,* mainly occurs in 'oilfields', is *dealt with by injecting water into the ground.* The re-injection of excess water is done at some distance from a ground fault.

2. **Liquid-dominated systems or wet steam fields:**

In such a system water temperature is above the normal boiling point (100°C). Due to the pressure inside the reservoir, water does not boil but remains in liquid state. When the water comes on the earth surface its *pressure reduces* resulting in *rapid boiling* and the liquid water *'flashes into a mixture of hot-water and steam'*. The *steam is separated from mixture and used to generate electricity.*

(a) **Liquid-dominated high temperature systems:**

For such systems, the following two methods are used :

  (*i*) The flash steam open system.

  (*ii*) The binary cycle system.

(i) **The flash steam open system:**

Fig. 7.5 shows a schematic diagram of flash steam open system.

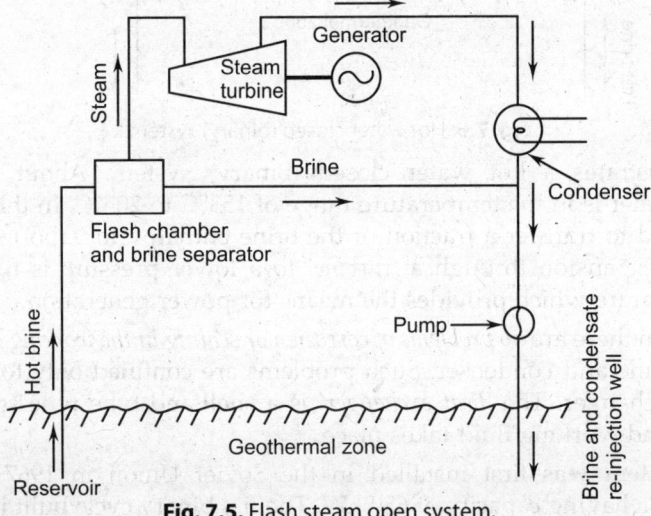

Fig. 7.5. Flash steam open system.

— Hot brine from the reservoir reaches the well head at lower pressure by *throttling process*. This low quality mixture is then throttled in *flash separator* which *improves the quality of mixture*. Now steam is separated as a dry saturated steam and supplied to the *'steam turbine'*, which *produces electric power* through a *'generator'*.

The *'power generation' from such system can be made more economical by associating chemical industry with power plant to make use of brine and gases effluent.*

- This system is used in Cerro Prietol Mexico, Otake (Japan).

  Limitations. Following are the *limitations* of flash steam open system as compared to vapour-dominated system:

  1. *Much larger total mass flow rates* through the well required.

  2. Owing to large amount of flows, there is a great degree of ground surface subsidence.

  3. A greater degree of *precipitation* of minerals from the brine results in the necessity for design of valves, pumps, separator internals, and other equipment for operation under *scaling conditions*.

(ii) The binary cycle system:

The binary cycle concept *isolates the steam turbine from corrosive or non-corrosive materials and/or to accommodate higher concentration of non-condensable gases.*

This is basically a **Rankine cycle** with an organic working fluid.

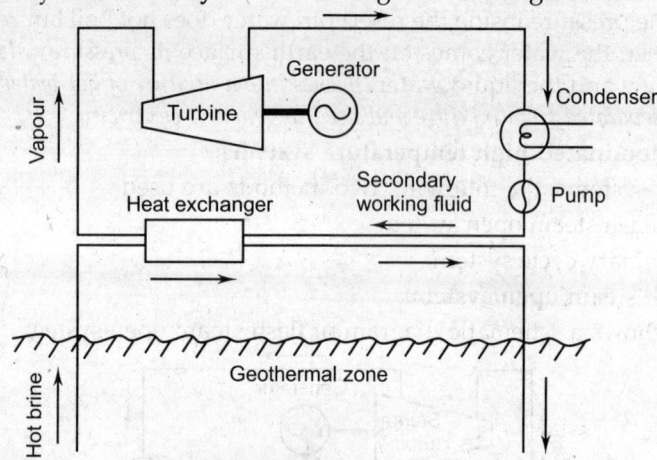

Fig. 7.6. Hot water closed (binary) system.

Fig. 7.6 illustrates a hot water closed (binary) system. About 50 per cent of hydrothermal water is in the temperature range of 153°C to 205°C. In this system, a *'heat exchanger'* is used to transfer a fraction of the brine enthalpy to vaporize the secondary working fluid. Expansion through a *'turbine'* to a lower pressure is fixed by the heat rejection temperature which provides the means for power generation.

In this system there are *no problems of corrosion or scaling in the working cycle components,* such as the turbine and condenser. Such problems are confined only to the well casing and the heat exchanger. The *'heat exchanger'* is a shell-and-tube type so that *no contact* between brine and working fluid takes place.

- This system was first installed in the Soviet Union in 1967 on Kamchatka Peninsula having capacity of 680 kW. The firs binary cycle built in U.S.A. is of 11 MW capacity in California and second one at Raft-river-Idalio, is of a capacity of 10 MW.

Total flow concept system:

In such a system, both *'kinetic energy'* and *'heat energy of the steam-liquid mixture, produced by flashing the geothermal brine, are utilised.* The *overall efficiency for conversion into 'eletrical energy'* should be *greater* than other methods (described earlier) in which *only the heat content of the brine is utilised.*

This system *utilises the principle of the 'Lysholm machine',* known in this connection as the *helical (or screw) expander* or *mixed phase expander.*

Fig. 7.7 shows the schematic diagram of a liquid-dominated total flow concept.

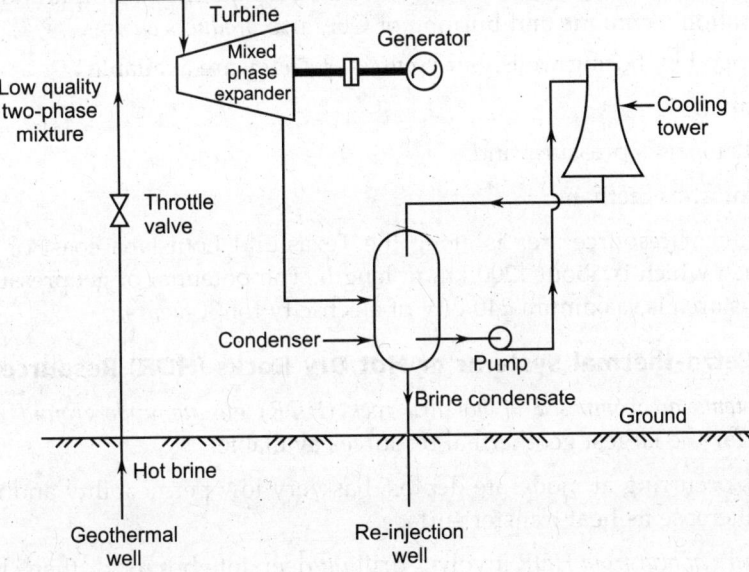

Fig. 7.7. Schematic diagram of a liquid-dominated total flow concept system.

The *hot brine* from geothermal well is throttled where it becomes a *two-phase mixture of low quality.* The two phases at this point are *not separated* and the *full flow* is expanded in the *'mixed phase expander' (turbine)* which is coupled with generator which generates *electrical power.* The mixture from the expander/turbine is discharged into condenser. Then the brine condensate is re-injected into the well.

Following are the *requirements* of mixed phase expanders:

(*i*) They should be able to *overcome the losses* associated with the impingement of liquid droplets on blades (the efficiency of the turbine decreases with the decrease in quality).

(*ii*) They must be able to *withstand the corrosive and erosive effects* of the significant quantities of dissolved solids in the brine.

(b) **Liquid-dominated low temperature systems (Geothermal fluids):**

The hydrothermal reservoirs of this system are available at moderate temperature range of 90°C to 175°C.

Due to low temperature, *little mineral water is extracted.* If there is danger of corrosion then it can be passed through a *heat exchanger* to transfer heat from the natural hot water.

The main *uses* of this system include the following:

(*i*) To provide heat for homes, commercial and agricultural buildings including greenhouses and animal shelters.

(*ii*) Hot water may also be used for air-conditioning and refrigeration.

7.8.2 Geopressured Resources

The geopressured resources contain moderate temperature brines (160°C) containing dissolved methane. These are trapped under high pressure (nearly 1000 bar) in a deep sedimentary formation sealed between impermeable layers of shale and clay at depths of 2 to 10 km.

At geopressure, dissolved methane gas is usually 1.9-3.8 m^3 per cubic metre of water. The methane gas is released from water on the earth surface because pressure at earth surface is lower. Therefore, methane gas is separated from brine by simple and economical gravity separation technique and burning of CH_4 also *produces energy*.

When tapped by boring wells, *three sources of energy* are available :

(*i*) Thermal;

(*ii*) Mechanical-as pressure; and

(*iii*) Chemical-as methane.

- The major resource area is along the Texas and Louisiana coast of the Gulf of Mexico which is about 1200 km in length. The potential of geopressured energy in this area is maximum 240 GW of electricity for 30 years.

7.8.3 Petro-thermal Systems or Hot Dry Rocks (HDR) Resources

These systems are composed of hot dry rock (HDR) but no underground water. They represent by far the largest geothermal resources available.

The rock, occurring at moderate depths, has very low permeability and needs to be fractured to increase its heat transfer surface.

The *recovery of heat from HDR* involves drilling deep into hot rocks, then cracking it to form cavity or fractures. This can be achieved by: (*i*) using high explosives at the bottom of the man-made well, (*ii*) using nuclear explosion, and (*iii*) by hydraulic fracturing (pumping water at high pressure into the rock).

The *thermal energy* of the HDR is *extracted by pumping water or fluid* through a well at the lower part of the fractured rock and *withdrawn by another well at a distance*. The temperature of the rock at a depth of 5 km is about 200°C. To achieve steady flow of high temperature water, the injection and extracting wells are joined to form a circulating loop. When heat is extracted through water, the rock cools down and due to temperature gradient between rocks, new cracks are developed.

- HRD technique is in operation near Valles Caldera, U.S.A., where fractures are made at a depth of 2.76 km and temperature at the location is about 185°C.

7.8.4 Magma Resources (Molten-rock-chamber Systems)

At some places, especially in the vicinity of relatively recent volcanic activity, molten or partially molten rock (*i.e. magma*) occur at a moderate depth (less than 5 km). The very high temperature above 650°C and the large volume make magma a substantial geothermal resources.

This resource has *not* been used yet due to the reason that the existing technology does not allow recovery of heat from these resources (Magma technology requires *special* manufacturing technology).

7.8.5 Hybrid Geothermal Fossil Systems

Such a system *utilises the relatively low-temperature heat of geothermal resources in the low temperature end of a conventional cycle and the high temperature heat from fossil-fuel combustion in the high temperature end of that cycle.*

Fig. 7.8 shows the schematic diagram of a geothermal **preheat hybrid system**, in which *low-temperature geothermal energy is used to heat feed water of a conventional fossil fuel system.*

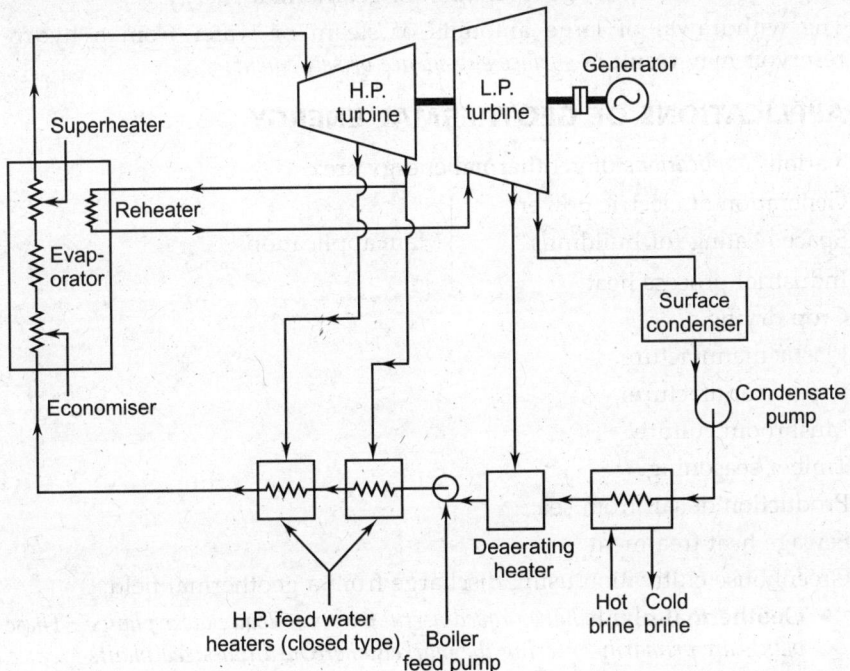

Fig. 7.8. Geothermal preheat hybrid with conventional plant.

In this system geothermal heat heats the feed water throughout the low-temperature steam end prior to an open-type deaerating heater. It is followed by a boiler feed pump and two closed-type feed water heaters with drains cascaded backward. These receive heat from steam bled from H.P. (high pressure) stages of turbine. *No steam is bled from L.P. (low pressure) stages because 'geothermal brine' fulfills this function.*

7.9 ADVANTAGES AND DISADVATAGES OF GEOTHERMAL ENERGY OVER OTHER ENERGY FORMS

Advantages of Geothermal Energy:

1. Geothermal energy is *cheaper*:
2. It is *versatile* in its use.
3. It is the *least polluting* as compared to other conventional energy sources.
4. It is amenable for *multiple uses* from a single resource.
5. Geothermal power plants have the *highest annual load factors* of 85 per cent to 90 per cent compared to 45 per cent to 50 per cent for fossil fuel plants.
6. It *delivers greater amount of net energy* from its system as compared to other alternative or conventional systems.

7. Geothermal energy from the earth's interior is *almost as inexhaustible as solar or wind energy*, so long as its sources are actively sought and economically tapped.

Disadvantages:

1. *Low overall power production efficiency* (about 15% as compared to 35 to 40% for fossil fuel plants).

2. Drilling operation is *noisy.*

3. *Large areas are needed* for exploitation of geothermal energy.

4. The withdrawal of large amounts of steam or water from a hydro-thermal reservoir may result in *surface subsidence or settlement.*

7.10 APPLICATIONS OF GEOTHERMAL ENERGY

The various *applications* of geothermal energy are:

1. Generation of electric power.
2. Space heating for buildings.   } Main applications
3. Industrial process heat
4. Crop drying.
5. Plastic manufacture.
6. Paper manufacture.
7. Mushroom culture.
8. Timber seasoning.
9. Production of salt from sea.
10. Sewage heat treatment
11. Greenhouse cultivation using discharge from a geothermal field.

- **Geothermal plants** *have proved useful for "base-load power plants". These kinds of plants are primarily entering the market where medium-sized plants are needed with low capital cost, short construction period and life-long fuel (i.e. geothermal heat).*

7.11 ENVIRONMENTAL PROBLEMS

Geothermal power plants *create some environmental problems* which are peculiar to them alone. The effluent will be salty and may contain sodium and potassium compounds. Additionally, in some cases lithium, fluorine, boron and arsenic compounds may be present. Such effluents cannot be discharged into the existing water courses unless properly treated without risking severe pollution problems. Some effluents contain boron, fluorine and arsenic. All these are *very harmful to plants and animal life in concentrations as low as two parts per million. Suitable waste treatment plants to prevent degradation of water quality will have to be installed to treat these new and increased sources of pollution.*

7.12 GEOTHERMAL ENERGY IN INDIA AND ABROAD

Some progress has been made in India on tapping geothermal energy on a commercial scale. Engineers from the Geological Survey of India have drilled about 50 shallow wells for steam in the Puga valley of the Ladakh region in Jammu and Kashmir. It may be possible to operate a 5 MW power station at the site. The Puga valley at an altitude of 4500 metres above sea level has the most promising geothermal field. The area extends to about 40 square kilometres out of which 5 sq. km is active. A combination of wet and dry steam to the tune of 170 tonnes of hot water per hour and 20 tonnes/hour of dry steam (superheated steam suitable for running steam turbines) is available. This is enough to run a small power station to light the homes of local population. The geothermal heat can also be used for space heating in the Puga valley as the temperature in this area, especially during winter months, goes down to 35 degrees below freezing point. There are no other energy sources in Ladakh region and coal, petroleum etc. have to be transported from Srinagar. It can also be used for poultry farming, mushroom cultivation and pashmina wool processing which need a warmer climate. In addition, there are good deposits of borax and sulphur in this area. Sulphur in elemental form is found only in this region in the whole of India.

There are many *hot water springs* in India. Hot water springs represent heat energy coming out of the earth from a large body of molten rock that has been pushed up into upper crust of the earth by geological forces. In North they occur in *Ladakh* and *Himachal Pradesh*. In western parts they are found in the *Cambay region of Gujarat and Maharashtra.* They are also found in the *Singhbhum region of Bihar* while there are some in *Assam.* The water from a hot spring at *Garampani, near Jawai, in Assam* is so hot in summer that rice kept in muslin bag gets worked in no time.

The Geological Survey of India has so far identified about *350 hot spring sites* which can be explored as sources of geothermal energy. The engineers have commissioned an experimental 1 kW generator running on geothermal energy in the Puga area. This is the first production of electricity from a hot water spring in India.

- Many countries with hot springs in their territories have realised their potential for power and heat production. Countries like *Italy, Iceland, New Zealand, the USA and the USSR have achieved remarkable progress in the application of geothermal energy.*

The Italian power plant at Landerello was started in 1904 on a small scale but now it produces 540 MW of electricity. This is equivalent to burning 1.5 million tonnes of oil in a year. New Zealand started exploration in 1950 and the Wairakei power station now produces 175 MW, which is equal to 0.7 million tonnes of oil per year. The power production in California, USA, began in 1960 and has already touched 50 MW. In the Philippines the drillers struck high pressure, high temperature steam at about 200 m only at Tiwi, a tiny sleepy village nestled at the base of the volcano Malino, in 1967. By January, 1976, the first geothermal power plant at Tiwi began producing 55 MW. A geothermal plant with a capacity of 11 MW has been in operation for nearly 20 years in USSR. The construction of another power plant at Mutnovsky with a capacity of 200 MW is in progress.

- At present, *35 countries* of the world use about 15000 MWe geothermal energy for *space heating, industrial and agricultural applications* whereas 21 countries utilise geothermal energy for electricity generation.

The following countries have installed generating **geothermal units** *(above 20 MWe)* up to the year 2005:

Country	Installed up to 2005 (MWe)
Mexico	953
Indonesia	793
Italy	790
Costa Rica	163

El Salvador	161
Kenya	127
Russia	79
Nicaragua	77
Guatemala	334
Portugal	20
Turkey	20

THEORETICAL QUESTIONS

1. Is geothermal energy renewable? Explain briefly.
2. What are the various means by which heat transfer from earth's interior takes place?
3. Explain briefly the various parts of the earth's interior.
4. Explain with a neat diagram the following parts of the earth's interior:
   (i) Crust;   (ii) Mantle;   (iii) Core.
5. What do you mean by 'Heat flux'?
6. Describe with a neat sketch the 'Hot Spring structure'.
7. What is geothermal system? Explain briefly.
8. Discuss briefly 'Earthquakes and Volcanoes'.
9. What is thermal gradient? Explain briefly.
10. How are geothermal resources classified?
11. Give the examples of hydrothermal resources.
12. Describe within a diagram the 'Dry-steam open system'. State its environmental aspects.
13. Explain with the help of a schematic diagram the 'Flash steam open system' used for power generation.
14. Explain the principle of the Binary cycle system' employed for power generation.
15. Draw a schematic diagram of a liquid-dominated 'total flow concept system' and explain it briefly.
16. Write a short note an 'liquid-dominated low temperature system (geothermal fluids)'
17. What are 'Geopressured resources'? Explain briefly.
18. What are 'Hot dry rocks (HDR) resources? Explain.
19. Write a short note on 'Magma resources'.
20. Explain clearly with a neat schematic diagram the working of a 'Hybrid geothermal fossil system.
21. List down the advantages and disadvantages of geothermal energy over other energy forms.
22. What are the applications of geothermal energy?
23. Write a short note on 'Geothermal energy in India and abroad'.
24. What are the environmental problems associated with geothermal energy?

CHAPTER

8

OCEAN ENERGY

8.1. Introduction; 8.2. Ocean energy sources; 8.3. Tidal energy – Introduction – Tidal range (R) – Production of tides – Origin of ocean tides (Tidal phenomenon) – Estimation of energy potential for a tidal power project – Energy power in a single-basin single-effect/cycle scheme, – Energy and power in a double-effect/ cycle system – Power generation (yearly) from tidal plants – Site requirements for a tidal power scheme – Components of a tidal power plant – Type of tidal power plants – Single-basin single-effect plant – single-basin double-effect plant – Double basin tidal plants – Double-basin, linked-basin scheme – Double-basin, paired-basin scheme – Advantages and disadvantages of tidal power – Global scenario of tidal energy – Economic aspects of tidal energy conversion; 8.4. Wave energy – Introduction – Advantage and disadvantages – Factors affecting wave energy – Parameters of ocean waves – Energy and power from waves – Characteristics of waves in real oceans – Wave energy potential – Wave energy conversion – *Wave energy conversion machines:* Classification of wave machines – Float wave-power machine – High level reservoir wave machine – Dolphine-type wave-power machine (generator) – Hydraulic accumulator wave machine – Nodding/oscillating ducks wave machine – Oscillating water column surge device – Wave power development; 8.5. Ocean Thermal Energy Conversion (OTEC) – Introduction – Solar energy absorption by water (Lambert's law of absorption) – Working principle of OTEC – Efficiency of OTEC – *Types of OTEC plants*: closed cycle (Anderson cycle, vapour cycle) system – Open cycle (Claude cycle, steam cycle) system – Modified open cycle OTEC plant – Hybrid OTEC system – Thermoelectric OTEC system – Bio-fouling – Advantages, disadvantages/limitations and applications of OTEC – Development of OTEC plants – Environmental impacts of wave power; 8.6. Mini and microhydel power plants – Introduction – Advantages and disadvantages/limitations of small hydro power potential – Clasification of small hydropower (SHP) stations – Classification of water turbines – Major components of small hydropower projects. *Highlights – Theoretical Questions.*

8.1. INTRODUCTION

The *oceans, large lakes and bays are huge reservoirs* of various useful and *renewable energy sources*. World's total estimated ocean energy reserves are about 130×10^6 MW. However, only a *fraction* can be recovered *economically*.

Due to rapidly depleting fossil fuel sources, the *ocean energy* is likely to gain a significant importance during coming decades. However, its present use is very limited.

Oceans receive water from the rivers in various parts on the earth. Ocean is a great collection of salt water that covers approximately 70 percent of earth's surface. Five principal oceans are:

1. Indian ocean;
2. Pacific ocean;
3. Atlantic ocean;
4. Arctic ocean;
5. Antarctic ocean.

Oceanography *is the science which deals with the environment in the oceans including the waters, depths, beds, biomass, energy resources etc.*

The various ocean energy technologies are presently in infant stage and their commercialisation is likely to take several more decades.

8.2. OCEAN ENERGY SOURCES

Ocean energy sources may be broadly divided into the following *four* categories:

1. *Tidal energy.*
2. *Wave energy.*
3. *Ocean thermal energy conversion (OTEC).* This concept was proposed as early as 1881 by the French physicist Jacques d'Arsonal.
4. *Hydroelectric energy*—Energy emanated from the *sun-ocean system* from the mechanism of surface water evaporation by solar heating *i.e. hydrological cycle".*

1. **Ocean tidal energy.** It refers to the *hydroenergy in ocean tides.* Ocean tides occur *due to gravitational attractive forces from sun and moon.* The level of the ocean water rises periodically during high tides and drops during low tides. The difference in head of water during high tide and low tide is used for rotating hydro turbine-generator units installed within barrages (dams) to obtain electrical energy.

2. **Ocean wave energy.** It refers to the *waves of water from ocean to the shore.* Ocean waves occur *due to rotation of earth and the winds over ocean surface.* Waves have an interval of 4 to 12 seconds and crest of a few centimetres to about 10 m. Locations having waves with crest height of *3 m and above have higher energy density.*

   Ocean *wave machines* are installed on *floating power plants* or *on-shore power plants.* The rotor of the wave machine is rotated by wave energy. Wave machine drives generator rotor or pumps water to the reservoir at higher level.

3. **Ocean thermal energy.** It refers to the *thermal energy acquired by the ocean water from solar radiation.* The warm water from upper levels of ocean (at about 25°C) is pumped through *heat exchangers.* Thermal energy is extracted and converted to electrical energy by steam turbine-generator or vapour turbine-generator. Cold water from the bottom of the sea (at about 10°C) is used for condenser.

   - **Ocean current energy.** *Energy from ocean currents* refers to *hydro-energy in water currents through the large rivers terminating in the ocean.* The currents have kinetic energy which is converted into electrical energy by turbine generators.
   - *Ocean wind energy* refers to off-shore wind energy resources over oceans.
   - *Ocean biomass energy* refers to organic matter from oceans e.g. aquatic vegetation, algae and animals. Rapidly growing varieties of ocean algae, ocean kelp are harvested periodically. The ocean biomass may be converted into *methane rich biogas* by *wet anaerobic digestion process.* Alternatively, biomass may be *dried* and *burnt.*

- *Ocean nuclear energy* resources refer to nuclear energy resources obtainable from ocean water or ocean beds.

4. **Hydroelectric energy.** The hydrogical cycle results in rainfall, which causes river flows that can be trapped behind *dam to even out the variations in the river flows* and thus become source of either *low-head or high-head (dam) hydroelectric energy. Small scale hydroelectric facilities* can supply in principle significant amounts of electricity for irrigation or, portable water pumping, lighting, health and educational purposes. The total amount of such a resource is very poorly documented but apt to be large.

8.3. TIDAL ENERGY

8.3.1. Introduction

The periodic rise and fall of the water level of sea which are carried by the action of the sun and moon on water of the earth is called the 'tide'.

The daily variation in tidal level is mainly due to the *changing position of the moon.*

- Tidal energy can furnish a significant portion of all such energies which are renewable in nature. The large scale up and down movement of sea water represents an *unlimited source of energy*. If some part of this vast energy can be converted into electrical energy, it would be an important source of hydropower.

- The *main feature of the tidal cycle is the difference in water surface elevations at the high tide and at the low tide*. If this differential head could be utilized in operating a hydraulic turbine, the tidal energy could be converted into electrical energy by means of an attached generator.

8.3.2. Tidal Range (R)

The tidal range is the difference between consecutive high tide and low tide water level. It is denoted by R and is measured in metres.

Tidal energy *refers to the potential energy in the tidal range.*

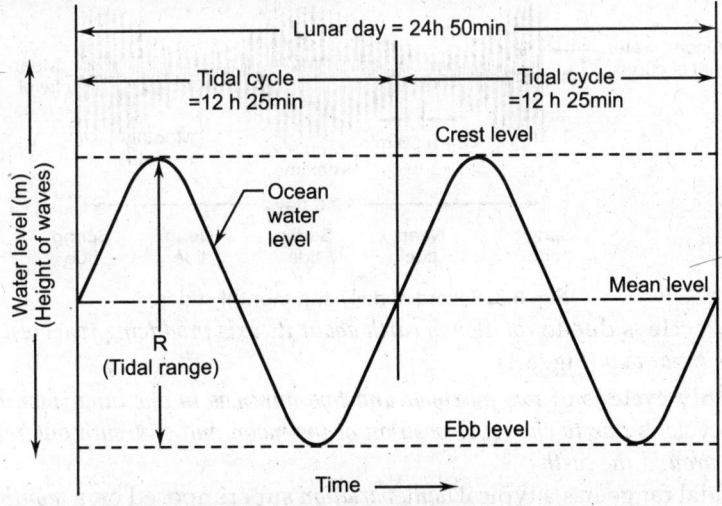

Fig. 8.1. Tidal range [Daily (diurnal) tides].

Fig. 8.1 shows the *time versus water-level characteristics* of ocean tide for a *lunar day*. The tidal curve against time is approximately sinusoidal. Range (R) is the difference in water level of high-tide crest and low-tide crest. This difference is utilised to obtain the head of water between ocean-side and basin-side of the barrage (dam).

- *Tidal range (amplitude)* varies widely depending upon *geographical location, contour of ocean bed, depth of oceans, distance from coasts etc.* It is *insignificant in the middle of ocean* and *significant near coast.* Tidal ranges of 0.25 m to 17 m have been recorded in different locations.

Tidal range is not constant at the same collection but *varies* with lunar days in the month. A lunar month is of 29.5 days.

8.3.3. Production of Ocean-tides

Fig. 8.2 explains how ocean-tides are produced.

— **Spring tides** *are those in which the tidal range is maximum on full moon and new moon.*

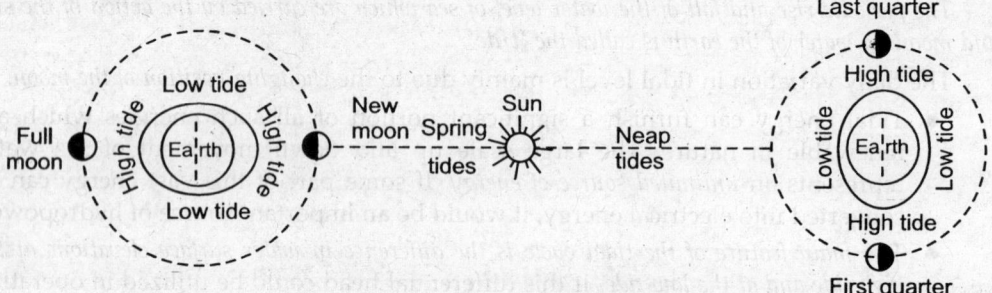

Fig. 8.2. Production of ocean-tides.

— *The tides in which the tidal range is minimum* on first quarter and third quarter are called **Neaptides.**

Fig. 8.3 shows a record of daily and monthly tides in a complete lunar month.

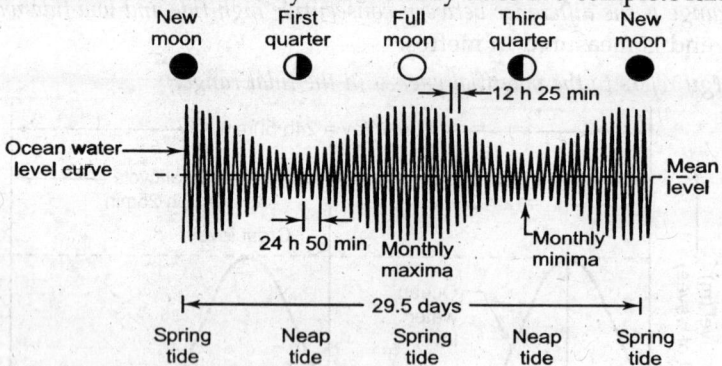

Fig. 8.3. Record of daily and monthly tides.

- **Daily cycle** is due to *rotation of earth about its axis producing two crests and two ebbs in one lunar day* (Fig. 8.1)
- **Monthly cycle** is of *two maximas and two minimas in one lunar month of 29.5 days.* This cycle is *due to changing position of the moon and sun with one revolution of the moon around the earth.*

The tidal range has a typical *daily variation* superimposed on a *monthly variation.*

8.3.4. Origin of Tides (Tidal Phenomenon)

The tidal energy *is due to the gravitational force of attraction between the earth and sun and between earth and moon.*

The gravitational force F between two bodies, say between sun and a molecule on earth, is given by:

$$F = \frac{KMm}{d^2} \qquad \qquad ...(8.1)$$

where,
$\qquad \qquad \qquad M$ = Mass of sun,
$\qquad \qquad \qquad m$ = Mass of water molecule,
$\qquad \qquad \qquad d$ = Distance between sun and water molecule, and
$\qquad \qquad \qquad K$ = Gravitational constant.

The gravitational force between moon and a water molecule will also be given by a similar equation. Since distance between moon and earth is lesser than that between sun and earth, the attraction between moon and water molecule is about 2 to 3 times that between sun and molecule.

- The moon revolves around the earth with a period of 24 hours 50 minutes per one revolution. The earth's surface facing the moon experiences greater attractive force than the surface away from the moon. Thus ocean water on the moon side experiences a swell (hide tide) and the other side experiences low tide. Such daily tides are called the *diurnal tides* (Fig. 8.3)

- The relative positions of the sun, the moon and the earth have hourly variation, daily variation. Hence tides are affected by these relative positions. *Spring tides* and *Neap tides* are caused by relative positions of the moon and the sun with respect to the position of ocean on the earth surface (Fig. 8.3).

8.3.5. Estimation of Energy Potential For a Tidal Power Project

To utilise tidal energy, water must be trapped at high side behind a dam or barrage and then made to drive turbine as it returns to sea during low side. The *available energy is proportional to the square of the amplitude of tide.* As such the available energy tends to be concentrated around regions of high side. *The amount of generation depends only on the tidal phenomenon and can be predicted fairly accurately.*

Because of the *variations in tidal pattern, the power output shows some variations as under:*

1. *Two bursts of generation activity per day,* beginning about 3 hours before high tide and lasting from 4 to 6 hours.
2. The power output in each tidal cycle will *increase* with the difference between high and low tides. Thus the power output curve will *display a 14 day cycle.*
3. The high tide time *shifts by about one hour every day* and the *power output will show a similar shift.*
4. Spring tide high water always occurs at the same time. Thus the maximum availability will *not be disturbed evenly during the day.*

Tidal power schemes are of two types:

1. Single basin scheme.
2. Two basin scheme.

8.3.5.1. Energy and power in a single-basin single-effect/cycle scheme

A single basin scheme is *cheaper.* However it is not very useful due to the above mentioned reasons in power output. This scheme produces power output, which *follows phases of sun and moon and not the load demand of the system.* Therefore, a single basin tidal power scheme would always *need a standby plant.*

Fig. 8.4 shows a single-basin single-effect tidal power scheme. It is the case of basin *beginning at high-tide level,* **emptying** *through the turbine to ocean, which is at 'low tide';* head (h) varying from (+R) to 0.

In case of a *'Double cycle system',* power is generated while *"filling"* the basin and *also while "emptying"* the basin. Hence double cycle scheme gives *"double energy"* per tide cycle of one crest and one ebb.

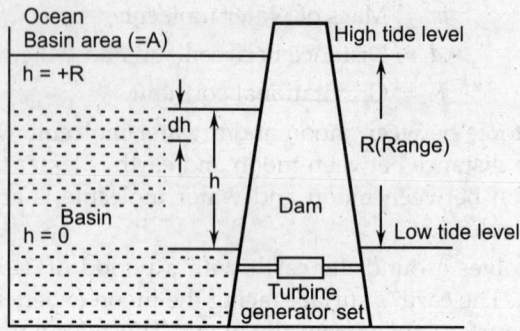

Fig. 8.4. Single-basin single-effect tidal system–"Emptying only".

Let, A = Area of basin, considered constant, m^2,

ρ = Density of water, kg/m^3,

g = Gravitational constant,

m = Mass flowing through the turbine, kg,

h = Head, m, and

W = Work done by water flowing through turbine, J.

For tidal range (amplitude) R, and certain head (h) at the given time during the flow from the ocean to basin, the *differential work done (dW)* is equal to the *change in potential energy* due to change in mass (dm) of water. Hence,

$$dW = dm.g.h, \text{ J}$$

But, $dm = -\rho.A.dh$

(–ve sign indicates *decrease* in the mass of water during in *emptying operation*)

So, that $dW = -\rho.A.dh.g.h, \text{ J}$

The *total work done (W)* by water while *emptying* the basin is obtained by integrating dW from R to 0,

$$W_{emp.} = \int_R^0 dW = \int_R^0 -\rho.A.dh.g.h = -g\rho A \int_R^0 h \, dh$$

$$= -g\rho A \left|\frac{h^2}{2}\right|_R^0 = \frac{1}{2} g\rho AR^2$$

i.e., $W_{emp.}$ (or Energy) $= \dfrac{1}{2} g\rho AR^2, \text{ J}$ ...(8.2)

where, ρ = Sea density in kg/m^3 = 1025 kg/m^3; g = 9.81 m/s^2

Eqn. (8.2) indicates that *work is proportional at to square of the tidal range.*

● The **power** *is the rate of doing work.*

The power is generated during emptying (or filling) and no power is generated during rest of the time.

The average power (theoretical) delivered $= \dfrac{W}{\text{time}}$

The duration of time for single effect is 6h 12.5 min which is equal to 22,350 seconds.

$\therefore \qquad\qquad P_{av.} = \dfrac{W}{\text{time (s)}} = \dfrac{g\rho AR^2}{2 \times 22{,}350} = \dfrac{1}{44{,}700} \times 9.81 \times 1025\, AR^2$

i.e., $\qquad\qquad P_{av} = 0.225\, AR^2$ ...(8.3)

and $\qquad\qquad \dfrac{P_{av.}}{A} = 0.225\, R^2\, \text{W/m}^2 = 0.225\, R^2\, (\text{MW/km}^2)$ ...[8.3 (a)]

The average power generated is calculated based on average operating head of $\dfrac{R}{2}$ which is available only for limited period under a single-basin emptying operation. There are friction losses, conversion efficiencies of the turbine and generator that *reduce the power*. Studies have shown that the optimal annual energy production is *about 30 percent* of the average theoretical power.

Example 8.1. *A tidal power plant of single-basin type, has a basin area of 24 km². The tide has a range of 10 m. The turbine stops operation when the head on it falls below 3m. Calculate the average power generated during one filling/emptying process in MW if the turbine-generator efficiency is 75 percent. Density of sea water = 1025 kg/m³; g = 9.8 m/s².*

Solution. *Given:* $A = 24$ km² $= 24 \times 10^6$ m²; $R = 10$ m; $r = 3$ m (the heat before turbine stops operating); $\eta = 0.75$; $\rho = 1025$ kg/m³; $g = 9.81$ m/s².

Average power generated, P_{av} :

Work done, $\qquad\qquad W = \displaystyle\int_{R}^{r} -g\rho Ah\,dh = -g\rho A \int_{R}^{r} h\,dh$

$\qquad\qquad\qquad\qquad = \dfrac{1}{2} g\rho A(R^2 - r^2)$ ...(1)

Thus, average power generated,

$\qquad\qquad P_{av.} = \dfrac{W}{\text{Time}}$

$\qquad\qquad\qquad = \dfrac{g\rho A(R^2 - r^2)}{2 \times 22350}$ .... (2)

$\qquad\qquad\qquad = \dfrac{1}{44{,}700} \times 9.81 \times 1025 \times (24 \times 10^6)\,(10^2 - 3^2) \times 10^{-6}\ \text{MW}$

or, $\qquad\qquad P_{av.} = 491.3$ MW

$\therefore$ Power generated $= 491.3 \times 0.75 = \textbf{368.5 MW}$ **(Ans.)**

8.3.5.2. Energy and power in a single-basin double-effect/cycle scheme

Fig. 6.5 shows a single-basin double-effect/cycle tidal power system. In this system energy is converted into electrical energy during flood tide (*rising tide*) when the basin is *filled* and *also* during the ebb tide (*falling tide*) when the basin is *emptied*. Since water flows through the turbine during rising and falling tides in *apposite directions*, therefore, a **reversible turbine** is used, which acts as a turbine for either direction of flow.

The average theoretical power from a double-cycle/effect scheme is **twice** that of a similar single-effect scheme. Thus, the theoretical average power generated (using eqn. 8.3),

$$(\text{Pav.})_{\text{double-effect}} = 2 \times 0.225 \ AR^2 = 0.45 \ AR^2 \qquad ...(8.4)$$

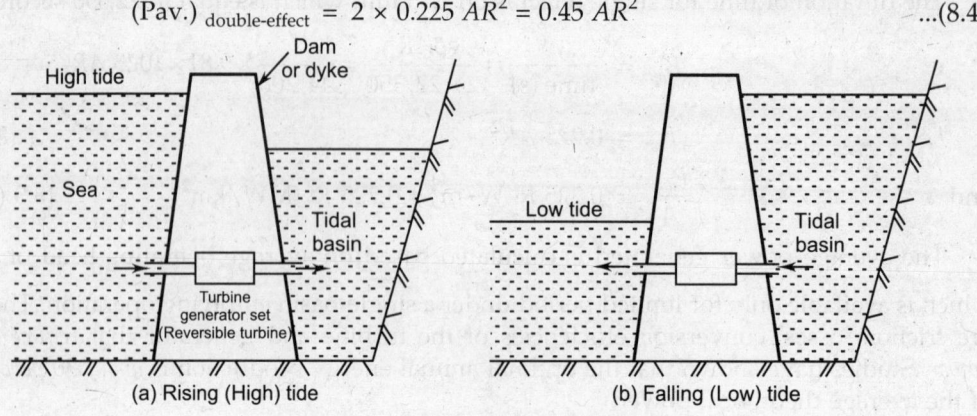

Fig. 8.5. Single-basin double-effect/cycle system.

This system is 100 percent more efficient that single-effect system/plant because it generates *double* energy per cycle.

8.3.6. Power Generation (Yearly) From Tidal Plants

The energy available from a tidal plant depends on the following *two* factors:

(*i*) The tidal range.

(*ii*) The volume of water accumulated in the basin.

Tidal energy is slowly-increasing hydro-energy during filling of the basin, and after a period of nearly 3 hours it attains its *peak value*. When the tide *recedes*, water is allowed to flow from basin to sea; it is then *slowly-decreasing hydro-energy* and attains its *lowest value* when the turbine stops after a period of 3 hours. Thus, the energy available for a tidal point can be calculated in a similar way as for an hydropower plant; *i.e., considering the average discharge and available head at any instant.*

Let, $\qquad\qquad\qquad A$ = Average cross-sectional area of the basin, m^3

$\qquad\qquad\qquad\quad H$ = Difference between maximum and minimum water levels, m, and

$\qquad\qquad\qquad\quad V$ = Volume of basin, m^3.

Now, $\qquad\qquad\qquad V = AH$

∴ Average discharge, $Q = \dfrac{AH}{t}$

where t is the total duration of generation in one filling/emptying operation in seconds.

Now, power generated at any *instant*,

$$P_{\text{inst.}} = \frac{\rho Q h}{75} \times \eta_0 \ \text{H.P.} \qquad ...(8.5\ a)$$

or, $\qquad\qquad P_{\text{inst.}} = \dfrac{\rho Q h}{75} \times \eta_0 \times 0.736 \ \text{kW} \qquad ...(8.5\ b)$

where, $\qquad\qquad\qquad h$ = available heat at that instant, m,

$\qquad\qquad\qquad\qquad \rho$ = 1025 kg/m^3 for sea water,

$$\eta_0 = \text{Overall efficiency of the system, and}$$
$$1 \text{ H.P.} = 75 \text{ kgm/s.}$$

Then, total energy $= \int_0^t P dt = \int_0^t \frac{\rho Q h}{75} dt \times \eta_0 \times 0.736 \text{ kW } per \text{ tideal cycle} \qquad ...(8.6)$

On average, there are *705 tidal cycle in a year*.

Then, *yearly* power generation from a tidal project

$$P_{yearly} = \int_0^t \frac{\rho Q h}{75} \, dt \times \eta_0 \times 0.736 \times 705 \text{ kWh/year} \qquad ...(8.7)$$

Example 8.2. *For a proposed tidal site, the observed difference between high and low water tide is 9 m. The basin area is about 0.45 sq. km which can generate power for 3 hours in each cycle. The average available head is assumed to be 8.5 m, and overall efficiency of the generation is 72 percent. Assume density of sea water as 1025 kg/m³. Calculate:*

(i) *Power at any instant.*

(ii) *Yearly power output.*

Solution. *Given:* $H = 9$ m; $A = 0.45 \times 10^6$ m²; $t = 3$ hours; $h = 8.5$ m; $\eta_0 = 72\%$, $\rho = 1025$ kg/m³.

Volume of basin $= AH$
$$= 0.45 \times 10^6 \times 9 = 4.05 \times 10^6 \text{ m}^3$$

$\therefore$ Average discharge, $Q = \dfrac{\text{Volume}}{\text{Time period}} = \dfrac{4.05 \times 10^6}{3 \times 3600} = 375 \text{ m}^3/\text{s}$

(i) **Power at any instant ($P_{inst.}$):**

$$P_{inst.} = \frac{\rho Q h}{75} \eta_0 \times 0.736 \text{ kW} \qquad ...[\text{Eqn. (8.5b)}]$$

or, $\qquad P_{inst.} = \dfrac{1025 \times 365 \times 8.5}{75} \times 0.72 \times 0.736 = \mathbf{23085 \text{ kW (Ans.)}}$

(ii) **Yearly power output:**

Energy generated per tidal cycle
$$= 23085 \times 3 = 69{,}255 \text{ kWh}$$

Total number of tidal cycle in a year $= 705$

$\therefore$ Year power output, $P_{yearly} = 69{,}255 \times 705 = \mathbf{488.25 \times 10^5 \text{ kWh/year (Ans.)}}$

8.3.7. Site Requirements for a Tidal Power Scheme

A favourable *site* for tide power scheme should meet with the following *requirements*:

1. The site should have a *large tidal range*.
2. Capable of *storing a large quantity of water* for energy production with minimum dam and dyke construction.
3. To achieve a high storage capacity, the site should be located in an *estuary or a creek.*
4. It should be near to a load centre to minimise the transmission requirements.
   - The following *points* need to be considered prior to the development of a tidal power scheme:

1. **Pre-feasibility study.** It pertains to the collection of data such as tides, local topography, infrastructure, etc. During this study following should be collected: (*i*) Local land area map, survey of India map and hydrographic charts, (*ii*) Historical data on tides and tidal currents; (*iii*) Geotechnical properties of sea bed and coastal region in the study area; (*iv*) Typical weather conditions, rainfall, wind and wave data; (*v*) Nearest high voltage substation for connecting the generated electric power with the state grid.

2. **Feasibility study.** This phase of the development of tidal power scheme consists of the following:
   — Mathematic modelling;
   — Preliminary energy computation;
   — Foundation investigations;
   — Hydraulic model studies;
   — Detailed analysis of various modes of operation.

3. Detailed design.

4. Preparation of specifications and tender documents.

5. Plant construction.

8.3.8. Components of a Tidal Power Plant

A tidal power plant consists of the following *three components:*

1. Dam or dyke (low wall);
2. Sluice ways;
3. Power house.

1. **Dam or dyke (barrage).** The function of dam or dyke is to form a barrier between the sea and the basin or between one basin and the other in case of multiple basin schemes.
   — It should be constructed by the material available at site or from a nearby place.
   — As the barrage has to withstand the force of sea waves, so the design should be suitable to the site conditions and to economic aspect of development.
   — The crest and slopes of the barrage should be armoured for protection against waves.

2. **Sluice ways.** These are used to *fill* the basin during the *high tide* or '*empty*' the basin during the low tide, as per operational requirement. These devices are controlled through gates.

   There are two types of sluice ways:

   (*i*) *Crest gates*: These are more prone to damage by wave action and masses carried by the flow.

   (*ii*) *Submerged gates with venturi type*: Vertical gates are the natural choice and can be fabricated from stainless steel.

3. **Power house.** A power house has *turbines, electric generators* and other auxiliary equipment. As far as possible, the power house and sluice ways should be in *alignment* with the dam or dyke.

   According to the suitability, for low heads the following turbines may be used; (*i*) Bulb turbine; (*ii*) Tube turbine, and (*iii*) Straight flow rim type turbine.

8.3.9. Types of Tidal Power Plants

Tidal power plants are *classified* as follows:

1. *Single-basin arrangement*:
   (i) Single-effect plant.
   (ii) Double-effect plant.
2. *Double-basin arrangement*:
   (i) Double-basin, paired-basin plant.
   (ii) Double-basin, linked-basin plant.

8.3.10. Single-Basin Tidal Plants

Fig. 8.6 shows the layout of a single-basin tidal plant.

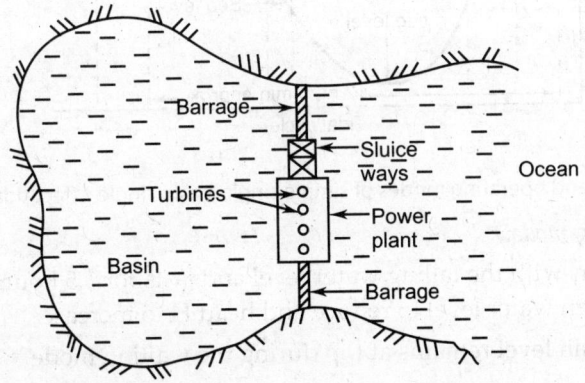

Fig. 8.6. Layout of a single-basin tidal plant.

— A barrage (Dyke, dam) separate, the basin from the sea.
— The sluice way is opened during high tide to fill the basin.
— The turbine-generator units are mounted within the ducts inside the barrage.
— Power house is built on the barrage.

  • A single-basin tidal plant *cannot generate power continuously.*
  • In these plants, *pumped storage capacity* may be incorporated to fill water to a higher level and empty the basin to lower level. A pumped storage plant is useful *when the load it supplies fluctuates considerably.*

8.3.10.1. Single-basin single-effect plant

Refer to Fig. 8.4. Such a plant generates power only with *one-way* flow of water through the turbines, *i.e.*, during low tide.

— The *basin is filled during the 'high-tide'* by opening the gates in the barrage. The water level in the basin reaches the crest level of the tide. The energy is stored in the form of tidal range. Tidal range gives the *head of water during the 'low tide;.*
— Initially, during the *low tide*, the level in the basin is R is *zero*. Water is released during the *generating mode* through the turbines, spillways located within the barrage, into the sea.

— The turbines are started only when the head reaches (R) *i.e.*, water in the basin is at high tide level and water in the sea is at low tide level. The turbines are designed for *single way operation. Only fixed blade turbines are necessary.*

Power generation is *intermittent* and mostly during off-peak load periods on daily load curves.

Fig. 8.7. indicates the flow intervals in a simple single-basin and single-effect plant.

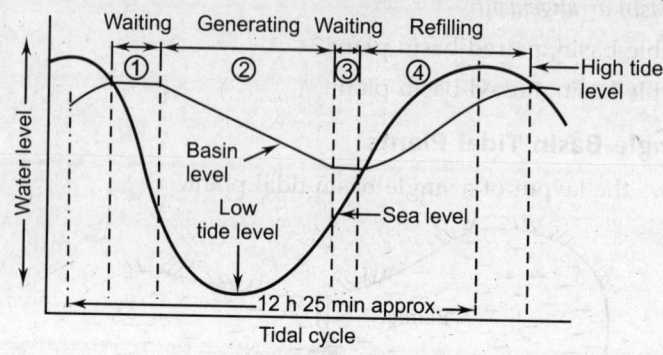

Fig. 8.7. Time and operating modes of simple single-basin single-effect tidal power plant.

1. *First waiting mode*:
   — It begins with the falling water level and lasts for 1.5 hours (approx.) to allow the ocean water level to reduce and head H to increase.
   — The basin level remains at top during the waiting mode.

2. *Generating mode*:
   — This mode begins, when the ocean level reduces to required mark and the head H reaches required value.

   Head = Basin water level-sea water level
   — The turbine-gates are opened, and the pool water flows through the turbines (during this mode).
   — This mode continues during 'emptying' of the basin and last for 4.5 hours (approx.)

3. *Second waiting mode*: The generating mode is followed by another waiting mode, during the rising tide.

4. *Refilling mode*: Its duration is 4.0 hours (approx.) during the rising tide.

8.3.10.2. Single-basin double-effect plant

Refer to Figs. 8.5 and 8.8. This is a *two-way scheme, Ebb and flood operation. **Reversible turbines** are installed and power is generated during 'filling' and 'emptying' of the basin.*

— The basin *fills* in during *rising tide (flooding)*. The water flows from the sea to the basin and drives the reversible turbines.

— The basin is *emptied* during *lowering-tide (ebbing)*.

Water flows from the basin into the sea and drives the reversible turbines.

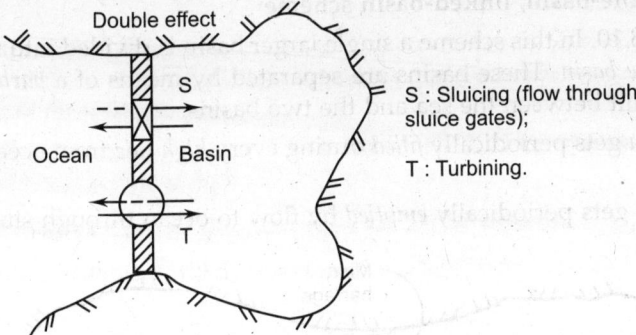

Fig. 8.8. Single-basin double-effect tidal plant, sluicing and turbining flow directions.

Fig. 8.9 shows the *operating modes* of single-basin double-effect tidal power plant.

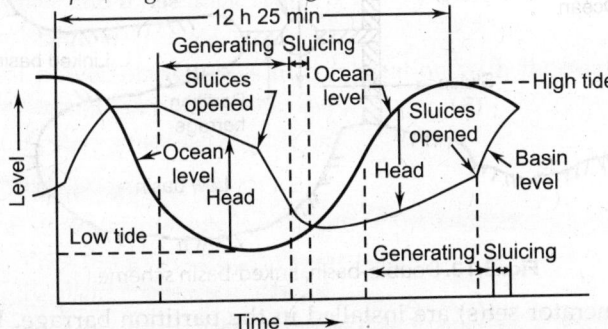

Fig. 8.9. Operating modes of single-basin double-effect tidal power plant (Fig. 8.5).

The operating modes during 12 h 25 min half-tidal cycle are:

1. *Generating mode* during *emptying*.
2. *Sluicing*—letting out water through sluice during low net heads.
3. *Generating period* during *filling*—sluices are closed, water is let out from the basin through the turbines.
4. *Second sluicing period*.

Energy conversion in half tidal period of 12h 25 min is *twice* that of a single-basin single-effect plant.

In a *single-basin, double-effect, 'pump turbine plant' in addition*, the turbines are used as pumps to increase or reduce the water level of the basin/reservoir some energy is spent to drive the pumps.

- Although the double-cycle system has *only short duration interruptions in turbine operation*, yet a *continuous generation of power is still not possible*. Furthermore the *periods of power generation coincide only occasionally with periods of peak demand*. These problems are overcome to some extent in the double-basin scheme.

8.3.11. Double-Basin Tidal Plants

A *'double-basin scheme' can produce continuous power output*. The *drawback* is that the civil works become *more extensive*.

Several tidal energy conversion projects under planning are of *double-basin type*. These plants may be of the following *two types*:

1. Double-basin, linked-basin plant.
2. Double-basin, paired-basin plant.

(*iii*) Diamond harbour;

(*iv*) Ganga Sagar.

The *basin in Kandla in Gujrat has been estimated to have a capacity of 600 MW.*

The *total potential of Indian coast is around 9000 MW,* which does not compare favourably with the sites in the American continent states. The technical and economic difficulties still prevail.

8.3.14. Economic Aspects of Tidal Energy Conversion

The cost of a tidal energy conversion scheme includes: (*i*) Cost of barrages, (*ii*) Cost of land of basins and development of basins, and (*iii*) Cost of power plant. The *capital cost per* kWh of energy is therefore *very high.* The running cost and maintenance cost are, however, *low.*

- The mini hydroprojects are more favoured than tidal power plants,
- The tidal power plants may be *economically comparable,* in future, when *cost of conventional fuels becomes more prohibitive.*

Inspite of the fact that tidal power plants are costly, they have the following *fringe benefits:*

1. Renewable energy, free of cost for entire period of time.
2. Performance is pollution free.
3. Development of regions on both the sides of the barrage and on the banks of the basin.
4. With pumped storage facility, continuous, dependable large power can be obtained. The rating of tidal power plant is in the range of several tens of MW.
5. Technology of bulb turbines developed for tidal power plants is useful in mini-hydro and pumped hydro-power plants.
6. Road on the top of the barrage eliminates the need of a separate bridge.
7. Tourist attraction in tidal power plants and development of tourism.

8.4. WAVE ENERGY

8.4.1. Introduction

Wave energy *comes from the interaction between the winds and surfaces of oceans.* The energy available varies with the *size and frequency of waves.* It is estimated that about 50 kW of power is available for *every metre width of true wave front.*

Ocean wave energy is due *to the periodic to-and-fro, up-and-down motion of water particles in the form of progressive waves.* The period of ocean waves is the order of a few seconds. Ocean waves are superimposed on ocean water. *Ocean water surface level varies with ocean tidal cycle.*

Ocean waves possess potential energy (P.E.) and kinetic energy (K.E.). The ocean waves originate in different parts of the ocean surface due to the surface winds. The waves travel in the direction of the wind to the shore. The waves may be due to the local winds or the planetary winds. The *height of the waves depends upon the wind velocities, depth of the ocean, contour of the shore etc.*

The typical ranges of ocean waves are:

(*i*) *Wind height* = 2 × amplitude = 0.2 m to 4 m

(*ii*) *Wave period* = 4 s to 12 s.

Very dangerous and destructive waves occur during storms and gusts. They may reach heights of 10 *m* and may topple ships and damage the ocean energy plants.

- Wave energy when active is very concentrated, therefore, wave energy conversion into useful energy can be carried out at *high power densities*. A large variety of devices (*e.g. hydraulic accumulator wave machine; high-level reservoir machine; Dolphin-type wave-power machine; Dam-Atoll wave machine*) have been developed for harnessing of energy but these are *complicated and fragile in face of gigantic power of ocean storms*.

8.4.2. Advantages and Disadvantages

Following are the *advantages* and *disadvantages* of wave energy:

Advantages:

1. It is relatively pollution free.
2. It is a free and renewable energy source.
3. After removal of power, the waves are in placed state.
4. Wave-power devices do not require large land masses.
5. Whenever there is a large wave activity, a string of devices have to be used. The system not *only produces electricity but also protects coast lines from the destructive action of large waves, minimises erosion and help create artificial harbour.*

Disadvantages:

1. Lack of dependability.
2. Relative scarcity of accessible sites of large wave activity.
3. The construction of conversion devices is relatively complicated.
4. The devices have to withstand enormous power of stormy seas.
5. There are unfavourable economic factors such as large capital investment and costs of repair, replacement and maintenance.

Problems associated with wave energy collection:

The collection of wave energy entails the following *problems:*

1. The variation of frequency and amplitude makes it an unsteady source.
2. Devices, installed to collect and to transfer wave energy from far off oceans, will have to withstand adverse weather conditions.

- Until now no major development programme for taming wave energy has been carried but successfully by any country. Small devices are available, however, and are in limited use as power supplies for buoys and navigational aids. From the engineering development point of view, wave energy development is not nearly as far long as wind and tidal energy.

8.4.3. Factors Affecting Wave Energy

Wave energy is affected by the following *three major factors:*

1. **Wind speed.** With the increase in wind speed, there is an increase in wind energy.
   — The amplitude of the waves depends on wind speed.
   — During storms and gusts, big ocean waves occur, which prove highly detrimental even to ships.
2. **Effective pitch value.** It is the *uninterrupted distance* on the ocean over which the wind can below before reaching the point of reference. The *larger* the distance, the *higher* the wave energy. This distance may vary from 5 km to 45 km.
3. **Depth of ocean water.** The greater the depth of ocean water the higher the wave velocity. Very large energy fluxes are available in deep ocean waves.

8.4.4. Parameters of Ocean Waves

A **progressive wave** (*travelling wave*) is a *wave whose crest line moves in the direction of propagation of wave.*

The important parameters and their notations are:

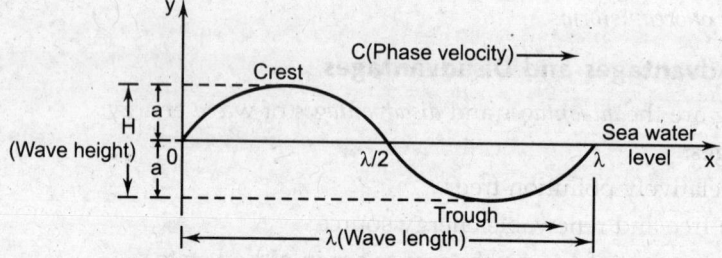

Fig. 8.12: Parameters of a progressive ocean wave

λ = Wavelength (= $C \times T$, where C is the phase velocity and T is the period), m;

α = Amplitude = $\dfrac{H}{2}$ (where H is the wave height), m;

B = Width along crest line, m;

A = Area of wave = $\lambda \times B$, m^2 (see Fig. 13);

f = Frequency, number of periods per second

$= \dfrac{1}{T}$.

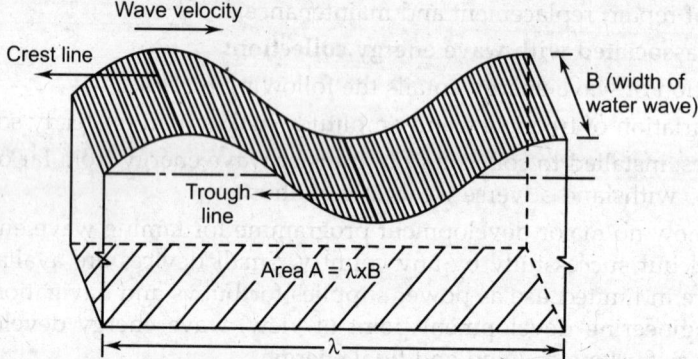

Fig. 8.13: Water wave length λ and width B.

The relation between λ and T is given by:

$$\lambda = 1.56 \; T^2 \text{ metre}$$

• The energy available in random sea is expressed as:

$$\rho = 0.96 \; H^2 T \; \text{kW}/m \text{ of wave crust} \qquad \qquad ...(8.8)$$

where, H = The wave heigh, m and

              T = The wave period, s.

Waves in ocean are not regular sine waves but random in nature.

8.4.5. Energy and Power from Waves

Fig. 8.14 shows a two-dimensional progressive wave, represented by the sinusoidal simple harmonic wave shown at time, $\tau = 0$, and at time τ. Although the sea waves are highly irregular yet, such a wave is assumed to be of sinusoidal harmonic wave shape for the purpose of mathematical analysis. The wave is moving in the direction of x–axis.

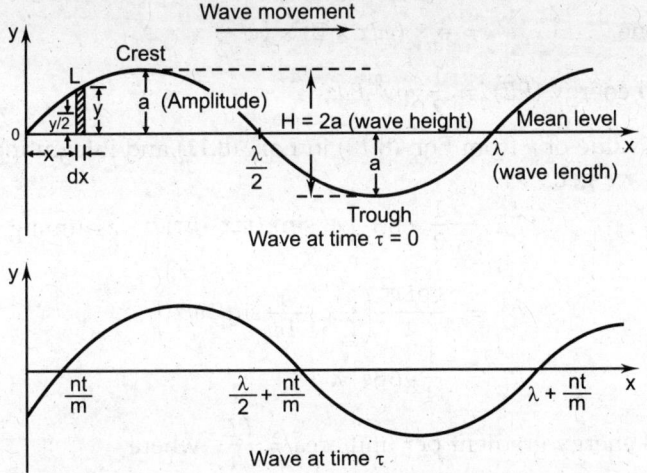

Fig. 8.14: A typical progressive wave, at time $\tau = 0$ and time τ.

Consider a point L (x, y) on the wave surface with an element of thickness dx along the x-axis with a coordinate y on the y-axis. This sine wave may be expressed by the following relation:

$$y = a \sin \left(\frac{2\pi}{\lambda} x - \frac{2\pi}{T} \tau \right) \qquad\qquad ...(8.9)$$

where,

y = Height above mean level, m,

a = Amplitude, m,

λ = Wave length, m,

τ = Time, s, and

T = Period, s.

Let,

$$m = \frac{2\pi}{\lambda}, \text{ and } n = \frac{2\pi}{T}$$

Then, eqn. (8.9) reduces to:

$$y = a \sin (mx - n\tau) \qquad\qquad ...(8.10)$$

where, $(mx - n\tau) = 2\pi \left(\dfrac{x}{\lambda} - \dfrac{\tau}{T} \right)$ = Phase angle, dimensionless.

Potential energy (PE):

The potential energy depends on the elevation of water above mean level (i.e., $y = 0$). In order to calculate the potential energy of elevated water, let us find out the work to be done in raising the quantity of water to the elevated height as given below:

$$\text{Work done} = \text{Force} \times \text{distance} = mgh \text{ joules}$$

where, m = Mass of water element dx

$= \rho \times ydx \times B$, kg

$h = \dfrac{y}{2}$ (i.e., mean height), m,

ρ = Density of sea water, kg/m^3, and

B = Wave width

$\therefore$ Work done   = $\rho \times (ydx \times B) \times gx \dfrac{y}{2}$

or,   Potential energy (PE) = $\dfrac{1}{2} g\rho y^2 \, Bdx$ ...(8.11)

Inserting the value of y from Eqn (8.10) in Eqn. (8.11) and integrating from 0 to λ for wave area $X \times B$, we get:

$$PE = \frac{1}{2} \, g\rho B \int_0^\lambda a^2 \sin^2 (mx - n\tau) dx \text{ , assuming } \tau = 0$$

$$= \frac{g\rho Ba^2}{2} \left(\frac{1}{2} x - \frac{1}{4m} \sin 2mx \right)$$

i.e., $PE = \dfrac{1}{4} g\rho Ba^2 \lambda$ ...(8.12)

The potential energy gradient per unit area = $\dfrac{PE}{A}$, where

A = λB, J/m^2.

$\therefore$ $\dfrac{PE}{A} = \dfrac{1}{4} g\rho a^2$ ...(8.13)

Kinetic energy (KE):

When the *amplitude a* of the wave is *small compared to its wave length*, then the PE and KE are *equal*.

$\therefore$ $KE = \dfrac{1}{4} g\rho Ba^2 \lambda$ ...(8.14)

The density of kinetic energy is given by:

$\dfrac{KE}{A} = \dfrac{1}{4} g\rho a^2$ ...(8.15)

Total energy (E):

$E = PE + KE$ joules

i.e., $E = \dfrac{1}{4} g\rho Ba^2 \lambda + \dfrac{1}{4} g\rho Ba^2 \lambda = \dfrac{1}{2} g\rho Ba^2 \lambda$ ....(8.16)

Now,   Energy density = $\dfrac{E}{A} = \dfrac{1}{2} g\rho a^2$ ...(8.17)

($\because$ Area, A = λB)

Wave power, P:

Power, $P = \dfrac{\text{Energy supplied}}{\text{Time taken}}$, J/s

                      = Energy $\times$ frequency, watts (W)

                      = $\dfrac{1}{2} g\rho B\lambda a^2 \, f$, watts (W),

Power density = $\dfrac{P}{A} = \dfrac{1}{2} g\rho a^2 f$, W/m^2 ...(8.18)

Empirical formulae on wave energy:

1. Scripps formula:

The scripps formula proposed by the Scripps Institution of Oceanography in La Jalla California gives a relationship between wave height and wind velocity as:

$$H = 0.085 \, U^2 \qquad \qquad ...(8.19)$$

where,
H = The wave height, m, and
U = The wind speed in knots (1 knot = 1.4 km/h)

2. Zinder Zee formula:

Zinder Zee formula is given by:

$$H = \frac{KV^2 F \cos a}{D} \qquad \qquad ...(8.20)$$

where,
H = Rise in water level above the normal, m,
K = 6.08×10^{-3} (constant),
F = Fetch *i.e.* unobstructed largest dimension of the lake, m,
V = Wind speed, km/h,
D = Average water depth, m, and
a = Angle between the wind direction and the fetch.

Example 8.3: *Ocean waves on an Indian coast had an amplitude of 1.2 m with a period of 6 s measured at the surface water 110 m deep. Taking water density as 1025 kg/m³, calculate the following:*

(i) The wavelength,

(ii) The wave velocity,

(iii) The energy density, and

(iv) The power density of the wave.

Solution. *Given:* a = 1.2 m; T = 6 s; ρ = 1025 kg/m³

(i) Wave length, λ:

$$\lambda = 1.56 \, T^2$$
$$= 1.56 \times 6^2 = \textbf{56.16 m (Ans.)}$$

(ii) Wave velocity, C:

We know that, $\quad C = \dfrac{\lambda}{T} \dfrac{56.16}{6} = \textbf{9.36 m/s (Ans.)}$

(iii) Energy density:

$$\text{Energy density} = \frac{E}{A} = \frac{1}{2} g \rho a^2 \qquad \qquad ...[\text{Eqn.}(8.17)]$$

$$= \frac{1}{2} \times 9.81 \times 1025 \times 1.2^2$$

$$= \textbf{7239.8 J/m}^2 \textbf{ (Ans.)}$$

(iv) Power density:

$$\text{Power density, } \frac{P}{A} = \frac{1}{2} g \rho a^2 \times f \qquad \qquad ...[\text{Eqn.}(8.18)]$$

$$= \frac{1}{2} g \rho a^2 \times \left(\frac{1}{T}\right) \qquad \qquad \left(\because f = \frac{1}{T}\right)$$

$$= \frac{1}{2} \times 9.81 \times 1025 \times 1.2^2 \times \frac{1}{6}$$

$$= \textbf{1206.6 W/m}^2 \textbf{ (Ans.)}$$

8.4.6. Characteristics of Waves in Real Oceans

In real oceans, final waves are made up by the combination of component waves; each component wave has different amplitude, wavelength, direction, period, etc. depending upon wind and ocean bed along the routes.

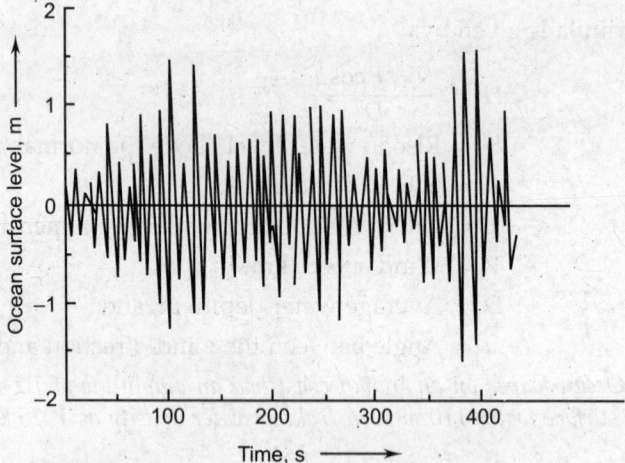

Fig. 8.15: Record of amplitudes of real wave with time.

- The waves generated in deep ocean by distant storms are very high and dangerous, called *swells*. These have *longer wavelengths*.
- The waves in shallow waters and by local winds are *not high* and have *shorter wavelengths*.

8.4.7. Wave Energy Potential

Collection of wave data:

The wave data is collected (by Institutes of Oceanography) by measuring the *parameters* of real ocean waves over a period of 1 to 3 years at selected locations on shore and high oceans.

— The measuring instruments (e.g. accelerometer, integrator, recorder etc.) are mounted on a ship or a buoy. The vertical displacement and vertical acceleration of the ship or the buoy are measured and recorded.

— The *data is scanned* at an internal of every 3 hours over a period of about 20 minutes with duration of 0.5 second per measurement.

— The data is *telemetered to the observatory* on the shore by means of radio communication channels. The data is *recorded* and *analysed*.

From the *analysis of wave data* the following *characteristics are determined:*

1. The significant *height* of the wave.
2. The *period* of significant wave.
3. The energy period.
4. The energy density.
5. The power density.
6. The power per unit width.

From these characteristics of ocean waves, a wave power plant in planned.

- These days the *computer-based methods are used for the analysis of wave data.* Such analysis of wave data helps *to estimate the potential of maximum, minimum and mean power at desired locations.*

Wave energy potential of Indian coast:

India has ample wave energy potential. Some data relating to ocean wave energy, for different locations obtained through measurements, are given in the tables 8.2, 8.3 and 8.4 respectively.

Table 8.2: Wave energy potential on Indian coast

S. No.	Location	North-east monsoon			South-west monsoon		
		Mean wave height (m)	Mean wave period (s)	Wave power (kW/m)	Mean wave height (m)	Mean wave period (s)	Wave power (kW/m)
1.	Near Calcutta 20-25° N and 85-95° E	1.33	8.00	13.85	1.95	7.65	28.80
2.	Near Vishakapatnam 20-25° N and 85-95° E	1.60	6.25	15.70	2.05	8.25	33.65
3.	Near Chennai 10-15° N and 85° E	1.55	5.85	13.45	1.70	5.80	16.60
4.	Near Cape Camorin 10-15° N and 70° E	1.20	5.35	7.80	1.80	6.30	19.55
5.	Near Mumbai 15-25° N and 70° E	1.00	5.00	4.90	2.65	6.95	47.00

Table 8.3: Wave energy potential near a Tamil Nadu coast

S. No.	Month	Mean wave height (m)	Maximum wave height (m)	Mean wave period (s)	Mean power (kW/m)	Maximum power (kW/m)
1.	Jan.	1.1	3.0	3.4	2.3	16.8
2.	Feb.	0.8	2.0	3.6	1.3	7.9
3.	March	0.7	1.5	2.1	0.6	2.6
4.	Apr.	0.8	2.0	2.8	1.0	6.2
5.	May	1.1	2.5	3.8	2.5	13.1
6.	June	1.7	4.5	4.0	6.4	44.6
7.	July	1.3	3.0	4.4	4.1	21.8
8.	Aug.	1.2	4.0	3.9	3.1	34.3
9.	Sept.	1.1	2.0	3.8	2.5	8.4
10.	Oct.	0.8	1.5	3.7	1.3	4.6
11.	Nov.	1.1	2.5	3.4	2.3	11.7
12.	Dec.	1.3	3.0	4.7	4.4	23.3

Location: *Latitude 12°-15° N; Longitude 81°-84° E.*

Table 8.4: Significant wave height and wave period for Nayachara Island

S. No.	Wind speed (km/h)	Mean wave height at depth		Mean wave period(s) at depth	
		6 m	10 m	6 m	10 m
1.	50	1.08	1.34	4.25	4.50
2.	60	1.24	1.56	4.50	5.0
3.	70	1.36	1.77	4.75	5.25
4.	80	1.47	1.96	5.00	5.50
5.	90	1.58	2.13	5.25	5.75
6.	100	1.69	2.29	5.50	6.0
7.	125	1.93	2.65	6.0	6.60

Location: *Latitude 22°; Longitude 88°7′.*

8.4.8. Wave Energy Conversion

The *wave energy is in the form of motion of water particles.* The potential energy and kinetic energy in water are converted to mechanical energy in the *'wave machine'* (wave energy converter) in the following *two* ways:

1. Wave machine *drives gears and electrical generator.*
2. Wave machine drives *air compressor or hydraulic pump to store energy in tanks, drive another machine for energy conversion.*

The former method/route is preferred as the energy can be transmitted to the shore.

8.4.9. Wave Energy Conversion Machines

Wave machines (or devices) are basically employed for energy conversion from the wave. The fluctuating mechanical energy from wave machine is *modified* to drive a generator.

8.4.9.1. Classification of wave machines:

Wave machines are *classified* as follows:

A. On the basis of location:
1. Off-shore or deep water device.
2. Shoreline devices.
- In *"off-shore wave machines"*, the *cost* of installation, operation and power transmission is *quite large.*
- In *"shoreline machines"* the cost of installation and maintenance is low.

B. On the basis of position with respect to sea level:
1. Floating devices.
2. Submerged devices.
3. Partially submerged devices.
- Submerged and partially submerged machines have the benefit of easier maintenance but have poor or worst storm conditions.

C. On the basis of actuating motion used in capturing the wave power:
1. Heaving float type.
2. Pitching type.
3. Heaving and pitching float type.

4. Oscillating water column type.

5. Surge devices.

8.4.9.2 Float wave-power machine

Wave motion is primarily 'horizontal', but the *motion of the water is primarily 'vertical'*. Mechanical power is obtained by floats making use of the motion of water. The concept visualises a large float that is driven up and down by the water within relatively stationary guides. This reciprocating motion is converted into mechanical and then electrical power is generated.

Fig. 8.16. shows Martin's float wave-power machine.

Construction. It consists of a *'square float'* which moves up and down.

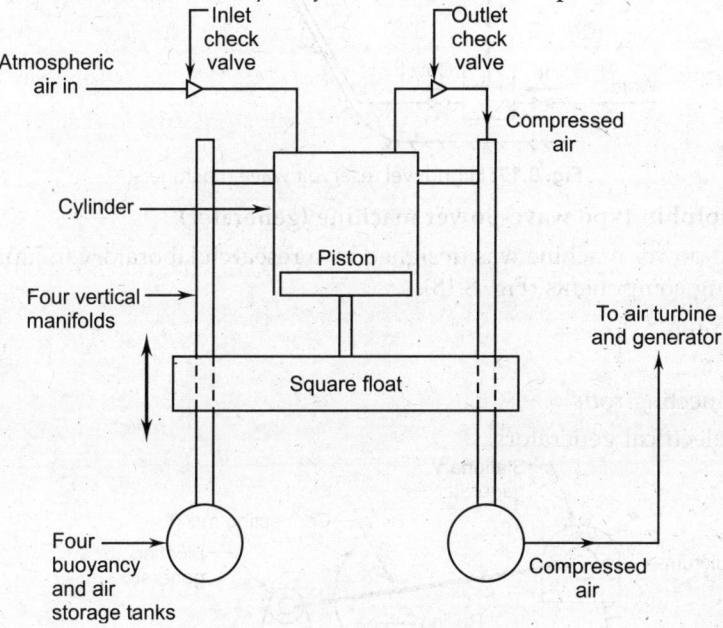

Fig. 8.16: Float wave-power machine.

'Four vertical manifolds' that are part of a platform guide the floats. A *'piston'* attached to the float compresses air in a *'cylinder'* which is stationary with the platform. The piston-cylinder arrangement acts as a *reciprocating air compressor*.

Working:

— The *downward motion* of the piston *draws air into cylinder* via inlet check valve.

— The *upward motion compresser the air* and send it through an outlet check valve to the four *underwater floatation tanks* via the four manifolds. These tanks serve the dual purpose of buoyancy and air storage tanks and the four manifolds serve the purpose of float guides and discharge air pipes.

— The compressed air in the buoyancy - storage tanks is then used to driver an *air turbine* which in term runs an electric *generator*, and the electricity is transmitted to the shore via underwater cable.

Limitations:

1. Waves are not perfectly sinusoidal.

2. Linear arrays of several kilometres are required to produce 100 MW.

3. Aspiration of water into intake systems and submersion by large waves.

4. Water entering turbine.

5. Design to withstand storms.

6. Transmission of power to shore.

7. Marine growth.

8.4.9.3. High-level reservoir wave machine

Fig. 8.17 shows a high-level reservoir machine. In this system an *hydraulic pump* is operated by the motion of *buoy to raise water to onshore reservoir*. The water in the reservoir is made to flow through a *turbo-generator to generate electricity*, and then back to sea level.

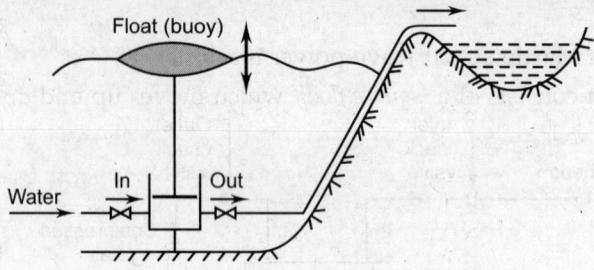

Fig. 8.17: High-level reservoir wave machine.

8.4.9.4. Dolphin-type wave-power machine (generator)

This wave-power machine was designed by a research laboratory in Japan. It consists of the following components (Fig. 8.18):

1. A dolphin;

2. A float;

3. A connecting rod;

4. Two electrical generators.

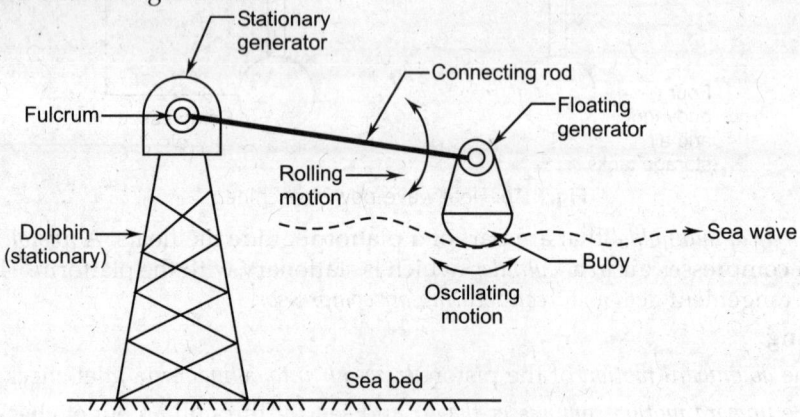

Fig. 8.18: Dolphin-type wave-power machine.

— A *"stationary generator"*, installed on the top of the structure, collects wave energy from the *"connecting rod"* with *rolling motion*. This generator has a *gear arrangement* which *rotates the rotor to generate electric power*.

— The *"buoy"* which is at the other end of the connecting rod float has two motions, namely *rolling motion* and *oscillatory motion*. The *"floating generator"* collects wave energy from the buoy through a gear arrangement and generates electric power continuously.

The power density (P/B) is given by the relation:

$$\frac{P}{B} = 1740a^2 \; T \; W/m \hspace{3cm} ...[Eqn. (8.21)]$$

where,
$\quad P$ = Power, W

$\quad B$ = Wave width, m

$\quad a$ = Amptitude of the wave, m, and

$\quad T$ = Wave period, s.

- The capacity of one dolphin type wave energy generator, normally, is 100 kW.

Example 8.4: *Calculate the installed capacity of a plant consisting of an array of dolphin type wave energy generators installed along a width 450 m. The mean amplitude of the wave is 2.2 m with a period of 9 s.*

Solution. *Given:* B = 450 m; a = 2.2 m; T = 9 s

Using the relation: $\dfrac{P}{B} = 1740\ a^2T$, we get: $\qquad$ ...[Eqn. (8.21)]

$$P = B \times 1740\ a^2T = 450 \times 1740 \times (2.2)^2 \times 9 \times 10^{-6}\ MW$$
$$= \textbf{34.1 MW (Ans.)}$$

8.4.9.5. Hydraulic accumulator wave machine

In this type of machine *low-pressure water* (at wave crust) is *pressurized and stored in a 'high-pressure accumulator'*, from which it flows through a water-turbine electric generator.

— A composite piston is composed of a large-diameter main system and a small-diameter piston at its centre. As the wave water enters through the opening, the main piston moves up and down. A closed water loop exists above the small piston. During the *upstroke*, the pressure on the main piston is *magnified*.

— The high-pressure water then moves through a one-way valve to a hydraulic accumulator at the top of the generator.

— High-pressure water flows through a hydraulic turbine that drives an electrical generator and is then discharged to a storage chamber *below the turbine*.

8.4.9.6. Nodding / Oscillating ducks wave machine

Refer to Fig. 8.19. This wave machine was designed by Stephen Salter at Edinburgh university in Scotland.

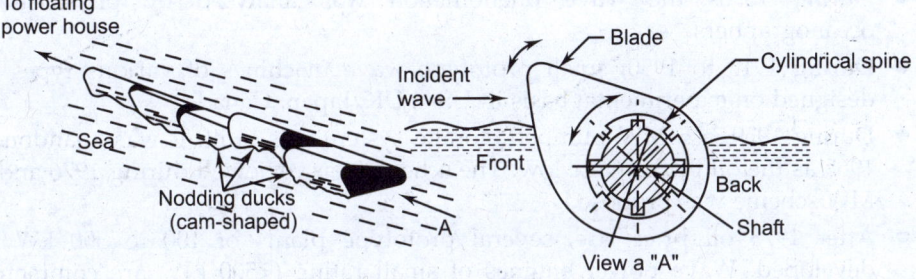

Fig. 8.19: Nodding /Oscillating wave machine

It consists of several duck-shaped devices (each 25 m long), installed in a *linear width-wise array along a line* which is *perpendicular to the direction of wave*.

— When the forward moving wave front strikes the head on the face of the ducks, wave energy is passed on and ducks start to *oscillate*.

— The ratchet and wheel mechanism conduct the oscillating motion of the ducks to the axial shaft. The wave energy is converted to mechanical energy (by the relative motion of the ducks). The shaft drives linkages, gears and the rotor of electric generator (the overall length of the cylindrical spine varies between 100 to 500 *m*).

8.4.9.7. Oscillating water column surge device

Refer to Fig. 8.20. When a moving wave is constricted, a surge is produced raising its amplitude. Such a device is known as *tapered channel device* (TAPCHAN). It comprises gradually narrowing channel with heights typically 3 to 5 m above sea level.

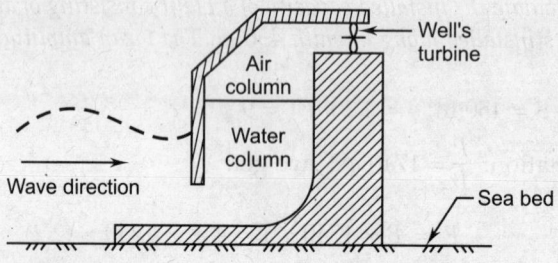

Fig. 8.20: Oscillating water column surge device.

The waves enter from the wide end of the channel and then propagate towards narrower region. The wave heights get *amplified* until the crests spills over the walls to a reservoir, which provides a stable water supply to a low head turbine.

The arrangement can be implemented successfully at *low tide sites only*.

- An off shore boating wave power vessel having TAPCHAN plant on a steel platform is suggested to *make the system insensitive to tidal range.*

8.4.10. Wave Power Development

- The history of wave machines begins with the first patent filed in parts in 1799 by Girard. The proposed plant envisaged a floating moored ship, connected to a mechanism and a pump mounted on the shore by means of a long rod. The ship oscillated vertically and the horizontal movement was prevented by mooring. The oscillating shaft motion was converted to rotary motion by mechanism at the shore; the mechanism drove the pumps.

- During 1890s the wave phenomenon was analysed by physicists and oceanographers.

- During 1910 to 1950, small prototype wave machines of various types were designed on experimental basis in USA, UK, Japan, Canada.

- During 1959, a 5 MW pumped storage system was built. It was abandoned in 1965 as the oil prices were low. The scheme was reviewed during 1976 and a 20 MW scheme was proposed.

- After 1973 oil price rise, several prototype plants of 100 to 500 kW were developed. Wave power engines of small rating (<500 kW) are commercially marketed by Japan for installing on lighthouses, buoys and remote Islands etc. Large pumped storage schemes operated by hydro-turbine pumps by wave energy were commercially successful in Mauritius (1976) for large base load power plant. Such schemes require suitable geographical contours for installing hydro-electric plants and reservoirs.

Wave power development in India:

The National Institute of Oceanography Goa has divided the Indian coastline into six zones, namely A, B, C, D, E, and F (Fig. 8.21), to expedite and identify high wave energy areas suitable for development of power.

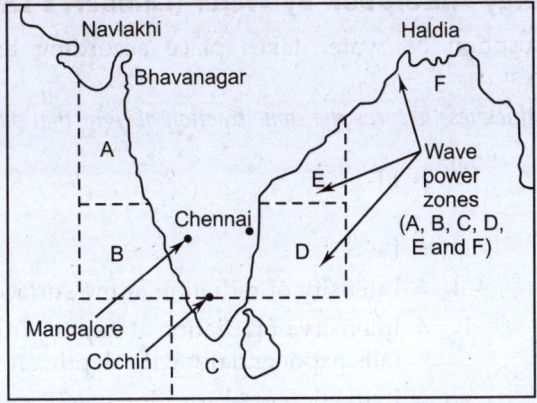

Fig. 8.21: Map of wave power zones in India.

The variation (estimated power in kW/m) in wave regime in different zones during different months is given in the table 8.5.

Table 8.5: Analysis of wave power (kW/m)

S. No.	Zone	Jan.	Feb.	Mar.	Apr.	May	Jun.	Jul.	Aug.	Sept.	Oct.	Nov.	Dec.
1.	A	3.02	3.73	3.91	4.47	6.98	26.77	39.57	24.84	10.03	2.69	3.58	4.74
2.	B	5.13	5.05	2.24	1.56	6.31	17.21	27.04	17.14	8.15	4.55	3.52	5.40
3.	C	9.26	4.45	4.05	5.50	11.44	18.85	17.69	15.34	10.11	7.21	6.67	7.52
4.	D	5.78	5.13	3.30	3.58	10.60	16.67	14.79	12.57	8.49	7.94	10.98	14.05
5.	E	4.03	1.69	2.35	3.69	11.14	17.24	17.45	16.16	9.18	6.90	9.71	5.62
6.	F	1.24	1.39	3.28	12.34	14.31	11.90	13.24	16.67	16.07	6.28	2.80	1.85

— On the estimates of the distribution of wave energy (*kW/m*) of sea frontage, the potential is seen to vary from 39 kW on the West coast to 15 kW on the East coast. On the basis of an average estimated wave power potential of 15 kW/m and total coastline of about 6000 *m* the total power potential is of the order of 90,000 MW, which is an enormous source of renewable energy which can be harnessed commercially.

8.5. OCEAN THERMAL ENERGY CONVERSION (OTEC)

8.5.1. Introduction

The oceans cover about 70% of the global surface and are particularly *extensive in the tropical zones*. Therefore, most of the sun's radiations are *absorbed by sea water*. Thus '*warm water*' (low density due to higher temperature) on the ocean's surface flows from *tropics towards poles*. '*Cold water*' (high density due to low temperature) circulates at the ocean bottom from the *poles to tropics*. Hence, in *tropical regions*, the water temperature is around **5°C** at a depth of 1000 *m*, whereas at the surface, it remains *almost constant at 25°C* (range being 24°C to 27°C) for the first few metres because of *mixing*; subsequently it *decreases and asymptotically approaches the value at the lower level*.

- Power obtained from OTEC plant is *renewable and eco-friendly*. The plant *can operate in remote islands and sea shore continuously*. According to MNRE, the overall potential of ocean energy in India may be in excess of 50,000 MW, and as such there is an enormous opportunity to this renewable source of energy.

8.5.2. Solar Energy Absorption by Water (Lambert's Law of Absorption)

Solar energy absorption by water takes place according to **Lambert's law of absorption** which states:

Each layer of equal thickness absorbs the same fraction of light that passes through it.

Mathematically, $\dfrac{dI_{(y)}}{dy} = \mu I$

or, $I_{(y)} = I_0 e^{-\mu y}$ ...(8.21)

where, I_0 = Intensity of radiation at the surface ($y = 0$),

$I_{(y)}$ = Intensity of radiation at depth y from water surface and falls exponentially with depth, and

μ = Extinction or absorption coefficient.

For very fresh clear water, $\mu = 0.05$ m^{-1}

For turbid fresh water, $\mu = 0.27$ m^{-1}

For very salty water, $\mu = 0.50$ m^{-1}

Almost all of the absorption occurs very close to the surface of deep waters. *Owing to heat and mass transfer at the surface itself, the maximum temperatures occur first below the surface.*

- With the *increase in temperature, the density of water decreases; pure water at 3.98°C has the 'maximum density'.*

- There will be *no convective currents* between the *'warmer'*, lighter water *at the top* and *'cooler'*, heavier water *at a depth.*

Similarly, heat transfer by thermal conduction between water layers at the surface and a depth is *too low* to alter the picture and the *mixing is retarded.* The 'warm water' stays at the *top* and 'cool water at' the *bottom.*

- In tropical waters there are essentially *"two infinite reservoirs"*:

(i) A *'heat source'* at about 27°C, and

(ii) A *'heat sink'* at about 4°C, at some depth of 1000 *m.*

Both these reservoirs are maintained annually by solar incidence. These temperatures vary with *latitude* and *season.*

Fig. 8.22 and 8.23 illustrate the variation of ocean surface water temperatures with latitude and season.

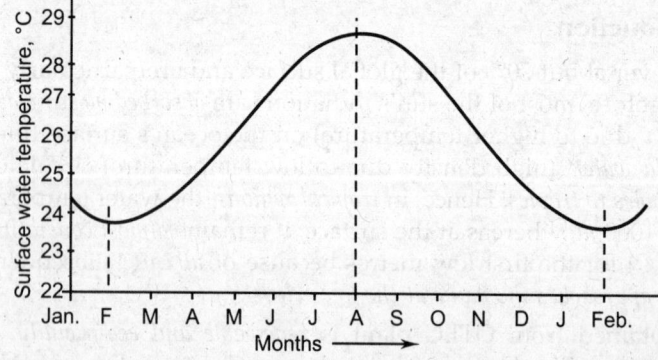

Fig. 8.22: Monthly variation of surface temperature of tropical ocean water.

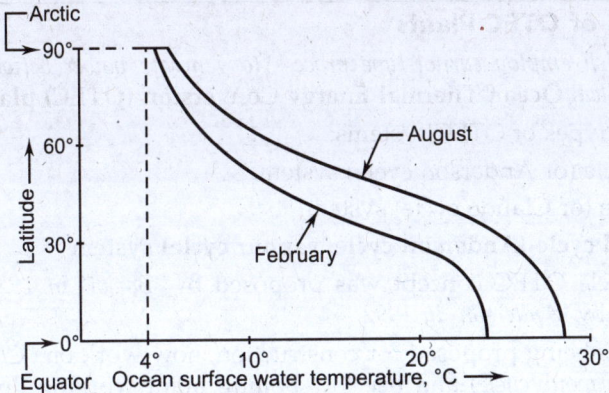

Fig. 8.23: Variation of surface temperature of tropical ocean water with latitude and season.

8.5.3. Working Principle of Ocean Thermal Energy Conversion (OTEC)

The *principle of OTEC is that "there is a temperature difference between water at the bottom of the sea and water at the top", this temperature difference can be used to operate a heat engine.* Most of the radiation is being absorbed at the surface layer of water. The mixing between hot and cold water is prevented because *no thermal convection occurs* between hot and cold water layers. This means that the *surface layer* will act as a *"source"* and *cold layers* act as a *"sink"*. Therefore, it is essential to connect the reversible heat engines between source and cold sink to produce work, that can be converted into required applications.

8.5.4. Efficiency of OTEC

The maximum efficiency of a heat engine working between two temperature limits *cannot be more than that of a Carnot cycle operating between the same temperature limits.*

The Carnot cycle efficiency, in the range of temperatures of warm water (T_1) in the upper surface layer and cold water (T_2) in the depth of the tropical ocean, is given by:

$$\eta_{carnot} = \frac{T_1 - T_2}{T_1} \qquad \qquad ...(8.22)$$

If, $t_1 = 27°C$ or $T_1 = 27 + 273 = 300$ K, and

$t_2 = 5°C$ or $T_2 = 5 + 273 = 278$ K, then

$$\eta_{carnot} = \frac{300 - 278}{300} = 0.0733 \text{ or } \textbf{7.33\%}$$

It is *only a theoretical value, but in actual practical is about 2% only.*

Example 8.4: *Determine the overall efficiency of an ocean thermal energy conversion plant if the temperature of water in the surface layer is 26°C and the temperature of cold water in the depth of the tropical ocean is 8°C. Assume the relative efficiency factor of the power plant 0.35.*

Solution. *Given:* $T_1 = 26 + 273 = 299$ K; $T_2 = 8 + 273 = 281$ K;

Efficiency factor, $EF = 0.35$.

Overall efficiency, η_{OTEC}:

$$\eta_{carnot} = \frac{T_1 - T_2}{T_1} = \frac{299 - 281}{299} = 0.06 \text{ or } 6\%$$

∴ $\eta_{OTEC} = EE \times \eta_{carnot} = 0.35 \times 6 = \textbf{2.1\% (Ans.)}$

8.5.5. Types of OTEC Plants

The plants which employ carnot-type process to generate power between the two steady temperatures are called. **Ocean Thermal Energy Conversion (OTEC) plants.**

The two basic types of OTEC systems:

1. Closed cycle (or Anderson cycle) system.
2. Open cycle (or Claude cycle) system.

8.5.5.1. Closed cycle (Anderson cycle, vapour cycle) system

The closed cycle OTEC concept was proposed by *'Barjot' in 1926.* The cycle was further developed by *'Anderson' in 1992.*

All the systems being proposed for construction, now work on *"Closed Rankine cycle'* (Anderson cycle, *vapour cycle*) and use low boiling point working fluids like *ammonia, propane, freon (R-12, R-22)* etc. These systems will be *located off-sore on large floating platforms* or *inside floating hulls.*

The *warm surface water* is used for supplying the *heat input* in the *'boiler'*, while *cold water* brought up from ocean depths is used for *extracting* the heat in the *'condenser'.*

Fig. 8.24 shows a schematic diagram of a closed or Anderson cycle OTEC plant.

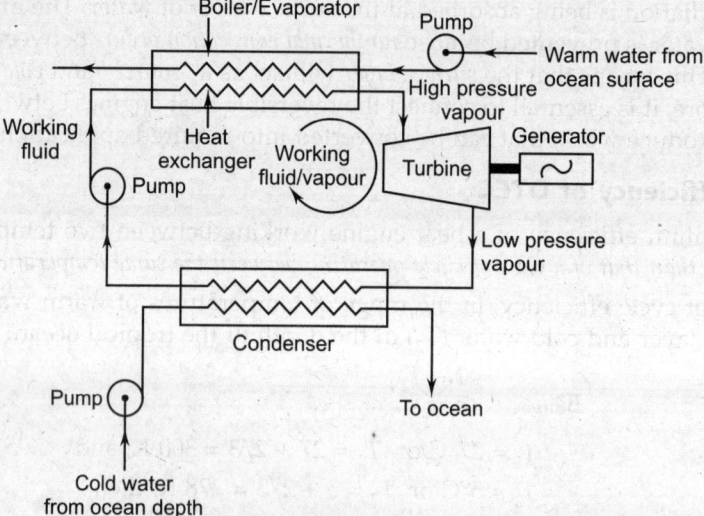

Fig. 8.24: Schematic layout of the closed or Anderson cycle OTEC plant.

Working:

— Warm water from ocean surface is circulated through a pump to a *'heat exchanger'* which acts as a *'boiler'* to generate working fluid ammonia vaour at high pressure.

— This vapour expands in the *'turbine'* to develop mechanical power, which in turn runs an electric *'generator'* to produce electric power.

— The working vapour from turbine at low pressure is condensed in the *'condenser'* with the help of cold water drawn from the depth of ocean through a pump.

The *overall efficiency* of such a plant is *very low* in the range of 2 to 3% only. Inspite of this, the concept of an OTEC system seems to be economically attractive because both the *collection and storage of solar energy is being done free by nature.*

● The *"major advantage"* of this system is that fluid evaporates at around 25°C and *does not require vacuum pumps.* The pressure at the turbine will be of the order of 9 to 6 bar resulting in compact turbines (Example: In a 1MW plant, the NH_3 turbine will have about 1.1 *m* diameter which is easy to fabricate technically).

The following points about OTEC are worth noting:

1. Each of the possible working fluids *i.e. ammonia* and *propane* has the following *advantages* and *disadvantages:*

   — "*Ammonia*" has better operating characteristics than propane and it is much less inflammable. On the other hand ammonia forms *irritating* vapour and probably could not be used with copper heat exchanger.

   — "*Propane*" is compatible with most heat-exchanger materials, but is highly flammable and forms an explosive mixture with air.

   ● *Ammonia* has been used as the working fluid in successful tests of the OTEC concept with closed cycle systems.

2. Because of the *low cycle efficiency* the heat to be transferred in the boiler and condenser is *large*. In addition, the temperature difference between the sea water and the working fluid in these heat exchangers has to be restricted to very small values. For these reasons, *very high flow rates are required for the sea water both in the boiler/evaporator and the condenser.* This results in *high pumping power requirements* and is reflected in the gross power outputs which are 20-50 percent higher than net power outputs.

   ● A second important consequence is that both the evaporator and condenser are *much larger in size than similar components in conventional practice.*

   ● The materials suggested for these heat exchangers are *titanium* or an *alloy of copper and nickel*. This is necessitated because of the *corrosive nature of the sea water.*

3. An examination of the break up of the OTEC system costs shows that the *cost of heat exchangers* plays an important role in costing; they contribute about *30 to 40 percent of the total.*

8.5.5.2. Open cycle (Claude cycle, steam cycle) system

In this system, the warm water is converted into '*steam*' in an evaporator. The steam drives steam-turbine generator to deliver electrical energy.

Fig. 8.25 shows the schematic layout of open or claude cycle OTEC plant.

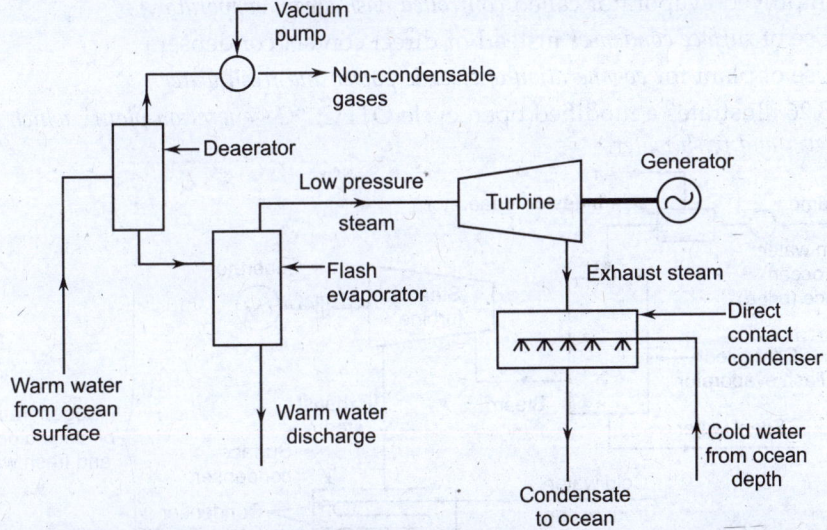

Fig. 8.25: Schematic layout of the open or claude cycle OTEC plant.

Working:

— The warm water from ocean surface is admitted through a '*deaerator*' to the '*flash evaporator*' which is maintained under *high vacuum*. As a result, *low pressure steam is generated* due to *throttling effect* and the remainder warm water is discharged back to the ocean at high depth.

- The deaerator also removes the dissolved non-condensable gases from water before supplied to the evaporate.

— The low pressure steam having very high specific volume is supplied to *'turbine'* where it expands and the mechanical power so developed is converted into electric power by the *'generator'*.

— The exhaust steam, from turbine is discharged into a direct contact type condenser where it mixes with cold water drawn from ocean at a depth of about 1 to 2 km. The mixture of condensed steam and ocean cold water are discharged into the ocean.

Advantages:

1. The warm ocean water is flash evaporated and the need for having a *surface heat exchanger is eliminated.*

2. The *portable water is obtained* when the exhaust steam from the turbine is condensed.

Disadvantages:

1. As the steam is generated at *very low pressure* (0.02 bar approx.) the *volume of steam* to be handled is *very large,* leading to a *very large diameter for the steam turbine.* (*Example:* A 1 MW OTEC plant requires a steam turbine of 12 *m* in diameter.)

2. To maintain the vacuum in the flash evaporator, massive vacuum pumps will be required.

3. The plant cost is high.

4. The plant is subjected to extremely reverse stresses.

8.5.4.3. Modified open cycle OTEC plant

The efficiency of the open cycle OTEC plant can be increased by making the following modifications:

1. Improved evaporator called *controlled flash-steam evaporator.*

2. Use of *surface condenser* instead of direct contact condenser.

3. Use of plant for *cogeneration of electric power and fresh water.*

Fig. 8.26 illustrates a modified open cycle OTEC, *"Cogeneration plant" which generates electric power and fresh water.*

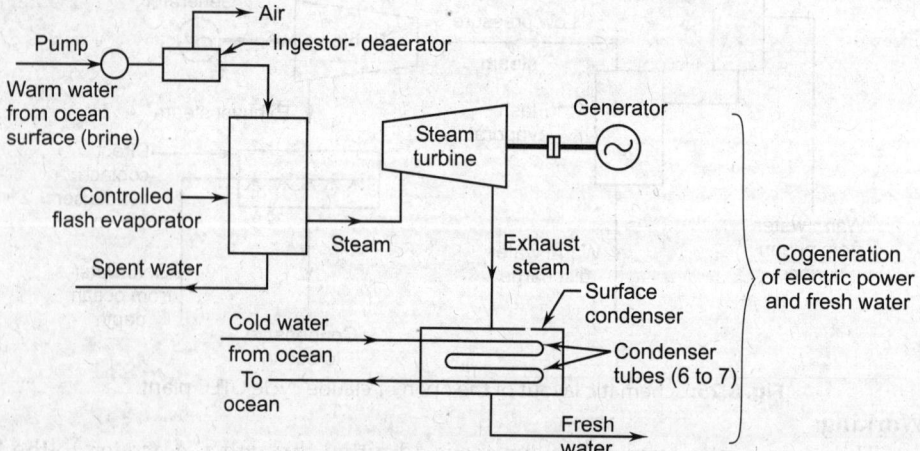

Fig. 8.26: Modified open cycle OTEC plant.

Working:

— Warm ocean water is fed into the *controlled flash evaporator* via the *ingestor-deaerator*.

— Steam collected from flash steam evaporator is supplied to the *steam turbine*. *Spent water* from the evaporator is returned to the ocean.

— Steam turbine drives the electric generator and delivers power to the network.

— Exhaust steam from the steam turbine is condensed in *'surface condenser'* and *'fresh water'* is supplied to users.

— As the *plant supplies electric power and fresh water, it is a* **"cogeneration plant"**.

— Cold water from condenser is drawn from deep ocean and after circulating through the condenser tubes the spent water is discharged into the ocean.

8.5.5.4. Hybrid OTEC system

This system combines the best features and avoids the worst features of the open and close cycle systems.

Working:

— The warm ocean water is *flash evaporated to steam, as in open cycle.*

— The heat is then transferred to *ammonia based closed Rankine cycle system.*

— The ammonia gas is then fed to the turbine, which is coupled to the generator to generate electricity.

— The ammonia goes to condenser unit and finally pumped to evaporator to repeat the cycle.

8.5.5.5. Thermoelectric OTEC system

This OTEC system was developed by Solar Energy Research Institute Colorado USA, during 1979.

It operates on the *"thermoelectric principle"* which is *simple in construction and economical.*

Fig. 8.27. shows a thermoelectric OTEC equipment. It consists of two separate packs, made of semiconductors, and covered by a thin thermal conducting sheet.

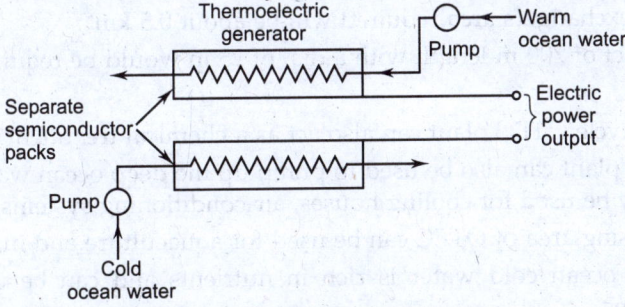

Fig. 8.27: Thermoelectric OTEC equipment.

Working:

— The *warm water* from the surface of ocean is circulated over *one device* and *cold water* pumped from the depth of ocean is allowed to flow over *other device.*

— The temperature difference of two waters with the solid state semiconductor devices generate the electric power using thermoelectric effect. The *economy of the OTEC depends on large variation of water temperature used for the surface and the deep ocean (minimum 20°C).*

8.5.5.6. Bio-fouling

Bio-fouling (the deposition and growth of micro-organisms) *is the biological impurity of sea water that deposits and grows on evaporator and condenser metal surfaces, creating thermal resistance for heat transfer.*

The total formulation of bio-fouling, corrosion and so on is referred to a **fouling** (or *scaling*) and will tend to inhibit heat transfer through it.

The *fouling factor* is a measure of the thermal resistance of a fauling film.

Provisions must be made to *inhibit* the formation of fouling layers on the surfaces of OTEC heat exchangers etc. This can be accomplished by periodically cleaning the heat exchanger surfaces through *mechanical, chemical or other means*.

8.5.6. Advantages, disadvantages / Limitations and Applications of OTEC

Following are the *advantages, disadvantages, limitations* and *applications* of OTEC:

Advantages:

1. It is clean form of energy conversion.
2. It does not occupy land areas.
3. No payment for the energy required.
4. It can be a steady source of energy since the temperatures are almost steady.

Disadvantages / Limitations:

1. About 30 percent of the power generated would be used to pump water.
2. The system would have to withstand strong convective effect of sea water; hurricanes and presence of debris and fish contribute additional hazard.
3. The materials used will have to withstand the highly corrosive atmosphere and working fluid.
4. Construction of floating power plants is difficult.
5. Plant size is limited to about 100 MW due to large size of components.
6. Very heavy investment is required.

   As an **example** for a 150 MW plant:

— A flow of 500 m^3/s would be required;

— The heat exchangers area required will be about 0.5 km^2;

— A cold duct of 700 m length with a dia. of 25 m would be required.

Applications:

1. A closed cycle OTEC plant can also act as a chemical treatment plant.
2. An OTEC plant can also be used to pump up the deep ocean water and this cold water may be used for cooling houses, air-conditioning systems etc.
3. The enclosing area of OTEC can be used for aquaculture and mariculture.
4. The deep ocean cold water is rich in nutrients and can be used for various applications.
5. OTEC plants are quite suitable for cogeneration of electricity and fresh water.
- OTEC power generation is a multipurpose project producing and supplying several useful products, like the river valley multipurpose projects.

8.5.7. Development of OTEC Plants

- In 1979, the first OTEC power plant was installed in the Hawai state of the USA:
  — It was a prototype 50 kW floating plant operated on *closed Rankine cycle* principle with NH_3 as the working fluid.
  — It was designed with the ocean water temperature difference of 21°C.
  — The *net power available was only 15 kW*, as 35 kW was consumed in pumping the warm and cold water.

- In 1981, another plant was installed in Nauru (Japan) in the Central Pacific ocean.
  — It was on-land plant and was economical in construction.
  — It required a 945 m long pipeline for pumping cold water.
  — Its gross power capacity was 100 kW, operated with a sea water temperature difference of 21.7°C. The net output power was 31.5 kW.
  — The turbine used was an axial flow type with 3000 rpm, the generator was directly coupled and supplied power at 415 V, 50 Hz.

Development of OTEC in India:

The National Institute of Ocean technology is implementing the world's first 1 *MW floating OTEC technology demonstration project off the Tutocorin coast in Tamil Nadu.*

— The various subsystems for the plant have been configured, designed and integrated on an OTEC floating barge.
— A 1 km long cold water pipe has been towed to the site and deployed vertically with an *anchoring system at a depth of 1.2 km.*
- The resource potential of India is estimated around 180×10^3 MW.

8.5.8. Environmental Impacts of Wave Power

The *"Wave power"* is *essentially non-polluting*. No appreciable environmental effects are forseen from island floating wave power devices.

— Onshore wave energy installations may *change visual landscape and degrade scenic ocean front views.* It may also *cause disturbance to marine life* including changes in distribution and types of life near the shore.
— There is *possible threat to navigation* from *collision due to low profile floating wave devices.* It would usually be both possible and necessary to avoid hazards to, or from, marine traffic by judicious planning and by the *provision of navigation aids.*

"Tidal energy" is a renewable source of energy, which *does not result in emission of gases responsible for global warming or acid rain associated with use of fossil fuels.*

— Changing tidal flow by *damming a bay or estuary could*, however, *result in negative impacts on aquatic and shoreline ecosystem, as well as navigation and recreation.*
- Studies have shown that the *environment impact at each site is different*, and *depends amply upon local geography.*

8.6. MINI AND MICRO HYDEL POWER PLANTS

8.6.1. Introduction

Small hydropower is *covered in the renewable energy power.* It can supply, in principle, significant amounts of electricity for:

(*i*) Irrigation; (*ii*) Portable water pumping; (*iii*) Lighting; (*iv*) Health; (*v*) Educational purposes, etc.

In order to meet with the present energy crisis *partly*, a solution is to develop **mini** *(5 m to 20 m head)* and **micro** *(less than 5 m head)* hydel potential in our country.

The low head hydropotential is scattered in our country and estimated potential from such sites could be as much as 20,000 MW.

By proper planning and implementation, it is possible to commission a small hydro-generating set-up of 5 MW with a period of one and half year against the period of a decade or two for large capacity power plants. Several sets upto 1 MW each have been already installed in Himachal Pradesh, UP, Arunachal Pradesh, West Bengal and Bhutan.

To reduce the cost of micro-hydel stations than that of the cost of conventional installation the following *considerations* are kept in view:

1. The *civil engineering work* needs to be kept to a *minimum* and designed to fit in with already existing structures e.g. irrigation, channels, locks, small dams etc.

2. The machines need to be manufactured in a *small range of sizes of simplified design*, allowing the use of unified tools and aimed at reducing the cost of manufacture.

3. These installations must be *automatically controlled to reduce attending personnel.*

4. The *equipment must be simple and robust, with easy accessibility to essential parts for maintenance.*

5. The *units must be light and adequately subassembled for ease of handling and transport* and to *keep down erection and dismantling costs.*

- **Microhydel plants** *(micro-stations) make use of standardised "bulb sets" with unit output ranging from 100 to 1000 kW working under heads between 1.5 to 10 metres.*

8.6.2. Advantages and Disadvantages / Limitations of Small Hydropower (SHP) Potential

Following are the *advantages and disadvantages / limitations of small hydropower potential:*

Advantages:

1. *Non-polluting* and *release no heat.*

2. *Non-cosumptive generator of electrical energy utilising a renewable resource,* which is made continually available through the hydrologic cycle of environment.

3. *Low gestation period* ranging from 8 to 24 months (as compared to other conventional energy generation machines).

4. *Low* operation cost.

5. An *ideal source of decentralised energy generation.* A small hydroplant can supply energy to rural feeders thereby cutting distribution losses to a large extent.

6. Can be *synchronised with grid;* synchronisation with utility grids *improves voltage profile.*

Disadvantages:

1. *Higher capital costs on per kW installation basis.*

2. Relatively low utilisation.

3. Higher managerial and administrative costs.

4. *Unstable operation* of isolated systems due to low inertia.

5. *Several problems crop up due to the disbursed nature of projects* in site identification, preparation of reports, proper construction planning etc.

8.6.3. Classification of Small Hydropower (SHP) Stations

According to the Central Electricity Authority (CEA) and Ministry of New and Renewable Energy (MNRE), SHPs are *classified* as follows (MNRE report 2005):

1. **Based on capacity range:**

	Category	Unit size
(i)	Micro	upto 100 kW
(ii)	Mini	101 to 1000 kW
(iii)	Small	1 to 25 MW.

2. **Based on available head:**

   (*i*) Ultra low head................... Below 3 *m*

   (*ii*) Low head............................. Above 3 *m* and upto 40 *m*

   (*iii*) Medium / high head .... Above 40 *m*.

"Small hydel schemes", on the basis of available head, can be further categorised as given below:

1. *Independent scheme:*

In such schemes a *stream flow is captured, regulated and developed* mainly for *"power generation only"*. For such a scheme, the *low heads are not found economical* for using as independent generation.

   — These schemes are successful in *Himachal Pradesh, Jammu and Kashmir, Nagaland and Manipur* etc.

2. *Sub ordinate schemes:*

These schemes are primarily meant for *irrigation* and *driking water;* the power generation being the secondary aim.

For such schemes, the feasible sites are *indogagetic plains (canals) and Peninsular India (reservoirs).*

Classification of Water Turbines

The water turbines are *classified* on the following bases:

   (*i*) Action of flowing water on turbine blades,

   (*ii*) The existing head and the quality of water available,

   (*iii*) The direction of water flow on turbine blades, and

   (*iv*) The name of the invertor.

Water Turbines:

  1. **Impulse turbine (high head):**

   (*i*) *Pelton wheel:* 60 to 700 *m* head

   (*ii*) *Turgo:* 30 to 210 *m* head

   (*iii*) *Ossberger crossflow:* 1 to 200 *m* head.

  2. **Reaction turbine:**

   (*a*) *Propeller turbines:*

   (*i*) Low head (Axial flow)

   (*ii*) *Kaplan:* 25 to 40 *m*

   (*iii*) *Tube:* 2 to 15 *m*

   (*iv*) *Bulb:* 1.3 to 25 *m*

   (*v*) *Strafflo:* 2 to 25 *m*

   (*b*) *Francis turbine:* Medium head (40 to 300 *m*) mixed flow (radial + axial).

   • For *details of water turbines* etc. refer to *"Chapter 15"*.

8.6.4. Major Component of Small Hydropower Projects

Fig. 8.28 shows a typical arrangement of small hydropower station. The basic and common components of a hydroelectric scheme are:

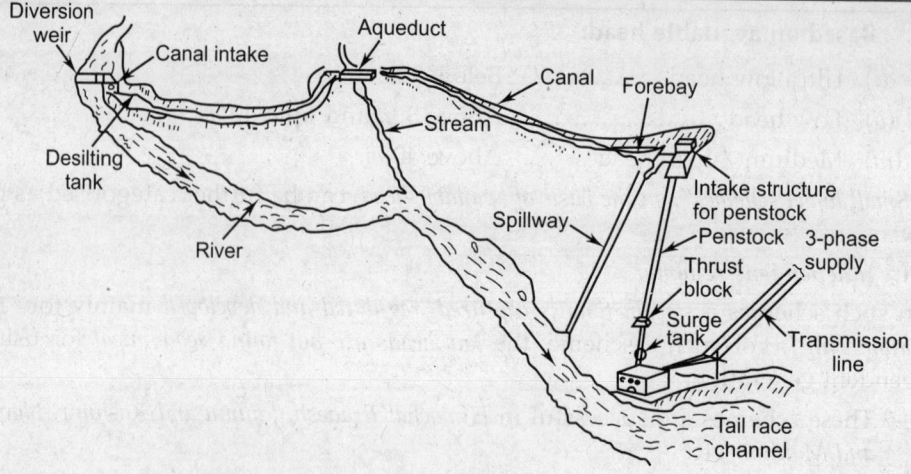

Fig. 8.28: A typical arrangement of small hydropower station.

1. Diversion and intake;
2. Desilting chamber;
3. Water conductor system;
4. Forbay / balancing reservoir;
5. Surge tank (if necessary);
6. Penstock, and
7. *Power house.*

● *Some small hydro electric projects in India are:*

1. Micro hydroelectric projects in Shansha (Keylong in Lahaul and Spiti valley) HP.
2. Micro hydel project in Kakori (Sonepat in Haryana)
3. Western Yamuna canal hydroelectric project in Yamuna nagar (Haryana).

HIGHLIGHTS

1. *Ocean energy sources:* Tidal energy; Wave energy; OTEC; Hydroelectric energy emanated from sun-ocean system from mechanism of surface water evaporation by solar heating *i.e.* hydrological cycle.

2. The periodic rise and fall of water level of sea which are carried by action of the sun and moon on water of the earth is called *tide.*

3. The *tidal range (R)* is the difference between consecutive high tide and low tide water.

4. *Energy* and *power* in a *single-basin single-effect/cycle* scheme:

Energy, $W = \frac{1}{2} g\rho AR^2$ joules

Power (Average), $P_{av} = 0.225 \, AR^2$ watts

Power density, $\frac{P_{av}}{A} = 0.225 \, R^2 \, W/m^2$,

where, A = Area of basin, m^2,

                       ρ = Density of water, kg/m^3,

                       g = Gravitation constant,

m = Mass flowing through the turbine, kg,

h = Head, m, and

W = Work done by water flowing through turbine, J.

5. *Yearly power generation* from a tidal project:

$$P_{yearly} = \int_0^t \frac{\rho Q h}{75} \, dt \times \eta_0 \times 0.736 \times 705 \text{ kWh/year}$$

(On average, there are 705 tidal cycles in a year)

6. *Wave energy* comes from the interaction between the winds and surfaces of oceans. The typical ranges of ocean waves are:

● Wave height = 2 × amplitude = 0.2 m to 4 m

● Wave period = 4 s to 12 s

7. The relation between λ (wavelength) and T (wave period) is given by:

$$\lambda = 1.56 \, T^2$$

8. The energy available in random sea is expressed as:

$$P = 0.96 \, H^2 T \text{ kW/m of wave crust}$$

where, H = The wave height, m, and

T = The wave period.

9. *Energy* and *power* from waves:

Potential energy, $PE = \dfrac{1}{4} g\rho Ba^2\lambda$

Kinetic energy, $KE = \dfrac{1}{4} g\rho Ba^2\lambda$

Total energy, $E = PE + KE = \dfrac{1}{2} g\rho Ba^2\lambda$

Energy density $= \dfrac{E}{A} = \dfrac{1}{2} g\rho a^2$ $(\because \text{Area, } A = \lambda B)$

Wave power, $P = \dfrac{1}{2} g\rho B\eta a^2 f$ watts

Power density, $\dfrac{P}{A} = \dfrac{1}{2} g\rho a^2 f \text{ W/m}^2$

where, ρ = Density of water, kg/m^3,

a = Amplitude, m,

λ = Wavelength, m,

B = Wave width, and

$f \left(= \dfrac{1}{T} \right)$ = Frequency of waves, T is the wave period in seconds.

10. *Wave motion* is primarily horizontal, but the motion of water is primarily vertical.

11. The plants which employ carnot-cycle process to generate power between the two steady temperatures are called Ocean Thermal Energy Conversion (OTEC) Plants.

12. Microhydel plants (micro-stations) make use of standardised 'bulb set' with unit output ranging from 100 to 1000 kW working under heads between 1.5 to 10 metres.

THEORETICAL QUESTIONS

1. Define the term 'Oceanography'.
2. How can the ocean energy sources categorised? Explain briefly.
3. What do you understand by tidal energy?
4. How can the energy be obtained from the tides of the sea?
5. What is tide?
6. What is the main feature of the tidal cycle?
7. What is tidal range? Explain briefly.
8. Explain how tides are produced.
9. What is the origin of tides?
10. Derive expression for 'energy' and 'power' in a single-basin, single-effect/cycle system.
11. Derive an expression for yearly power generation from a tidal project.
12. What are the site requirements for a tidal power plant?
13. Explain briefly the various components of a tidal power plant.
14. Give the classification of tidal power plants.
15. Explain briefly the 'single-basin double-effect tidal plant.
16. What is the difference between single-basin and double-basin tidal plants.
17. Explain with a neat diagram the working of double-basin, paired-basin scheme. What are its limitations?
18. List the advantage, disadvantages and limitations of tidal power.
19. Write a short note on the global scenario of tidal energy.
20. Discuss briefly the economic aspects of tidal energy conversion.
21. What is wave energy? Give the typical ranges of ocean wave.
22. What are the advantages and disadvantages of wave energy?
23. What problems are associated with wave energy?
24. Explain the factors which affect wave energy?
25. What is a progressive wave? Discuss its parameters.
26. Derive the expressions for energy and power from waves.
27. Discuss the characteristics of waves in real oceans.
28. Write a short note on 'Collection of wave data'.
29. Give the classification of wave energy conversion machines.
30. Explain briefly the following wave-power machines:
    (i) Float wave-power machine.
    (ii) High-level reservoir wave machine.
31. Explain with a neat diagram the 'Dolphin-type' wave-power machine.
32. Describe the working of a hydraulic accumulator wave machine.
33. Explain with a neat sketch the working of a 'Nodding/oscillating ducks machine.
34. Describe the working of an oscillating water column surge device.
35. Write a short note on 'Wave power development'.
36. What is the working principle of Ocean Thermal Energy Conversion (OTEC)?
37. What is the efficiency of OTEC?
38. What are the types of OTEC systems? Explain any one of them briefly.
39. Describe with a neat diagram the working of a 'closed cycle OTEC system'.
40. Explain briefly with a neat sketch the construction and working of a 'open cycle OTEC system'. State its advantages and disadvantages also.
41. Describe with a sketch the working of a modified open cycle OTEC plant used for getting electric power as well as fresh water.
42. What is a hybrid OTEC system?
43. Write short note on "Thermoelectric OTEC".
44. What is 'Bio-fouling'?
45. Give the advantages, disadvantages and applications of OTEC?
46. Write a short note on 'Development of OTEC plants'.
47. What are the 'Environment impacts of wave power'?

9

DIRECT ENERGY CONVERSION (DEC)

9.1. Introduction; 9.2. Thermoelectric effect – Seebeck effect-Peltier effect-Thomson effect - Joule effect; **9.3. Thermoelectric generator** – Construction and working - Thermoelectric materials and their selection – Advantages and disadvantages of thermoelectric power generator – Power output and efficiency; **9.4. Thermionic generator/converter** – Introduction – Thermionic generator – Thermionic convertor materials – Advantages, disadvantages/limitations and applications of thermionic converter; **9.5. Magneto Hydro Dynamic (MHD) generation** – Introduction – Principle of MHD power generation – *MHD systems:* Classification – open cycle MHD system – close cycle MHD systems – Materials for MHD generators – Advantages and disadvantages/drawbacks of MHD system – Voltage and power output of MHD generator – Parameters governing power output; **9.6. Electro Gas-dynamic (EGD) generator: 9.7 Fuel cells** Introduction – Advantages, disadvantages and applications of fuel cell-Components and working theory of a fuel cell – Classification of fuel cells – Requirements of electrolyte and electrode – Desirable characteristics of a fuel cell – Hydrogen-oxygen fuel cell (Hydrox cell) – Alkaline fuel cell (AFC) – Phosphoric acid fuel cell (PAFC) – Polymer electrolyte membrane fuel cell (PEMFC) – Molten carbonate fuel cell (MCFC) – Solid oxide fuel cell (SOFC) – Regenerative cell – Performance analysis of a fuel cell – Fuel cell power plant – Advantages of fuel cell power plants; **9.8. Photovoltaic power system** – Photovoltaic generators – historical background – Photovoltaic cell – Basic photovoltaic system for power generation – Fabrication of cells – Advantages and disadvantages of photovoltaic solar energy conversion; **9.9. Electrostatic Mechanical generators; 9.10. Nuclear batteries.** *Highlights – Theoretical Questions.*

9.1. INTRODUCTION

The devices which convert naturally available energy into electricity, without an intermediate conversion, into mechanical energy; energy source may be thermal, solar or chemical are called **Direct energy conversion (DEC) devices.**

Until now, their use has been *confined to small scale, special purpose applications* since the *voltage output available* with them is *rather small* and no inexpensive device that is reliable like a turbine or alternator has been built.

Under this topic the following *devices* will be discussed:

1. Thermoelectric generator.
2. Thermoionic generator.

3. Magneto Hydro Dynamic (MHD) power generation.

4. Electro Gas-Dynamic generator (EGD).

5. Fuel Cells.

6. Electrostatic mechanical generators.

7. Solar cells.

8. Nuclear batteries.

9.2. THERMOELECTRIC EFFECTS

The quest for a *reliable, silent, energy converter* with *no moving parts* that transforms heat to electrical power has led engineers to reconsider a set of phenomena called the *"Thermoelectric effects"*. These effects, known for over a hundred years, have permitted the development of *small, self contained electrical power sources*.

The various thermoelectric effects, in relation to the thermodynamic conversion of heat to work, are as follows:

1. Seebeck effect.

2. Peltier effect. ⎫

3. Thomson effect. ⎬ Reversible processes

4. Joule effect. ⎭

9.2.1. Seebeck Effect

*Seebeck Effect states:

If *two dissimilar materials are joined to form a loop and two junctions maintained at different temperatures, an e.m.f. will be set up around the loop.* The magnitude of current depends on the *materials* (1 and 2) and the *temperature difference between the junctions* (Fig. 9.1.)

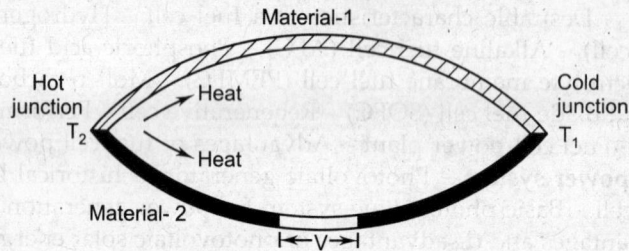

Material-1

Hot junction T_2 Heat   Heat   Cold junction T_1

Material- 2

|◄—V—►|

Fig. 9.1. Seebeck effect—The principle of thermocouple.

The magnitide of e.m.f. (*E*) is given by:

$$E = \alpha_s \, \Delta T \qquad \qquad ...(9.1)$$

where, α_s = Seebeck coefficient, and

             ΔT = Temperature difference between hot and cold junctions.

● This effect has been used in *"Thermocouples"* to *measure temperatures*.

9.2.2. Peltier Effect

Refer to Fig. 9.2.

Peltier effect states:

"When an electric current flows across an isothermal junction of two dissimilar materials, their is either an evolution or absorption of heat at the junction".

---

* Seebeck effect was discovered by the German scientist seebeck in 1822.

Peltier's coefficient $\alpha_{1\text{-}2}$ is defined as the *heat evolved or absorbed as the junction per unit current flow, per unit time*.

Mathematically, $\alpha_{p1\text{-}2} = \alpha_{p1} - \alpha_{p2} = \dfrac{Q_p}{I}$ ...(9.2)

and, $Q_p = \alpha_{p1\text{-}2}I$ ...(9.3)

where, I = Peltier heat per unit time.

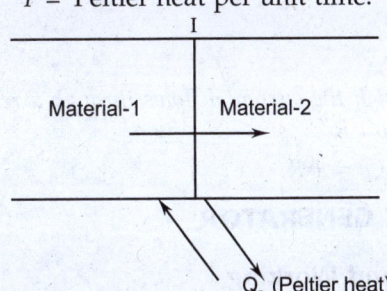

Fig. 9.2. Peltier effect.

9.2.3. Thomson Effect

Refer to Fig. 9.3.

Thomson effect states:

"Any current carrying conductor with a temperature difference between two points will either absorb or emit heat, depending upon the material".

The *'Thomson's coefficient'* is defined as *the heat absorbed (or evolved) per unit time per unit electric current and per unit temperature gradient.*

Mathematically, $\sigma = \dfrac{dQ_t/dx}{dT/dx}$ ...(9.4)

where, $\dfrac{dQ_t}{dx}$ = Heat interchange per unit time per unit length of conductor, and

    $\dfrac{dT}{dx}$ = Temperature gradient.

Hence, the *Thomson heat per unit time* is given by:

$$\dfrac{dQ_t}{dx} = \sigma I \dfrac{dT}{dx}$$ ...(9.5)

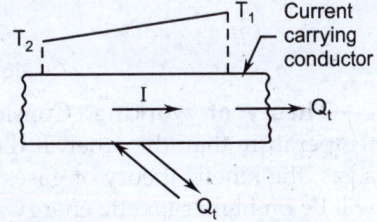

Fig. 9.3. Thomson effect.

- The Seebeck, Peltier and Thomson effects are more *pronounced* in "semiconductors" as compared to metals.

The relations between seebeck, peltier and thomson coefficients as derived by Kelvin are as follows:

1. The relation *between seebeck effect and peltier effect:*

$$\left. \begin{array}{l} \alpha_p = \alpha_s . T \\ \alpha_{p1\text{-}2} = \alpha_{s1\text{-}2} \, T \end{array} \right\}$$ ...(9.6)

2. The relation between *seebeck effect and thomson effect:*

$$\sigma = T \cdot \frac{d\alpha_s}{dT} \Bigg\}$$

$$\text{or,} \qquad\qquad \sigma_{1-2} = T \frac{d\alpha_{s1-2}}{dT} \Bigg\} \qquad\qquad\qquad ...(9.7)$$

9.2.4. Joule Effect

Joule effect states:

"In a closed electric circuit if the current I flows through a resistance R, the heat generated (Q) by the resistance is equal to I^2R":

Mathematically, $\qquad Q = I^2R \qquad\qquad\qquad\qquad\qquad ...(9.8)$

9.3. THERMOELECTRIC GENERATOR

9.3.1. Construction and Working

Fig. 9.4 shows a schematic diagram of a *thermoelectric power generator*. It uses seebeck effect to produce electrical energy directly from the available heat input. Its thermal efficiency is *very low*, of the order of 1 to 3%. In any heat engine the efficiency of thermoelectric generator *depends upon the temperatures of hot and cold junctions*.

The thermocouple materials A and B are joined at the hot end, but the other ends are kept cold; an electric voltage or electromotive force is then generated between the cold ends. A direct current (D.C.) will flow in a circuit or load connected between these ends. The flow of current will continue as long as the heat is supplied to the hot *junction and removed from the cold ends*. For a given *thermocouple*, the voltage and electric power output are *increased by increasing the temperature difference between the hot and cold ends*.

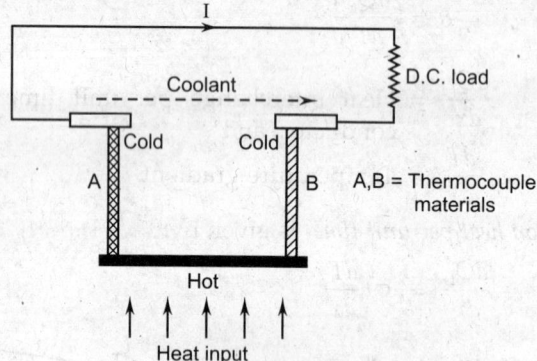

Fig. 9.4. Thermoelectric generator.

Theory of working. Consider a metal bar where one side is kept at a higher temperature than the other. If the free electrons in the metal are considered to behave as a gas, the kinetic theory of gases predicts that the free electrons in the hot side of the bar will be on higher kinetic energy and will be moving at *greater speed* than those in the cold side of the bar. As the fast moving electrons flow from the hot side to the cold side of the bar, it results in an *accumulation of negative charge at the cold side, preventing further build up until circuit is closed. In a closed circuit the current will flow to reduce the charge built up and will continue as long as the temperature is maintained.*

In a practical thermoelectric converter, several thermocouples are convected in series to *increase both voltage and power* as shown in Fig. 9.5. If the output voltage is insufficient to operate a particular device or equipment, it can be *increased*, with little loss of power, by an *inverter transformers combination*. The direct current generated by the thermocouples is first *changed into alternating current (A.C.)* of essentially the same average by means of an *"inverter"*. The alternating current and voltage is then increased to the desired value with the help of *a transformer*. The high voltage alternating current *can be reconverted into direct current if required, by the use of a rectifier*.

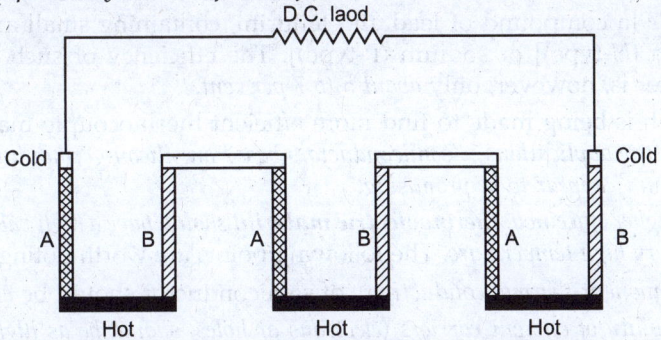

Fig. 9.5. Thermocouples in series (to increase voltage).

The source of heat for a thermoelectric generator may be a small oil or gas burner, a radio-isotope or direct solar radiation.

A typical couple operating with hot and cold junction temperatures of 600°C and 200° C could be designed to give about 0.1 V and 2 A *i.e.*, about 0.5 W, so that a 1kW device could require about 5000 couples in series.

- Taking into account mechanical characteristics, stability under operating conditions and ease of fabrication, **Bismuth telluride** *appears to be most suitable material*. It can be *alloyed* with such materials as Bismuth selenide, Antimony telluride, lead selenide and tin telluride to give improved properties.

- Research is being carried out on the possibility of using thermoelectric devices within the core of a nuclear reactor. The hot junction would be located on the fuel element and the cold junctions in contact with the coolants.

Note. A thermoelectric converter is a form of *heat engine* which takes up heat at an upper temperature *(hot junction)* converts it partly into electrical energy and discharges the remaining part at a *lower temperature* (cold junction). The efficiency of a thermocouple, as is the case with other heat engines, increases by increasing the upper temperature and decreasing the lower temperature. Since the lower temperature is usually that of environment the efficiency of a thermocouple, practically, depends upon the *hot junction temperature*.

9.3.2. Thermoelectric Materials and their Selection

The following materials find use in the making of thermoelectric elements:

Material	Formula	Figure of merit, Z (°K⁻¹)
1. *Lead telluride*	$PbTe$	1.5×10^{-3}
2. *Bismuth telluride* (doped with Sb or Se)	$Bi_2 Te_3$	4×10^{-3}
3. *Germanium telluride* (with bismuth)	$GeTe$	1.5×10^{-3}
4. *Cesium sulphide*	CeS	1.0×10^{-3}
5. *Zinc antimonide* (doped with silver)	$Zn Sb$	1.5×10^{-3}

where Z is an *index used in rating thermoelectirc converters*. It depends on the *properties of thermoelectric materials used*. A high value of Z (Figure of merit) is obtained by using materials of:

    (*i*) *Large seebeck co-efficient*   (*ii*) *Small thermal conductivity*

 (*iii*) *Small electrical resistivity*.

- In recent times, the *most commonly used material for thermoelectric converters* is **lead telluride** [a compound of lead and tellurim, containing small amounts of either bismuth (*N*-type)] or sodium (P-type)]. The efficiency of such a thermoelectric converter is, however, only *about 5 to 7 per cent*.

- Research is being made to find more efficient thermocouple materials. For '*high temperature applications*', "*semiconductors based on silicon-germanium and compounds of selenium*" appear to be promising.

To *achieve higher efficeincy*, **thermoelectric material** *should have a high value of Z and be able to operate upto very high temperature*. The following points are worth noting in this regard:

1. The *component thermal conductivity* of semiconductor should be *as low as possible*.
2. The *mobility of current carriers* (*electrons or holes, should be as high as is compatible with condition* 1).
3. One of the arms should consist of a *purely hole type* and the other of a *purely electronic type* semiconductor.
4. In the low temperature zone the impurity concentration should be *lower* than in the higher temperature zone.
5. So that the thermoelectric material *may not crack* under the effect of stresses it should possess the following properties:
   - (*i*) It should be able to *resist chemical influences* such as oxidation etc.
   - (*ii*) It should have *good mechanical strength*.
   - (*iii*) It should be amply elastic.

9.3.3. Advantages and Disadvantages of Thermoelectric Power Generator

Following are the *advantages and disadvantages* of thermoelectric power generator.

Advantages:

1. *Highly reliable*.
2. *Free from noise* due to the absence of parts.
3. *Compact* and *durable*.
4. *Minimum maintenance*.
5. *Portable* (can be used at any location).
6. *Uses low grade thermal energy*.

Disadvantages:

1. Low output.
2. Low efficiency.
3. High cost.

9.3.4. Power Output and Efficiency of Thermoelectric Power Generator

Refer to Fig. 9.6.

Let, T_1 = Temperature of the source, K

                             T_0 = Temperature of the sink, K,

L = Length of thermoelectric material, m,

A = Area of thermoelectric materials, m^2

k = Thermal conductivity of material(s) W/mK,

ρ = Electrical resistivity of materials, Ωm,

K = Thermal conductance of material(s) $\left(\dfrac{kA}{L}\right)$, W/K,

R = Electrical resistance of elements(s), $\left(\dfrac{\rho L}{A}\right)$, Ω,

α = Absolute value of seebeck coefficient(s), $\left(\dfrac{V}{K}\right)$, and

π = Absolute value of peltier coefficient,

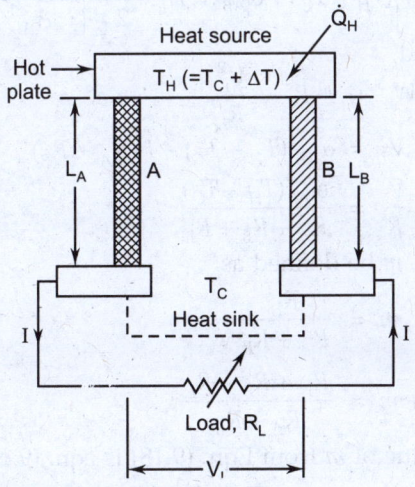

Fig. 9.6. Analysis of thermoelectric generator.

$\dot{Q}_H$ = The heat rate into hot plate,

$\dot{Q}_k$ = Heat conducted into two legs,

$\dot{Q}_j$ = Joulean heat rate, and

$\dot{Q}_p$ = Peltier heat rate at the junction due to flow of current in the circuit.

Consider hot plate at control volume.

The seebeck voltage (V) generated as a result of current flow I in the circuit when the load R_L is placed in the circuit is given by :

$$V = \alpha_{AB} \cdot (T_H - T_C) \qquad \qquad \text{...(9.9)}$$

According to first law of thermodynamics, the *energy balance equation* can be written as:

$$\underbrace{\dot{Q}_H}_{\substack{\text{Heat rate into the} \\ \text{hot plate}}} + \underbrace{\frac{\dot{Q}_j}{2}}_{\substack{\text{Joulean heat rate flowing into } each \\ \text{junction (Assuming half the} \\ \text{heat appears at each junction)}}} = \underbrace{\dot{Q}_k}_{\substack{\text{Heat rate} \\ \text{conducted} \\ \text{into two legs}}} + \underbrace{\dot{Q}_p}_{\substack{\text{Peltier heat rate} \\ \text{at the junction due} \\ \text{to current flow} \\ \text{in the circuit}}}$$

i.e.
$$\dot{Q}_H = \dot{Q}_k + \dot{Q}_p - \frac{\dot{Q}_j}{2} \qquad \qquad \qquad ...(9.10)$$

where,
$$\dot{Q}_p = \pi_{AB}.I = \alpha_{AB}. I. T_H \qquad \qquad ...(9.11)$$

$$\dot{Q}_j = I^2 (R_A + R_B) = I^2 \left[\frac{\rho_A \cdot L_A}{A_A} + \frac{\rho_B \cdot L_B}{A_B} \right] \qquad ...(9.12)$$

$$\dot{Q}_k = (K_A + K_B) (T_H - T_C)$$

$$= \left[\frac{k_A \cdot A_A}{L_A} + \frac{k_B \cdot A_B}{L_B} \right] (T_H - T_C) \qquad ...(9.13)$$

Inserting these values, we obtain:

$$\dot{Q}_H = \left[\frac{k_A \cdot A_A}{L_A} + \frac{k_B \cdot A_B}{L_B} \right] (T_H - T_C) + \alpha_{AB}.I.T_H - \frac{1}{2} I^2 \left[\frac{\rho_A \cdot L_A}{A_A} + \frac{\rho_B \cdot L_B}{A_B} \right] \qquad ...(9.14)$$

Useful power generated,

$$\dot{W}_L = I^2.R_L = \frac{V_L^2}{R_L} \qquad \qquad ...(9.15)$$

where,
$$V_L = \alpha_{AB} (T_H - T_C) - I (R_A + R_B) \qquad \qquad ...(9.16)$$

or,
$$I = \frac{V}{R} = \frac{\alpha_{AB}(T_H - T_C)}{R_A + R_B + R_L} \qquad \qquad ...(9.17)$$

Let the resistance ratio, *m* be defined as:

$$m = \frac{R_L}{R_A + R_B} \qquad \qquad ...(9.18)$$

Hence,
$$1 + m = \frac{R_A + R_B + R_L}{R_A + R_B}$$

On substituting the value of *m* from Eqn. (9.18) is eqn. (9.17), we get:

$$I = \frac{\alpha_{AB}(T_H - T_C)}{(1 + m)(R_A + R_B)} \qquad \qquad ...(9.19)$$

Now, *power output* from Eqn. (9.15), (9.18) and (9.19) can be written as:

$$\dot{W}_L = \frac{[\alpha_{AB}(T_H - T_C)]^2}{(1 + m)^2 (R_A + R_B)^2} \times m(R_A + R_B)$$

or,
$$\dot{W}_L = \frac{m}{(1 + m)^2} \frac{[\alpha_{AB}(T_H - T_C)]^2}{R_A + R_B} \qquad ...(9.20)$$

From Eqn. (9.14), the **rate of heat input** $\left(\dot{Q}_H \right)$ becomes:

$$\dot{Q}_H = (K_A + K_B) (T_H - T_C) + \alpha_{AB} \left[\frac{\alpha_{AB}(T_H - T_C)}{(1 + m)(R_A + R_B)} \right].T_H$$

$$- \frac{1}{2}(R_A + R_B) \left[\frac{\alpha_{AB}^2(T_H - T_C)^2}{(1 + m)^2 (R_A + R_B)^2} \right]$$

or,
$$\dot{Q}_H = (K_A + K_B) (T_H - T_C) + \frac{\alpha_{AB}^2 \cdot T_H(T_H - T_C)}{(1 + m)(R_A + R_B)}$$

$$- \frac{1}{2} \frac{\alpha_{AB}^2}{(1 + m)^2} \times \frac{(T_H - T_C)^2}{R_A + R_B} \qquad ...(9.21)$$

Efficiency of generator (η_{gen}) can be written as:

$$\eta_{gen} = \frac{W_L}{Q_H}$$

On substituting the values of W_L and Q_H from Eqns. 9.20 and (9.21), and on solving, we get:

$$\eta_{gen} = \left(\frac{T_H - T_C}{T_H}\right)\frac{m}{(1+m) - \frac{1}{2}\left(\frac{T_H - T_C}{T_H}\right) + \frac{(K_A + K_B)(R_A + R_B)(1+m)^2}{\alpha_{AB}^2 \cdot T_H}} \quad ...(9.22)$$

We define *figure of merit* (Z) as:

$$Z = \frac{\alpha_{AB}^2}{(R_A + R_B)(K_A + K_B)}$$

— Z consists of material properties of semiconductors A and B only. Thus *efficiency will increase with increase in Z.*

— The seebeck effect can be +ve or –ve. High magnitude of Z is required with high value of seebeck coefficient (opposite polarity), low resistivity and low thermal conductivity.

— For high efficiency, ZT should be as high as possible.

The maximum power output and maximum efficiency are given by the relations:

$$Z_{opt} = \left[\frac{\alpha_{AB}}{\sqrt{\rho_A \cdot k_A} + \sqrt{\rho_B \cdot k_B}}\right]^2 \quad ...(9.23)$$

Maximum power output,

$$(W_L)_{max} = \frac{1}{4}\left[\frac{\alpha_{AB}^2 (T_H - T_C)^2}{(R_A + R_B)}\right] \quad ...(9.24)$$

Maximum efficiency, η_{max}

$$\eta_{max} = \frac{T_H - T_C}{T_H}\left[\frac{1}{2 - \frac{1}{2}\left(\frac{T_H - T_C}{T_H}\right) + \frac{4}{Z_{opt} \cdot T_H}}\right] \quad ...(9.25)$$

9.4. THERMIONIC GENERATOR/CONVERTER

9.4.1. Introduction

A **Thermionic converter (or generator)** *converts heat energy directly to electrical energy by utilising thermionic emission effect.*

In this device, *electrons act as the working fluid* (or gas) *and electrons are emitted from the surface of the heated metal.*

A thermionic converter can be analysed from at least *three* different points of view:

1. In terms of *thermodynamics,* it may be viewed as a heat-engine that uses an electron gas as a working substance.

2. In terms of *electronics,* it may be viewed as a *diode* that transforms heat to electricity by the law of thermionic emission.

3. In terms of *thermoelectricity,* it may be viewed as a thermocouple in which an evacuated space or a plasma has been substituted for one of the conductors.

Regardless of the point of view adopted in analysis, a *thermionic converter works because of the phenomenon of "thermionic emission". Thermionic emission implies emission of electrons from the metal when it is heated.*

Work function (ϕ):

It is *defined as the energy required to extract an electron from the metal*. It is measured in *electron volts*. The value of work function varies with the *nature of the metal and its surface condition.*

A thermionic converter, in principle, consists of *two metals or electrodes with different work functions sealed into an evacuated vessel*. The electrode with a *large work function* is maintained at a *higher temperature* than one with the smaller work function.

9.4.2. Thermionic Generator

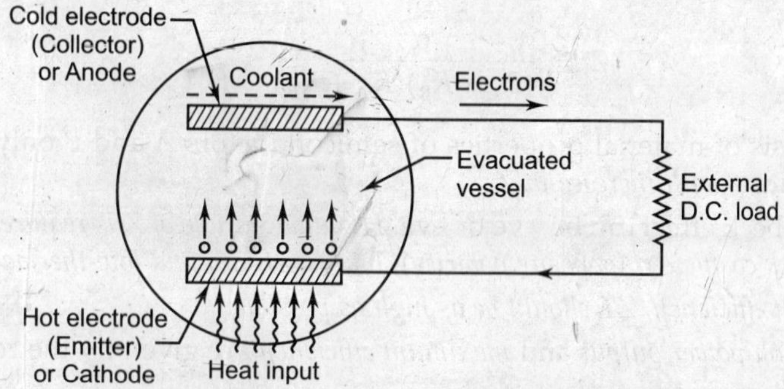

Fig. 9.7. Thermionic generator.

Refer to Fig. 9.7.

Construction and working:

— A thermionic converter/generator comprises a *heated* **cathode** (electron *emitter*) and an **anode** (electron *collector*) *separated by 'vacuum'*, the electrical output circuit being connected between the two as shown in the Fig. 9.7.

— The heat which is supplied to the cathode raises the energy of its electrons to such a level that it enables than, to *escape from the surface and flow to anode. At the 'anode' the energy of electrons appears partially as heat, removed by cooling and partially as electrical energy delivered to the circuit.*

— Although the distance between anode and cathode is only about 1mm. The negative space charge with such an arrangement hinders the passage of the electrons and *must be reduced*, this can be achieved by *introducing positive ions* into the inter electrode space, *'cesium vapour' being valuable source of such ions.*

- In order to materialise a substantial electron emission rate (per unit area of emitter), and hence a significant current output as well as a high efficiency, the *emitter temperature in a thermionic converter containing cesium should be at least* 1000°C, the efficiency is *then 10 per cent*. Efficiency as high as 40 per cent can be obtained by operating at *still higher temperatures*. Although temperature has little effect on the voltage generated, the increase in current (per unit emitter area) associated with a temperature increase results in increase in power. Electric power (P) is the product of voltage (E) and current (I) *i.e.*, $P = EI$.
- *Anode materials* should have a *low work function e.g. barium and strontium oxides* while that of the ***cathode*** should be *considerably higher, tungsten impregnated with a*

barium compound being a suitable material. Even with these materials temperatures upto 2000°C will be required to secure for the generator itself, efficiencies of 30–35 per cent. Electrical outputs of about 6 W/cm^2 of anode surface are envisaged with about 13 W/cm^2 removed by coolant.

- A thermionic generator, in principle, *can make use of any fuel (may be fossil fuel, a nuclear fuel or solar energy) subject to the condition that* sufficiently high temperatures are obtainable. The thermionic conversion can be utilized in several different situations—*remote locations on the earth and in space.*

9.4.3. Desirable of Properties of Thermionic Converter Materials

The problem of developing materials suitable for use in thermionic converters ranks next to the space charge control problem in the development of efficient thermionic generators.

Following *properties are desirable in materials suitable for converters:*

Emitter:

A good emitter should have the following *properties:*

1. High-electron emission capability coupled with a low rate of deterioration.
2. *Low emissivity*, to reduce heat transfer by radiation from the emitter.
3. High mechanical strength.
4. High melting point.
5. It should be such that in the event some of it vapourizes and subsequently condenses on the collector it will not poison the collector (that is, change the collector properties, thereby making it less effective).

The relative importance of these properties is dependent upon the type of converter being designed. It should be noted that efficiency is a much slower rising function of electron emission capability if space charge is present than if there is no space charge.

The work function may be reduced considerably by an absorbed single layer of foreign atoms. This comes about by the establishment of a dipole layer at the surface. The layer can be formed by atoms or molecules. This is essentially what happens in a cesium converter, which is designed so that cesium condenses on the emitter or collector.

Collector:

The *main criteria for choosing a 'collector material' is that it should have as low a work function as possible.*

— Because the collector temperature is held below any temperature that will cause significant electron emission, its actual emission characteristics are of no consequence. The lower the collector work function (ϕ_c), however, the less energy the electron will have to give up as it enters the collector surface.

— In practice the lowest value of ϕ_c that can be maintained stably is *about 1.5 eV.*

— For applications in which it is desirable to maintain the collector at elevated temperatures (greater than 900°K) such as space applications, an *optimum value of ϕ_c may be determined.*

— *Molybdenum* has been *widely used as a collector*; it is frequently *assumed* to have a work function of *1.7 eV.*

9.4.4. Advantages, Disadvantages/Limitations and Applications of Thermionic Converter

Following are the *advantages, disadvantages/limitations and applications* of thermionic converter:

Advantages:

1. Compact device.
2. High conversion efficiency.
3. Quiet in operation and durable.
4. Low cost and less maintenance required.
5. Absence of rotating parts.
6. Operates at high temperatures.
7. Can be developed for very low power to very high power generation.
8. Can operate in remote areas and harsh environments.

Disadvantages/Limitations :

1. Metal is costly as it has to withstand high temperatures.
2. Needs high operating temperatures at anode.
3. Needs special seal to protect the cathode from corrosive gases.
4. Needs cesium vapour in the tube to reduce the space charge.

Applications:

1. Power generation in both centralised and distributed systems.
2. Residential and Commercial purposes.
3. Automobiles, marine and air vehicles.
4. Electronics and telecommunications.
5. Aerospace and military systems.

9.4.5. Analysis of Thermionic Generator

Fig. 9.8, shows the characteristic curve of thermionic converter.

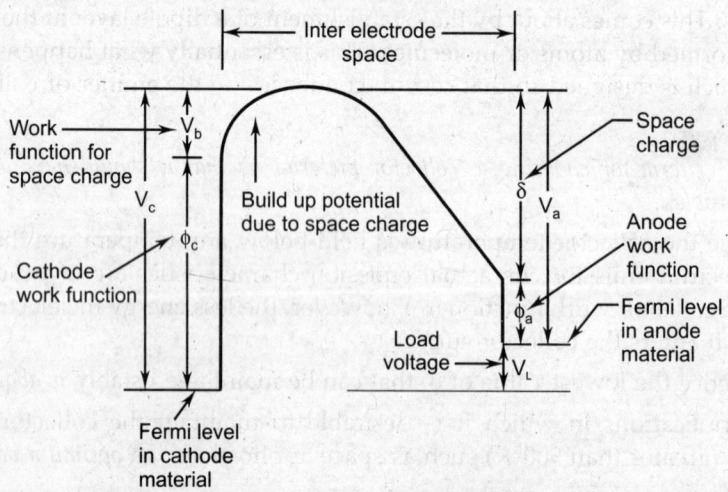

Fig. 9.8. Characteristic curve of thermionic converter.

Let,
$$\phi_c = \text{Cathode work function,}$$
$$\phi_{a'} = \text{Anode work function,}$$

V_b = Kinetic energy barrier at cathod (Work function for space charge)

V_L = Load voltage,

δ = Space charge,

T_c = Temperature of cathode, and

T_a = Temperature of anode.

The total energy required at cathode = $\phi_c + V_b$

Current density of cathode (J_c) is given by:

$$J_c = AT_c^2 \exp\left(\frac{-V_c}{KT_c}\right) \qquad ...(9.26)$$

Similarly, current density of *anode* (J_a) is given by:

$$J_a = AT_a^2 \exp\left(\frac{-V_a}{kT_a}\right) \qquad ...(9.27)$$

Load voltage (output), $V_L = V_c - V_a = \phi_c - \phi_a$

Energy required for the electrons to reach at *anode*,

$$Q_1 = (\phi_c + V_b) \times J_c = V_c J_c \qquad ...(9.28)$$

The electrons also carry the energy in the form of *kinetic energy* ($2KT_c$), given by:

$$Q_2 = J_c \cdot \frac{2KT_c}{e} \qquad ...(9.29)$$

$\therefore$ Total energy needed for electrons to reach at anode,

$$Q_c = Q_1 + Q_2$$

$$= J_c V_c + J_c \cdot \frac{2KT_c}{e}$$

or, $$Q_c = J_c\left(V_c + \frac{2kT_c}{e}\right) \qquad ...(9.30)$$

Similarly *back emission from anode carries the energy*, given by:

$$Q_a = J_a V_a + J_a \frac{2KT_a}{e} \qquad ...(9.31)$$

i.e. $$Q_a = J_a\left(V_a + \frac{2KT_a}{e}\right) \qquad ...(9.32)$$

The *net energy* supplied to the *cathode*,

$$Q_{net} = J_c\left(V_c + \frac{2KT_c}{e}\right) - J_a\left(V_a + \frac{2KT_a}{e}\right) \qquad ...(9.33)$$

Power output, $P = V_L (J_c - J_a)$ $\qquad ...(9.34)$

Efficient of converter, $\eta = \dfrac{P}{Q_{net}}$

$$= \frac{V_L(J_c - J_a)}{J_c\left(V_c + \dfrac{2KT_c}{e}\right) - J_a\left(V_a + \dfrac{2KT_a}{e}\right)}$$

or,
$$\eta = \frac{(V_c - V_a)(J_c - J_a)}{J_c\left(V_c + \frac{2KT_c}{e}\right) - J_a\left(V_a + \frac{2KT_a}{e}\right)} \qquad ...(9.35)$$

$$(\because V_L = V_c - V_a)$$

Let,
$$\frac{V_c}{KT_c} = \gamma_c; \quad \frac{V_a}{KT_a} = \gamma_a; \quad \frac{T_a}{T_c} = \beta$$

Substituting the values in Eqn. (9.35) and similifying, we get:

or
$$\eta = \frac{(\gamma_c - \beta\gamma_a) \times [1 - \beta^2 \exp(\gamma_c - \gamma_a)]}{(\gamma_c + 2) - \beta^2(\gamma_a + 2\beta)\exp(\gamma_c - \gamma_a)} \qquad ...(9.36)$$

If,
$$\gamma_a = \gamma_c = \gamma, \text{ then}$$

$$\eta_{max} = (1 - \beta)\frac{\gamma}{\gamma + 2}\left[\frac{1 - \beta^2}{1 - \frac{\beta^2(\gamma + 2\beta)}{\gamma + 2}}\right] \qquad ...(9.37)$$

But,
$$\frac{1 - \beta^2}{1 - \frac{\beta^2(\gamma + 2\beta)}{\gamma + 2}} \simeq 1$$

$$\therefore \qquad \eta_{max} = (1 - \beta)\frac{\gamma}{\gamma + 2} \qquad ...(9.38)$$

9.5. MAGNETO HYDRO DYNAMIC (MHD) GENERATION

9.5.1. Introduction

Magnetohydrodynamics is *concerned with the flow of a conducting fluid in the presence of magnetic and electric fields.*

MHD generator *is a device which converts heat energy of fuel directly into electrical energy without a conventional electric generator.*

— *MHD* converter system is a *heat engine* whose efficiency, like all engines, is increased by supplying the heat at the highest practical temperature and rejecting it at the lowest practical temperature.

— MHD generation looks the most promising of the direct energy conversion techniques for *large scale production* of *electrical energy.*

9.5.2. Principle of MHD Power Generation

Faraday's law of electromagnetic induction states:

"When a conductor and a magnetic field move in respect to each other, an electric voltage is induced in the conductor".

The conductor need not be solid — it may be a *gas or liquid.*

MHD generator uses this principle by *forcing a high-pressure high-temperature combustion gas through a strong magnetic field.*

Fig. 9.9, shows the comparison between a *turbo generator and the MHD generator:*

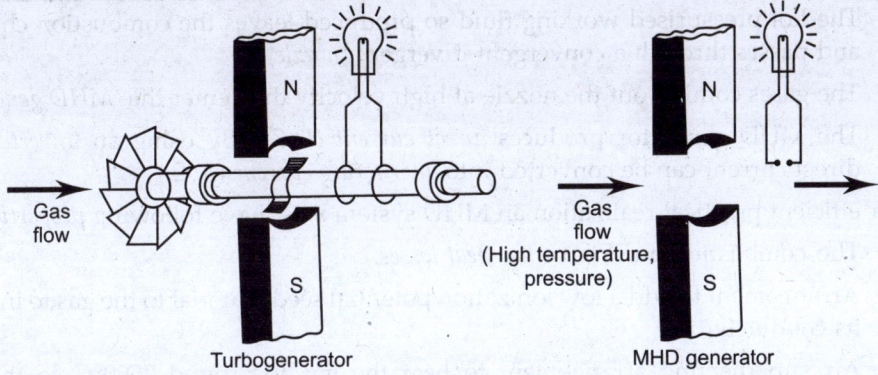

Fig. 9.9. Comparison between the conventional turbogenerator and the MHD generator

9.5.3 MHD Systems

9.5.3.1. Classification

The *broad classification* of the MHD systems is as follows:

1. Open-cycle systems.
2. Closed-cycle systems :
   (*i*) Seeded inert gas systems
   (*ii*) Liquid metal systems.

9.5.3.2. Open-cycle MHD system

Fig. 9.10, shows the schematic diagram of an open-cycle MHD system.

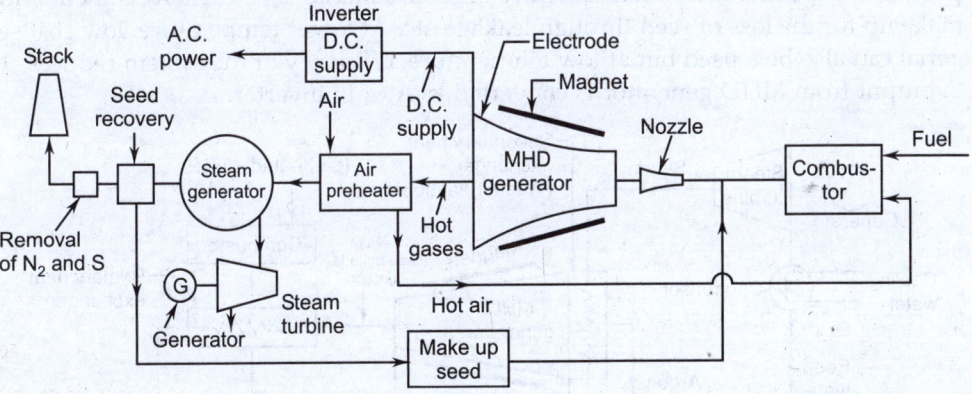

Fig. 9.10. Open-cycle MHD system

Construction and Working:

— The fuel (*e.g. Oil, coal, natural gas*) is burnt in *'combustion chamber'*, air required for combustion is supplied from *'air preheater'*.

— The hot gas produced by the combustion chamber are then *seeded* with a small amount of an ionized *alkali metal* (cesium or potassium) to *increase* the electrical conductivity of the gas.

— The *ionization of potassium* (generally *potassium carbonate* is used as *seed material*) takes place due to gases produced at temperature of about 2300-2700°C by combustion.

— The hot pressurised working fluid so produced leaves the combustion chamber and passes through a convergent-divergent 'nozzle'.

— The gases coming out the nozzle at high velocity then enter the 'MHD generator'.

— The MHD generator produces *direct current* (D.C.). By using an 'inverter' this direct current can be converted into *alternating current* (A.C.)

For efficient practical realization an MHD system *must* have following *properties:*

1. The combustion must have *low heat losses.*

2. Arrangement to add a low ionization potential seed material to the gas to increase its conductivity.

3. Air superheating arrangement to heat the gas to around 2500°C so that the electrical conductivity of the gas is increased.

4. Seed recovery apparatus — necessary for both environmental and economic reasons.

5. A magnet capable of producing high magnetic flux density (5 to 7 teslas).

6. A water cooled but electrically insulating expanding duct with long life electrodes.

9.5.3.3. Closed-cycle MHD systems

1. Seeded inert gas closed-cycle MHD system:

This MHD system works on *Brayton cycle* with inert carrier gas (Argon/helium).

Refer to Fig. 9.11. In seeded inert gas system, electrical conductivity is maintained in the working fluid by ionization of a *seed material.* The inert gas is *compressed* and heated in primary heat exchanger. Small quantity of seed material (*e.g.,* cesium) is them added to make up for the loss of seed through leakage etc. At lower temperature, low cost seed material can also be a used but at low temperature, efficiency of the system reduces. The D.C. output from MHD generator is converted to A.C. in inverter.

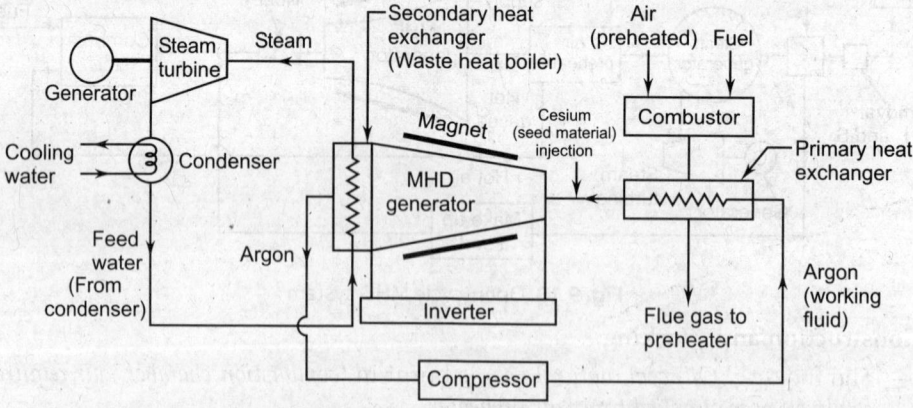

Fig. 9.11. Seeded inert gas closed-cycle MHD generator.

2. Close-cycle liquid metal MHD system

Fig. 9.12 shows the schematic of a *closed-cycle liquid metal system.* In this system, a liquid metal (*Potassium*) is used *as a working fluid.*

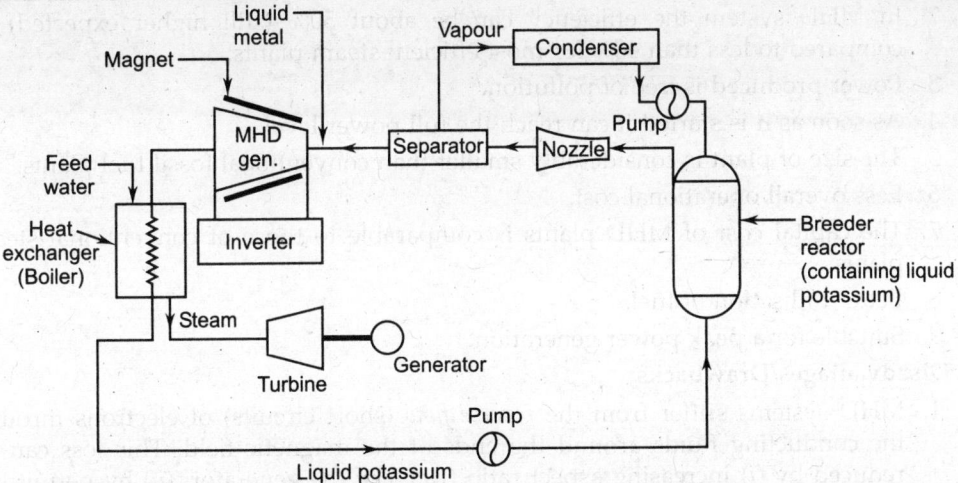

Fig. 9.12. Closed-cycle (liquid metal) system.

Construction and Working:

— The liquid potassium after being heated in the *'breeder reactor'* is passed through the *'nozzle'* where its velocity is increased.

— The vapours formed due to nozzle action are separated in the *'separater'* and *condensed* and them pumped back to the reactor.

— Then the liquid metal with high velocity is passed through *'MHD generator'* to produce D.C. power.

— The liquid possium coming out of MHD generator is passed through the *heat exchanger'* (boiler) to use its remaining heat to run a turbine-generator set and then pumped back to the reactor.

● The *system entails many constructional and operational difficulties.*

9.5.4. Materials for MHD Generators

Owing to high temperature of plasma (2700°C), refractory materials are required in various parts of MHD generator like electrodes, channel or duct wall.

The materials for various parts are given in Table 9.1 below:

Table 9.1. Materials for MHD generators

S. No.	Components	Required properties	Materials
1.	*Electrodes*	High conductivity, stable at high temperature	Zirconia based materials; chromite based materials, Aluminate based materials, Nickel oxide based materials.
2.	*Channel or duct*	Thermal and electrical insulator	$MgAl_2O_3$, Beta alumina ($KAl_{11}O_{17}$), MgO
3.	*Magnet*	Stronger than permanent magnet	Tin, Niobium, Niobium 33, Niobium-titanium
4.	*Insulator*	Withstand high temperature	Oxides, Nitrides, Zirconates, etc.

9.5.5. Advantages and Disadvantages/Drawbacks of MHD Systems

Following are the *advantages* and *disadvantage/drawbacks* of MHD system:

Advantages:

1. More reliable since there are *no* moving parts.

2. In MHD system the efficiency can be about 50% (still higher expected) as compared to less than 40% for most efficient steam plants.

3. Power produced is free of pollution.

4. As soon as it is started it can reach the full power level.

5. The size of plant is considerably smaller than conventional fossil fuel plants.

6. Less overall operational cost.

7. The capital cost of MHD plants is comparable to those of conventional steam plants.

8. Better utilisation of fuel.

9. Suitable for a peak power generation.

Disadvantages/Drawbacks:

1. MHD systems suffer from the *reverse flow* (short circuits) of electrons through the conducting fluids around the ends of the magnetic field. This loss can be reduced by (*i*) increasing aspect ratio (L/D) of the generator, (*ii*) by permitting the magnetic field poles to extend beyond the end of electrodes, and (*iii*) by using insulated vanes in the fluid ducts and at the inlet and outlet of the generator.

2. There will be *high friction losses* and *heat transfer losses*. The friction loss may be as high as 12% of the input.

3. The MHD system operates at very high temperatures to obtain high electrical conductivity. But the electrodes must be relatively at low temperatures and hence the gas in the vicinity of the electrodes is cooler. This *increases the resistivity of the gas* near the electrodes and hence there will be a very large voltage drop across the gas film. By *adding the seed material, the resistivity can be reduced.*

4. The MHD system needs very *large magnets* and *this is a major expense.*

5. Coal, when used as a fuel, poses the problem of molten ash which may *short circuit the electrodes.* Hence, *oil or natural gas are considered to be much better fuels for this system.* This restriction on the use of fuel makes the operation *more expensive.*

9.5.6. Voltage and Power Output of MHD Generator

When a charged particle (having charge q) moving at a velocity v towards right is subjected to a perpendicular magnetic field (pointing into the paper), a magnetic force F acts on the charged particle (Fig. 9.13). This force is given by:

$$\vec{F} = q(\vec{v} \times \vec{B}) \qquad \qquad ...(9.39)$$

Force F, velocity v and magnetic flux density B are vectors.

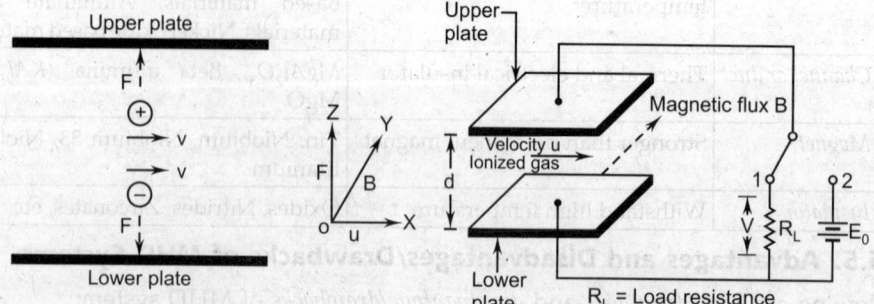

Fig. 9.13. Motion of a charged particle in magnetic field.

Fig. 9.14. Directions of magnetic flux, gas velocity and force in MHD system.

If an electric field E is also present in the region, the total force is:

$$\vec{F} = q(\vec{E} + \vec{v}\,\vec{B})$$ ...(9.40)

The velocity to be used in the Eqn. (9.40) is the *vector sum* of gas velocity u and the particle drift velocity v. Thus, force is given by:

$$\vec{F} = q(\vec{E} + \vec{v}\,\vec{B} + \vec{u}\,\vec{B})$$ ...(9.41)

Eqn. (9.41) can be written as:

$$\vec{F} = q(\vec{E'} + \vec{u}\,\vec{B})$$ ...(9.42)

where,

$$\vec{E'} = \vec{E} + \vec{v}\,\vec{B}$$

Refer to Fig. 9.14. The motion of the gas (v) is in x-direction, magnetic field (B), is in the y-direct and the force (F) on the particle is in z-direction.

A resistance R (load) is connected between the two plates and current I flows, developing a voltage V across the load.

The electric intensity between the plates is:

$$\vec{E_z} = -\frac{V}{d}$$ ...(9.43)

where, d is the distance between the plates. The total electric field is:

$$\vec{E_z'} = \vec{E_z} + \vec{B}u = -\frac{V}{d} + \vec{B}u$$

$$= \frac{1}{d}(Bud - V)$$ ...(9.44)

The electromagnetic field E_z and B acting on the moving gas produce the same force on the ions as the electromagnetic field E_z' and B produce on a gas with *zero average velocity*. Evidently the open circuit voltage, E_0 is;

$$E_0 = Bud$$ ...(9.45)

Let,

R_G = Internal resistance of the generator

$$= \frac{d}{\sigma A}$$ ...(9.46)

where,

σ = Conductivity of the gas, and

A = Plate area.

Maximum power output (P_{max}) is obtained when:

Internal resistance = Load resistance

i.e.

$$R_G = R_L$$

$$P_{max} = E_0 I$$

$$= I \times R_G \times I = I^2 R_G$$

$$= \left(\frac{E_0}{R_L + R_G}\right)^2 R_G$$

or,

$$P_{max} = \left(\frac{E_0}{R_a + R_G}\right)^2 R_G$$

i.e. $$P_{max} = \frac{E_0^2}{4R_G} = \frac{B^2u^2d^2}{4R_G} \qquad (\because E_0 = Bud) \qquad ...(9.47)$$

Inserting the volue of R_G from eqn. (9.46), we get:

$$P_{max} = \frac{B^2u^2d^2}{4 \times \left(\dfrac{d}{\sigma A}\right)} = \frac{B^2u^2d^2\sigma A}{4d}$$

i.e. $$P_{max} = \frac{1}{4}\sigma B^2 u^2 d.A \qquad ...(9.48)$$

and, *Maximum power per unit volume* $= \dfrac{1}{4}\sigma B^2 u^2$ ...(9.49)

9.5.6. Parameters Governing Power Output

The parameters which govern the power output, for *constant plate area and spacing* are:

1. Gas velocity,
2. Flux density, and
3. Gaseous conductivity.

1. **Gas velocity.** The power output *varies as square of the gas velocity*, therefore it is advantageous to obtain as high a velocity as possible. This is achieved by *converting thermal energy* of the gas emerging from the furnace into *direct kinetic energy*. This conversion is carried out through the *conventional nozzle expansion techniques*.

It is possible to achieve velocities of the order of 1000 m/s.

2. **Flux density.** The power output also *varies as square of magnetic flux density*. In conventional magnets flux density upto 3 *Wb/m²* can be obtained.

— Magnets made of superconducting materials are being developed. With such magnets flux densities around 10 *Wb/m²* can be obtained. The *cost of the magnetic system of an MHD system is *quite enormous*.

3. **Gaseous conductivity.** The power output *varies directly as conductivity*. The gas must have a reasonable value of conductivity between 10 and 1000 *mho/m* to obtain a reasonable power output from MHD generator.

The conductivity can be *increased* by *introducing a seeding material into the gas*. The *ionization potential* of the seeding material is *lower* than that of gas atoms and thus this material *ionizes more readily* than the gas and *increases the conductivity*. 'Cesium' is the *best seeding material* having an *ionization potential of 3.89 eV*, but it is *very costly*. The next best is 'Potassium' which has an ionization potential of 4.34 eV.

9.6. ELECTRO GAS-DYNAMIC GENERATOR (EGD)

The EGD generator *uses the potential energy of a high pressure gas to carry electrons from a low potential electrode to a high potential electrode, thereby doing work against an electric field.*

A schematic diagram of EGD is shown in Fig. 9.15.

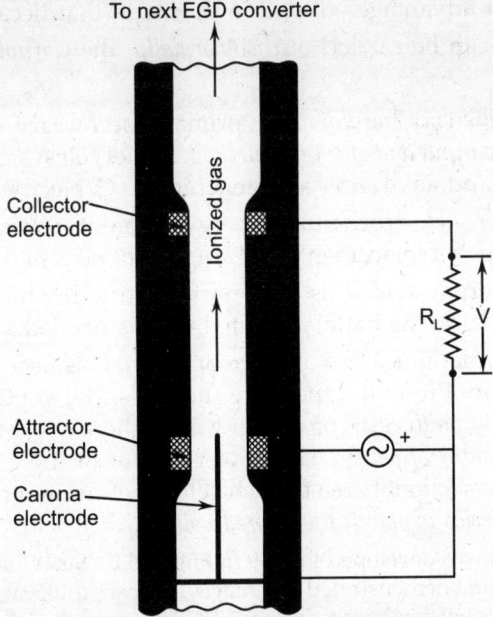

Fig. 9.15. Gas duct in an EGD converter.

'*Carona electrode*' at the entrance of the duct generates electrons. This ionised gas particles are carried down the duct with the neutral atoms and the ionized particles are neutralised by the '*collector electrode*', at the end of the insulated duct. The working fluid in these systems are commonly, either combustion gases produced by burning of fuel at high pressures, or it is a pressurised reactor gas coolant. The maximum power output from EGD is about 10 to 30 W per channel. Hence, several thousand channels are connected in series and parallel. The voltage produced is very high, of the order of 1,00,000 to 2,00,000 V. Thus, it needs *very good high voltage insulators*(Beryllium Oxide is generally used).

- EGD can produce a *high efficiency equal to MHD-steam combination.*

Advantages of EGD over MHD systems:

1. EGD systems operate at relatively low temperatures.
2. No need for injection and recovery of seed material.
3. This system is self contained since it does not need a steam generator.
4. Energy can be extracted till the gases reach almost the stack temperature.
5. It does not need large quantities of condenser cooling water.

EGD and MHD hold the prospects of offering the best solutions for high efficiency, large capacity systems for the production of electricity.

9.7. FUEL CELLS

9.7.1. Introduction

A **fuel cell** *is an electrochemical device in which the chemical energy of a conventional fuel is converted directly and efficiently into low voltage, direct current electrical energy.*

- Fuel cell systems generally *operate on pure hydrogen and air to produce electricity.*

- One of the chief advantages of such a device in that because the conversion, at least in theory, can be carried out *isothermally*, the *Carnot limitation on efficiency does not apply.*

- The essential *difference* between the primary/secondary cell and *fuel cell* is of *continuous energy input and output of fuel cell.* A fuel cell system requires continuous supply of a *fuel* and an *oxidizer* and generates D.C. electric power continuously.

- **A battery** has *stored* electrochemical energy within its container. After discharge it needs recharging or replacement. *Fuel cells do not need such recharging replacement.* A fuel cell is often described as a primary battery in which the fuel and oxidize are stored external to the battery and fed to it as needed.

- Fuel cells can be manufactured as large or as small as necessary for the particular power application. Presently, there are fuel cells that are the *size of a pencil eraser* and generate few *milliwatts* of power while there are others large enough to provide *large amount of power.* The power output of fuel cells is fully *scalable* by varying the cross-sectional area of each cell to get desired current and by *stacking multiple cells in series to obtain the desired voltage.*

Note. The first fuel cell was developed in 1839 in England by Sir William Grove. However, the application of fuel cell was first demonstrated by Francis T. Bacon in 1959 when his model generated 5kW at 24V. Its practical application began during the 1960s when the US space programme chose fuel cells over nuclear power and solar energy. Fuel cells provided power to the Gemini, Apollo and Skylab spacecraft, continue to be used to provide electricity and water to space shuttles.

9.7.2. Advantages, Disadvantages and Applications of Fuel Cells

Following are the *advantages, disadvantages* and *applications of fuel cells*:

Advantages:

1. Conversion efficiencies are high.
2. Require little attention and less maintenance.
3. Can be installed near the use point, thus reducing electrical transmission requirements and accompanying losses.
4. Fuel cell is odourless and does not make any noise.
5. A little time is needed to go into operation.
6. Space requirement considerably less in comparison to conventional power plants.
7. Simple and safe.
8. Pollution free.
9. No cooling water needed.
10. Capacity can be increased as the demand grows.
11. Long life.

Disadvantages:

1. High initial cost.
2. Low service life.
3. Problems for refilling in vehicles.

Applications:

The applications of fuel cell relate to:

1. Domestic use.
2. Automotive vehicles.

3. Central power stations.
4. Defence applications.
5. Space projects.

● *Fuel cells are primarily suited for low voltage and high current applications.*

Note. The human body functions essentially like a fuel cell. Living things take in food (fuel) and oxygen to produce both thermal energy and work output. They are *not heat engines.*

9.7.3. Components and Working Theory of a Fuel Cell

Components. The main components of a cell are:

1. Anode (Fuel electrode)   2. Cathode (oxidant electrode)
3. Electrolyte   4. Container
5. Separators   6. Sealings
7. Fuel supply   8. Oxidizer.

Working theory:

Fig. 9.16 shows a schematic diagram of a fuel cell.

The *'fuel gas' diffuses through the anode and is oxidized, thus releasing electrons to the external circuit.*

The *'oxidizer' diffuses through the cathode and is reduced by the electrons that have come from the anode by way of the external circuit.*

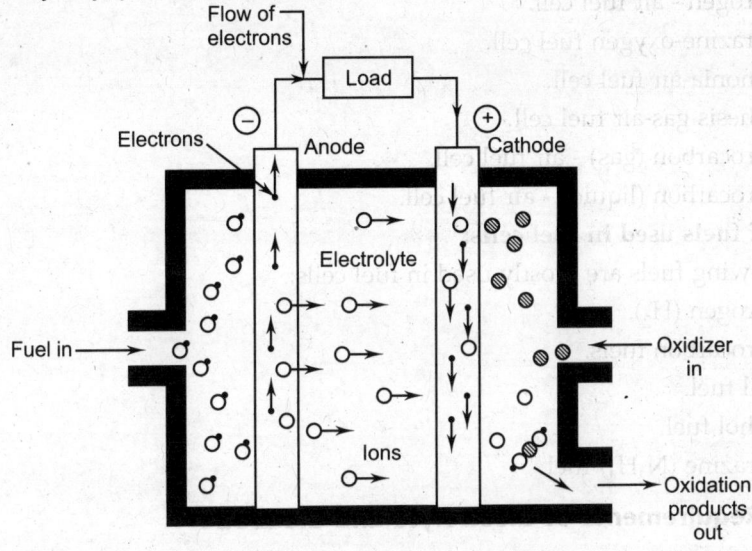

Fig. 9.16. Schematic of a fuel cell.

— The *fuel cell is a device that keeps the fuel molecules from mixing with the oxidizer molecules, permitting,* however, *the transfer of electrons by a metallic path that may contain a load.*

● Of the available fuels, **hydrogen** has so far given the most *promising results,* although cells consuming coal, oil or natural gas would be economically much more useful for large scale applications.

Some of the possible reactions are:

Hydrogen/oxygen   1.23 V $2H_2 + O_2 \longrightarrow 2H_2O$

Hydrazine   1.56 V $N_2H_4 + O_2 \longrightarrow 2H_2O + N_2$

Carbon (coal)	1.02 V	$C + O_2$	$\longrightarrow$	CO_2
Methane	1.05 V	$CH_4 + 2O_2$	$\longrightarrow$	$CO_2 + 2H_2O$

9.7.4. Classification Fuel Cells

Fuel cells may be *classified* as follows:

A. Based on the type of electrolyte:

1. Alkaline fuel cell (AFC).
2. Phosphoric acid fuel cell (PAFC).
3. Polymer electrolytic membrane fuel cell (PEMFC).
4. Molten carbonate fuel cell (MCFC).
5. Solid oxide fuel cell (SOFC).

B. Based on operating temperature:

1. Low temperature fuel cell 25–100°C.
2. Medium temperature fuel cell (below 100–500°C).
3. High temperature fuel cell (500–1000°C).
4. Very high temperature fuel cell (Above 1100°C).

C. Based on the types of fuel and oxidant:

1. Hydrogen-oxygen fuel cell.
2. Hydrogen - air fuel cell.
3. Hydrazine-oxygen fuel cell.
4. Ammonia-air fuel cell.
5. Synthesis gas-air fuel cell.
6. Hydrocarbon (gas) - air fuel cell.
7. Hydrocarbon (liquid) - air fuel cell.

Types of fuels used in fuel cells:

The following fuels are mostly used in fuel cells:

1. Hydrogen (H_2).
2. Hydrocarbon fuels.
3. Fossil fuel.
4. Alcohol fuel.
5. Hydrazine (N_2H_4) fuel.

9.7.5. Requirements of Electrolyte and Electrode

Following are the *requirements* of electrolyte and electrode:

Electrolyte:

1. It should be *conductive to ions*.
2. It should be *electrically non-conductive*.
3. *Ions should be free to move through the electrolyte*.
4. The composition of electrolyte should *not get changed during operation*.

Electrode:

1. It should be electrically *conductive*.
2. It should *not react with electrolyte* to prevent corrosion.

3. It should be *able to withstand high temperature*.

4. It should also *act as a catalyst* to convert hydrogen and oxygen molecules into their ions.

9.7.6. Desirable Characteristics of a Fuel Cell

The *fuel cell* should have the following *characteristics*:

1. It should have high *energy conversion efficiency*.

2. It should produce *low chemical pollution*.

3. It should be *flexible to choose any fuel*.

4. It should have cogeneration capability and *rapid load response*.

9.7.7. Hydrogen-Oxygen Fuel Cell (Hydrox Cell)-Alkaline Fuel Cell (AFC)

In this cell **hydrogen** and **oxygen** are used as the *'fuel'* and *'oxidant'* respectively as these elements are most *reactive with least complications*. The *'electrolyte'* is potassium hydroxide (20 to 40% concentration) which has *high electrical conductivity* and is less *corrosive than acids*.

Fig. 9.17 shows structure of a hydrox cell (alkaline cell).

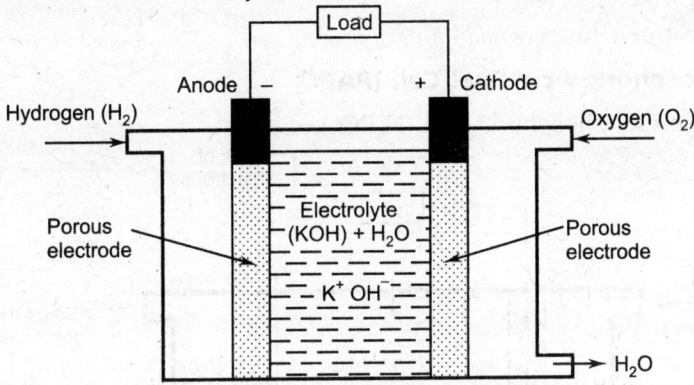

Fig. 9.17. Hydrogen-oxygen (Alkaline) fuel cell.

Construction. It has three chambers separated by two porous nickel electrodes, the *anode* and *cathode*. The middle chamber between the electrodes is filled with a strong solution of potassium hydroxide (KOH). The surfaces of the electrodes are chemically treated to repel the electrolyte, so that there is minimum leakage of potassium hydroxide into the outer chamber.

Working. The gases diffuse through the electrodes, undergoing *reactions* as shown below:

$$KOH \longrightarrow K^+ + (OH)^-$$

Anode: $\qquad H_2 + 2(OH)^- \longrightarrow 2H_2O + 2e^-$

The *electrons* so produced build up a negative potential and move towards the cathode through an externally connected circuit.

Cathode: At cathode, the electrons are picked up by oxygen atoms available there, react with water present in the electrolyte to form hydroxide $(OH)^-$ ions, the *reaction* being:

$$\frac{1}{2}O_2 + H_2O + 2e^- \longrightarrow 2(OH)^-$$

(OH)⁻ ions combine with hydrogen ion (H⁺) to form water as per the following reactions,

$$H^+ + OH^- \longrightarrow H_2O$$

Overall cell reaction:

$$2H_2 + O_2 \longrightarrow 2H_2O + \text{electrical energy}$$

The water formed is drawn off from the side.

The electrolyte provides the (OH)⁻ ions needed for the reaction, and *remains unchanged at the end*, since these are *regenerated*. The electrons liberated at the anode find their way to the cathode through the external circuit. This transfer is *equivalent to the flow of current from the cathode to anode*.

- Such cells when properly designed and operated, have an open circuit voltage of about 1.1V. Unfortunately, their *life is limited* since the water formed continuously *dilutes* the electrolyte. The working temperature is 90°C. Fuel efficiencies and high as 60 to 70% may be obtained.

- When pure H₂ and O₂ reactants are available, as in *rocket* and *spacraft*, there is *no fuel cell that can compete with the high power densities offered by alkaline fuel cells* (AFCs).

- The *Low-temperature cell* and *High-pressure cell* are typical developments of the hydrogen-oxygen cell.

Applications. In *space and military works*.

9.7.8 Phosphoric Acid Fuel Cell (PAFC)

Fig. 9.18 shows a phosphoric acid (H₃PO₄) cell.

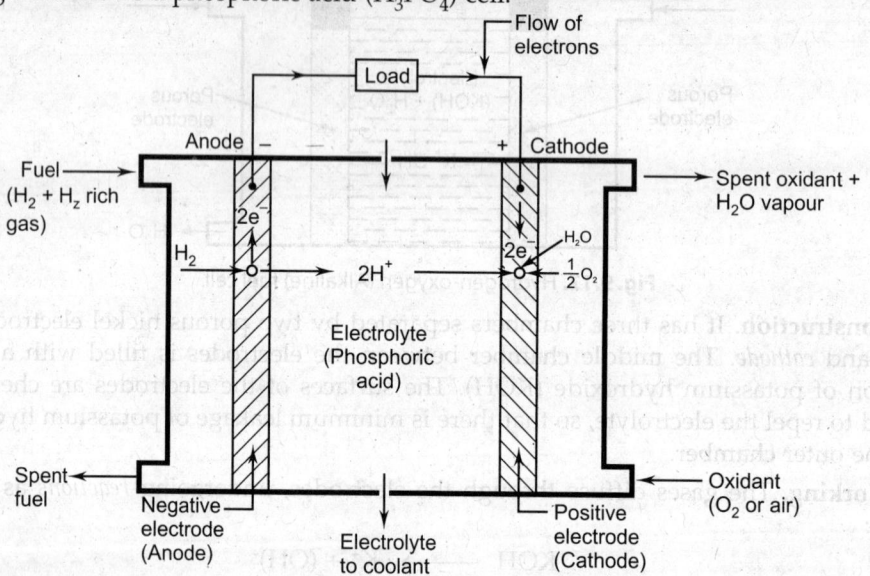

Fig. 9.18 Phosphoric acid cell.

The phosphoric acid cell consists of *two electrodes* of porous conducting material (*e.g. nickel*) to collect charge and, *'phosphoric acid' used as electrolyte*.

At anode, hydrogen molecule is split into *hydrogen ions* (protons) and *electrons*. The electrons flow through external circuit and produce electric power while protons travel through electrolyte and combine with oxygen, usually from air, at the *cathode* to form water. The electrochemical reaction is very slow, so a *catalyst* is required in the electrode to accelerate the reaction. The catalysts used are platinum, nickel (for anode) and silver (for cathode). *Platinum is the best catalyst for both electrodes*.

Reactions in this fuel cell produce electricity and by-product heat. The *reactions* are given below:

Anode: $\qquad\qquad\qquad\qquad$ $H_2 \longrightarrow 2H^+ + 2e^-$

Cathode: $\qquad$ $\dfrac{1}{2}O_2 + 2H^+ + 2e^- \longrightarrow 2H_2O$

Overall cell reaction:

$$2H_2 + O_2 \longrightarrow 2H_2O + \text{Electrical energy}$$

- At atmospheric pressure PAFC produces an ideal emf of 1. 23V at 25°C which reduces to 1.15 V at operating temperature between 150 to 200°C.

Application. These cells are *used commercially having the plant capacity in the range of 50 kW to 200 kW.*

9.7.9. Polymer Electrolyte Membrane Fuel Cell (PEMFC)

Fig. 9.19. illustrates a polymer electrolyte membrane fuel cell.

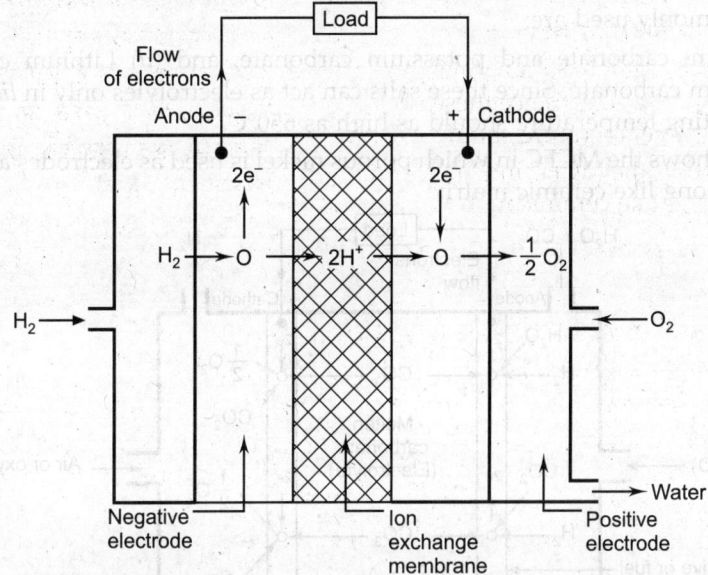

Fig. 9.19. Polymer electrolyte membrane fuel cell (PEMFC).

In PEMFC cell, *electrolyte* is a *solid polymer membrane of an organic material* such as polystyrene sulphonic acid and this is *permeable* to protons (H^+) when it is saturated with water but it does not conduct electrons.

The fuel is *hydrogen* and charge carriers are hydrogen ions (protons). At the *anode*, the hydrogen molecule is split into hydrogen ions and electrons. The hydrogen ions permeate across the electrolyte to *cathode* while the electrons flow through an external circuit and produce electric power. *Oxygen* is supplied to the *cathode* and combines with electrons and hydrogen ions to produce water.

The *reactions* at anode and cathode are given below:

Anode: $\qquad\qquad\qquad\qquad$ $H_2 \longrightarrow 2H^+ + 2e^-$

Cathode: $\qquad$ $\dfrac{1}{2}O_2 + 2H^+ + 2e^- \longrightarrow H_2O$

Overall reaction: $\qquad$ $2H_2 + O_2 \longrightarrow 2H_2O + \text{Electrical energy}$

The membrane is coated on both sides with finely powdered platinum which acts as a *catalyst*.

This cell are also called *"Ion-exchange membrane cell"*.

The desired properties of the electrolyte of an ideal ion-exchange membrane cell:

   (*i*)   Low permeability of fuel and oxidants.

  (*ii*)   High ionic conductivity.

 (*iii*)   Zero electronic conductivity.

 (*iv*)   Low degree of electro-osmosis.

  (*v*)   Mechanical stability.

 (*vi*)   High resistance to hydration.

(*vii*)   High resistance to the oxidation or hydrolysis.

Applications. Residential, portable laptops, cellular phones, video cameras, buses, cars, railway locomotives.

9.7.10. Molten Carbonate Fuel Cell (MCFC)

This type of cell uses an *electrolyte*, which is a *molten mixture of carbonate salts*. Two mixtures commonly used are:

   (*i*)   Lithium carbonate and potassium carbonate, and (*ii*) Lithium carbonate and sodium carbonate. Since these salts can act as electrolytes only in *liquid phase*, the operating temperature should as high as 650°C.

Fig. 9.20 shows the MCFC in which porous nickel is used as electrodes and electrolyte is held in a spong like ceramic matrix.

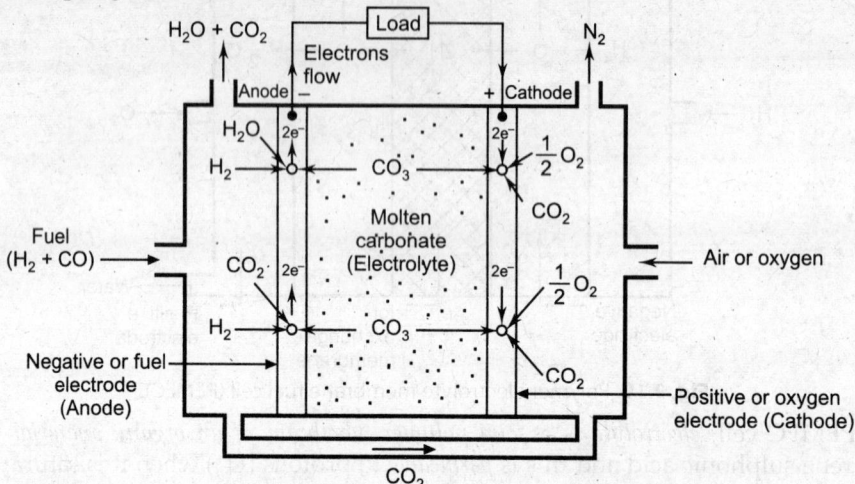

Fig. 9.20. Molten carbonate fuel cell (MCFC).

A hydrocarbon fuel, such as methane or kerosene, is used. The fuel is reacted inside the cell to produce H_2O and CO. At the *fuel electrode* (anode) H_2O and CO react with CO_3 ions in the electrolyte, releasing electrons to the electrode, and forming H_2O and CO_2. At the *oxygen electrode* O_2 reacts with the returning electrons and CO_2 diverted from the fuel electrode to form CO_3 ions. These CO_3 ions then migrate through the electrolyte to the fuel rod.

The *reactions* are given below:

Anode: $H_2 + CO_3^{--} = H_2O + CO_2 + 2e^-$

               $CO + CO_3^{--} = CO_2 + 2e^-$

These released electrons circulate through external resistance (load), forming load current and reach at oxygen electrode (cathode).

Cathode: $CO_2 + \dfrac{1}{2}O_2 + 2e^- = CO_3^{--}$

The CO_3^{--} ions are responsible for transportation of charge from cathode to anode within the electrolyte.

Overall reaction:

$$H_2 + CO + O_2 \longrightarrow H_2O + CO_2 + \text{Electrical energy}$$

● The theoretical *e.m.f.* produced by each cell is 1 V and actual e.m.f. being *0.8 V at 700° C*. The expected *efficiency is about 60%*.

Applications. *Dispersed power and utility power.*

9.7.11. Solid Oxide Fuel Cell (SOFC)

Fig. 9.21 Illustrates a solid oxide fuel cell (tubular shape)

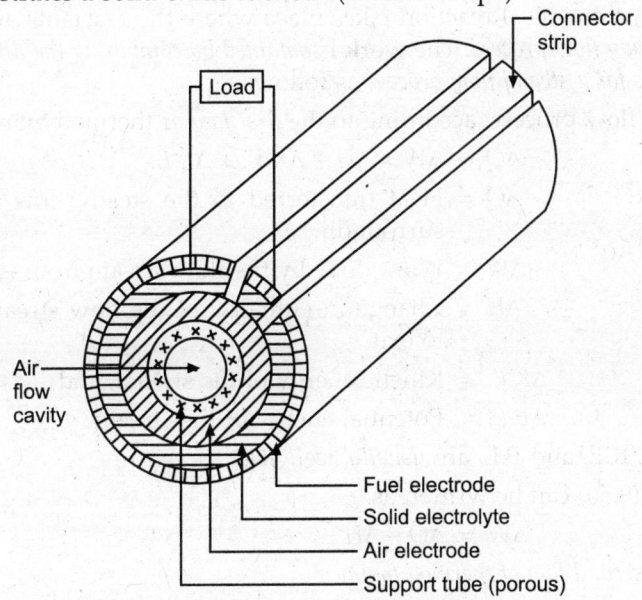

Fig. 9.21. Solid oxide fuel cell (tubular shape).

— This cell is based on a *solid metal oxide 'electrolyte'* (zirconium dioxide) called *zirconia*. It allows ionic conductivity of oxygen ions from cathode to anode. The *electrodes* are electric conductors with a *high porosity*. The operating temperature (800–1000°C) is *high enough for internal reforming of natural gas* in the anode chamber. The water gas shift reaction takes place at the anode, thus enabling H_2 and CO mixtures to be used as fuel feedstock.

— The construction materials used are *metal oxides* and *ceramics*. The central hollow space is for *air flow that acts as an oxidant*. Its operation is efficient at 1000°C and 1 atmospheric pressure.

— Fuel gas flows through the outermost layer of the fuel electrode. Next to it is the *electrolyte layer*. The *fuel gas* permeates through the porous electrodes and is *oxidized by air containing oxygen*. The *air electrode* is next to electrolyte and air *flows axially through the central hollow space*.

— *Both fuel gas and oxidant are fed into the cell continuously which get consumed and the cell delivers electrical energy.*

● The solid oxide fuel cell fueled by *natural gas* can attain a high electrical efficiency upto 55 percent. Each cell delivers 25 A current at 0.7V and a pack of 50 fuel cells give an output of 1 kW.

- These cells promise a vast potential in utilisation of low grade high ash, graded coals through Fluidized Bed Gasification.

Applications. *Domestic and commercial utility power; mobile applications for railways.*

9.7.12. Regenerative Cell

A *regenerative cell operates in a closed loop*. Fuel cell generates the electricity, heat and water from hydrogen and oxygen. The hydrogen would be generated from the electrolysis of water *i.e.* water can be split into hydrogen and oxygen by the use of renewable energy sources such as sun, wind etc. Oxygen can be used as oxidant and water produced is re-circulated for electrolysis.

9.7.13. Performance Analysis of a Fuel Cell

In a fuel cell, a chemical reaction takes place where the reactants are converted into products in a *steady flow process*. The **work** is *obtained by combining the first and second laws of thermodynamics for a steady flow process* as follows:

For a steady flow process, according to the *first law* of thermodynamics,

$$\Delta Q - \Delta W = \Delta H + \Delta K.E. + \Delta P.E. \qquad ...(9.50)$$

where, ΔQ = Heat transferred to the steady flow stream from the surrounding,

ΔW = Work done by the flow stream from entrance to exit,

ΔH = Change in enthalpy of the flow stream from entrance to exit.

$\Delta K.E.$ = Kinetic energy of the stream, and

$\Delta P.E.$ = Potential energy of the stream.

In a fuel cell, K.E. and P.E. are *usually negligible.*

Hence, Eqn. (9.50) can be written as:

$$\Delta W = \Delta Q - \Delta H \qquad ...(9.51)$$

According to second law of thermodynamics

$$\Delta Q = T. \Delta S \qquad ...(9.52)$$

Combining Eqn. (9.51) and (9.52), we obtain:

$$\Delta W_{max} = T.\Delta S - \Delta H \qquad ...(9.53)$$

The Gibbs free energy is given by:

$$\Delta G = \Delta H - \Delta T. \Delta S \quad \text{(where } T \text{ is constant)} \qquad ...(9.54)$$

From Eqns. (9.53) and (9.54), we get:

$$\Delta W_{max} = - \Delta G$$

Efficiency of fuel cell (η_{FC}):

The *efficiency (η_{FC}) of energy conversion of a fuel cell is defined as the ratio of the useful work to the heat of combustion of the fuel"*.

Mathematically, $\eta_{FC} = \dfrac{\Delta W_{max}}{-(\Delta H)} = \dfrac{-(\Delta G)}{-(\Delta H)}$ $\qquad ...(9.55)$

- For hydrogen oxygen fuel cell, $\Delta G = (- 237191)$ kJ/kg mole at 25°C while its heat of reaction ΔH is about (-285838) kJ/kg mole, then,

$$(\eta_{FC})_{max} = \frac{237191}{285838} \simeq 0.83 \text{ or } \textbf{83\%}.$$

E.M.F. of a fuel cell:

The e.m.f. (electromotive force) that will drive electrons through the external load is *proportional* to Gibbs free energy change.

Mathematically, $\qquad E = \dfrac{-\Delta G}{nF}$ ...(9.56)

where, $\qquad E$ = Electromotive force,

ΔG = Change in Gibbs free energy (J/mole),

n = Number of electrons per mole of fuel, and

F = Faraday's constant (=96487 coulombs/mole).

From Eqns. (9.55) and (9.56), we get

$$(\eta_{FC})_{max} = \dfrac{-nFE}{\Delta H}$$ ...(9.57)

The *overall efficiency of 'reversible fuel cell'* is given by:

$$(\eta_{FC})_{rev.} = \eta_{FC} \times \text{loss factor}$$ ...(9.58)

The *power output of a reversible fuel cell* ($P_{rev.}$) is given by:

$$P_{rev.} = \dfrac{\Delta G_{max}}{\text{Molar mass of hydrogen}}$$ ...(9.59)

where, Molar mass of hydrogen = 2.016 kg/mole

Actual electrical power output,

$$P_{actual} = P_{rev.} \times (\eta_{FC})_{overall}$$ ...(9.60)

The *rate of heat released,*

$$Q = P_{rev.} - P_{actual}$$ ... (9.61)

Voltage efficiency (η_v):

All the losses in a fuel cell may be included under *voltage efficiency* and it can be expressed as:

$$\eta_v = \dfrac{\text{Operating voltage}}{\text{Theoretical voltage}} = \dfrac{V_c}{E}$$ ...(9.62)

where, $\qquad V_c$ = Operating voltage of the fuel cell at a given current density, and

E = Theoretical open circuit voltage (e.m.f.).

V.I and P.I. characteristics of a fuel cell:

The *various characteristics of a fuel cell* are:

1. Current-Voltage (V–I) characteristics.
2. Power-current (P-I) characteristics.

Current-voltage (V-I) characteristics:

The performance of a fuel cell is evaluated by the cell voltage V_c versus electrode current density I_d curve (Fig. 9.22). V_c drops with increase in I_d due to *'polarization'* within the cell. Hence, the curve is also called the *polarization curve* of the fuel cell.

Polarisation is internal, chemical, electrical, thermal effects within the fuel cell resulting in inefficiencies. Polarization, *the cause of internal energy loss,* is measured in terms of polarization voltage V_p.

$$\Delta V_p = E - V_c$$ ...(9.63)

where, E = E.m.f. of the cell, and

                   V_c = On-load voltage of the cell.

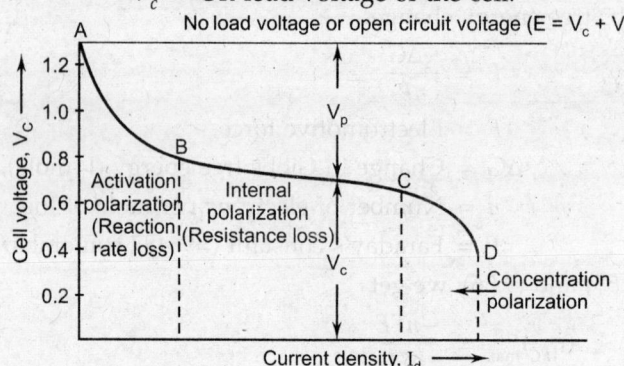

Fig. 9.22. V-I (V_c vs. I_d) characteristics of fuel cell.

With increase in load of the cell, the internal losses increase, resulting in the drop of cell terminal voltage (V_c).

Power-current (P-I) characteristics:

Power per cell, $P_c = V_c \times I_C$ ...(9.64)

The **power** of *a cell increases with the increase in current density till saturation point is reached*; thereafter it *decreases due to polarization effects.*

         Output power = Input power – polarization losses

Now, Efficiency of fuel cell,

$$\eta_{FC} = \frac{\text{Output power}}{\text{Input power}} \qquad ...(9.65)$$

The simple method of calculating the *efficiency* is as follows:

$$\eta_{FC} = \frac{\text{On-load cell voltage}}{\text{No-load cell voltage}} = \frac{V_c}{E} = \frac{E - V_p}{E} \qquad ...(9.66)$$

Fig. 9.23 shows the power and efficiency characteristics of a fuel cell.

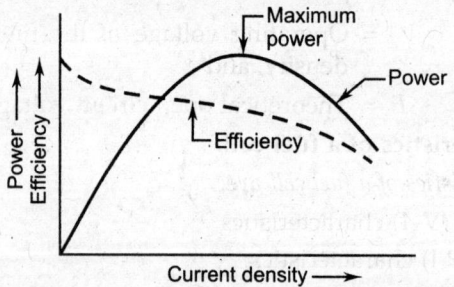

Fig. 9.23. Power and efficiency characteristics of a fuel cell.

— The efficiency of a fuel cell *varies* with the current density at electrode surface due to the polarizationeffect.

— The *power losss is converted to waste heat* and *released to atmosphere.*

Example 9.1. *A hydrogen-oxygen cell operates at 25°C. Given:* $\Delta H°_{298K}$ = – 285838 kJ/kg mole., $\Delta G°_{298K}$ = –237191 kJ/kg mole, and molar mass of hydrogen = 2.016.

Calculate:

(i) Efficiency of the fuel cell.

(ii) Electrical work output per mole H_2 consumed and per mole of H_2O produced.

(iii) Heat transfer to the surroundings.

Solution. *Given :* $T = 25 + 273 = 298$ K; $\Delta H^\circ_{298} = -285838$ kJ/kg mole; $\Delta G^\circ_{298K} = -237191$ kJ/kg mole.

(i) **Efficiency of the fuel cell, η_{FC}:**

$$\eta_{FC} = \frac{-\Delta G^\circ}{-\Delta H^\circ} = \frac{237191}{285838} \qquad \text{...[Eqn. (9.55)]}$$

$$\simeq 0.83 \text{ or } \textbf{83\% (Ans.)}$$

(ii) **Electrical work output/mole of H_2O produced, P_{rev}:**

$$P_{rev.} = \frac{\Delta G_{max}}{\text{Molar mass of hydrogen}} \qquad \text{...[Eqn. (9.56)]}$$

$$= \frac{237191}{2.016} \times 10^{-3} \text{kW}$$

$$= \textbf{117.65 kW (Ans.)}$$

(iii) **Heat transfer to the surroundings, Q:**

$$Q = T \Delta S = (\Delta H^\circ)_{298K} - (\Delta G^\circ)_{298K} \qquad \text{...(Eqn. (9.54))}$$

$$= -285838 - (237191)$$

$$= \textbf{-48647 kJ/kg mole (Ans.)}$$

The *-ve sign* indicates that *heat is transferred from the system to surroundings.*

9.7.14. Fuel Cell Power Plant

The *primary fossil fuels* are used to *generate electrical energy in fuel cell power plant.*

Fig. 9.24. shows the schematic of fuel cell based electrical power generation scheme:

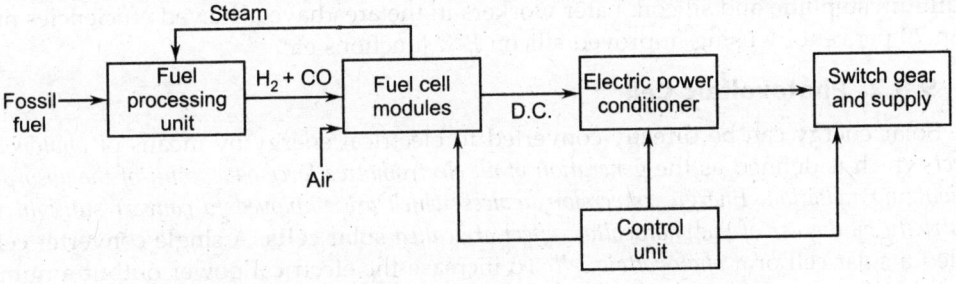

Fig. 9.24. Fuel cell based electrical power generation system.

— The fossil fuel is supplied to the *'fuel processing unit'*, where fuel is purified and then supplied to *'fuel cell modules'.*

— The fuel cell modules convert fuel energy electrochemically into D.C. power.

— A number of fuel cells are stacked to form a module and several modules are interconnected to form a *power producing unit.*

— The *power conditioning unit* converts D.C. output to A.C output using *'inverter'*; and the standard rated supply being 3-phase, 400 V, 50 Hz/60 Hz or single phase, 230 V/110 V, 50 Hz/60Hz.

● Modules of size 200–250 kW are commonly available.

9.7.15. Advantages of Fuel cell Power Plants

Following are the *advantages of fuel cell power plants*:

1. Besides *electric power,* fuel cell plants also supply *hot water, space heat* and *steam.*

2. These plants are *eco-friendly* and *noiseless* (since they don't have rotating parts.)

3. Fuel cell plants can attain *high efficiency upto 55%,* whereas conventional thermal plants operate at around 30% efficiency.

4. These plants have *cogeneration capabilities.*

5. A large degree of modularity is available with capacity ranging from *5 kW to 25 MW.*

6. There is a *wide choice of fuels* for fuel cells.

7. It is a *decentralised plant,* can be operated in isolation for military installations and hospitals where noise and smoke are prohibited.

9.8. PHOTOVOLTAIC POWER SYSTEM

9.8.1. Photovoltaic Generators - Historical Background

Edmond Becquerel in 1839 noted that a voltage was developed when light was directed onto one of the electrodes in an electrolytic solution. The effect was first observed in a solid in 1877 by W.G. Adams and R.E. Day, who conducted experiments with selenium. Other early workers with solids included Schottky, Lange and Grandahl, who did pioneering work in producing photovoltaic cells with selenium and cuprous oxide. This work led to the development of photoelectric exposure meters. 1954 researchers turned to the problem of utilizing the photovoltaic effect as a source of power. In that year several groups including the workers at Bell Telephone Laboratories achieved conversion efficiencies of about 6 per cent by means of junctions of *P*-type and *N*-type semiconductors. These early junctions, commonly called *P-N* junctions, were made of cadmium sulphide and silicon. Later workers in the area have achieved efficiencies more than 20 per cent by using improved silicon *P-N* junctions etc.

9.8.2. Photovoltaic Cell

Solar energy can be directly converted to electrical energy by means of *photovoltaic effect* which is defined as the *generation of an electromotive force as a result of the absorption of ionizing radiation. Energy conversion devices which are employed to convert sunlight into electricity by the use of the photovoltaic effect are called* **solar cells.** A single converter cell is called a solar cell or a *photovoltaic cell.* To increase the electrical power output a number of such cells are combined and the combination is called a *'solar array'* or *'solar module'.*

In a photovoltaic cell sensitive element is a *semiconductor* (not metal) which generates voltage in proportion to the light or any radiant energy incident on it. The most commonly used photovoltaic cells are barrier layer type like iron-selenium cells or $Cu-CuO_2$ cells.

Fig. 9.25 shows a typical widely used photovoltaic cell–"*Selenium cell*". It consists of a metal electrode on which a layer of selenium is deposited; on the top of this a barrier layer is formed which is coated with a very thin layer of gold. The latter serves as a translucent electrode through which light can impinge on the layer below. Under the influence of this light, a negative charge will build up on the gold electrode and a positive charge on the bottom electrode.

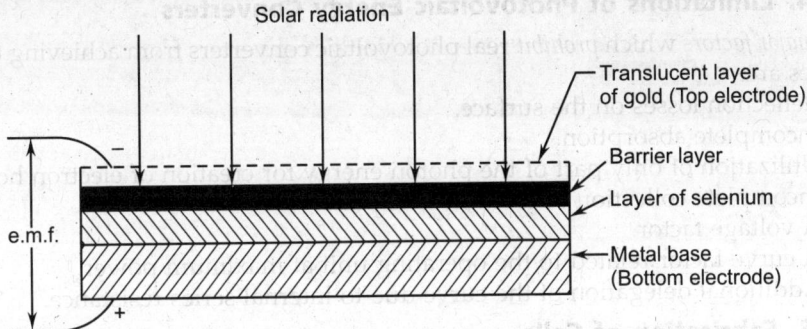

Fig. 9.25. Photovoltaic cell.

Photovoltaic cells *are widely used in the following fields*:

(*i*) Automatic control systems.

(*ii*) Television circuits.

(*iii*) Sound motion picture and reproducing equipment.

9.8.3. Basic Photovoltaic System for Power Generation

Fig. 9.26. shows a basic photovoltaic system integrated with the utility grid. With the help of this system the generated electrical power can be delivered to the local load.

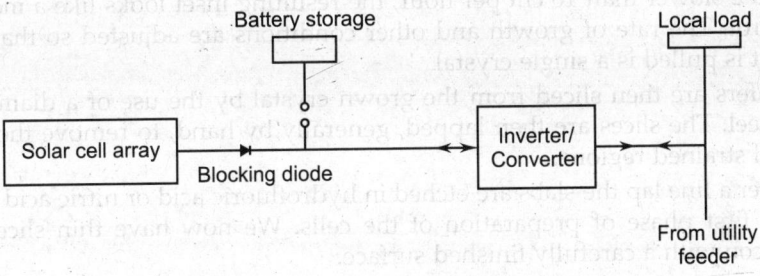

Fig. 9.26. Basic photovoltaic system integrated with power grid.

This system consists of the following:

1. Solar array   2. Blocking diode

3. Battery storage   4. Inverter/converter

5. Switches and circuit breakers.

— The *solar array* (large or small) converts the insolation to useful D.C. electrical power.

— The *blocking diode* confines the electrical power generated by the solar array to flow towards the battery or grid only. In the absence of blocking diode the battery would discharge back (through the solar array) during the period when there is no insolation.

— *Battery storage* stores the electrical power generated through solar array.

— *Inverter/converter* (usually solid state) converts the battery bus voltage to A.C. of frequency and phase to match that needed to integrate with the utility grid. Thus it is typically a D.C, A.C. inverter.

— *Switches and circuit breakers* permit isolating parts of the system, as the battery.

9.8.4. Limitations of Photovoltaic Energy Converters

The *major factors* which *prohibit* real photovoltaic converters from achieving the higher efficiencies are:

1. Reflection losses on the surface.
2. Incomplete absorption.
3. Utilization of only part of the photon energy for creation of electron hole pairs.
4. Incomplete collection of electron-hole pairs.
5. A voltage factor.
6. A curve factor related to the operating unit at maximum power.
7. Additional delegation of the curve due to internal series resistance.

9.8.5. Fabrication of Cells

A. Silicon cells:

Silicon cells are most widely used. Next to oxygen, silicon is the most abundant element on earth. The pure silicon used in cell manufacture is extracted from sand which is mostly silicon dioxide (SiO_2). The silicon required for solar cell use, because of its high purity, is expensive.

The fabrication of silicon cells include the following **steps:**

(*i*) The pure silicon is placed in an induction furnace where boron is added to melt. This turns the crystal resulting from the melt into *P*-type material.

(*ii*) A small seed of single crystal silicon is dipped into the melt and withdrawn at a rate slower than 10 cm per hour, the resulting inset looks like a medium sized carrot. The rate of growth and other conditions are adjusted so that the crystal that is pulled is a single crystal.

(*iii*) Wafers are then sliced from the grown crystal by the use of a diamond cutting wheel. The slices are then lapped, generally by hand, to remove the saw marks and strained regions.

(*iv*) After a fine lap the slabs are etched in hydrofluoric acid or nitric acid to complete the first phase of preparation of the cells. We now have thin slices of *P*-type silicon with a carefully finished surface.

(*v*) The wafers are then sealed in a quartz tube partly filled with phosphorous pentoxide and the arrangement is placed in a diffusion furnace where temperature is carefully controlled; this process causes the phosphorous to diffuse into the *P*-type silicon to a depth of about 10^{-4} cm to 10^{-5} cm.

(*vi*) The cells are then etched in a concentrated acid to remove unwanted coatings that formed during manufacture. Wax or Teflon masking tape is used to protect the surfaces not to be etched.

B. Thin film solar cells:

These cells have the following *advantages*:

(*i*) The material cost is low.

(*ii*) The manufacturing cost is low (possibly avoiding the need for single crystal growth).

(*iii*) High power-to-weight ratios.

(*iv*) Low array costs, because the number of connections needed will be greatly reduced.

The example of this type of cell is *cadmium sulphide (CdS) cells*. CdS cells having areas of 50 cm^2 have been made by evaporating the semiconductor on to a flexible substrate such as kapton, a metallized plastic substrate. A barrier layer of copper sulphide is then deposited on top of the CdS. Power to weight ratios of 200 watts/kg are claimed for such cells. These cells have low *efficiency and instability*.

9.8.6. Advantages and Disadvantages of Photovoltaic Solar Energy Conversion

Advantages:

(i) There are no moving parts.

(ii) Solar cells are easy to operate and need little maintenance.

(iii) They have longer life.

(iv) They are highly reliable.

(v) They do not create pollution problem.

(vi) Their energy source is unlimited.

(vii) They can be fabricated easily.

(viii) They have high power to weight ratio.

(ix) They can be used with or without sun tracking, making possible a wide range of application possibilities.

(x) They have ability to function unattended for long periods as evident in space programme.

Disadvantages:

(i) The cost of a solar cell is quite high.

(ii) The output of a solar cell is not constant, it varies with the time of day and weather.

(iii) Amount of power generated is small.

9.9. ELECTROSTATIC MECHANICAL GENERATORS

Electrostatic mechanical generators *convert mechanical energy, usually mechanical potential energy of a fluid, directly into electrical energy.*

Fig. 9.27 shows the principle of working of liquid drop electrostatic mechanical generator. In this, *the gravitational potential energy of water droplets is directly converted into electrical energy.* The electric charge is transferred from one electrode to another by an insulated belt. All these electrostatic devices are having the characteristic of fairly *low currents and very high voltages.* This is yet only a laboratory model and commerical power generation has yet to be done.

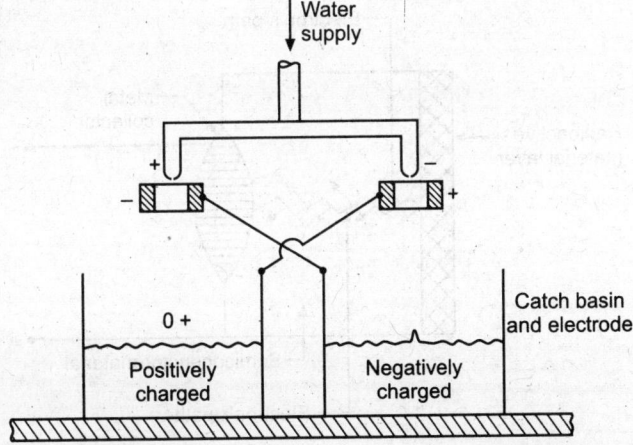

Fig. 9.27. Liquid drop electrostatic mechanical generator.

9.10. NUCLEAR BATTERIES

A **nuclear battery** *works on the principle* that *beta emitter can produce the electrical energy.* Nuclear batteries *are of the following two kinds:* (*i*) High voltage atomic battery and (*ii*) *Low voltage* atomic battery.

9.10.1. High Voltage Atomic Battery

Fig. 9.28 shows the schematic diagram of a high voltage atomic battery.

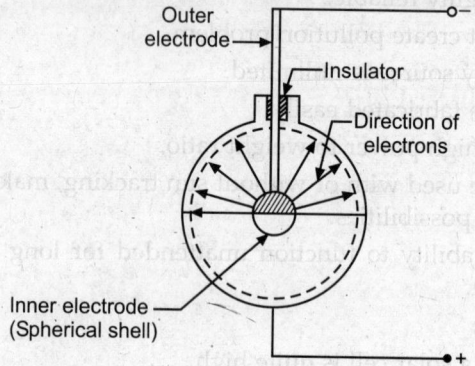

Fig. 9.28. High Voltage atomic battery.

It consists of an inner spherical electrode on the surface of which is deposited a powerful beta emitting (*i.e.,* fast electrons) substance. This is surrounded by another spherical condenser and the inner surface of this condenser becomes negatively charged. This acts as the outside electrode and is properly insulated at the opening. The inner and outer electrodes become the – ve and + ve terminals of the battery.

Sr^{90} isotope can be used since its half life is 28 years. These batteries perform independent of the temperature unlike the accumulators whose electrolyte freezes at low temperatures. These also supply a *very steady constant potential.*

This has yet to be developed on a commercial scale.

9.10.2 Low Voltage Atomic Battery

Fig. 9.29 shows a low voltage battery.

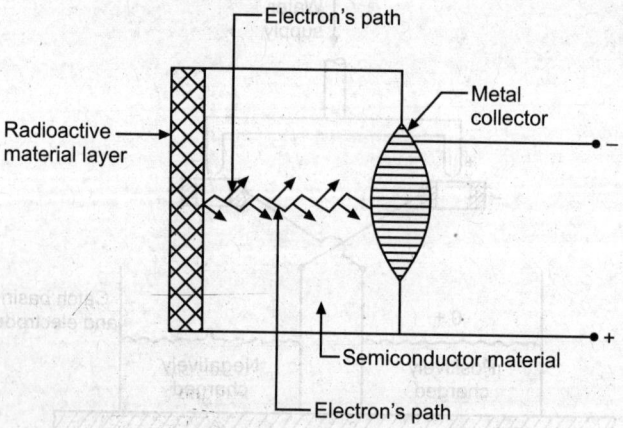

Fig. 9.29. Low voltage atomic battery.

In this Sr^{90} (beta emitter) is deposited on the surface of a semiconductor (germanium or silicon) at one end and the other end is having a metallic collector. The semiconductor has the characteristic of undirectional current flow. The fast electrons pass through the semiconductor and strike the metallic disc at the other end and the two ends of the semiconductor become the – ve and + ve terminals of the battery. The power produced by this battery is a few microwatts at a potential difference of 1/10 volt. This power is sufficient to *feed a small radio*.

HIGHLIGHTS

1. The devices which converts naturally available energy into electricity, without an intermediate conversion, into mechanical energy; energy source may be thermal, solar or chemical, are called *"Direct energy conversion devices"*.

2. To achieve higher efficiency, *thermoelectric material* should have a *high value of Z* and be able to operate upon very high temperatures.

3. *Power output and efficiency of a thermoelectric power generator:*

$$\eta_{gen} = \left(\frac{T_H - T_C}{T_H}\right) \times$$

$$\frac{m}{(1+m) - \frac{1}{2}\left(\frac{T_H - T_C}{T_H}\right) + \frac{(K_A + K_B)(R_A + R_B)(1+m)^2}{\alpha_{AB}^2 \cdot T_H}}$$

The figure of merit, $Z = \dfrac{\alpha_{AB}^2}{(R_A + R_B)(K_A + K_B)}$

and, $Z_{opt} = \left[\dfrac{\alpha_{AB}}{\sqrt{\rho_A \cdot k_B} + \sqrt{\rho_B \cdot k_B}}\right]^2$

Maximum power output,

$$(W_L)_{max} = \frac{1}{4}\left[\frac{\alpha_{AB}^2 (T_H - T_C)^2}{(R_A + R_B)}\right]$$

Maximum efficiency,

$$\eta_{max} = \left(\frac{T_H - T_C}{T_H}\right)\left[\frac{1}{2 - \frac{1}{2}\left(\frac{T_H - T_C}{T_H}\right) + \frac{4}{Z_{op} \cdot T_H}}\right]$$

where, T_H = Temperature of the source, K,

T_C = Temperature of the sink, K,

A = Area of thermoelectric material (s) m^2,

k = Thermal conductivity of materials W/mK,

ρ = Electrical resistivity of material(s), Ωm,

K = Thermal conductance of material (s) $\left(= \dfrac{kA}{L}\right)$, W/K,

R = Electrical resistance of element (s), Ω, and

α = Absolute value of seebeck coefficient(s), $\dfrac{V}{K}$.

4. A *"thermionic converter/generator"* converts heat energy directly to electrical energy by utilizing thermionic emission effect.

Power output, $\qquad P = V_L \, (J_c - J_a)$

Efficiency, $\qquad \eta = \dfrac{(\gamma_c - \beta\gamma_a) \times \left[1 - \beta^2 \exp(\gamma_c - \gamma_a)\right]}{(\gamma_c + 2) - \beta^2(\gamma_a + 2\beta)\exp(\gamma_c - \gamma_a)}$

Maximum efficiency,

$$\eta_{\mathbf{max}} = (1-\beta)\frac{\gamma}{\gamma+2} \qquad \text{(when } \gamma_a = \gamma_c = \gamma)$$

where, $\qquad\qquad V_L$ = Load voltage,

$\qquad\qquad\qquad\quad J_c$ = Current density of cathode,

$\qquad\qquad\qquad\quad J_a$ = Current density of anode,

$\qquad\qquad\qquad\quad T_c$ = Temperature of cathode,

$\qquad\qquad\qquad\quad T_a$ = Temperature of anode, and

$$\gamma_c = \frac{V_c}{KT_c}; \gamma_a = \frac{V_a}{KT_a}; \beta = \frac{T_a}{T_c}$$

5. *MHD generator* is a device which converts heat energy of fuel directly into electrical energy without a conventional electric generator.

Open circuit voltage,

$$E_0 = \text{Bud}$$

Maximum power output,

$$P_{max} = \frac{1}{4}\sigma\beta^2 u^2 d \cdot A$$

where, $\qquad\qquad B$ = Flux density,

$\qquad\qquad\qquad\quad u$ = Velocity of the charged particle,

$\qquad\qquad\qquad\quad d$ = Distance between the two plates, and

$\qquad\qquad\qquad\quad A$ = Area of the plate.

6. A *'fuel cell'* is an electrochemical device in which the chemical energy of a conventional fuel is converted directly and efficiently into low voltage, direct current electrical energy.

Efficiency of fuel cell,

$$\eta_{FC} = \frac{-(\Delta G)}{-(\Delta H)}$$

E.m.f. of a fuel cell,

$$E = \frac{-\Delta G}{nF}$$

Maximum efficiency of fuel cell,

$$(\eta_{FC})_{max} = -\frac{nFE}{\Delta H}$$

$$(\eta_{FC})_{rev.} = \eta_{FC} \times \text{loss factor}$$

$$P_{rev.} = \frac{\Delta G_{max}}{\text{Molar mass of hydrogen}(= 2.016 kg/\text{mole})}$$

Actual *electrical power output,*

$$P_{actual} = P_{rev.} \times (\eta_{FC})_{overall}$$

The rate of heat released,

$$Q = P_{rev.} - P_{actual}$$

Where, E = Electromotive force,

ΔG = Change in Gibbs energy,

ΔH = Change in enthalpy of the flow stream,

n = Number of electrons per mole of fuel, and

F = Faraday constant (= 96487 coulombs/mole).

THEORETICAL QUESTIONS

1. What is "Direct energy conversion device?
2. What are the 'Thermoelectric effects'?
3. Explain the following:
    (*i*)   Seebeck effect,
    (*ii*) Peltier effect, and
    (*iii*) Thomson effect.
4. Describe the working of a 'Thermoelectric generator'.
5. Analyse the working of a thermoelectric generator. Derive an expression for its power output.
6. What are the advantages and disadvantages of thermoelectric power generator?
7. Explain the principle on which a thermoelectric generator works.
8. What are the different types of semiconductors used in thermoelectric power generation? What are the criteria for the selection of these materials?
9. What is a thermionic converter?
10. What is work function?
11. Explain with a neat diagram the construction and working of a thermionic generator.
12. What are the desirable properties of thermionic converter materials? Explain briefly.
13. Give the advantages, disadvantages and applications of thermionic converter.
14. Derive expressions for power output and efficiency of a thermionic converter.
15. What is a MHD generator? How does it work?
16. What is the principle of MHD power generation?
17. How are MHD systems classified?
18. Explain with a neat sketch the working of an open cycle MHD system.
19. What are the properties which a MHD system most have to achieve efficient practical realization?
20. Describe closed-cycle MHD system.
21. List the materials used for MHD generators.
22. What are the advantages and disadvantages/drawbacks of MHD systems?
23. Derive expressions for voltage and power output of MHD generator.
24. Explain briefly the parameters which govern power output of a MHD system.
25. Explain briefly Electro gas-dynamic generator. What are its advantages over MHD systems?

26. What is a fuel cell?
27. What are the advantages, disadvantages and applications of fuel cell?
28. List the main components of a fuel cell.
29. Describe briefly the working of a fuel cell.
30. How are fuel cell classified?
31. Name the fuels which are used in fuel cells.
32. What are the requirements of electrolyte and electrode in a fuel cell?
33. Explain with a neat diagram the working of hydrogen-oxygen fuel cell. What are its fields of applications?
34. Describe briefly the any two of the following fuel cell:
    - (*i*)   Phosphoric acid fuel cells (PAFC)
    - (*ii*) Polymer electrolyte membrane fuel cell (PEMFC)
    - (*iii*) Molten carbonate fuel cell (MCFC).
35. What is a regenerative cell?
36. Derive the expressions for the overall efficiency of a fuel cell.
37. Draw the P-I. and V-I characteristics of a fuel cell.
38. Explain briefly a fuel cell based electrical power generation system.
39. What are the advantages of fuel cell power plants?
40. What are the limitations of photovoltaic energy converters?
41. Give the advantages and disadvantages of photovoltaic solar energy conversion.
42. Explain briefly the following:
    - (*i*)   Electrostatic mechanical generator.
    - (*ii*) Nuclear batteries.

CHAPTER 10

HYDROGEN ENERGY

10.1. INTRODUCTION

Hydrogen energy (*"energy carrier" like electricity*) *is an alternative energy of future which can also be stored in addition to other qualities of electrical energy.* But it is *highly inflammable* and special handling precautions are needed during its production, transportation, storage and utilisation.

- Hydrogen energy has a *tremendous potential* because it can be produced from *water* which is available in *abundance in nature*. It can be produced from water by using *solar energy. All plants and hydrocarbons (fossil fuels) are sources of hydrogen.*

- Hydrogen has the *highest energy content per unit mass*. Its specific energy content is almost *three times that of hydrocarbon fuels*. Therefore, it can be directly used as aircraft fuel for air transport. It has been used as a fuel for space crafts. A $H_2 - O_2$ fuel cell *liberates energy and also water as sole material product* for the use of space craft passengers.

- The simplest way to obtain hydrogen from water is its *"electrolysis"* using electricity. The latter can be generated from renewable energy sources like solar energy, wind energy and geothermal energy.

- The *main issues* associated with the *use of hydrogen at energy* source are: (*i*) *Production;* (*ii*) *Storage and transportation;* (*iii*) *Utilisation;* (*iv*) *Safety* and *Management,* and (*v*) *Economy.*

- One of the potential advantages of *hydrogen as a secondary "fuel"* is that it can be *transmitted and distributed by pipeline* in much the same way as natural gas. *Hydrogen can serve as a means of carrying energy from the place where a primary source is available to a distant load centre where the energy is used.*

- Hydrogen is an *efficient and clean fuel*. It has *minimum carbon content* compared to other fuels. A *carbon-rich fuel* produces more CO_2 which contributes to *global warming*. By adopting a *leaner carbon and richer hydrogen content, it is a step towards better "environmental friently" source of fuel.*

- Hydrogen has *huge market*, however, *enormous capital investment is required* for its production, distribution and storage. For the *safe operation* of equipment and systems, special *design precautions* are needed.

  Sources of hydrogen:

  Hydrogen is found only in *compound form with other* elements. H_2 combines with O_2 to form water. Hydrogen combines with carbon to form different compounds such as coal, petroleum, methane gas etc.

10.2. PROPERTIES OF HYDROGEN

The *properties* of hydrogen are:

1. The burning process of hydrogen is *pollution free*.
2. The standard heating value of *hydrogen gas* is 12.1 MJ/m^3 compared with 38.3 MJ/m^3 for natural gas.
3. The heating value of *liquid hydrogen* is 120 MJ/kg or 8400 MJ/m^3 as compared to 44 MJ/kg or 32000 MJ/m^3 of aviation petrol.
4. Hydrogen is a *light gas* at *room temperature and pressure*. Its density is $\frac{1}{4}$ th of that air and $\frac{1}{9}$ th that natural gas.
5. Hydrogen can be *liquefied at –253°C at atmospheric pressure*. The liquid hydrogen has a *specific gravity of 0.07* which is $\frac{1}{10}$ th that of gasoline.
6. Mixture of hydrogen and air are *combustible over wide range of composition*. The flammability limits are from 4 to 74% by volume of hydrogen in air at ordinary temperatures.
7. The flame speed of hydrogen when burning in air is *much greater* than for natural gas.
8. The ignition energy to initiate combustion is *less* for hydrogen than for natural gas.
9. The combustion of hydrogen with oxygen from air results in release of energy and water as byproduct.
10. *Detonation* can occur between hydrogen-air mixture between 18 and 59%. The *I.C. engines working on hydrogen fuel can work from very rich(excess fuel) to very lean (excess air) mixture.* The adjustment of air-fuel ratio is *less critical* than for gasoline engine.

10.3 ADVANTAGES OF HYDROGEN AS FUEL

Following are the *advantages* of hydrogen as fuel:

1. *Very* high energy content.
2. Burning is *non-polluting*.
3. Hydrogen produced from biomass and supplied to the consumers in the transport sector *costs only 50%* compared to hydrogen produced electrolytically.
4. For fuel-cell operated bus, hydrogen produced from biomass can *compete well* with gasoline-operated vehicles.
5. It is a *superior fuel for turbojet aircraft* due to *greater economy, lower noise level* and *little pollution*.

6. Hydrogen as a velicular fuel can reduce dependence on fossil fuel which is increasing in cost every year.

7. Hydrogen can easily be transported and distributed through pipelines.

8. Hydrogen being a high density fuel, its low transport cost compensates for its high product cost to make it an *economically viable fuel.*

9. Hydrogen can be used for generating electricity for domestic appliances, in domestic cooking as a fuel, in automobiles etc.

10.4. APPLICATIONS OF HYDROGEN (ENERGY)

Following are the *applications* of hydrogen energy:

1. It can be used for $H_2 - O_2$ *fuel cell* for production of electrical energy.

2. Hydrogen is used as fuel in *aircrafts and rockets in liquid form.*

3. *Used in cooking, water heaters* and *air-conditioning.*

4. Can also be used in *furnaces.*

5. Can also be used in *generators.*

6. Widely used in petroleum refining.

7. Widely used in manufacture of vanaspati, fertilizers and alcohols.

• Hydrogen-based vehicles have been developed by Mazda Motor Corporation, BMW Germany, Toyota hybrid highlander and Taiwanese scooter. In India, Tatas are working on the modification of I.C. engines in the present vehicles that can be run on hydrogen fuel.

• National Hydrogen Energy Board (NHEB) has prepared a workable plan to make hydrogen as a *commercial fuel. In the near future, large amounts of hydrogen could be produced in remote wind farms, solar stations and ocean power plants, and stored underground.*

'Hydrogen utilisation' in the near future:

Following are the *possible areas of utilisation of hydrogen in the near future:*

1. Use of hydrogen in the *processing of heavy oil.*

2. *Reduction of iron oxides* by means of hydrogen in the steel industry.

3. Using hydrogn to *manufacture synthetic liquid* or *gaseous fuels.*

4. *Direct addition of hydrogen to the existing natural gas distribution network.*

5. *Direct use of hydrogen as an aircraft fuel in air transport.*

6. *Direct use of hydrogen as a motor vehicle fuel* in urban transport, particularly where air pollution problems are already critical.

7. *Production of electrolytic hydrogen,* for *full-load exploitation of nuclear power stations.*

Problems associated with hydrogen use:

The use of hydrogen entails the following *problems*:

1. *Commercial production* of hydrogen at *cheap cost.*

2. *Effective energy utilisation.*

3. *Difficulty in storage,* since it is *highly explosive.*

4. *Lack* of safety and management.

10.5. PRODUCTION OF HYDROGEN

The *most commonly used methods of production of hydrogen* are enumerated and described as follows:

1. Electrolysis of water.
2. Steam reformation.
3. Coal gasification.
4. Methane gas reformation.
5. Biological production of hydrogen.
6. Photo-electrolysis

The methods of producing hydrogen may be *classified* according to the *immediate source of addition of energy to decompose* as follows:

1. *Electrical energy* .... Electrolysis
2. *Fossil fuels* ..... Steam reformation; Coal gasification
3. *Heat energy* ..... Thermo-chemical methods
4. *Solar energy*:
   (*i*) Bio-photolysis.
   (*ii*) Photo-electrolysis.

10.5.1. Electrolysis of Water

It is the process in which hydrogen splits from water by means of direct electrical current, by using two electrodes and electrolyte.

The reaction of electrolysis is as follows:

$$2H_2O \xrightarrow{\text{Direct current}} \underbrace{2H_2}_{\text{At cathode}} + \underbrace{O_2}_{\text{At anode}} + \text{Heat energy released}$$

Fig. 10.1 shows the *principle of operation of an alkaline water electrolysis cell.*

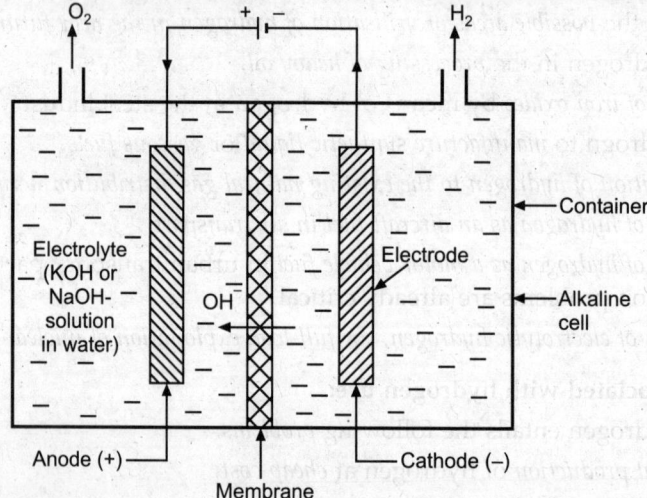

Fig. 10.1. Alkaline water electrolysis.

When the direct current flows through the electrolyte from the anode (+ve electrode), to the cathode (–ve electrode), the water in the electrolyte solution is *decomposed* into H_2

which is released at the *cathode*, and O_2 released at the *anode*. Although only the water is split, an *electrolyte (e.g. KOH solution) is required because water itself is a very poor conductor of electricity.*

The reactions at cathode and anode are given below:

Cathode: $4H_2O + 4e^- \longrightarrow 2H_2\uparrow + 4OH^-$

Anode: $4OH^- \longrightarrow O_2\uparrow + 2H_2O + 4e^-$

Overall reaction: $\underbrace{2H_2O}_{\text{Electrolyte}} + Energy \longrightarrow \underbrace{2H_2}_{\text{At cathode}} + \underbrace{O_2}_{\text{At anode}}$

The energy required to produce hydrogen is $3.5\ kWh/m^3$.

The *available electrolysis* processes are:

1. Alkaline electrolysis,
2. Membrane electrolysis, and
3. High temperature steam electrolysis.

10.5.2. Steam Reformation (Fossil Fuels)

In this process, steam is passed over hot sponge iron sheets at a suitable temperature where hot iron and steam react to produce *ferric oxide* (Fe_2O_3), hydrogen (H_2), carbon dioxide (CO_2) and carbon monoxide (CO) in small quantities. The gases are passed through a *scrubber* where dilute NaOH absorbs CO_2 and CO.

Reaction: $3H_2O + 2Fe \rightarrow Fe_2O_3 + 3H_2$

10.5.3. Coal Gasification

In the gasification of coal, there is *complete conversion of the organic part of the coal into gas*, so that ash alone remains. This is done by reacting the coal with a gasifying agent *e.g.* steam above 700°C.

In this process, the carbon in coal reacts with steam to form CO and H_2. This low energy gas mixture is submitted to water gas shift reaction with *steam*. The CO is then converted into CO_2 with the formation of *additional H_2 gas.*

The *reactions* for coal gasification are:

$$C + H_2O \rightarrow CO + H_2$$
$$CO + H_2O + H_2 \rightarrow 2H_2 + CO_2$$

10.5.4. Methane Gas Reformation

In this process, *methane mixed with steam is passed over a nickel oxide catalyst at elevated temperature.* The *reforming reaction* being *endothermic*, it is usually carried out in fire tube reformed in a fire to be reformer where the catalyst is loaded in the tubes.

The reaction for methane gas reforming:

$$CH_4 + H_2O \rightarrow CO + 3H_2 + Heat\ energy\ released.$$

After steam reforming, the gas products contain considerable amount of CO which may further undergo reaction with *additional steam*, and it *increases the H_2 production.*

The reaction of CO with steam:

$$CO + H_2O \rightarrow CO_2 + H_2 + Heat\ energy\ absorbed.$$

The above reaction is known as water gas shift reaction which is *exothermic*.

10.5.5 THERMO-CHEMICAL METHOD

The thermo-chemical method *involves thermal chemical reactions between primary energy, water and specific chemicals to produce hydrogen at temperatures range from 700°C to 1000°C.*

General thermo-chemical *reactions*:

$$ZO_x + H_2O \rightarrow ZO_{x+1} + H_2$$

$$ZO_{x+1} + Heat \rightarrow ZO_x + \frac{1}{2}O_2$$

where Z represents a *metallic ion or a complex radical.*

10.5.6. Solar Energy Methods

These methods are described below:

10.5.6.1. Bio-photolysis (Biological production)

In this process, Hydrogen is produced with the help of photosynthetic micro-organisms. Blue green algae as well as other anaerobic bacteria are capable of splitting water into hydrogen and oxygen by light driven process and such a process is called *bio-photolysis.*

10.5.6.2. Photo-electrolysis.

In this process, the *decomposition of water into hydrogen and oxygen takes place with the help of electric current which is generated by exposing electrode to sunlight.*

In this process at least one of the electrodes is usually *semiconductor; a catalyst may be included to facilitate the electrode process.*

10.6. HYDROGEN STORAGE

The need for storage arises due to the almost inevitable mismatch between the optimum production rate of energy and fluctuations in demand for energy by users.

The following *three* methods are used for *storage of hydrogen*:

1. Compressed gas storage.
2. Liquid storage.
3. Solid state storage.

10.6.1. Compressed Gas Storage

Hydrogen can be stored in compressed gaseous state in *underground reservoirs* similar to natural gas or can be stored in *high pressure cylinders.*

This method of storage is costly as large quantity of steel is required to store a small amount of hydrogen. The gaseous storage of hydrogen, for industrial use is *economically not viable as a fuel.*

10.6.2. Liquid Storage

Liquid hydrogen fuel is used as a *rocket propellent* in space vehicles as it has the *highest energy density* which is almost *three times the conventional fuels.*

It boils at –253°C (*i.e.*, 20 K) and therefore must be maintained at or below this temperature in storage unless pressure build up can be tolerated. It is necessary to use insulated cylinder to *avoid air condensation* over its surface.

The main *problems associated with the storage of hydrogen are*:

1. Flammabily danger from the fact that liquefied atmospheric gases, rich in oxygen, would concentrate in the vicinity of hydrogen tank.

2. Considerable amount of energy (25-30% of the heating value of hydrogen) is required to convert hydrogen gas into the liquid phase.

Thus, a liquid hydrogen plant requires some kind of primary refrigeration, such as a *liquid nitrogen plant, to precool hydrogen.*

10.6.3. Solid State Storage

A number of metals and alloys form solid compounds, called metal "*hydrides*", by *direct reaction with hydrogen gas.*

In a solid storage, the hydrogen is stored in the form of metallic hydrides. The **metal hydride system** is *based on the principle that a few metals absorb hydrogen in an 'exthermonic reaction' when treated with gas and the absorbed gas is released when the metal hydride is heated.*

The *Chemical reactions* given below:

$$H_2 + \text{Metal} \xrightarrow{\text{Charge}} \textit{Metal hydride} + \text{Heat}$$

$$\textit{Metal hydride} + \textit{heat} \xrightarrow{\text{Discharge}} \text{Metal} + \mathbf{H_2}$$

For hydrogen storage, the metal hydride should have the following *properties*:

1. *Large amount of hydrogen per unit volume per unit mass.*
2. Fairly *inexpensive.*
3. Release of gas at a significant pressure from the hybrid at a moderately high temperature, preferably below 100°C.
4. Easy formation of hybride (by reaction of metal with H_2 gas) should be stable at room temperature.

10.7. GAS HYDRATES

Gas hydrates have been identified by ONGC (Oil Natural Gas Commission) as one of the non-conventional energy sources, to be studied for exploration to ensure energy security for the country. These are naturally occurring *ice like compounds* of *methane and are water formed under low temperature and pressure conditions.* These ice formations consist of water molecules that trapped gas molecules in a *cage-like structure*, found at varying depths in areas of low temperature.

The gas hydrates are *important* due to following *reasons.*

1. Total energy contained in hydrogen is estimated to be *double the amount of the total fossil fuels.*
2. Contain a *great volume of methane*, which is a source of cleaner fuel.
3. One volume of gas hydrates produces 164 volume of gas at standard temperature and pressure.
4. Methane made by drilling around these gas hydrates, can be *captured, stored* and *fed into pipelines* for further use.

Methane can be used to extract hydrogen and use it to power fuel cells.

10.8. HYDROGEN TRANSPORTATION

Hydrogen can be *transported* by the following *three methods*:

1. By pipelines.
2. By insulated tanks.
3. By metal hydride transportation.

10.8.1. By Pipelines

Long distance gas transmission lines of lengths greater than about 90 km must be supplied with pipeline compressors at fairly regular intervals. These compressors must handle a *considerably greater volume* of the gas–somewhere between 3 to 4 times the number of m^3 to the same energy capacity. This would require a considerably higher horse power to drive a hydrogen compressor (in comparison then that needed to drive a natural gas compressor for the same energy throughout).

Therefore, the cost of hydrogen transportation by pipelines must include the cost of piping, compressors and power consumption by compressors.

10.8.2. By Insulated Tanks

Hydrogen in bulk can be transported in well insulated cryogenic tanks having liquid hydrogen either by trucks or rails. The transportation cost, however, is very high.

10.8.3. By Metal Hydrides

Hydrogen can also be transported as a solid metal hydride. The main drawback is the heavy weight of hydride relative to its hydrogen yield. The transportation cost is very high.

10.9. SAFETY PRECAUTIONS

Hydrogen is *highly inflammable* and *explosive* and can lead to fire and serious accidents. The production, storage and distribution of hydrogen require special precautions, as given below:

1. The system should be designed to withstand the explosion pressures.
2. The system should be designed to withstand pressure surges.
3. Proper explosion relief system must be provided.
4. Flame traps, flame suppressors, explosion-relief devices and rapid closing devices must be used.
5. The design, manufacture and storage methods/system should follow Petroleum Act.

HIGHLIGHTS

1. Hydrogen energy (energy carrier like electricity) is an alternative energy of future which can also be stored in addition to other qualities of electrical energy.
2. Problems associated with hydrogen use are:
   (*i*) Commercial production of hydrogen at cheap cost.
   (*ii*) Energy utilisation.
   (*iii*) Difficulty in storage, since it is highly explosive.
   (*iv*) Lack of safety and management.
3. The most commonly used methods of production of hydrogen are:
   (*i*) Steam reformation;   (*ii*) Electrolysis of water;
   (*iii*) Coal gasification;   (*iv*) Methane gas reformation;
   (*v*) Thermo-chemical methods;   (*vi*) Solar energy method.

4. Methods of hydrogen storage are:

   (*i*) Compressed gas storage;   (*ii*) Liquid storage;

   (*iii*) Solid state storage.

THEORETICAL QUESTIONS

1. Discuss the main issues associated with the use of hydrogen as energy source.
2. What are the properties of hydrogen?
3. What are the advantages of hydrogen as fuel.
4. What are the applications of hydrogen (energy)?
5. List the possible areas of utilisation of hydrogen in near future.
6. Discuss briefly the various method of producing hydrogen.
7. Explain clearly the "electrolysis of water" method used for producing hydrogen.
8. How can hydrogen be produced by the following methods:
      (*i*)   Methane gas reformation.
      (*ii*) Solar energy method.
9. Explain any two of the following methods of hydrogen storage:
      (*i*)   Compressed gas storage.
      (*ii*) Liquid storage.
      (*iii*) Solid state storage.
10. Write a short note on 'Gas hydrates'.
11. Explain briefly any two of the following methods employed for hydrogen transportation:
       (*i*)   By pipelines.
       (*ii*) By insulated tanks.
       (*iii*) By metal hydrides.

NUCLEAR POWER PLANT

> **11.1 General aspects of nuclear engineering** – Atomic structure – Atomic mass weight – Isotopes – Radioactivity – Radioactive decay – Nuclear reactions – Fertile materials – Fission and fusion; **11.2 Nuclear reactors** – Definition and classification – Essential components of a nuclear reactor – Power of a nuclear reactor; **11.3 Main components of a nuclear power plant; 11.4 Brief description of reactors** – Pressurised water reactor (PWR) – Boiling water reactor (BWR) – CANDU reactor – Gas cooled reactor – Liquid metal cooled reactor – Breeder reactor; **11.5 Advantages and disadvantages of nuclear power plants; 11.6 Nuclear-plant site selection; 11.7 Nuclear power plants in India.** *Highlights – Theoretical Questions.*

11.1. GENERAL ASPECTS OF NUCLEAR ENGINEERING

11.1.1 Atomic Structure

- An element is defined as a substance *which cannot be decomposed into other substances.* The *smallest particle* of an element which takes part in chemical reaction is known as an *'atom'*. The word atom is derived from Greek word 'Atom' which means indivisible and for a long time the atom was considered as such. *Dalton's atomic theory states* that (*i*) all the atoms of one element are precisely alike, have the same mass but differs from the atoms of other elements (*ii*) the chemical combination consists of the union of a small fixed number of atoms of one element with a small fixed number of other elements.

- Various atomic models proposed by scientists over the last few decades are: 1. Thompson's plum puddling model, 2. Rutherford's nuclear model, 3. Bohr's model, 4. Sommerfeld's model, 5. Vector model, 6. Wave-mechanical model.

- The complex structure of atom can be classified into *electrons* and *nucleus*. The nucleus consists of *protons* and *neutrons* both being referred as *nucleons*. *Protons are positively* charged and *neutrons are neutral*, thus making complete nucleus as positively charged.

- The *electrons* carry *negative* charge and circulate about the nucleus. As the positive charge on proton particle is equal to the negative charge on electron particle, and the *number of electrons is equal to the number of protons*, atom is a neutral element. Any addition of the number of electrons to the neutral atom will make it negatively charged. Similarly any subtraction of the electrons will make it positively charged. Such an atom is known as *ion* and the process of charging the atom is termed an *ionisation*.

- *"The nuclear power engineering"* is specially connected with *variation of nucleons in nucleus.* Protons and neutrons are the particles having the mass of about 1837 times and 1839 times the mass of an electron.

- The modern atomic theory tells that the atom has a diameter of about 10^{-7} mm. In a neutral atom the electrons are bound to the nucleus by the electrostatic forces, which follows the Coloumb's law of forces, *i.e.,* like charges repel and unlike charges attract each other. The function of electrostatic force is similar to the gravitational force.

- The atomic spectrum study has revealed that every electron in an atom is in one group of specific states of motion which is corresponding to its total energy. In an atom the electrons are spinning around the nucleus in orbits. These orbits are called *shells*, which represent the energy levels for the electrons. *All the electrons having very nearly the same total energy are said to be in the same shell.* The shells have been named as *K, L, M, N* etc. Each shell consists of the specific maximum number of electrons. The *K* shell (inner shell) contains **2** electrons, *L* shell has **8** electrons, *M* shell is limited to **18** and the *N* shell possesses **32** electrons. In fact, the number of electrons in any orbit is equal to $2n^2$ where *n* is the serial number of the orbit taking first orbit nearest to the nucleus, *with the exception that the outermost orbit cannot have more than eight electrons.* In a given atom all orbits may not be complete. It is obvious from the study that *amplitude difference in energy between two shells is much more than the difference in between energy levels in one shell.* In a *shell less than the specified number of electrons may exist but not a large number.* The inner shell is filled up first and then the other successive shell are completed.

- *The chemical properties of the atom varies with composition of number of electrons in various shells and the state of energies within the shells determine the electrical characteristics of the atom.* For example, *Hydrogen* (H_2) consists of one electron in the first shell, *Helium* (He) has two electrons in the first shell, *Lithium* (Li) has two electrons in first shell and one is second shell, *Carbon* (C) consists of two electrons in first and four in second shell.

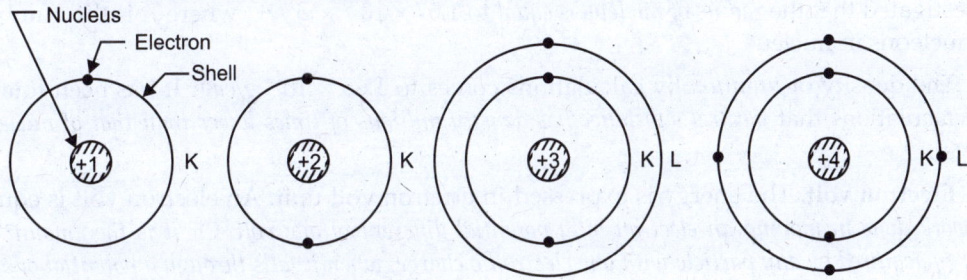

Fig. 11.1: Atomic structure of H_2, He, Li and C.

- The electrons lying in the outermost shell are termed *valence electrons.* If the outermost shell is completely filled, the atom is stable and will not take any electron to fill up the gap. However, the incomplete outer shell will try to *snatch* the required number of electrons from the adjacent atom in a matter. The binding force between the electron and nucleus is the electrostatic force of attraction. *To emit one electron, energy required is more than the electrostatic force of attraction. When the energy is supplied, the electron jumps from one discrete energy level to another permissible level. The process starts from outer shell.* The electron possesses the energy in two forms, *i.e.,* kinetic energy due to its motion and potential energy due to its position with respect to the nucleus. It is obvious that *electrons cannot exist in between the permissible orbits.*

● The charge of nucleus is represented by the *number of protons* present. This number is known as *atomic number* and designated by the letter Z. It also shows the position of atom in the periodic table. 'Hydrogen' has *only one* number but natural 'uranium' has *ninety two.* The atoms having *higher atomic number* have been developed artificially ranging from 93 to 102. These are einsteinium (Z = 99), Ferinium (Z = 100), and mendelevium (Z = 101). *Platonium (Z = 94) is an important element to the nuclear power field.*

The *mass number (A)* is the sum of total number of protons and neutrons in a nucleus. The number of electrons is represented by the letter *N, i.e., N = (A − Z).*

11.1.2. Atomic Mass Unit

The *mass of the atom is expressed in terms of the mass of the electron.* The unit of mass has been considered as $\frac{1}{16}$ th of the mass of neutral oxygen atom which contains 8 protons and 8 neutrons. The atomic mass unit (a.m.u.) is equal to $\frac{1}{16}$ th the mass of oxygen neutral atom.

One a.m.u. = 1.66×10^{-24} g

Mass of proton = 1837 me = $\dfrac{1837 \times 9.1 \times 10^{-28}}{1.66 \times 10^{-24}}$ = 1.00758 a.m.u.

Mass of neutron = 1839 me = $\dfrac{1839 \times 9.1 \times 10^{-28}}{1.66 \times 10^{-24}}$ = 1.00893 a.m.u.

It has been concluded that the density of matter in a nucleus is enormous. It has been investigated that the *radius of nucleus is equal to* $1.57 \times 10^{-3} \times 3\sqrt{A}$, where A is the number of nucleons in nucleus.

The density of *uranium* by calculations comes to 1.65×10^{14} g/cm³. It has been found by calculations that *natural substance has density millions of times lower than that of nuclear matter.*

Electron volt. The energy is expressed in electron volt unit. An electron volt is equal to *work done in moving an electron by a potential different of one volt. Or it is the amount of energy acquired by any particle with one electronic charge, when it falls through a potential of one volt.*

One electron volt = 1.602×10^{19} joule.

11.1.3. Isotopes

In any atom, the *number of electrons is equal to number of protons.* This is independent of neutrons in the nucleus. *Atoms having different number of neutrons than the number of protons are known as 'Isotopes'.*

Examples: Isotopes of hydrogen are shown in Fig. 11.2.

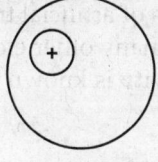 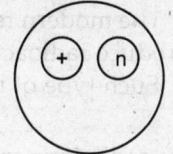 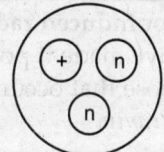

H₁-Hydrogen   H₂- Heavy hydrogen   H₃- Heavy hydrogen
(No neutron)   or Deuterium-D   or Tritium-T

Fig. 11.2

These isotopes have the *same chemical properties* and *have the same atomic number* and *occupy the same place in the periodic table*. But the *nuclear properties* of each of the isotopes are different *because of the different number of neutrons in the nucleus*.

The isotopes of oxygen vary from O_{14} to O_{19}. The change of number of neutrons in nucleus affect the mass of atom.

Example. Weight of heavy hydrogen is twice the weight of simple hydrogen. This means a volume of H_2O weighs less than the same volume of D_2O.

The isotopes can be represented with mass number (A) as subscript and atomic number (Z) as subscript like $_ZH^A$.

Examples. Hydrogen isotopes are represented as $_1H^2$ (Deuterium), $_1H^3$ (Tritium) and Uranium isotopes as U^{234}, U^{235}, U^{238}.

— The isotopes are *not stable* and *disintegrate* at a certain rate. The isotope which disintegrates at a fixed rate is called *Radioactive isotope* or *Radio isotope*.

The instability of the nucleus can be either by the separation of parent nucleus into 2 or more nuclei or by the rearrangement of nucleons in the matter so that there is an emission of particles or energy in the form of rays or by rearrangement of electrons. During this transformation there is emission of particles at a very high velocity. This is known as *"radiation"*.

We must note that for *any specific isotope, the rate of radiation from a unit mass and also the energy distribution are fixed and cannot be changed by any method*. Thus, for any isotope, the quantity of radiation per unit time can be determined easily.

11.1.4. Radioactivity

Radioactivity was originally discovered by Becquerel in 1896. This phenomenon is *confined almost entirely to the heaviest element* from 83 to 106 in the periodic table.

The phenomenon of spontaneous emission of powerful radiations exhibited by heavy elements is called **radioactivity.** Radioactivity is essentially a *nuclear phenomenon* and is a *drastic process* because the element changes its kind. It is *spontaneous* and an *irreversible self-disintegrating activity* because the element breaks itself up for good. Those elements which exhibit this activity are called *radioactive elements*. Examples are: *Uranium, polonium, radium, radon, ionium, thorium, actinium* and *mesothorium*.

The radioactive radiations emitted by the radioactive elements are found to consist of the following:

(*i*) Alpha (α) rays or α-particles.

(*ii*) β rays or β-particles.

(*iii*) γ-rays or photons.

The radioactivity may be *natural* or *artificial*.

Natural radioactivity. It is that *which is exhibited by element as found in Nature*. It is always found in heavier elements in the periodic table.

Artificial or induced radioactivity. The modern techniques of artificial transmutation of elements have made it possible to produce radioactivity in many other elements much lighter than those that occur in Nature. Such type of radioactivity is known as *artificial or induced radioactivity.*

The general *properties* of radioactive radiations are:

1. These radiations are *highly penetrating, they affect photographic plates, ionise gases, cause scintillations on fluorescent screen, develop heat and produce chemical changes.*

2. As radiations are given out, *new elements are formed* in an irreversible process— the new elements themselves being usually radioactive.

3. To emission of radiations is *spontaneous* and is not affected by external agents.

4. The emission is *not instantaneous* but is *prolonged i.e.,* it is extented over a period of time otherwise it would not have been discovered at all.

5. Except for radioactivity, there is nothing abnormal about the radioactivity elements as regards their physical and chemical properties.

11.1.5. Radioactive Decay

It has been observed that the emission of the particles in the form of alpha, beta or gamma radiations is not an instantaneous process. For various elements the decay time is different, which follows a certain *law*. Obviously the process is independent of the physical and chemical properties of the given isotope at a particular temperature and pressure.

The *'law'* states: *"The small amount of disintegration of the isotope in a small period is directly proportional to the total number of radioactive nuclei and proportionality constant".*

If, N = Number of radioactive nuclei present at any time t,

N_0 = Initial number of such nuclei,

λ = Proportionality constant (also known as disintegration constant or the radioactive decay constant of the material),

Then, the above law can be stated in the form of equation as follows:

$$\Delta N = -\lambda N \Delta t \qquad \qquad ...(11.1)$$

or,
$$\frac{dN}{dt} = -\lambda N \qquad \qquad ...(11.2)$$

The *negative* sign represents that during disintegration the number of the nuclei is *decreasing.*

Integrating the above equation (11.2) after proper arrangement within the proper limits, we get:

$$\int_{N_0}^{N} \frac{dN}{N} = -\lambda \int_{0}^{t} dt \qquad \qquad ...(11.3)$$

or, $\log_e N - \log_e N_0 = -\lambda t,$ or, $\log_e \dfrac{N}{N_0} = -\lambda t$

or, $\dfrac{N}{N_0} = e^{-\lambda t},$ or, $N = N_0 e^{-\lambda t}$ $...(11.4)$

$$\frac{dN}{dt} = -\lambda N = -\lambda N_0 e^{-\lambda t} \qquad \qquad ...(11.5)$$

The eqn. (11.5) represents that the *decay scheme follows the exponential law.*

Activity:

The intensity of emitted radiation is termed **activity.**

This is directly dependent on the rate of disintegration of the element.

If, A = Activity at time t,

A_1 = Initial activity,

k = Detection coefficient,

Then, $A = k\left(-\dfrac{dN}{dt}\right) = k\lambda N = k\lambda N_0 e^{-\lambda t} = A_1\, e^{-\lambda t}$ ...(11.6)

Half-life:

Half-life represents the rate of decay of the radioactive isotopes. The half-life is the time required for half of the parent nuclei to decay or to disintegrate.

Putting $N = \dfrac{N_0}{2}$ and $t = t_{1/2}$ in eqn. (11.5), we get

$\dfrac{N_0}{2} = N_0\, e^{-\lambda t_{1/2}}$ $\therefore$ $e^{-\lambda t_{1/2}} = 1/2$

$\therefore$ $\lambda t_{1/2} = \log_e 2 = 0.693$ $\therefore$ $t_{1/2} = \dfrac{0.693}{\lambda}$ ...(11.7)

Here $t_{1/2}$ is the half-life of radioactive nuclei. After passing every half-life the number of nuclei is reduced to half and so is the activity. This process is repeated for the several half lives till the activity becomes negligible. The *variation of half-life is from fraction of seconds to million of years.*

Half-life of some of the metals is given below:

Metal	Half-life
Po-214	170 μ sec
I-137	25 sec
Carbon-14	5100 years
Th-232	1.4×10^{10} years
Uranium-238	4.525×10^9 years

Average (mean) life:

This indicates the average of total time for which the radioactive nuclei has disintegrated for several half lives. Hence this is *greater than half-life.* This is obtained by taking the sum of the decay time of the radioactive nuclei and then it is divided by the initial number of nuclei.

If T is the time of average life, then

$$T = \dfrac{-\int_0^\infty t\,dN}{N_0} = \dfrac{\lambda N_0 \int_0^\infty t e^{-\lambda t}\,dt}{N_0}$$ ...(11.8)

Now integrating by parts, we get:

$$T = \left[-t e^{-\lambda t} - \dfrac{e^{-\lambda t}}{\lambda}\right]_0^\infty = \dfrac{1}{\lambda}$$ ...(11.9)

But, $t_{1/2} = \dfrac{0.693}{\lambda}$

From the above eqns. it is clear that *mean life is 1.445 times greater than half-life.*

Note. Number of disintegrations per second is the unit of radioactivity and is termed *curie,* as this phenomenon was first discovered by Curie.

11.1.6. Nuclear Reactions

During a nuclear reaction, the *change in the mass of the particle represents the release or an absorption of energy.* If the total mass of the particle after the reaction is reduced, the process releases the energy, consequently, the increase in the mass of the resultant particle, will cause the absorption of energy.

The equations of nuclear reactions are connected with the resettlement of protons and neutrons within the atom. The equations are much similar to chemical reactions. The energy variation is also of the order of MeV. In simple term the equation shows the *balance of neutron and proton.*

A nuclear reaction is written as follows:

(*i*) The bombarded nuclei or the target nuclei is written first from left hand side.

(*ii*) In the middle within brackets, first is the incident particle and second one the ejected.

(*iii*) On the right hand side, the resultant nucleus is placed.

A *neutron* is written as: $_0n^1$ because it has unit mass and it does not have any charge.

Any *electron* is written as: $_{-1}e^0$ because its mass is negligible as compared to proton or neutron and its charge is equal but opposite to the charge of proton.

Some of the *examples of reactions* are given below:

(*i*) When $_{11}Na^{23}$ is bombarded with protons possessing high energy, it is converted to $_{12}Mg^{23}$

$$_{11}Na^{23} + {}_1H^1 \longrightarrow {}_{12}Mg^{23} + {}_0n^1 + q \qquad ...(11.10)$$

(where q = release or absorption of energy in the reaction)

(*ii*) When $_{13}Al^{27}$ is bombarded with high energy protons it is transformed to $_{14}Si^{27}$.

$$_{13}Al^{27} + {}_1H^1 \longrightarrow {}_{14}Si^{27} + {}_0n^1 + q \qquad ...(11.11)$$

(*iii*) When $_{13}Al^{27}$ is bombarded with deutrons, Al^{28} and proton may be produced.

$$_{13}Al^{27} + {}_1H^2 \longrightarrow {}_{13}Al^{28} + {}_1H^1 \qquad ...(11.12)$$

The eqns. (11.10), (11.11) and (11.12) may be written in the equation form as given below:

$$Na^{23}\ (p,\ n)Mg^{23} \qquad ...(11.13)$$
$$_{13}Al^{27}\ (p,\ n)Si^{27} \qquad ...(11.14)$$
$$Al^{27}\ (d,\ p)Al^{28} \qquad ...(11.15)$$

11.1.7. Fertile Materials

It has been found that some materials are not *fissionable by themselves* but they *can be converted to the fissionable materials,* these are known as **fertile materials.**

Pu^{239} and U^{233} are not found in Nature but U^{238} and Th^{232} can produce them by nuclear reactions. When U^{238} is bombarded with slow neutrons it produces $_{92}U^{239}$ with half-life of 23.5 days which is unstable and undergoes two beta disintegration. The resultant Pu^{239} has half-life of 2.44×10^4 yrs and is a good alpha emitter.

$$_{92}U^{238} + {}_0n^1 \longrightarrow {}_{92}U^{239} + \gamma \qquad ...(11.16)$$
$$_{92}U^{239} \xrightarrow{\ 23.5\ min\ } {}_{-1}e^0 + {}_{91}Np^{231} \qquad ...(11.17)$$

$$_{93}\text{Np}^{239} \xrightarrow{\text{2.3 days}} {}_{-1}e^0 + {}_{94}\text{Pu}^{239} \qquad\qquad ...(11.18)$$

During conversion the above noted reactions will take place. The other isotopes of neptunium such as 2.1 day Np^{238} and plutonium can also be produced by the bombardment of heavy particles accelerated by the cyclotron.

The nuclear transformations to convert $_{90}\text{Th}^{232}$ to U^{233} are given below:

$$_{90}\text{Th}^{232} + {}_0n^1 \xrightarrow{\hspace{3cm}} {}_{90}\text{Th}^{233} + \gamma \qquad\qquad ...(11.19)$$

$$_{90}\text{Th}^{233} \xrightarrow{\text{23.3 min}} {}_{91}\text{Pa}^{233} + {}_{-1}e^0 \qquad\qquad ...(11.20)$$

$$_{91}\text{Th}^{233} \xrightarrow{\text{27.4 days}} {}_{92}\text{U}^{233} + {}_{-1}e^0 \qquad\qquad ...(11.21)$$

U^{235} is the source of neutrons required to derive Pu^{239} and U^{233} from Th^{232} and U^{238} respectively. This process of conversion is performed in the *breeder reactors*.

Other fissionable materials: Th^{227}, Pa^{232}, U^{231}, Np^{238} and Pu^{241} are the other nuclides which are having high cross-sections for neutron thermal fission. *Pu^{241} is the important nuclide which is used in plutonium fueled power reactors.*

11.1.8. Fission and Fusion

Fission *is the process that occurs when a neutron collides with the nucleus of certain of the heavy atoms, causing the original nucleus to split into two or more unequal fragments which cary off most of the energy of fission as kinetic energy.* This process is *accompanied by the emission of neutron and gamma rays.*

The chain reaction:

A chain reaction is that process in which the number of neutrons keeps on multiplying rapidly (in geometrical progression) during fission till whole of the fissionable material is disintegrated. The chain reaction will become self-sustaining or self propagating only if, for every neutron absorbed, at least one fission neutron becomes available for causing fission of another nucleus. This condition can be conveniently expressed in the form of *multiplication factor or reproduction factor (K)* of the system which may be defined as:

$$K = \frac{\text{No. of neutrons in any particular generation}}{\text{No. of neutrons in the preceding generation}}$$

If $K > 1$, chain reaction will continue and if $K < 1$, chain reaction cannot be maintained.

Nuclear fusion *is the process of combining or fusing two lighter nuclei into a stable and heavier nuclide.* In this case also, *large amount of energy is released* because mass of the product nucleus is *less* than the masses of the two nuclei which are fused.

Several reactions between nuclei of low mass numbers have been brought about by accelerating one or the other nucleus in a suitable manner. These are often fusion processes accompanied by release of energy. However, reactions involving artificially-accelerated particles cannot be regarded as of much significance for the utilisation of nuclear energy. To have practical value, fusion reactions must occur in such a manner as to make them *self-sustaining, i.e., more energy must be released than is consumed in initiating the reaction.*

It is thought the energy liberated in the sun and other stars of the main sequence type is due to the nuclear fusion reactions occurring at the very high stellar temperature of 30 million °K. Such processes are called *thermonuclear reactions* because they are temperature-dependent.

Comparison of "Fission" and "Fusion" processes:

The comparison between *'Fission'* and *'Fusion'* processes is given below:

Fission	Fusion
1. When heavy unstable nucleon is bombarded with neutrons, the nucleus splits into fragments of equal mass and energy is released.	1. Some light elements *fuse together* with the release of energy.
2. About *one thousandth* of the mass is *converted into energy.*	2. It is possible to have *four thousandths* of mass *converted into energy.*
3. Nuclear reaction *residual problem is great.*	3. Residual problem is *much less.*
4. Amount of radioactive material in a fission reactor is *high.*	4. A possible advantage is that the total amount of radioactive material in a working fusion reactor is likely to be *very much less* than that in a fission reactor.
5. Because of higher radioactive material, *health hazards is high in case of accidents.*	5. Because of lesser radioactive material, *health hazards is much less.*
6. It is *possible to construct self-sustained fission reactors* and have positive energy release.	6. It is *extremely difficult to construct controlled fusion reactors.*
7. *Manageable temperatures* are obtained.	7. Needs *unmanageable temperatures* like 30 million degrees for fusion process to occur.
8. Raw fissionable material is *not available in plenty.*	8. Reserves of deuterium, the fusion element, *is available in great quantity.*

11.2. NUCLEAR REACTORS

11.2.1. Definition and Classification

Definition. *A nuclear reactor is an apparatus in which nuclear fission is produced in the form of a controlled self-sustaining chain reaction.* In other words, it is a controlled chain-reacting system supplying nuclear energy. It may be looked upon as a sort of nuclear furnace which burns fuels like U^{235}, U^{233} or Pu^{239} and, in turn, *produces many useful products like heat, neutrons and radioisotopes.*

Mechanism of heat production. Most of the energy is imparted to the two fission fragments into which the nucleus divides causing them to move at high speed. However, because they have taken birth in a dense mass of metal, they are rapidly slowed down and brought to rest by colliding with other atoms of the metal. In so doing, their energy is converted into heat in much the same way as energy given up by a slowing motor can be converted into heat in the brake lining. In this way, the mass of uranium metal gets heated up.

Nuclear reactors are *"classified"* according to the *chain reacting system, use, coolants, fuel material etc.*

1. **Neutron energies at which the fission occurs:**
    (*i*) Fast fission is caused by high energy neutrons   *Fast reactors*
    (*ii*) Intermediate or epithermal   *Intermediate reactors*
    (iii) Low energy i.e., thermal   *Slow reactors.*

On the basis of the energy of the neutrons to cause fission the reactors have been divided into *three groups—fast, intermediate and thermal.*

(a) In **"Fast reactors"** the high velocity neutrons produced by fission are *utilised directly* to cause fission of the fuel in the reactor. The velocity of the neutrons is *not reduced deliberately*.

(b) If in a reactor the fission process is maintained due to the slow neutrons capture, the reactor is known as **"Slow reactor"**. The minimum velocity to which neutrons *are slowed down* before the fission is equal to the thermal velocity which the slow neutrons may acquire in a state of thermal equilibrium with the medium. This velocity is of the order of 2150 m/s at room temperature which is equivalent to $\frac{1}{40}$ eV or neutron energy. The neutrons associated with the energy of this order are known as *thermal neutrons* and the reactor as **"Thermal reactors"**. With the *moderator the neutrons are slowed down*. The *main advantage is that the probability of reaction increases*.

(c) If the velocity of neutrons is kept between both the above noted limits, the reactors are termed as *"Intermediate reactors"*.

2. Fuel-moderator assembly:

    (i) Homogeneous reactors;   (ii) Heterogeneous reactors.

In *"Hemogeneous reactor"* the fuel and moderator are mixed to form a homogeneous material, i.e., uranium fuel salt forms a homogeneous solution in water which is a moderator, or fine particles of uranium and carbon give a mechanical mixture.

In *"Heterogeneous reactor"* the fuel is used in the form of rods, plates or wires etc. and the moderator surrounds the each fuel element in the reactor core.

3. Fuel state:

    (i) Solid   (ii) Liquid   (iii) Gas

The nuclear fuel is available in three states—solid, liquid and gas. In reactors the fuel is mostly used in solid state or in the form of solution dissolved in water. *The "liquid metal reactors" are in practical use.*

4. Fuel material:

    (i) Natural uranium with U^{235} contents (occurs in nature)

    (ii) Enriched uranium with more than 0.71 of U^{235}

    (iii) Pu^{239}, Pu^{241} or Pu^{239} (man made)

    (iv) U^{233} (man made).

Considering the necessary requirement of fission process and its availability economically the fuels used in reactors are *uranium, plutonium* and *thorium*. U^{235} is easily available in natural uranium (i.e., 0.7%) and its content increases upto 90% in enriched uranium.

5. Moderator:

    (i) Water (H_2O) (ii) Heavy water (D_2O)

    (iii) Graphite   (iv) Beryllium or beryllium oxide

    (v) Hydrocarbons or hydrides.

A moderator's function is to *absorb the part of the kinetic energy of the neutrons*. The neutrons collide directly with the moderator and thus slowed down. No ideal moderator is available in nature or has been produced artificially. The light weight nuclei materials are not suited at all as a moderator because they do not possess the property of absorption of neutrons.

Light water, heavy water and *graphite* are the *most common moderators* used in reactors.

6. **Principal product:**

(i) Research features to produce neutrons	*Research reactors*
(ii) Power reactor to produce heat	*Power reactors*
(iii) Breeder reactors to produce fissionable materials	*Breeder reactors*
(iv) Production reactors to produce isotopes	*Production reactors*

Research reactors. These are desgined to produce the high neutron flux for research work and these neutrons are used to determine the neutron properties of interaction with the nuclei and the effect of bombardment of neutrons on the materials. The reactors are operated at high neutron flux and low power level otherwise the cooling will be a problem. The unit is cooled constantly during operation. The by-products are heat and fission products which are removed during operation.

Power reactors. In these reactors the energy is produced in heat form which is carried away to the heat exchanger by circulating the coolant through the reactor and heat exchanger. In the heat exchanger the coolant converts the water into steam to run the turbine. The by-products are fission products, neutrons and other radiation particles.

These reactors are useful to produce *huge amount of power* and are *widely used in power plant stations.* In such reactors *consumption is very low.*

Breeder reactors. A breeder reactor *converts fertile materials into fissionable materials* such as U^{238} and Th^{232} to Pu^{239} and U^{233} respectively *besides the power production.* It is worth noting that the amount of fissionable material produced is *more* than its consumption of fissionable material. By-products are the same as those of power reactors.

Production reactors: The output of such reactors is radioactive materials which are used as sources of radiation and tracers in research in all areas of science. By-products are the same as those of power reactors.

7. **Coolant:**

    (i) Air, carbon or helium-cooled reactors

    (ii) Water or other liquid-cooled reactors

    (iii) Liquid metal-cooled reactors.

In *gas-cooled reactors* the amount of gas required to extract the heat is too much and therefore these reactors are expensive. Gases have poor heat carrying capacity. CO_2 and He have been used in the early reactors. *Mostly water is used as a coolant.*

Liquid metal cooled reactors are also suitable as the metal is having high boiled point and low steam pressure. These are the power reactors.

8. **Construction of core**

(i) Cubical	(ii) Cylindrical	(iii) Octagonal
(iv) Spherical	(v) Slab	(vi) Annulus.

The proper shape to the core is given on the practical consideration and can have cubical, cylindrical or ring type construction.

11.2.2. Essential Components of a Nuclear Reactor

The *essential components of a nuclear reactor* are as follows:

1. Reactor core	2. Reflector	3. Control mechanism
4. Moderator	5. Coolants	6. Measuring instruments
7. Shielding.		

1. **Reactor core:** Refer to Fig. 11.3.

The reactor core is that part of a nuclear power plant where fission chain reaction is made to occur and where *fission energy is liberated in the form of heat for operating power conversion equipment.* The *core* of the reactor consists of an assemblage of fuel elements, control rods, coolant and moderator. Reactor cores generally have a shape approximating to a right circular cylinder with diameters ranging from 0.5 m to 15 m. The pressure vessels which houses the reactor core is also considered a part of the core (Fig. 11.3). The fuel elements are made of plates or rods of uranium metal. These plates or rods are usually clad in a thin sheath of stainless steel, zirconium or aluminium to provide corrosion resistance, retention of radioactivity and in some cases, structural support. Enough space is provided between individual plates or rods to allow free passage of the coolant.

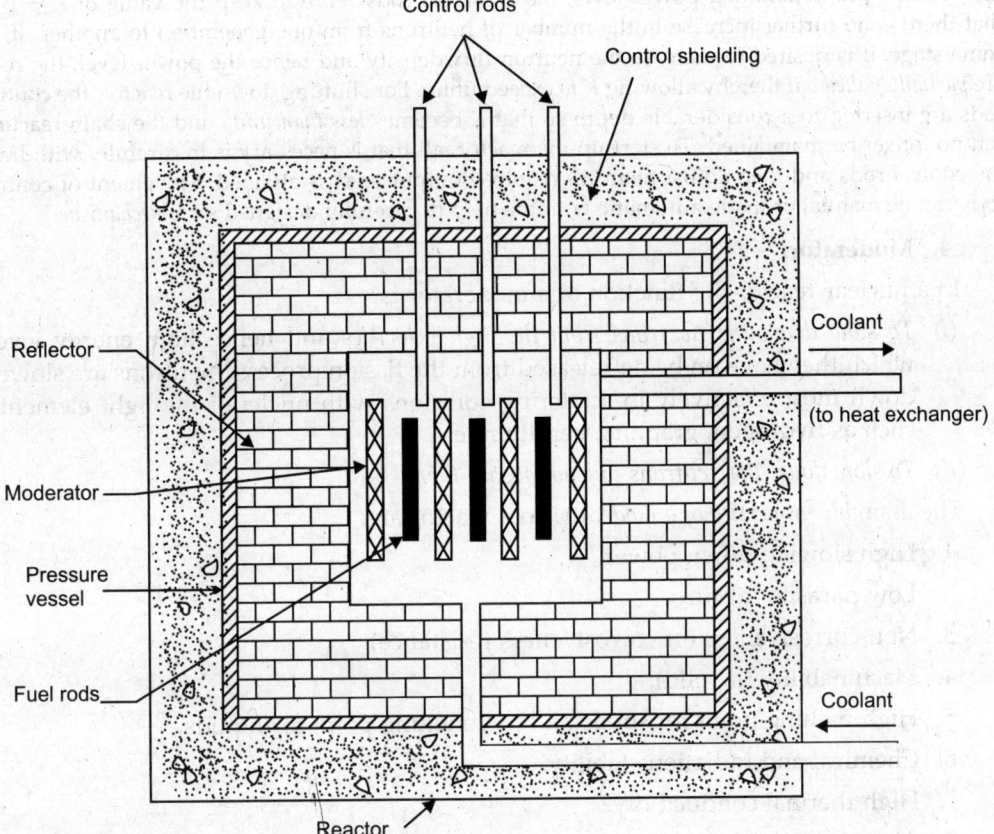

Fig. 11.3: Nuclear reactor.

2. **Reflector:**

A reflector is usually placed round the core to reflect back some of the neutrons that leak out from the surface of the core. It is generally made of the same materials as the moderator.

3. **Control mechanism:**

It is an essential part of a reactor and serves the following *purposes*:

(*i*) For starting the reactor *i.e.,* to bring the reactor up to its *normal operating level.*

(*ii*) For maintaining at *that level i.e.,* keep power production at a steady state.

(*iii*) For shutting the reactor down under normal or emergency conditions.

The control system is also necessary to prevent the chain reaction from becoming violent and consequently damaging the reactor. The effective multiplication factor of the reactor is always kept *greater than unity* in order that the number of neutrons keeps on increasing in successive generations. As the number of neutrons and hence the neutron flux density increases, the temperature also increases. Unless the growth is checked at some point, the reactor is likely to be damaged as a result of too rapid liberation of energy.

Note. The *control system* works on the simple *principle of absorbing the excess neutrons* with the help of control rods either made of boron steel or cadmium strips. Both these materials have very large cross-section for thermal neutrons *i.e.*, they are very good absorbers of slow neutrons and also have the advantage of not becoming radioactive due to neutron capture. By pushing these rods deeper into the central core, any amount of excess neutrons can be absorbed. Once the reactor has reached pre-determined power level, these control rods serve to keep the value of $K = 1$ so that there is no further increase in the number of neutrons from one generation to another. If, at some stage, it is desired to increase the neutron flux density and hence the power level, the rods are *partially pulled out* thereby allowing K *to* exceed unity. For shutting down the reactor, the control rods are inserted to a considerable depth so that K becomes *less than unity* and the chain-reaction can no longer be maintained. To start up the reactor, all that is necessary is to carefully withdraw the control rods and then adjust them till required output level is attained. Movement of control rods can be manual or made automatic with the help of carefully designed *servomechanism*.

4. Moderator:

In a nuclear reactor the function of a moderator is:

(*i*) *To slow down the neutrons from the high velocities* and hence high energy level, which they have on being released from the fission process. Neutrons are slowed down most effectively in scattering collisions with nuclei of the light elements, such as hydrogen, graphite, beryllium etc.

(*ii*) *To slow down the neutrons but not absorb them.*

The *desirable properties of a moderator* in a reactor are:

1. High slowing down power.
2. Low parasite capture.
3. Non-corrosiveness (or corrosiveness resistance).
4. Machinability (if solid).
5. High melting point for solids and low melting point for liquids.
6. Chemical and radiation stability.
7. High thermal conductivity.
8. Abundance in pure form.

H_2O, D_2O (heavy water), He (gas), Be and C (graphite) are the commonly used moderators.

As a moderator D_2O is *the best material available* (moderating ratio of D_2O is 12000 as compared to 72 for H_2O and 170 for carbon) because (*i*) it has excellant neutron slowing properties, (*ii*) it has very small cross-section for neutron capture, and (*iii*) it can be used as a coolant as well. Its *disadvantages* are: (*i*) it has low boiling point so that it *necessitates pressurisation*, and (*ii*) it is *very expensive*. But, the advantages of D_2O as moderator or moderator coolant outweigh its high cost.

5. Coolants:

The function of a coolant is to remove the intense heat produced in the reactor and to bring out for being utilised.

The desirable characteristics for a reactor coolant are:

1. Low parasite capture.
2. Low melting point.
3. High boiling point.
4. Chemical and radiation stability.
5. Low viscosity.
6. Non-toxicity.
7. Non-corrosiveness.
8. Minimum induced activity (short half-lives, low energy emission).
9. High specific heat (reduces pumping power and thermal stresses).
10. High density (reduces pumping power and physical plant size).

Commonly used coolants: Santiwax R (organic, Hg, He, CO_2).

The most widely-used gaseous coolant is CO_2 particularly in large-power reactors. It is (*i*) cheap, (*ii*) does not attack metals at reasonable temperatures, and (*iii*) has small cross-section for neutron capture.

6. Measuring instruments:

Main instrument required is for the purpose of measuring thermal neutron flux which determines the power developed by the reactor.

7. Shielding:

Shielding is necessary in order to:

(*i*) protect the walls of the reactor vessel from radiation damage, and also to

(*ii*) protect operating personnel from exposure to radiation.

The first known as *thermal shield* is provided through the *steel lining*, while the other called *external* or *biological* shield is generally made of *thick concrete surrounding the reactor installation.*

Among the nuclear radiations produced in a reactor the alpha and beta particles, thermal (slow) neutrons, fast neutrons and gamma rays are harmful ones and must be shielded against. Of these only the fast neutrons and gamma rays present some serious difficulty in designing the reactor shielding, since alpha and beta particles can be stopped by a fraction of an inch of a solid substance, while thermal neutrons can be automatically guarded against with a shield thick enough to provide protection against fast neutrons and gamma rays.

The effectiveness of a nuclear shield against gamma rays approximately depends upon its mass. A heavy material like lead will be a more effective shield per unit weight, than a light element such as carbon. On the other hand, light elements, particularly hydrogen are much more effective per unit weight than heavy elements for fast neutron shielding. Concrete is a material that offers a compromise between these two extreme characteristics of shielding material for both gamma rays and fast neutrons. It is a material which has low cost and is easily available.

The actual design of the shield, however, involves the following *considerations*:

(*i*) The total amount of radiation produced in the reactor.

(*ii*) The amount of radiation that can be permitted to leak through the shield.

(*iii*) The shielding properties of material.

11.2.3. Power of a Nuclear Reactor

The fission rate of a reactor, *i.e.*, total number of nuclei undergoing fission per second in a reactor is

$$= nC\sigma NV = \phi_{nu}\sigma NV$$

where,

n = Average neutron density *i.e.*, number per m³,

C = Average speed in m/s,

ϕ_{nu} = nC = Average neutron flux,

N = Number of fissile nuclei / m³,

σ = Fission cross-section in m², and

V = Volume of the nuclear fuel.

Since 3.1×10^{10} fission per second generates a power of one watt, the power P of a nuclear reactor is given by,

$$P = \frac{nC\sigma NV}{3.1 \times 10^{10}} \text{ watt}$$

$$= 3.2 \times 10^{-11} nC\sigma NV \text{ watt}$$

$$= 3.2 \times 10^{-11} \phi_{nu} \sigma NV \text{ watt}$$

Now,

NV = Total number of fissile nuclei in the reactor fuel

$$= m \times 6.02 \times 10^{26}/235$$

where, m is the mass of the U²³⁵ fuel. It is known that fission cross-section σ of U²³⁵ for thermal neutrons is 582 barns = 582×10^{-28} m².

$$\therefore \quad P = \frac{3.2 \times 10^{-11} \times \phi_{nu} \times 582 \times 10^{-28} \times m \times 6.02 \times 10^{26}}{235}$$

$$= 4.77 \times 10^{-12} m \, \phi_{nu} \text{ watt}$$

$$= 4.8 \times 10^{-12} mnC \text{ watt}.$$

11.3. MAIN COMPONENTS OF A NUCLEAR POWER PLANT

Fig. 11.4 shows schematically a nuclear power plant.

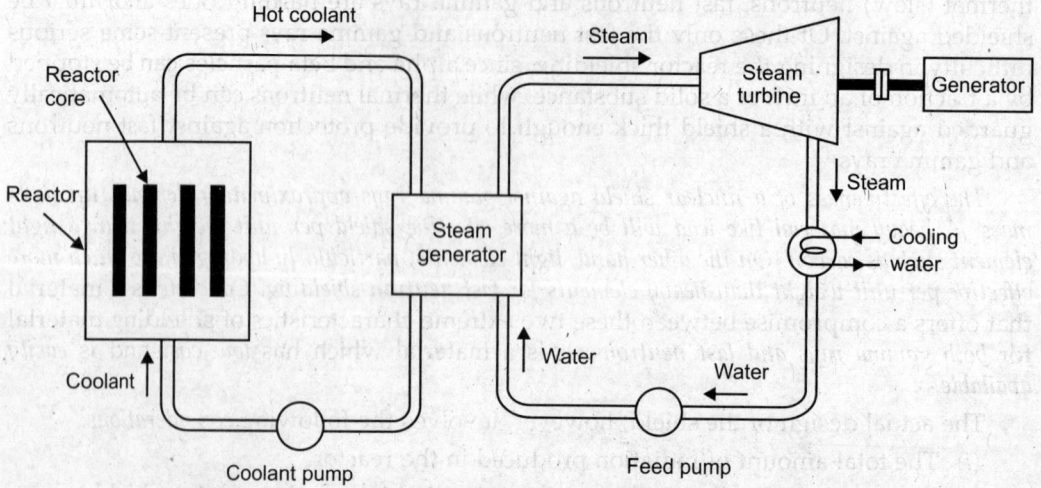

Fig. 11.4: Nuclear power plant.

The *main components of a nuclear power plant* are:

1. Nuclear reactor
2. Heat exchanger (steam generator)
3. Steam turbine
4. Condenser
5. Electric generator.

In a nuclear power plant the reactor performs the same function as that of the furnace of steam power plant (*i.e.* produces heat). The heat liberated in the reactor as a result of the nuclear fission of the fuel is taken up by the coolant circulating through the reactor core. Hot coolant leaves the reactor at the top and then flows through the tubes of steam generator and passes on its heat to the feed water. The steam so produced expands in the steam turbine, producing work and thereafter is condensed in the condenser. The steam turbine in turn runs an electric generator thereby producing electrical energy. In order to maintain the flow of coolant, condensate and feed water pumps are provided as shown in Fig. 11.4.

11.4. BRIEF DESCRIPTION OF REACTORS

11.4.1. Pressurised Water Reactor (PWR)

A pressurised water reactor, in its simplest form, is a light water-cooled and moderated thermal reactor having an unusual core design, using both natural and highly enriched fuel. The principal parts of the reactor are:

1. Pressure vessel
2. Reactor thermal shield
3. Fuel elements
4. Control rods
5. Reactor containment
6. Reactor pressuriser.

The components of the secondary system of pressurised water plant are similar to those in a normal steam station.

Refer to Fig. 11.5. In PWR, there are two circuits of water, one *primary circuit* which passes through the fuel core and is *radioactive*. This primary circuit then produces steam in a *secondary circuit* which consists of heat exchanger or the boiler and the turbine. As such the steam in the turbine is *not radioactive* and need not be shielded. The *pressure* in the primary circuit should be *high* so that the boiling of water takes place at high pressure. A *pressuring tank* keeps the water at about 100 kgf/cm² *so that it will not boil*. Electric heating coils in the pressuriser boil some of the water to form steam that collects in the dome. As more steam is forced into the dome by boiling, its pressure rises and pressurises the entire circuit. The pressure may be reduced by providing cooling coils or spraying water on the steam.

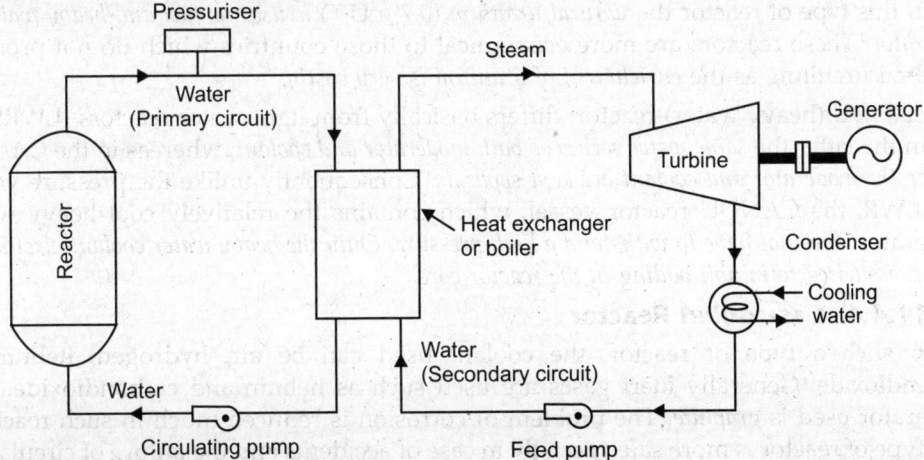

Fig. 11.5: Pressurised water reactor.

Water acts both as coolant as well as moderator. Either heavy water or the light water may be used for the above purpose.

A pressurised water reactor *can produce only saturated steam.* By providing a separate furnace, the steam formed from the reactor could be superheated.

11.4.2. Boiling Water Reactor (BWR)

In a boiling water reactor *enriched fuel* is used. As compared to PWR, the arrangement of BWR plant is simple. The plant can be safely operated using natural convection within the core or forced circulation as shown in the Fig. 11.6. For the safe operation of the reactor the pressure in the forced circulation must be maintained constant irrespective of the load. In case of *part load operation of the turbine some steam is by-passed.*

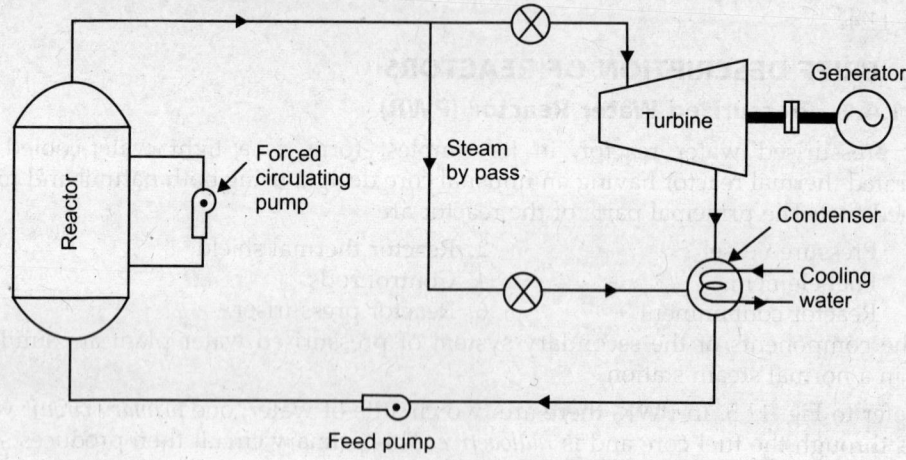

Fig. 11.6: Boiling water reactor.

11.4.3. CANDU (Canadian-Deuterium-Uranium) Reactor

CANDU is a thermal nuclear power reactor in which *heavy water* (99.8% deuterium oxide D_2O) *is the moderator and coolant, as well as the neutron reflector.* This reactor was developed in Canada and is being extensively used in this company. A few CANDU reactors are operating or under construction in some other countries as well.

In this type of reactor the *natural uranium* (0.7% U^{235}) *is used as fuel and heavy water as moderator.* These reactors are more economical to those countries which do not produce enriched uranium, as the *enrichment of uranium is very costly.*

CANDU (heavy water) reactor, differs basically from light-water reactors (LWRS) in that in the latter the *same water serves as both moderator and coolant,* whereas in the *CANDU reactor the moderator and coolant are kept separate.* Consequently unlike the pressure vessel of a LWR, the CANDU reactor vessel, which contains the relatively cool heavy water moderator, *does not have to withstand a high pressure. Only the heavy water coolant circuit has to be pressurised to inhibit boiling in the reactor core.*

11.4.4. Gas-cooled Reactor

In such a type of reactor, the coolant used can be air, hydrogen, helium or carbondioxide. Generally inert gases are used such as helium and carbondioxide. The moderator used is *graphite.* The problem of corrosion is reduced much in such reactors. This type of reactor is more safe specially in case of accidents and the failure of circulating pumps. The thickness of gas cooled reactor shield is much reduced as compared to the other types of reactors.

11.4.5. Liquid Metal-cooled Reactors

Sodium-graphite reactor (SGR) is one of the typical *liquid metal reactors*. In this reactor *sodium works as a coolant* and *graphite works as moderator*.

Sodium boils at 880°C under atmospheric pressure and freezes at 95°C. Hence *sodium* is first *melted* by electric heating system and is pressurised to about 7 bar. The liquid sodium is then circulated by the circulation pump. The reactor will have *two coolant circuits or loops*:

(i) The *primary circuit* has *liquid sodium* which circulates through the fuel core and gets heated by the fissioning of the fuel. This liquid sodium gets cooled in the intermediate heat exchanger and goes back to the reactor vessel.

(ii) The *secondary circuit* has an *alloy of sodium and potassium in liquid form*. This coolant takes heat from the intermediate heat exchanger which gets heat from liquid sodium of primary circuit. The liquid sodium-potassium then passes through a boiler which is once through type having tubes only. The steam generated from this boiler will be superheated. Feed water from the condenser enters the boiler, the heated sodium-potassium passing through the tubes gives it heat to the water thus converting it into steam. The sodium-potassium liquid in the second circuit is then pumped back to the intermediate heat exchanger thus making it a closed circuit.

The reactor vessel, primary loop and the intermediate heat exchanger are to be shielded for radio-activity. The liquid metal be handled under the cover of an inert gas, such as helium, to prevent contact with air while charging or draining the primary or secondary circuit/loop.

The arrangement of a sodium-graphite reactor (SGR) is shown in Fig. 11.7.

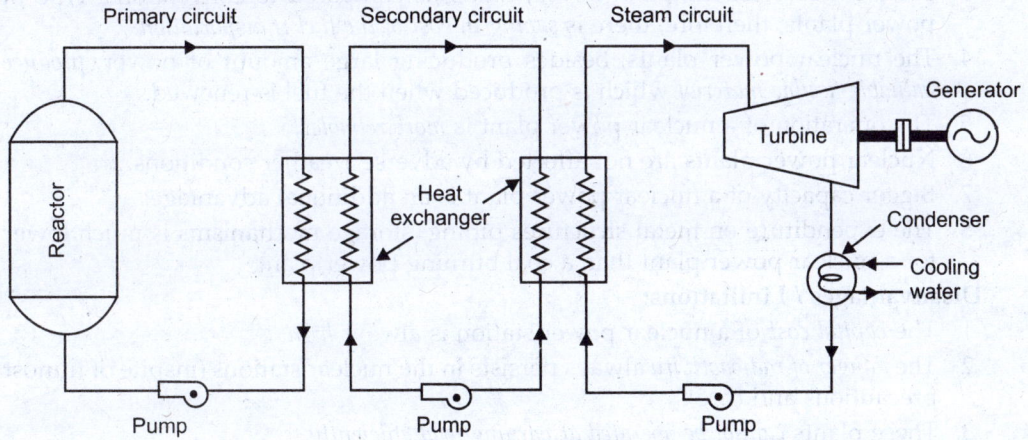

Fig. 11.7: Liquid metal-cooled reactor.

11.4.6. Breeder Reactor

In its simplest form a *fast breeder reactor* is a small vessel in which necessary amount of *enriched plutonium* is kept *without using moderator*. A fissible material, which absorbs neutrons, surrounds the vessels. The reactor core is cooled by liquid metal. Necessary *neutron shielding* is provided by the use of *light water, oil or graphite*. Additional shielding is also provided for gamma rays. (It is worth noting that when U^{235} is fissioned, it produces heat and additional neutrons. If some U^{238} is kept in the same reactor, part of the additional neutrons available, after reaction with U^{235}, convert U^{238} into fissible plutonium).

Fig. 11.8 shows a schematic diagram of a breeder reactor.

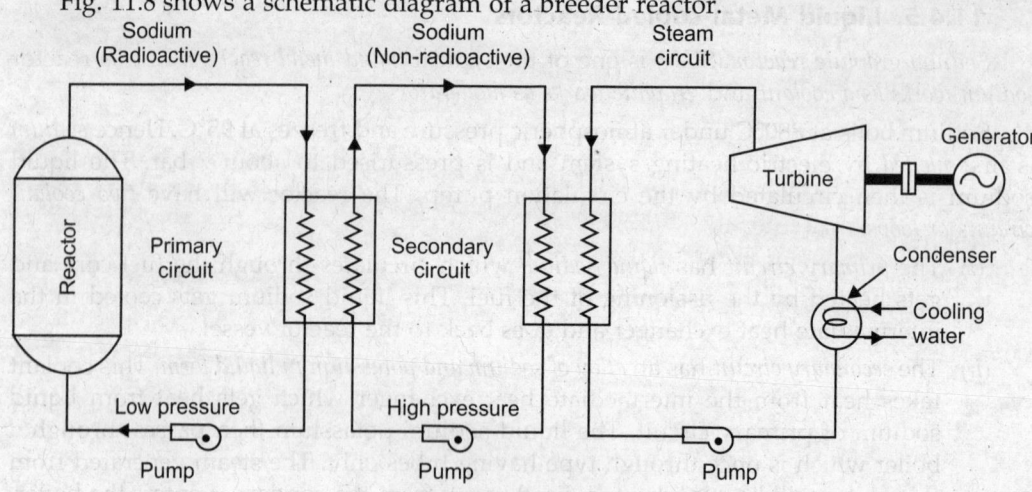

Fig. 11.8: Breeder reactor.

11.5. ADVANTAGES AND DISADVANTAGES OF NUCLEAR POWER PLANTS

Advantages:

Some of the *major advantages* of nuclear power plants are:

1. A nuclear power plant *needs less space* as compared to other conventional power plant of equal size.
2. Nuclear power plants are well suited to meet large power demands. They give better performance at high load factors (80 to 90%).
3. Since the fuel consumption is very small as compared to conventional type of power plants, therefore, there is *saving in cost of the fuel transportation.*
4. The nuclear power plants, besides producing large amount of power, *produce valuable fissible material* which is produced when the fuel is renewed.
5. The operation of a nuclear power plant is *more reliable.*
6. Nuclear power plants are not affected by adverse weather conditions.
7. Bigger capacity of a nuclear power plant is an additional advantage.
8. The expenditure on metal structures piping, storage mechanisms is much lower for a nuclear power plant than a coal burning power plant.

Disadvantages / Limitations:

1. The *capital cost* of a nuclear power station is always *high.*
2. The *danger of radioactivity* always persists in the nuclear stations (inspite of utmost precautions and care).
3. These plants *cannot be operated at varying load efficiently.*
4. The *maintenance cost is always high* (due to lack of standardisation and high salaries of the trained personnel in this field of specialisation).
5. The disposal of fission products is a big problem.
6. Working conditions in nuclear power station are always detrimental to the health of the workers.

11.6. NUCLEAR-PLANT SITE SELECTION

Nuclear power plants *must meet all the economic and technical and most of the legal criteria* that apply to the siting of conventional fossil-fuel-fired power plants. *In addition, the importance of site characteristics in the assessment of public safety results in greater concern in siting nuclear plants than with any other type of industrial facility.* Of particular concern are the population distribution with respect to the site and the natural factors which could affect the transport of radioactive material to the public, under normal operating conditions and in the highly unlikely event of an accident which could release radioactive material to the environment.

The various factors to be considered while selecting the site for a nuclear power plant are enumerated below:

1. Proximity to load centre   2. Population distribution
3. Land use   4. Meteorology
5. Geology   6. Seismology
7. Hydrology.

11.7. NUCLEAR POWER PLANTS IN INDIA

The various nuclear power plants situated in India are as follows:

1. Tarapur power plant   2. Rana Partap Sagar power plant
3. Kalpakkam power plant   4. Narora power plant
5. Kakrapar power plant.

The particulars of these plants are given in tabular form below:

Particulars	Power plants				
	Tarapur	*Rana Partap Sagar*	*Kalpakkam*	*Narora*	*Kakrapar*
1. *Location*	Maharashtra (65 miles north Mumbai)	Near Kota in Rajasthan	Near Chennai in Tamil Nadu state	U.P.	Surat district, Gujrat
2. *Capacity*	380 MW	400 MW (2 × 200 MW)	470 MW (2 × 235 MW)	2 × 235 MW (under construction) 40 bar at 250°C	4 × 235 MW (proposed)
3 *Steam pressure and temperature*	35 bar at 240°C	40 bar at 250°C	—	—	—

In view of increasing energy requirements of our country, more number of nuclear power plants are likely to be set up in near future.

- The most important emission from a nuclear plant are *heat and a trace of radioactivity.* It is necessary to ensure that the operating staff and people residing in the vicinity of nuclear plants *are not overexposed* to nuclear radiations. Risk to public can be minimised by *intense safety design* and by *auxiliary systems that control radioactive release from normal operations as well as accidents.* Also Refer to **Article 16.3.**

HIGHLIGHTS

1. Those pairs of atoms which have the same atomic number and hence similar chemical properties but different atomic mass number are called *isotopes.*
2. Those atoms which have the same mass number but different atomic numbers are called *isobars.* Obviously, these atoms belong to different chemical elements.

3. Those pairs of atoms (nuclides) which have the same atomic number and atomic mass number but have different *radioactive properties* are called *isomers* and their existence is referred to as nuclear isomerism.

4. Those atoms whose nuclei have the same number of neutrons are called *isotones*.

5. The phenomenon of spontaneous emission of powerful radiations exhibited by heavy elements is called *radioacivity*. The radioactivity may be natural or artificial.

6. The five types of nucelar radiations are:

   (*i*) Gamma rays (or photons): electromagnetic radiation.

   (*ii*) Neutrons: uncharged particles, mass approximately 1.

   (*iii*) Protons: + 1 charged particles, mass approximately 1.

   (*iv*) Alpha particles: helium nuclei, charge + 2, mass 4.

   (*v*) Beta particles: electrons (charge – 1), positrons (charge + 1), mass very small.

7. *Half life* represents the rate of decay of the radioactive isotopes. The half life is the time required for half of the parent nuclei to decay or to disintegrate.

8. *Nuclear cross sections* (or attenuation co-efficients) are measures of the probability that a given reaction will take place between a nucleus or nuclei and incident *radiation*.

9. It has been found that some materials are not fissionable by themselves but they can be converted to the fissionable materials, these are known as *fertile materials*.

10. *Fission* is the process that occurs when a neutron collides with the nucleus of certain of the heavy atoms, causing the original nucleus to split into two or more unequal fragments which carry-off most of the energy of fission as kinetic energy. This process is accompanied by the emission of neutrons and gamma rays.

11. A *chain reaction* is that process in which the number of neutrons keeps on multiplying rapidly (in geometrical progression) during fission till whole the fissionable material is disintegrated. The multiplication or reproduction factor (*K*) is given by:

$$K = \frac{\text{No. of neutrons in any particular generation}}{\text{No. of neutrons in the preceding generation}}$$

If $K > 1$, chain reaction will continue and if $K < 1$, chain reaction cannot be maintained.

12. *Nuclear fusion* is the process of combining or fusing two lighter nuclei into a stable and heavier nuclide. In this case large amount of energy is released because mass of the product nucleus is less than the masses of the two nuclei which are fused.

13. A *nuclear reactor* is an apparatus in which nuclear fission is produced is the form of a controlled self-sustaining chain reaction.

14. Essential components of a nuclear reactor are:

   (*i*) Reactor core   (*ii*) Reflector

   (*iii*) Control mechanism   (*iv*) Moderator

   (*v*) Coolants   (*vi*) Measuring instruments

   (*vii*) Shielding.

15. The main components of a nuclear power plant are:

   (*i*) Nuclear reactor   (*ii*) Heat exchanger (steam generator)

   (*iii*) Steam turbine   (*iv*) Condenser

   (*v*) Electric generator.

16. Some important reactors are:
    (*i*) Pressurised water reactor (PWR) (*ii*) Boiling water reactor (BWR)
    (*iii*) Gas-cooled reactor (*iv*) Liquid metal-cooled reactor
    (*v*) Breeder reactor.

17. Following factors should be considered while selecting the site for a nuclear power plant:
    (*i*) Proximity to load centre (*ii*) Population distribution
    (*iii*) Land use (*iv*) Meteorology
    (*v*) Geology (*vi*) Seismology
    (*vii*) Hydrology

18. Typically, all costs of nuclear power plants are broken down into the following categories:
    (*i*) Capital costs (total) (*ii*) Fuel costs (per year)
    (*iii*) Other operating and maintenance cost (per year).

THEORETICAL QUESTIONS

1. Explain the following terms:
    (*i*) Atomic model (*ii*) Atomic mass unit
    (*iii*) Isotopes (*iv*) Isobars
    (*v*) Isomers (*vi*) Isotones.

2. What do you mean by the term 'Radioactivity'?

3. What is the difference between 'Artificial radioactivity' and 'Natural radioactivity'?

4. Name five types of radiation of interest, in nuclear power technology.

5. Explain briefly the following:
    (*i*) Prompt-fission gamma rays (*ii*) Fission-product-decay gamma rays
    (*iii*) Capture gamma rays (*iv*) Activation gamma rays
    (*v*) Inelastic scattering gamma rays.

6. Explain briefly the following types of neutrons:
    (*i*) Prompt-fission neutrons (*ii*) Delayed neutrons
    (*iii*) Photoneutrons (*iv*) Activation neutrons
    (*v*) Reaction neutrons.

7. What do you mean by 'Binding Energy'? What are the total binding energy and binding energy per nucleon for the $_6C^{12}$ nucleus?

8. Explain briefly the following terms relating radioactive decay:
    (*i*) Activity (*ii*) Half life
    (*iii*) Average (mean) life.

9. What do you mean by the following:
    (*i*) Elastic scattering (*ii*) Inelastic scattering
    (*iii*) Capture (*iv*) Fission.

10. Write a short note on 'Fertile materials'.

11. What do you mean by 'Fission of nuclear fuel'?

12. What is a chain reaction?

13. What are the requirement of fission process?

14. How are the following defined?
    (*i*) Critical mass (*ii*) Critical size.

15. What is 'nuclear fusion'? How does it differ from *nuclear fission*?

16. What is a nuclear reactor?

17. How are nuclear reactors classified?

18. Enumerate and explain essential components of a nuclear reactor.

19. Explain with help of neat diagram the construction and working of a nuclear power plant.

20. What is a moderator? Name common moderators and discuss their advantages and limitations.

21. Give the functions and materials for the following:

    (*i*) Reflector   (*ii*) Control rods

    (*iii*) Biological shield.

22. Describe with the help of a sketch the construction working of a Pressurised Water Reactor (PWR). What are its advantages and disadvantages?

23. What is 'Boiling Water Reactor' (BWR)? How does it differ from 'Pressurised Water Reactor' (PWR)?

24. Give the construction and working of a 'Gas-cooled reactor'. What are its advantages and disadvantages?

25. What is a 'Liquid metal-cooled reactor'? Explain briefly a typical liquid metal reactor.

26. Describe a breeder reactor. What are its advantages and disadvantages?

27. What factors must be considered while selecting materials for the various reactor components?

28. List the advantages and disadvantages/limitations of nuclear power plants

29. Discuss the various factors which must be considered while selecting a site for a nuclear power plant.

30. Give the application of nuclear power plants.

31. What do you mean by 'Economics of nuclear power plants'?

32. List down some important safety measures for nuclear power plants.

33. What is the future of nuclear power?

12

STEAM THERMAL
POWER PLANTS

12.1. INTRODUCTION

A **steam power plant** *converts the chemical energy of the fossil fuels (coal, oil, gas) into mechanical/electrical energy.* This is achieved by raising the steam in the boilers, expanding it through the turbines and coupling the turbines to the generators which convert mechanical energy to electrical energy as shown in Fig. 12.1.

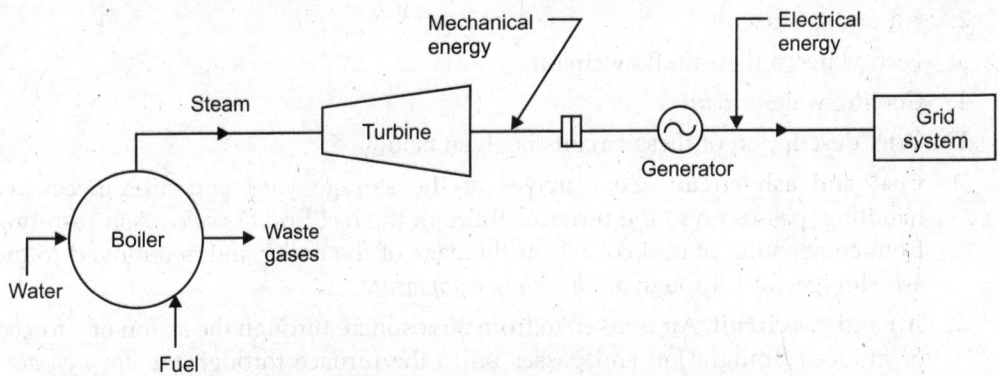

Fig. 12.1. Production of electric energy by steam power plant.

The following *two purposes* can be served by a steam power plant:

1. To produce electric power.

2. To produce steam for industrial purposes besides producing electric power. The steam may be used for varying purposes in the industries such as *textiles, food manufacture, paper mills, sugar mills* and *refineries* etc.

12.2. CLASSIFICATION OF STEAM POWER PLANTS

The steam power plants may be *classified* as follows:

1. Central stations.
2. Industrial power stations or captive power stations.

1. **Central stations.** The electrical energy available from these stations is meant for general sale to the customers who wish to purchase it. Generally, these stations are *condensing type* where the exhaust steam is discharged into a condenser instead of into the atmosphere. In the condenser the pressure is maintained below the atmospheric pressure and the exhaust steam is condensed.

2. **Industrial power stations or captive power stations.** This type of power station is run by a manufacturing company for its own use and its output is *not available for general sale*. Normally these plants are *non-condensing* because a large quantity of steam (low pressure) is required for different manufacturing operations.

In the *condensing steam power plants* the following *advantages* accrue:

   (*i*) The amount of energy extracted per kg of steam is *increased* (a given size of the engine or turbine develops more power).

   (*ii*) The steam which has been condensed into water in the condenser, can be recirculated to the boilers with the help of pumps.

In non-condensing steam power plants a continuous supply of fresh feed water is required which becomes a problem at places where there is a shortage of pure water.

12.3. LAYOUT OF A MODERN STEAM POWER PLANT

Refer to Fig. 12.2. The layout of a modern steam power plant comprises of the following *four circuits* :

1. Coal and ash circuit.
2. Air and gas circuit.
3. Feed water and steam flow circuit.
4. Cooling water circuit.

The brief description of these circuits is given below :

1. **Coal and ash circuit.** Coal arrives at the storage yard and after necessary handling, passes on to the furnaces through the *fuel feeding device*. Ash resulting from combustion of coal collects at the back of the boiler and is removed to the ash storage yard through *ash handling equipment*.

2. **Air and gas circuit.** Air is taken in from atmosphere through the action of a forced or induced draught fan and passes on to the furnace through the *air preheater*, where it has been heated by the heat of flue gases which pass to the chimney *via* the preheater. The flue gases after passing around boiler tubes and superheater tubes in the furnace pass through a *dust* catching device or precipitator, then

through the economiser, and finally through the air preheater before being
exhausted to the atmosphere.

3. **Feed water and steam flow circuit.** In the water and steam circuit condensate
   leaving the condenser is first heated in a closed feed water heater through
   extracted steam from the lowest pressure extraction point of the turbine. It then
   passes through the *deaerator* and a few more water heaters before going into the
   boiler through *economiser*.

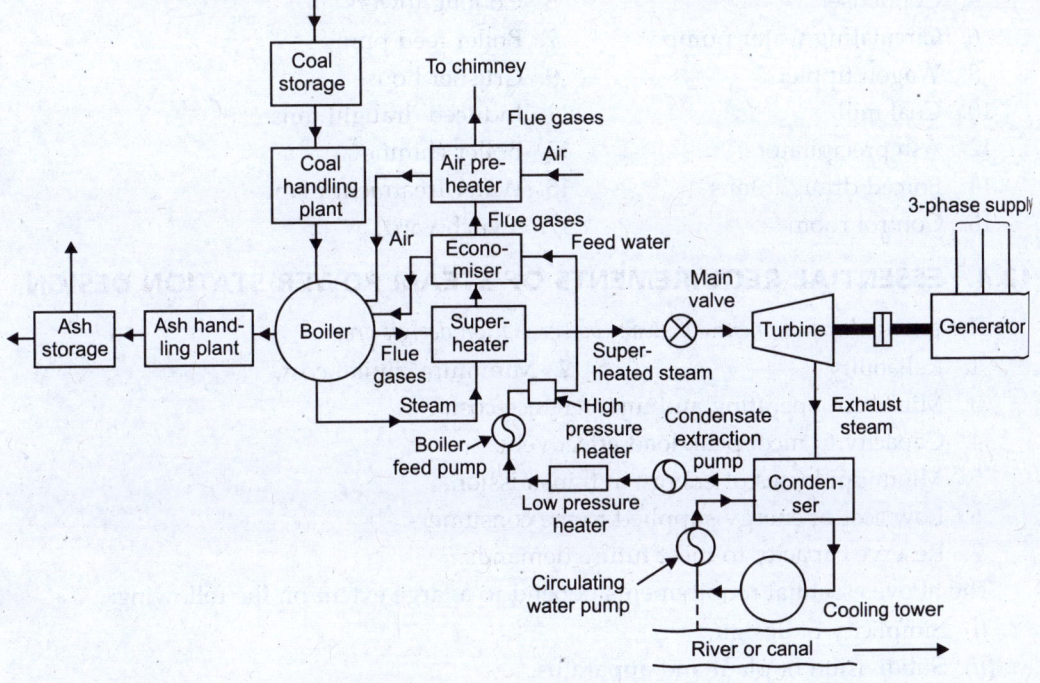

Fig. 12.2. Layout of a steam power plant.

In the boiler drum and tubes, water circulates due to the difference between the
density of water in the lower temperature and the higher temperature sections of the
boiler. Wet steam from the drum is further heated up in the superheater before being
supplied to the primemover. After expanding in high pressure turbine steam is taken to
the reheat boiler and brought to its original dryness or superheat, before being passed on
to the low pressure turbine. From there it is exhausted through the condenser into the hot
well. The condensate is heated in the feed heaters using the steam trapped (bled steam)
from different points of turbine.

A part of steam and water is lost while passing through different components and
this is compensated by supplying *additional feed water*. This feed water *should be purified*
before hand, to avoid the scaling of the tubes of the boiler.

4. **Cooling water circuit.** The cooling water supply to the condenser helps in
   maintaining a low pressure in it. The water may be taken from a natural source
   such as river, lake or sea or the same water may be cooled and circulated over
   again. In the latter case the cooling arrangement is made through spray pond or
   cooling tower.

Components of a Modern Steam Power Plant:

A modern steam power plant consists of the following *components*:

1. Boiler
    (i) Superheater
    (iii) Economiser
2. Steam turbine
4. Condenser
6. Circulating water pump
8. Wagon tippler
10. Coal mill
12. Ash precipitators
14. Forced draught fans
16. Control room

(ii) Reaheater
(iv) Air-heater
3. Generator
5. Cooling towers
7. Boiler feed pump
9. Crusher house
11. Induced draught fans
13. Boiler chimney
15. Water treatment plant
17. Switch yard.

12.4. ESSENTIAL REQUIREMENTS OF STEAM POWER STATION DESIGN

The *essential requirements of steam power station design are*:

1. Reliability
2. Minimum capital cost.
3. Minimum operating and maintenance cost.
4. Capacity to meet peak load effectively.
5. Minimum losses of energy in transmission.
6. Low cost of energy supplied to the consumers.
7. Reserve capacity to meet future demands.

The above essential requirements depend to a large extent on the following:

(i) Simplicity of design.
(ii) Subdivision of plant and apparatus.
(iii) Use of automatic equipment.
(iv) Extensibility.

12.5. SELECTION OF SITE FOR STEAM POWER STATION

The following points should be taken into consideration while *selecting the site for a steam power station* :

1. Availability of raw materials
2. Nature of land
3. Cost of land
4. Availability of water
5. Transport facilities
6. Ash disposal facilities
7. Availability of labour
8. Size of the plant
9. Load centre
10. Public problems
11. Future extensions.

1. **Availability of raw material.** Modern steam power stations using coal or oil as fuel require huge quantity of it per annum. A thermal power plant of 400 MW capacity requires 5000 to 6000 tonnes of coal per day. Therefore, it is necessary to locate the plant as far as possible near the coalfields in order to save the transportation charges. Besides transportation charges, a plant located away from the coalfields, cannot always depend on the coal deliveries in time as there may be failure of transportation system, and there may be strike etc. at the mines. For

these reasons a considerable amount of coal must be stored at the power stations, this results in:

(*i*) increased investment,

(*ii*) increased space required at the site of the plant for the storage,

(*iii*) losses in storage, and

(*iv*) additional staff requirement.

If it is not possible to locate the plant near the coalfields then the plant should be located as near the railway station as possible. Even if this is not possible then at least arrangement should be made for railway siding to the power plant so that the coal wagons can be shunted from the station to the site of power plant. This applies to plants using oil as fuel, as well.

2. **Nature of land.** The type of the land to be selected should have *good bearing capacity* as it has to withstand the *dead load* of the plant and the *forces transmitted to the foundation* due to the machine operations. The minimum bearing capacity of the land should be 1 MN/m^2.

3. **Cost of land.** Considerable area is required for the power stations. The cost of the land for that purpose should be reasonable. The large plants in the heart of big cities and near the load centre are not economical as the cost of land is very high.

4. **Availability of water.** Steam power stations use water as the working fluid which is repeatedly evaporated and condensed. Theoretically there should be no loss of water, but in fact some make up water is required. Besides this, considerable amount of water is required for condensers. A large quantity of water is also required for disposing the ash if hydraulic system is used. It is, therefore, necessary to locate the power plant near the water source which will be able to supply the required quantity of water through out the year.

5. **Transport facilities.** Availability of proper transport facilities is another important consideration in locating the thermal power station. It is always necessary to have a railway line available near the power station for bringing in heavy machinery for installation and for bringing the fuel.

6. **Ash disposal facilities.** The ash handling problem is more serious than coal handling because it comes out in hot condition and it is highly corrosive. Its effects on atmospheric pollution are more serious as the human health is concerned. Therefore, there must be sufficient space to dispose of large quantity of ash.

In a power station of 400 MW capacity 10 hectares area is required per year if the ash is dumped to a height of 6.5 metres.

7. **Availability of labour.** During construction of plant enough labour is required. The labour should be available at the proposed site at cheap rate.

8. **Size of the plant.** The expenses involved in electric transmission from a *small plant* are *relatively severe*, owing to the impracticability of using high voltages, so that the electric transmission feature alone becomes dominant in the location of the plant. It case of *large plants*, the costs of transporting enormous quantities of coal and water are considerably high; therefore, the plant must be located near the pit head provided the required water quantity must be available as near as possible. The large plants should be located close to the railroad offering adequate services. The economic significance of the large plant with small one is much greater than the mere ratio of size.

9. **Load centre.** A power station must be located *near the loads* to which it is supplying power. However, a plant cannot be located near all loads. As such C.G. of the loads is

determined with reference to two arbitrarily chosen axes, this C.G. is known as the *load centre* (see Fig. 12.3.)

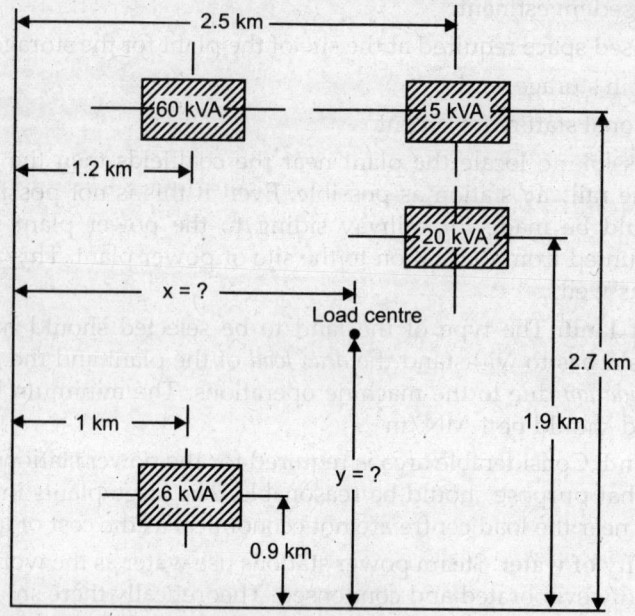

Fig. 12.3. Load centre.

The power plant should be located, as far as possible, near the load centre in order to *minimize the cost of transmission lines and also the losses occuring in them.*

10. **Public problems.** In order to avoid the nuisance from smoke, fly ash and heat discharged from the power plant, it should be located far away from the towns.

11. **Future extensions.** The choice of the site should allow for economical extensions consistent with the estimated growth of load.

12. Consent of Town Planning Department must be sought in case urban area is selected for the purpose.

12.6. FUEL HANDLING

12.6.1. Introduction

Three types of fuels can be burnt in any type of steam generating plant : 1. Solid fuel such as *coal,* 2. *Liquid* fuel as oil, and 3. *Gaseous* fuel as gas. Supply of these fuels to the power plants from various sources is one of the important considerations for a power plant engineer. The handling of these fuels is an important aspect. The following *factors* should be considered in *selecting the fuel handling system:*

1. Plant fuel rate.

2. Plant location in respect of fuel shipping.

3. Storage area available.

Fuel handling plant needs extra attention, while designing a thermal power station, as almost 50 to 60 percent of the total operating cost consists of fuel purchasing and handling. *Fuel system is designed in accordance with the type and nature of fuel.*

Continuously increasing demand for power at lower cost calls for setting up of higher capacity power stations. Rise in capacity of the plant poses a problem in coal supply stystem from coal mines to the power stations. The coal from coal mines may be transported by the following *means*:

1. Transportation by sea or river,
2. Transportation by rail,
3. Transportation by ropeways,
4. Transportation by road, and
5. Transportation of coal by pipeline.

The *'pipeline' coal transport system* offers the following *advantages*:

1. It provides simplicity in installation and increased safety in operation.
2. More economical than other modes of transport when dealing with large volume of coal over long distances.
3. This system is continuous as it remains unaffected by the vagaries of climate and weather.
4. High degree of reliability.
5. Loss of coal during transport due to theft and pilferage is totally eliminated.
6. Manpower requirement is low.

12.6.2. Requirements of Good Coal Handling Plant

1. It should need minimum maintenance.
2. It should be reliable.
3. It should be simple and sound.
4. It should require a minimum of operatives.
5. It should be able to deliver requisite quantity of coal at the destination during peak periods.
6. There should be minimum wear in running the equipment due to abrasive action of coal particles.

12.6.3. Coal Handling Systems

"Mechanical handling" of coal is *preferred* over *"manual handling"* due to the following *reasons*:

1. Higher reliability.
2. Less labour required.
3. Economical for medium and large capacity plants.
4. Operation is easy and smooth.
5. Can be easily started and can be economically adjusted according to the need.
6. With reduced labour, management and control of the plant becomes easy and smooth.
7. Minimum labour is put to unhealthy condition.
8. Losses in transport are minimised.

Disadvantages:

1. Needs continuous maintenance and repair.
2. Capital cost of the plant is increased.
3. In mechanical handling some power generated is usually consumed, resulting in less net power available for supply to consumers.

12.6.4. Coal Handling

Refer to Fig. 12.4.

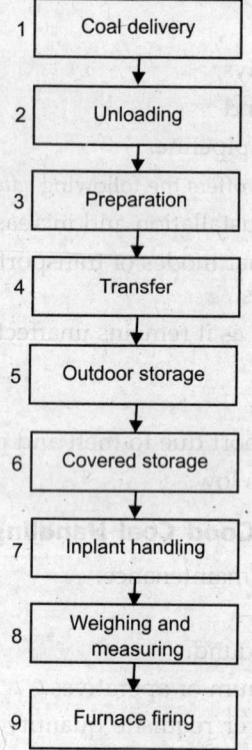

1	Coal delivery
2	Unloading
3	Preparation
4	Transfer
5	Outdoor storage
6	Covered storage
7	Inplant handling
8	Weighing and measuring
9	Furnace firing

Fig. 12.4. Various stages in coal handling

The following *stages/steps* are involved in handling the coal:

1. Coal delivery
2. Unlaoding
3. Preparation
4. Transfer
5. Storage of coal
6. In-plant handling
7. Weighing and measuring
8. Furnace firing.

Fig. 12.5 shows the outline of coal handling equipment.

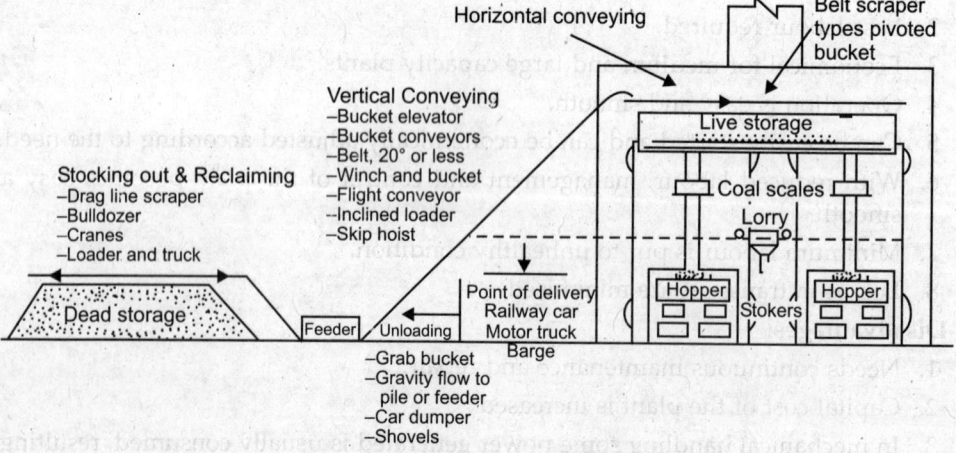

Fig. 12.5. Outline of coal handling equipment.

12.6.5 Layout of a Fuel Handling Equipment

Fig. 14.6. shows a schematic layout of a fuel handling equipment of a modern steam power plant where coal (a solid fuel) is used. Brief description is as follows:

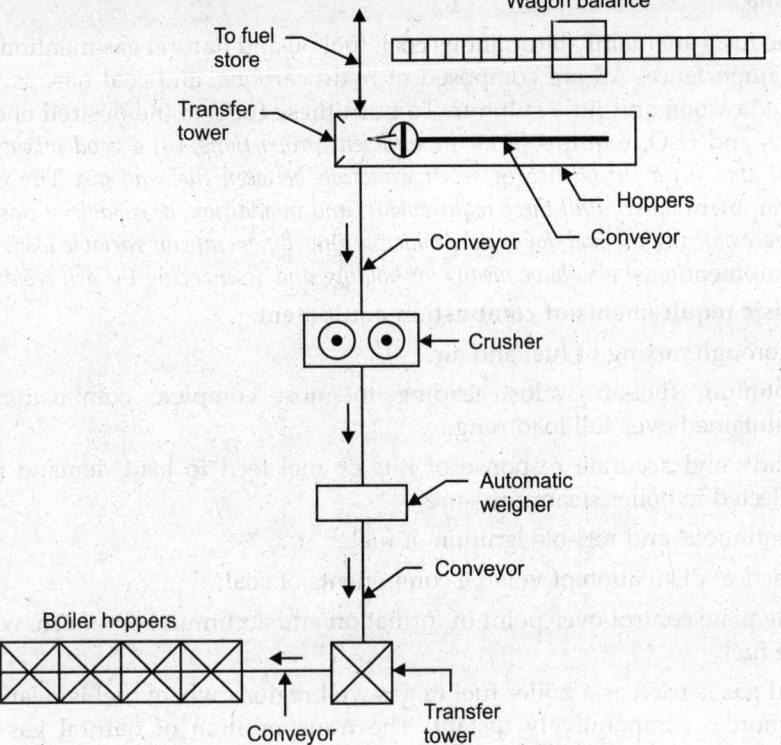

Fig. 12.6. Layout of a fuel handling equipment.

— Coal is supplied to the power plant in railway wagons.
— After weighing on wagon balance the coal is then unloaded into underground hoppers or bunkers. The wagon can be unloaded either manually or through rotary wagon tipplers.
— From the bunkers, the coal is lifted by conveyor to the transfer tower from where it can be delivered either to the fuel store or by a conveyor to a crusher.
— The coal is then passed through the magnetic separators and screens and crushed in crushers into pieces 25 to 30 mm in size for stoker firing and 10 to 20 mm when pulverised fuel is fired in boiler furnaces. The crushed coal in the later case is *milled to a fine powder* and then it is carried through automatic weigher to a transfer tower where fuel is lifted and distributed between boiler hoppers by a conveyor.

12.7. COMBUSTION EQUIPMENT FOR STEAM BOILERS

12.7.1. General Aspects

The combustion equipment is a component of the steam generator. Since the source of heat is the combustion of a fuel, a working unit must have, whatever, equipment is necessary to receive the fuel and air, proportioned to each other and to the boiler steam demand, mix, ignite, and perform and other special combustion duties, such as distillation of volatile from coal prior to ignition.

- *Fluid fuels are handled by burners; solid lump fuels by stokers.*
- In boiler plants hand firing on grates is *practically unheard* of nowdays in new plants, although there are many small industrial plants still in service with hand firing.
- The fuels are mainly bituminous coal, fuel oil and natural gas mentioned in order of importance. All are composed of hydrocarbons, and coal has, as well, much fixed carbon and little sulphur. To burn these fuels to the desired end products, CO_2 and H_2O, requires (*i*) *air in sufficient proportions*, (*ii*) *a good mixing of the fuel and air*, (*iii*) *a turbulence* or *relative motion between fuel and air*. The combustion equipment *must fulfill these requirements and in addition, be capable of close regulation of rate of firing the fuel, for boilers which ordinarily operate on variable load.* Coal-firing equipment *must also have means for holding and discharging the ash residue.*

The basic requirements of combustion equipment:

1. Thorough mixing of fuel and air.
2. Optimum fuel-air ratios, leading to most complete combustion possible, maintained over full load range.
3. Ready and accurate response of rate of fuel feed to load demand (usually as reflected in boiler steam pressure).
4. Continuous and reliable ignition of fuel.
5. Practical distillation of volatile components of coal.
6. Adequate control over point of formation and accumulation of ash, when coal is the fuel.

Natural gas is used as a boiler fuel in gas well regions where fuel is relatively cheap and coal sources comparatively distant. The transportation of natural gas over land to supply cities with domestic and industrial heat has made the gas in the well more valuable and the gas-fired steam generator more difficult to justify in comparison with coal, or fuel cost alone. Cleanliness and convenience in use are other criteria of selection, but more decisive in small plants in central power stations.

Transportation costs add less to the delivery price of oil than gas; also fuel oil may be stored in tanks at a reasonable cost, whereas, gas cannot. Hence although fuel oil is usually more costly than coal per kg of steam generated, many operators select fuel oil burners rather than stokers because of the simplicity and cleanliness of storing and transporting the fuel from storage to burner.

Depending on the type of combustion equipment, boilers may be *classified* as follows:

1. **Solid fuels fired:**
(*a*) Hand fired
(*b*) Stoker fired:
    (*i*) Overfeed stokers
    (*ii*) Underfeed stokers.
(*c*) Pulverised fuel fired:
    (*i*) Unit system
    (*ii*) Central system
    (*iii*) Combination of (*i*) and (*ii*).

2. **Liquid fuel fired:**

(a) Injection system

(b) Evaporation system

(c) Combination of (a) and (b).

3. **Gaseous fuel fired:**

(a) Atmospheric pressure system

(b) High pressure system.

12.7.2. Combustion Equipment for Solid Fuels—Selection Considerations

While selecting combustion equipment for solid fuels the following *considerations* should be taken into account:

1. Initial cost of the equipment.
2. Sufficient combustion space and its ability to withstand high flame temperature.
3. Area of the grate (over which fuel burns).
4. Operating cost.
5. Minimum smoke nuisance.
6. Flexibility of operation.
7. Arrangements for thorough mixing of air with fuel for efficient combustion.

12.7.3. Burning of Coal

The two most commonly used *methods for the burning of coal* are:

1. Stroker firing
2. Pulverised fuel firing.

The selection of one of the above methods depends upon the following *factors*:

(i) Characteristics of the coal available.

(ii) Capacity of the boiler unit.

(iii) Load fluctuations.

(iv) Station load factor.

(v) Reliability and efficiency of the various types of combustion equipment available.

12.7.3.1 Stoker Firing

A "**stoker**" *is a power operated fuel feeding mechanism and grate.*

Advantages of stoker firing:

1. A cheaper grade of fuel can be used.
2. A higher efficiency attained.
3. A greater flexibility of operations assured.
4. Less smoke produced.
5. Generally less building space is necessary.
6. Can be used for small or large boiler units.
7. Very reliable, maintenance charges are reasonably low.
8. Practically immune from explosions.
9. Reduction in auxiliary plant.
10. Capital investment as compared to pulverised fuel system is less.

11. Some reserve is gained by the large amount of coal stored on the grate in the event of coal handling plant failure.

Disadvantages:

1. Construction is complicated.

2. In case of very large units the initial cost may be rather higher than with pulverised fuel.

3. There is always a certain amount of loss of coal in the form of riddling through the grates.

4. Sudden variations in the steam demand cannot be met to the same degree.

5. Troubles due to slagging and clinkering of combustion chamber walls are experienced.

6. Banking and standby losses are always present.

7. Structural arrangements are not so simple and surrounding floors have to be designed for heavy loadings.

8. There is excessive wear of moving parts due to abrasive action of coal.

Classification of stoker firing:

Automatic stokers are *classified* as follows:

1. Overfeed stokers.

2. Underfeed stokers.

In case of overfeed stokers, the coal is fed into the grate *above* the point of air admission and in case of underfeed stokers, the coal is admitted into the furnace *below* the point of air admission.

1. Overfeed stokers:

Principle of operations. Refer to Fig. 12.7.

The principle of an overfeed stoker is discussed below:

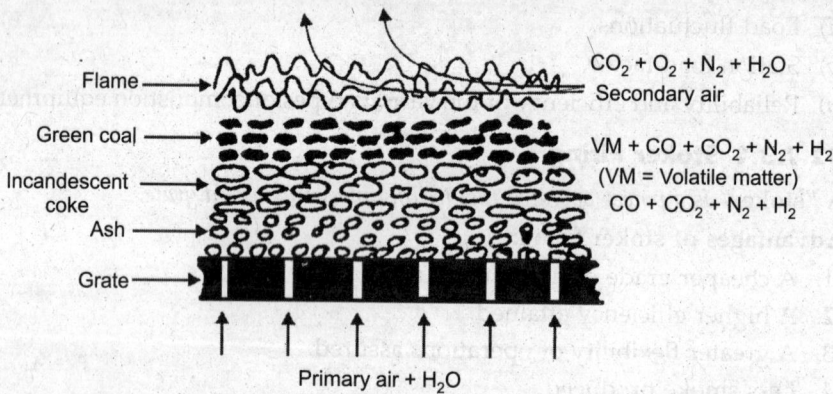

Fig. 12.7. Principle of overfeed stoker.

The fuel bed section receives fresh coal on top surface. The ignition plane lies between green coal and incandescent coke.

The air (with its water vapour content from atmosphere) enters the bottom of the grate under pressure. In flowing through the grate opening the air is heated while it cools the grate. The warm air then passes through a layer of hot ashes and picks up the heat energy.

The region immediately above the ashes contains a mixture of incandescent coke and ash, coke content increasing in upward direction. As the air comes in contact with incandescent coke, the oxygen reacts with carbon to form carbondioxide. Water vapour entering with the air reacts with coke to form CO_2, CO and free H_2. Upon further travel through the incandescent region some of the CO_2 reacts with coke to form CO. Hence no free O_2 will be present in the gases leaving the incandescent region.

Fresh fuel undergoing distillation of its volatile matter forms the top-most layer of the fuel bed. Heat for distillation and eventually ignition comes from the following *four sources*:

(i) By conduction from the incandescent coke below.

(ii) From high temperature gases diffusing through the surface of the bed.

(iii) By radiation from flames and hot gases in the furnace.

(iv) From the hot furnace walls.

The ignition zone lies directly below the raw fuel undergoing distillation.

To burn the combustible gases, additional *'secondary air' must be fed into the furnace to supply the needed oxygen*. The secondary air must be injected at considerable speed to create turbulence and to penetrate to all parts of the area above the fuel bed. The commbustible gases then completely burn in the furnace.

Fuel, coke and ash in the fuel bed move in direction opposite to that of air and gases. Raw fuel continually drops on the surface of the bed. The rising air cools the ash until it finally rests in a plane immediately adjacent to the grate.

Types of overfeed stokers

The "overfeed stokers" are used for large capacity boiler installation where the coal is burnt with pulverisation.

These stokers are mainly of following two types.

(i) Travelling grate stoker

    (a) Chain grate type   (b) Bar grate type

(ii) Spreader stoker.

(i) **Travelling grate stoker:**

These stokers may be *chain grate type or bar grate type*. These two differ only in the details of grate construction.

Fig. 12.8 shows a **"Chain grate stoker"**.

A chain grate stoker consists of flexible endless chain which forms a support for the fuel bed. The chain travels over two sprocket wheels one at the front and one at the rear of furnace. The front sprocket is connected to a variable speed drive mechanism. The speed of the stroker is 15 cm to 50 cm per minute.

The coal bed thickness is shown for all times by an index plate. This can be regulated either by adjusting the opening of fuel grate or by the speed control of the stoker driving motor.

The air is admitted from the underside of the grate which is divided into several compartments each connected to an air duct. The grate should be saved from being overheated. For this, coal should have sufficient ash content which will form a layer on the grate.

Since there is practically no agitation of the fuel bed, 'non-coking coals' are best suited for chain grate stokers.

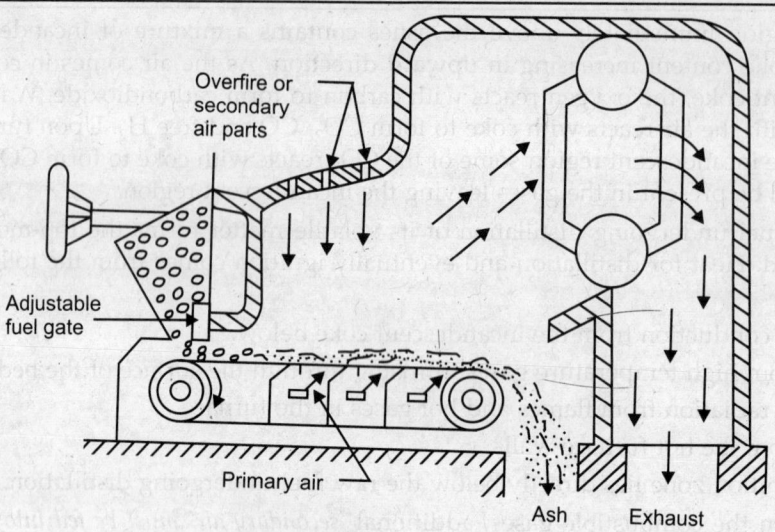

Fig. 12.8. Chain grate stoker

The rate of burning with this stoker is 200 to 300 kg per m² per hour when forced draught is used.

Advantages of chain grate stoker:

1. Simple in construction.
2. Initial cost low.
3. Maintenance charges low.
4. Self-cleaning stoker.
5. Gives high release rates per unit volume of the furnace.
6. Heat release rates can be controlled just by controlling the speed of chain.

Disadvantages:

1. Preheated air temperatures are limited to 180°C maximum.
2. The clinker troubles are very common.
3. There is always some loss of coal in the form of fine particles through riddlings.
4. Ignition arches are required (to suit specific furnace conditions).
5. This cannot be used for high capaicty boilers (200 tonnes/hour or more).

(*ii*) **Spreader stoker.** Refer to Fig. 12.9.

— In this type of stoker the coal is not fed into furnace by means of grate. The function of the grate is only to support a bed of ash and move it out of the furnace.

— From the coal hopper, coal is fed into the path of a rotor by means of a conveyer, and is thrown into the furnace by the rotor and *is burnt in suspension.* The *air for combustion* is supplied through the holes in the grate.

— The *secondary air* (or *overfire air*) to *create turbulence and supply oxygen for thorough combustion of coal* is supplied through *nozzles* located directly above the ignition arch.

— Unburnt coal and ash are deposited on the grate which can be moved periodically to remove ash out of the furnace.

● Spreader stokers *can burn any type of coal.*

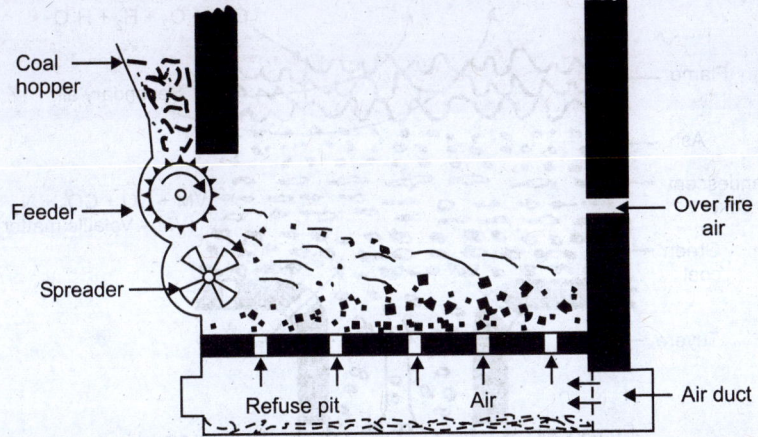

Fig. 12.9. Spreader stoker

- This type of stoker can be used for boiler capacities from 70000 kg to 140000 kg of steam per hour. The heat release rate of 10×10^6 kcal/m²-hr is possible with *stationary grate* and of 20×10^6 kcal/m²-hr is possible with *travelling grate*.

Advantages:

1. A wide variety of coal can be burnt.
2. This stoker is simple to operate, easy to light up and bring into commission.
3. The use of high temperature preheated air is possible.
4. Operation cost is considerably low.
5. The clinkering difficulties are reduced even with coals which have high clinkering tendencies.
6. Volatile matter is easily burnt.
7. Fire arches etc. are generally not required with this type of stokers.
8. As the depth of coal bed on the grate is usually limited to 10 to 15 cm only, fluctuating loads can be easily met with.

Disadvantages:

1. It is difficult to operate spreader with varying sizes of coal with varying moisture content.
2. Fly-ash is much more.
3. No remedy for clinkder troubles.
4. There is a possibility of some fuel loss in the cinders up the stack because of the thin fuel bed and suspension burning.

2. Underfeed feeders:

Principle of operation. Refer to Fig. 12.10 (*a*).

- *The underfeed principle is suitable for burning the semi-bituminous and bituminous coals.*
- Air entering through the holes in the grate comes in contact with the raw coal (green coal). Then it passes through the incandescent coke where reactions similar to overfeed system take place. The gases produced then pass through a layer of ash. The secondary air is supplied to burn the combustible gases.

The underfeed stokers fall into two main groups, the *single retort* and *multi-retort stokers.*

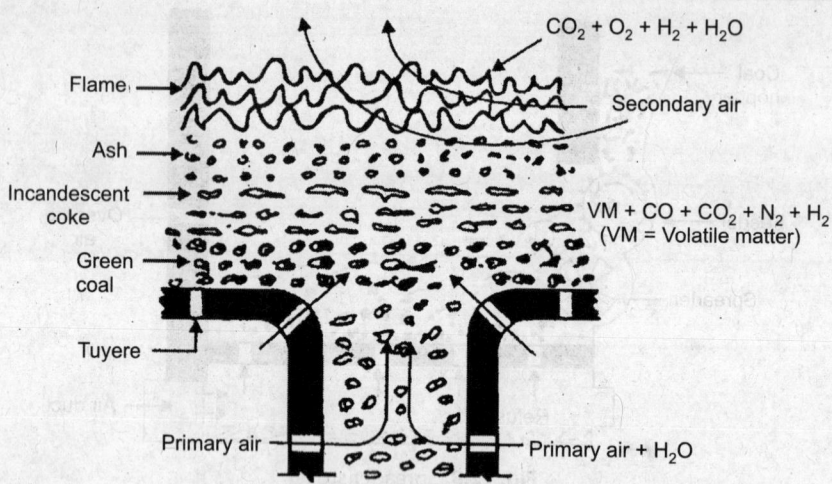

Fig. 12.10. (*a*) Principle of underfeed feeders.

Multi-retort underfeed stokers:

Refer to Fig. 12.10 (*b*)

Fig. 12.10. (b) Multi-retort underfeed stokers

— The stoker consists of a series of sloping parallel troughs formed by tuyere stacks. These troughs are called *retorts*. Under the coal hopper at the head end of the retorts, *feeding rams* reciprocate back and forth. With the ram in the outer position, coal from the hopper falls into space vacated by the ram. On the inward stroke the ram forces the coal into the retort.

— The height and profile of the fuel bed is controlled by *secondary*, or *distributing rams*. These rams oscillate parallel to the retort axes, the length of their stokes can be varied as needed. They slowly move the entire fuel bed down the length of the stoker.

— At the rear of the stoker the partly burned fuel bed moves onto an *extension grate* arranged in sections. These sections also oscillate parallel to the fuel-bed movement. The sharp slope of the stoker aids in moving the fuel bed. Fuel-bed movement keeps it slightly agitated to break up clinker formation. From extension grate the ash moves onto ash dump plate. Tilting the dump plate at long intervals deposits the ash in the ashpit below.

— *Primary air* from the wind box underneath the stoker enters the fuel bed through holes in the vertical sides of the tuyeres. The extension grate carries a much thinner fuel bed and so must have a lower air pressure under it. The air entering from the main wind box into the extension-grate wind box is regulated by a *controlling air damper*.

● In this stoker the number of retorts may vary from 2 to 20 with coal burning capacity ranging from 300 kg to 2000 kg per hour per retort.

● Underfeed strokers are *suitable for non-clinkering, high voltatile coals having caking properties and low ash contents.*

Advantages:
1. High thermal efficiency (as compared to chain grate stokers).
2. Combustion rate is considerably higher.
3. The grate is self cleaning.
4. Part load efficiency is high particularly with multi-retort type.
5. Different varieties of coals can be used.
6. Much higher steaming rates are possible with this type of stoker.
7. Grate bars, tuyeres and retorts are not subjected to high temperature as they remain always in contact with fresh coal.
8. Overload capacity of the boiler is high as large amount of coal is carried on the grate.
9. Smokeless operation is possible even at very light load.
10. With the use of clinker grinder, more heat can be liberated out of fuel.
11. Substantial amount of coal always remains on the grate so that the boiler may remain in service in the event of temporary breakdown of the coal supply system.
12. It can be used with all refractory furnaces because of non-exposure of stoker mechanism to the furnace.

Disadvantages:
1. High initial cost.
2. Require large building space.
3. The clinker troubles are usually present.
4. Low grade fuels with high ash content cannot be burnt economically.

12.7.3.2. Pulverised fuel firing

In pulverised fuel firing system the *coal is reduced to a fine powder with the help of grinding mill and then projected into the combustion chamber with the help of hot air current.* The amount of air required (known as *secondary air*) to complete the combustion is supplied separately to the combustion chamber. The resulting turbulence in the combustion chamber helps for uniform mixing of fuel and air and thorough combustion. *The amount of air which is used to carry the coal and to dry it before entering into the combustion chamber is known as* **Primary air** *and the amount of air which is supplied separately for completing the combustion is known as* **Secondary air.**

The *efficiency* of the pulverised fuel firing system mostly *depends upon the size of the powder*. The fineness of the coal should be such as 70% of it would pass through a 200 mesh sieve and 90% through 50 mesh sieve.

Fig. 12.11 shows elements of pulverised coal system.

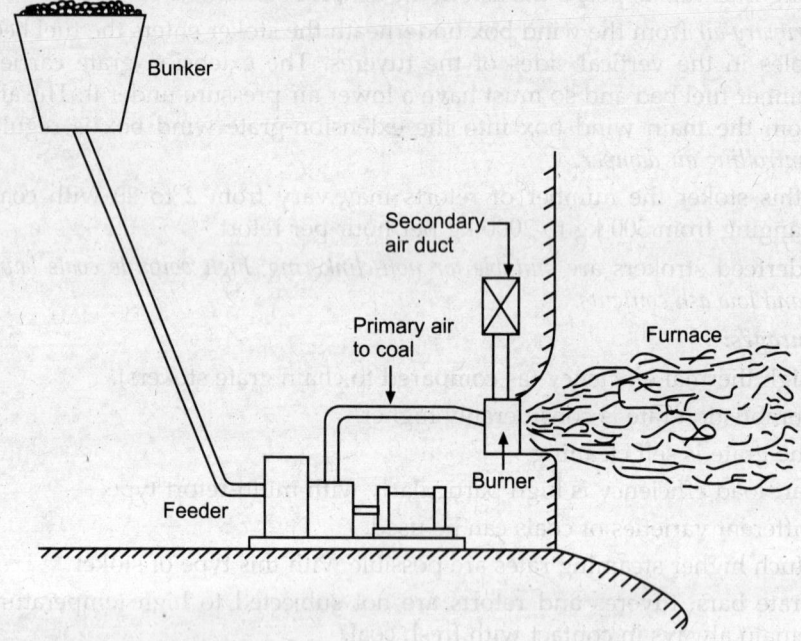

Fig. 12.11. Elements of pulverised coal system.

Advantages:

1. Any grade of coal can be used since coal is powdered before use.
2. The rate of feed of the fuel can be regulated properly resulting in fuel economy.
3. Since there is almost complete combustion of the fuel there is increased rate of evaporation and higher boiler efficiency.
4. Greater capacity to meet peak loads.
5. The system is practically free from sagging and clinkering troubles.
6. No standby losses due to banked fires.
7. Practically no ash handling problems.
8. No moving part in the furnace is subjected to high temperatures.
9. This system works successfully with or in combination with gas and oil.
10. Much smaller quantity of air is required as compared to that of stoker firing.
11. The external heating surfaces are free from corrosion.
12. It is possible to use highly preheated secondary air (350°C) which helps for rapid flame propagation.
13. The furnace volume required is considerably less.

Disadvantages:

1. High capital cost.
2. Lot of fly-ash in the exhaust, which makes the removing of fine dust uneconomical.
3. The possibility of explosion is more as coal burns like gas.
4. The maintenance of furnace brickwork is costly.
5. Special equipment is needed to start this system.
6. The skilled operators are required.
7. A separate coal preparation plant is necessary.
8. High furnace temperatures cause rapid deterioration of the refractory surfaces of the furnace.
9. Nuisance is created by the emission of very fine particles of grit and dust.
10. Fine regular grinding of fuel and proper distribution to burners is usually difficult to achieve.

Pulverised Fuel Handling:

Basically, pulverised fuel plants may be divided into the following *two systems*:

1. Unit system.
2. Central system.

Unit system:

A *unit system* is shown in Fig. 12.12.

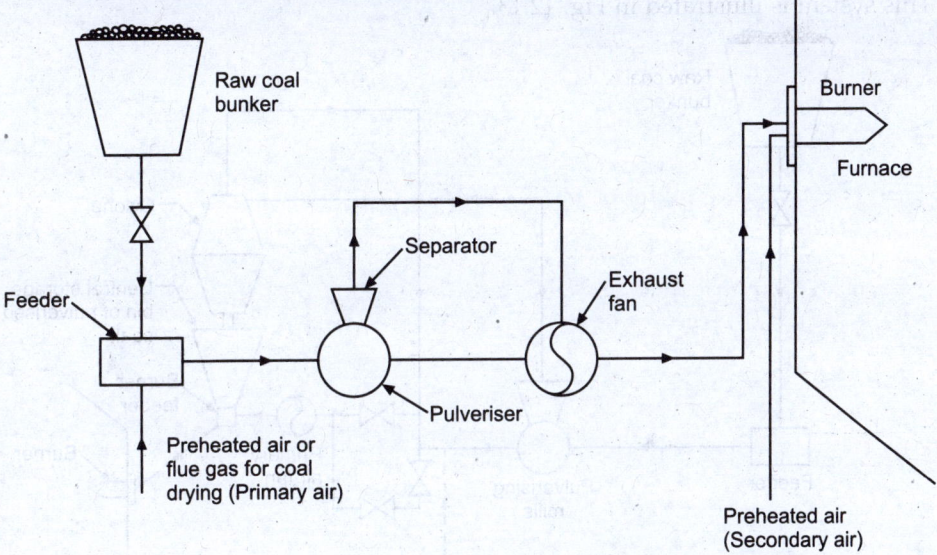

Fig. 12.12. Unit system.

Most pulverised coal plants are now being installed with unit pulveriser.

The *unit system* is so called from the fact that each burner or burner group and the pulveriser constitute a unit. Crushed coal is fed to the pulverising mill at a variable rate governed by the combustion requirements of the boiler and furnace. *'Primary air'* is admitted to the mill and *becomes the transport air* which carries the coal through the short delivery pipe to the burner. This air may be preheated if mill drying is desirable.

Advantages:

1. The layout is simple and permits easy operation.
2. It is cheaper than central system.
3. Less spaces are required.
4. It allows direct control of combustion from the pulveriser.
5. Maintenance charges are less.
6. There is no complex transportation system.
7. In a replacement of stoker, the old conveyor and bunker equipment may be used.
8. Coal which would require drying in order to function satisfactorily in the central system may usually be employed without drying in the unit system.

Disadvantages:

1. Firing aisle is obstructed with pulverising equipment, unless the latter is relegated to a basement.
2. The mills operate at variable load, a condition not especially conducive to best results.
3. With load factors in common practice, total mill capacity must be higher than for the central system.
4. Flexibility is less than central system.

Central system:

This system is illustrated in Fig. 12.13.

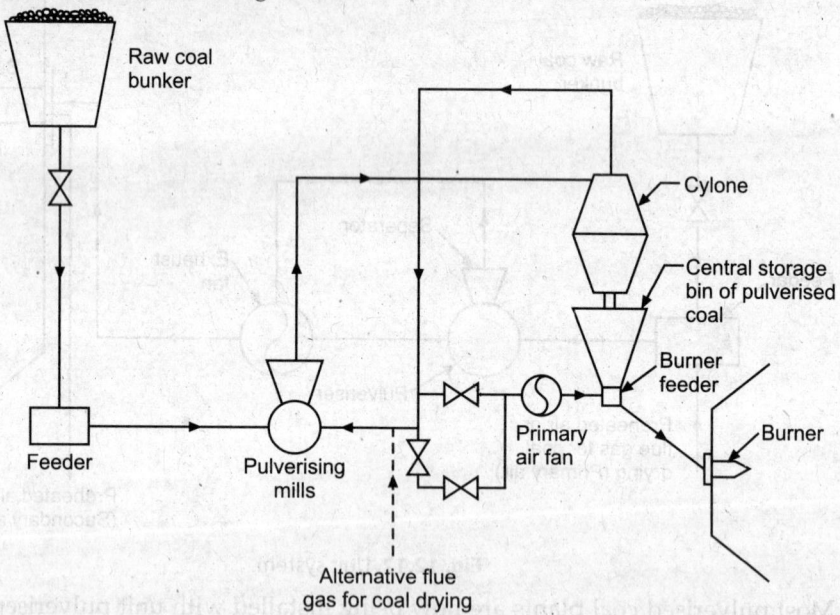

Fig. 12.13. Central system.

A central pulverising system employs a limited number of large capacity pulverisers at a central point to prepare coal for all the burners. Driers, if required, are conveniently installed at this point. From the pulverisers the coal is transported to a central storage bin where it is deposited and its transporting air vented from the bin through a *"cyclone"*.

This bin may contain from 12 to 24 hours supply of pulverised coal. From the bin the coal is metered to the burners by motor-driven feeders of varied design. Primary air, added at the feeders, floats the coal to the burners.

Advantages:

1. Offers good control of coal fineness.
2. The pulverising mill may work at constant load because of the storage capacity between it and the burners.
3. The boiler aisels are unobstructed.
4. More latitude in the arrangement and number of burners is allowed to the designers.
5. The large storage is protection against interruption of fuel supply to the burners.
6. Less labour is required.
7. Power consumption per tonne of coal handled is low.
8. Burners can be operated independent of the operation of coal preparation plant.
9. Fans handle only air, as such, there is no problem of excessive wear as in case of unit system, where air and coal both are handled by the fan.

Disadvantages:

1. Driers are usually necessary.
2. Fire hazard of quantities of stored pulverised coal.
3. Central preparation may require a separate building.
4. Additional cost and complexity of coal transportation system.
5. Power consumption of auxiliaries is high.

Pulveriser. Coal is *pulverised* in order *to increase its surface exposure*, thus promoting rapid combustion without using large quantities of excess air. A pulveriser is the most important part of a pulverised coal system. Pulverisers (sometimes called *mills*) are classified as follows:

1. *Attrition mills*:
   (*i*) Bowl mills
   (*ii*) Ball and race mills.
2. *Impact mills*:
   (*i*) Ball mills
   (*ii*) Hammer mills

Pulverisers are driven by electric motors with the feeders either actuated by the main drive or by a small d.c. motor, depending upon the control used.

12.7.4. Burners

Primary air that carries the powdered coal from the mill to the furnace is only about 20% of the total air needed for combustion. Before the coal enters the furnace, it must be mixed with additional air, known as *secondary air*, in burners mounted in the furnace wall. *In addition to the prime function of mixing, burners must also maintain stable ignition of fuel-air mix and control the flame shape and travel in the furnace.* Ignition depends on the rate of flame propagation. To prevent flash back into the burner, the coal-air mixture must move away from the burner at a rate equal to flame front travel. *Too much secondary air can cool the mixture and prevent its heating to ignition temperature.*

The *requirements of a burner* can be summarised as follows:

(*i*) The coal and air should be so handled that there is stability of ignition.

(*ii*) The combustion is complete.

(*iii*) In the flame the heat is uniformly developed avoiding any superheat spots.

(*iv*) Adequate protection against overheating, internal fires and excessive abrasive wear.

12.7.4.1. Burners

Pulverised fuel burners may be *classified* as follows:

1. Long flame burners.
2. Turbulent burners.
3. Tangential burners.
4. Cyclone burners.

12.7.4.2. Oil burners

Principle of oil firing. *The functions of an oil burner are to mix the fuel and air in proper proportion and to prepare the fuel for combustion.* Fig. 12.14 shows the principle of oil firing.

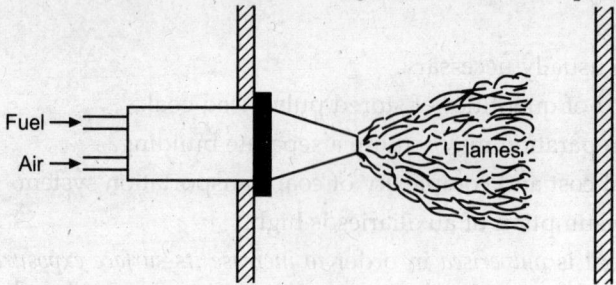

Fig. 12.14. Principle of oil firing.

Classification of oil burners. The oil burners may be *classified* as:

1. *Vapourising oil burnres*:

(*a*) Atmospheric pressure atomising burner.

(*b*) Rotating cup burner

(*c*) Recirculation burner

(*d*) Wick type burner

2. *Atomising fuel burners*:

(*a*) Mechanical or oil pressure atomising burner

(*b*) Steam or high pressure air atomising burner

(*c*) Low pressure air atomising burner.

12.7.4.3. Gas burners

Gas burning claims the following *advantages*:

(*i*) It is much simpler as the fuel is ready for combustion and requires no preparation.

(*ii*) Furnace temperature can be easily controlled.

(*iii*) A long slow burning flame with uniform and gradual heat liberation can be produced.

(*iv*) Cleanliness.

(v) High chimney is not required.

(vi) No ash removal is required.

For generation of steam, natural gas is invariably used in the following cases:

(i) Gas producing areas.

(ii) Areas served by gas transmission lines.

(iii) Where coal is costlier.

Typical gas burners used are shown in Figs. 12.15, 12.16 and 12.17.

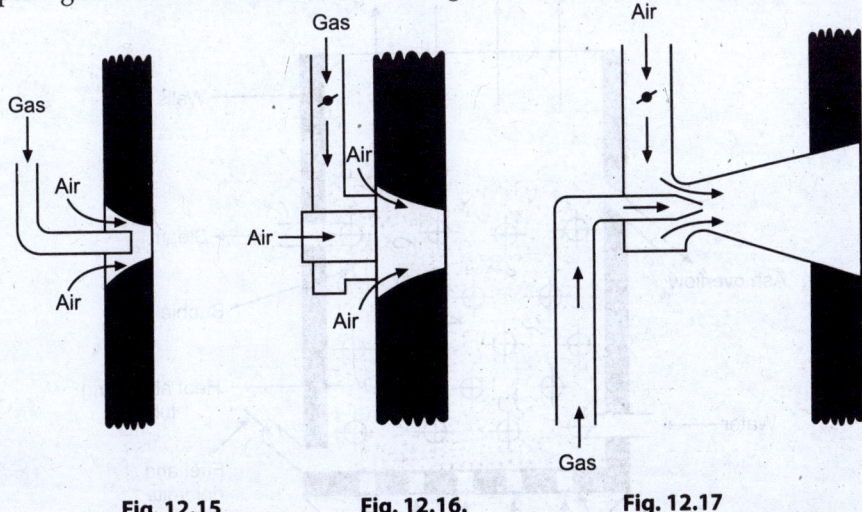

| Fig. 12.15 | Fig. 12.16. | Fig. 12.17 |

Refer to Fig. 12.15. In this burner the mixing is poor and a fairly long flame results.

Refer to Fig. 12.16. This is a *ring type burner* in which a short flame is obtained.

Refer to Fig. 12.17. This arrangement is used when both gas and air are under pressure.

In order to prevent the flame from turning back the velocity of the gas should be more than the *"rate of flame propagation"*.

12.8. FLUIDISED BED COMBUSTION (FBC)

A *fluidised bed* may be defined as the bed of solid particles behaving as a fluid. The *principle of FBC-system* is given below:

When a gas is passed through a packed bed of finely divided solid particles, it experiences a pressure drop across the bed. At low gas velocities, this pressure drop is small and does not disturb the particles. But if the gas velocity is increased further, a stage is reached, *when particles are suspended in the gas stream and the packed bed becomes a 'fluidised bed'*. With further increase in gas velocity, the bed becomes turbulent and rapid mixing of particles occurs. In general, the behavior of this mixture of solid particles and gas is *like a fluid*. Burning of a fuel in such a state is known as a *fluidised bed combustion*.

Fig. 12.18 shows the arrangement of the *FBC system*.

On the distributor plate are fed the fuel and inert material *dolomite* and from its bottom air is supplied. The high velocity of air keeps the solid feed material in suspending condition during burning. The generated heat is rapidly transferred to the water passing through the tubes immersed in the bed and generated steam is taken out. During the burning sulphur dioxide formed is absorbed by the dolomite and prevents its escape with the exhaust gases. The molten slag is tapped from the top surface of the bed.

The primary object of using the *inert material* is to control the bed temperature, it accounts for 90% of the bed volume. It is very necessary that the selection of an inert material should be done judiciously as it remains with the fuel in continuous motion and at high temperature to the tune of 800°C. Moreover, the inert material should not disintegrate coal, the parent material of the bed.

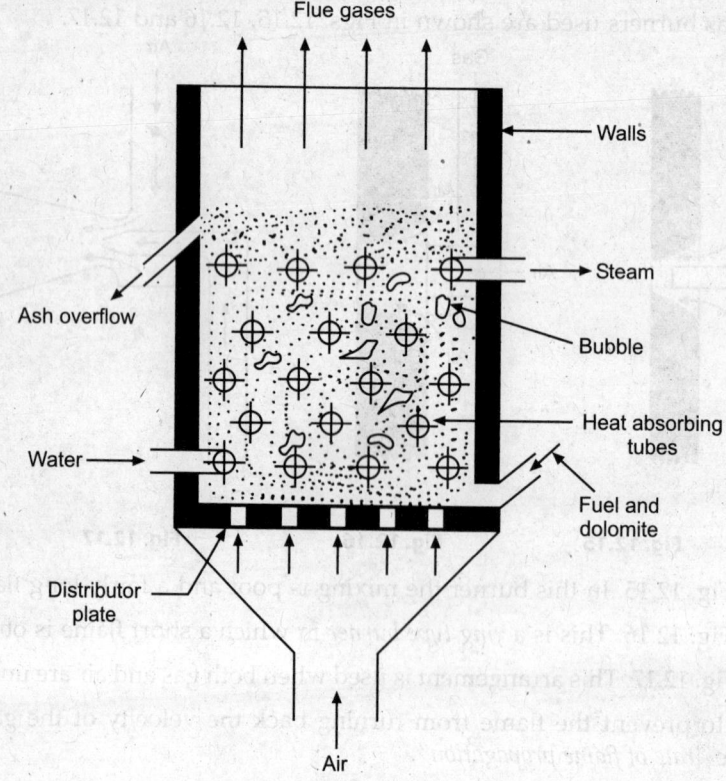

Fig. 12.18. Basic FBC system.

The cost economic shows that a saving of about 10% in operating cost and 15% capital cost could be achieved for a unit rating of 120 MW and it may be still higher for bigger units.

Advantages:

1. As a result of better heat transfer, the unit size and hence the *capital costs are reduced.*

2. It can respond rapidly to changes in load demand (since thermal equilibrium between air and coal particles in the bed is quickly established.)

3. Low combustion temperatures (800 to 950°C) inhibits the formation of nitrogen oxides like nitric oxide and nitrogen dioxide.

4. Since combustion temperatures are low the fouling and corrosion of tubes is reduced considerably.

5. As it is not necessary to grind the coal very fine as is done in pulverised fuel firing, therefore, the cost of coal crushing is reduced.

6. Pollution is controlled and combustion of high-sulphur coal is possible.

7. FBC system can use solid, liquid or gaseous fuel or mix as well as domestic and industrial waste. Any variety of coal can be used successfully.

8. Combustion temperature can be controlled accurately.

9. The system can be readily designed for operation at raised combustion pressure, owing to the simplicity of arrangement, small size of the plant and reduced likelihood of corrosion or erosion of gas turbine blades.

10. The combustion in conventional system becomes unstable when the ash exceeds 48%, but even 70% ash containing coal can be efficiently burned in FBC.

11. The large quantity of bed material acts as a thermal storage which reduce the effect of any fluctuation in fuel feed ratio.

12.9. ASH HANDLING

A huge quantity of ash is produced in central stations, sometimes being as much as 10 to 20% of the total quantity of coal burnt in a day. Hundred of tonnes of ash may have to be handled every day in large power stations and mechanical devices become indispensable. A station using low grade fuel has to deal with large quantities of ash.

Handling of ash includes:

(*i*) Its removal from the furnace.

(*ii*) Loading on the conveyers and delivery to the fill or dump from where it can be disposed off by sale or otherwise.

Handling of ash is a problem because ash coming out of the furnace is too hot, it is dusty and irritating to handle and is accompanied by some poisonous gas. Ash needs to be *quenched* before handling due to following *reasons*:

(*i*) Quenching reduces corrosion action of the ash.

(*ii*) It reduces the dust accompanying the ash.

(*iii*) It reduces temperature of the ash.

(*iv*) Ash forms clinkers by fusing in large lumps and by quenching clinkers will disintegrate.

12.9.1 Ash Handling Equipment

A *good ash handling plant* should have the following *characteristics*:

1. It should have enough capacity to cope with the volume of ash that may be produced in a station.

2. It should be able to handle large clinkers, boiler refuse, soot etc., with little personal attention of the workmen.

3. It should be able to handle hot and wet ash effectively and with good speed.

4. It should be possible to minimise the corrosive or abrasive action of ashes and dust nuisance should not exist.

5. The plant should not cost much.

6. The operation charges should be minimum possible.

7. The operation of the plant should be noiseless as much as possible.

8. The plant should be able to operate effectively under all variable load conditions.

9. In case of addition of units, it should need minimum changes in original layout of plant.

10. The plant should have high rate of handling.

The *commonly used equipment for ash handling* in large and medium size plants may comprise of:

    (*i*) Bucket elevator

    (*ii*) Bucket conveyor

    (*iii*) Belt conveyor

    (*iv*) Pneumatic conveyor

    (*v*) Hydraulic sluicing equipment

    (*vi*) Trollies or rail cars etc.

Fig. 12.19. shows the outline of ash disposal equipment.

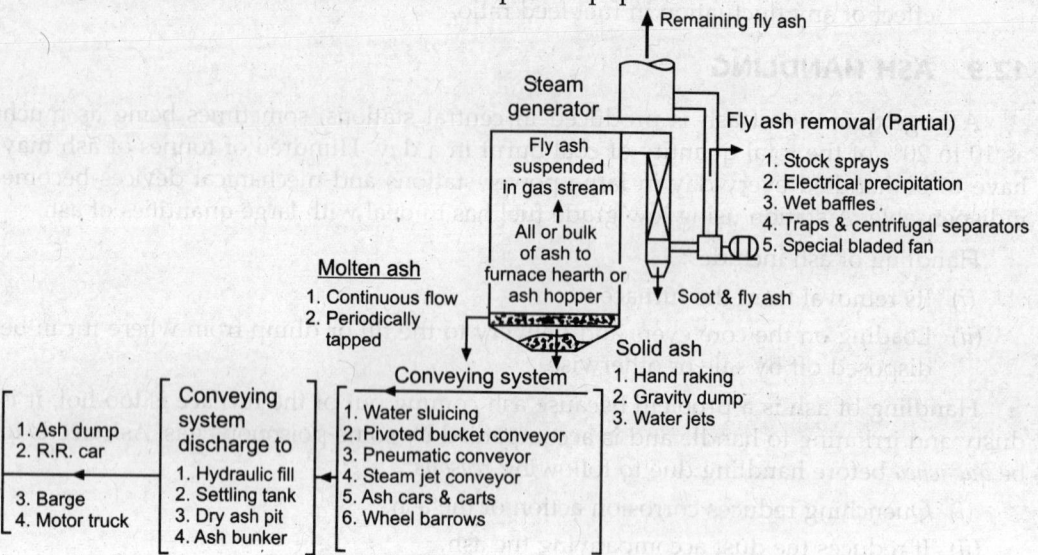

Fig. 12.19. Outline of ash disposal equipment.

The modern ash-handling systems are mainly *classified* into *four* groups:

    1. Mechanical handling system   2. Hydraulic system

    3. Pneumatic system   4. Steam jet system.

12.10. DUST COLLECTION

12.10.1 Introduction

The products of combustion of coal-fed fires contain particles of solid matter floating in suspension. This may be smoke or dust. If *smoke*, the indication is that combustion conditions are faulty, and the proper remedy is in the design and management of the furnace. If *dust*, the particles are mainly fine ash particles called *"Fly-ash"* intermixed with some quantity of carbon-ash material called *"cinder"*. *Pulverised coal and spreader stoker firing units are the principle types causing difficulty from this source.* Other stokers may produce minor quantities of dust but generally not enough to demand special gas cleaning equipment. The two mentioned are troublesome because coal is burned in suspension—in a turbulent furnace atmosphere and every opportunity is offered for the gas to pick up the smaller particles and sweep them along with it.

The size of the dust particles is measured in *microns*. The micron is one millionth of a metre. As an indication of the scale of this measure, the diameter of a human hair is approximately 80 microns. Typical classification of particles by name is given in

Fig. 12.20, but the limits shown are, for the most part, arbitrary. A critical characteristic of dust is its *"Settling Velocity"* in still air. *This is proportional to the product of the square of micron size and mass density.*

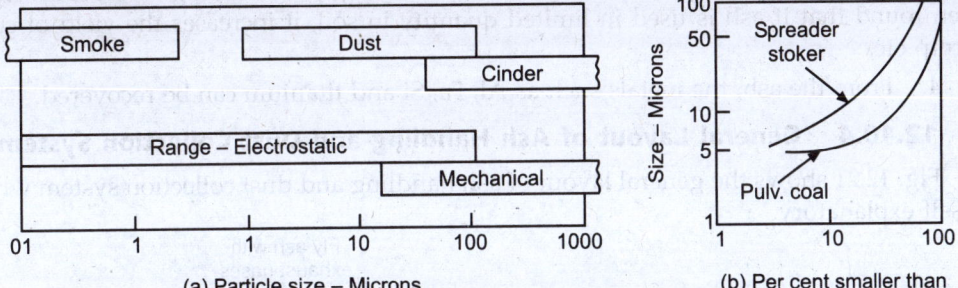

(a) Particle size – Microns   (b) Per cent smaller than

Fig. 12.20. Typical particle sizes: (*a*) Flue gas particles and ranges of collecting equipment.
(*b*) Typical distribution of particle size in products of combustion.

12.10.2 Removal of Smoke

Smoke *is produced due to the incomplete combustion of fuels.* Smoke particles are less than 1 micron in size. The smoke disposal to the atmosphere is not desirable due to the following *reasons*:

(*i*) Smoke is produced due to incomplete combustion of coal. This will create a big economic loss due loss of heating value of coal.

(*ii*) A smoky atmosphere is unhealthy.

(*iii*) Smoke corrodes the metals, darkens the paints and gives lower standards of cleanliness.

In order to check the nuisance of smoke the coal should be completely burnt in the furnace. *The presence of dense smoke indicates poor furnace conditions and a loss in efficiency and capacity of a boiler plant.*

12.10.3 Removal of Dust and Dust Collectors

The removal of dust and cinders from flue gas can usually be effected to the required degree by commercial *dust collectors*.

The dust collectors may be *classified* as follows:

1. *Mechanical dust collectors*:
(*i*) Wet type (Scrubbers)
    (*a*) Spray type
    (*b*) Packed type
    (*c*) Impingement type.
(*ii*) Dry type :
    (*a*) Gravitational separators
    (*b*) Cyclone separators.
2. *Electrical dust collectors*:
(*i*) Rod type
(*ii*) Plate type.

12.10.3 Uses of Ash and Dust

The *uses of ash and dust* are listed below:

1. Ash is widely used in the production of cement.

2.   Ash is used in the production of concrete, 20 percent fly-ash and 30 percent bottom ash are presently used constructively in U.S.A.

3.   Because of their better alkali values, they are used for treating acidic soils. It has been found that if ash is used in limited quantity in soil, it increases the yield of corn, turnip etc.

4.   From the ash, the metals such as Al, Fe, Si and titanium can be recovered.

12.10.4   General Layout of Ash Handling and Dust Collection System

Fig. 12.21 shows the general layout of ash handling and dust collection system which is self explanatory.

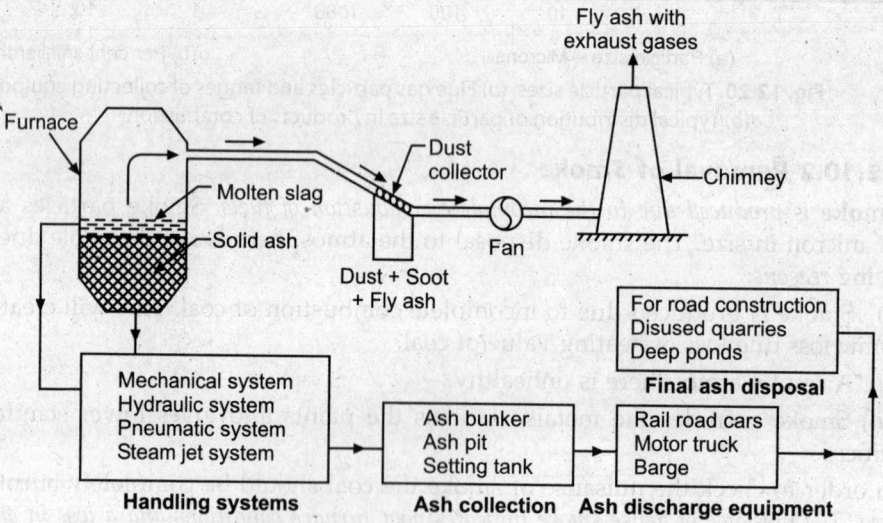

Fig. 12.21. General layout of ash handling and dust collection system.

12.11. HIGH PRRESSURE BOILERS

12.11.1   Introduction

In simple, a *boiler* may be defined as a closed vessel in which steam is produced from water by combustion of fuel.

According to American Society of Mechanical Engineers (A.S.M.E) a *'steam generating unit'* is defined as:

"A combination of apparatus for producing, furnishing or recovering heat together with the apparatus for transferring the heat so made available to the fluid being heated and vapourised".

The steam generated is employed for the following *purposes*:

(*i*)   For generating power in steam engines or steam turbines.

(*ii*)   In the textile industries for sizing and bleaching etc., and many other industries like sugar mills; chemical industries.

(*iii*)   For heating the buildings in cold weather and for producing hot water for hot water supply.

In applications where steam is needed at pressure, 30 bar, and individual boilers are required to raise less than about 30000 kg of steam per hour, *shell boilers are considerably cheaper than the water-tube boilers.* Above these limits, shell boilers (generally factory built) are difficult to transport if not impossible. There are no such limits to water-tube boilers.

These can be site erected from easily transportable parts, and moreover the pressure parts are of smaller diameter and therefore can be thinner. The geometry can be varied to suit a wide range of situations and furnace is not limited to cylindrical form. Therefore, *water-tube boilers are generally preferred for high pressure and high output, whereas, shell boilers for low pressure and low output.*

The modern high pressure boilers employed for power generation are for steam capacities 30 to 650 tonnes/h and above with a pressure up to 160 bar and maximum steam temperature of about 540°C.

12.11.2   Unique Features of the High Pressure Boilers

Following are the *unique features* of high pressure boilers:

1. Method of water circulation.
2. Type of tubing.
3. Improved method of heating.

1. **Method of water circulation**. The circulation of water through the boiler may be *natural circulation* due to density difference or *forced circulation*. In all modern high pressure boiler plants, the water circulation is maintained with the help of pump which forces the water through the boiler plant. The use of natural circulation is limited to sub-critical boilers due to its limitations.

2. **Type of tubing**. In most of the high pressure boilers, the water is circulated through the tubes and their external surfaces are exposed to the flue gases. In water-tube boilers, if the flow takes place through one continuous tube, the large pressure drop takes place due to friction. This is *considerably reduced by arranging the flow to pass through parallel system of tubing*. In most of the cases, several sets of the tubings are used. This type of arrangement helps to *reduce the pressure loss*, and better control over the quality of the steam.

3. **Improved method of heating**. The following improved methods of heating may be used to increase the heat transfer:

   (*i*) The saving of heat by *evaporation of water* above critical pressure of the steam.

   (*ii*) The heating of water can be made by *mixing the superheated steam*. The mixing phenomenon *gives highest heat transfer coefficient*.

   (*iii*) The overall heat transfer coefficient can be increased by *increasing the water velocity inside the tube and increasing the gas velocity above sonic velocity*.

12.11.3   Advantages of High Pressure Boilers

The following are the *advantages* of high pressure boilers:

1. In high pressure boilers pumps are used to maintain forced circulation of water through the tubes of the boiler. This *ensures positive circulation of water and increases evaporative capacity of the boiler and less number of steam drums will be required.*

2. The *heat of combustion is utilised more efficiently* by the use of small diameter tubes in large number and in multiple circuits.

3. *Pressurised combustion* is used which *increases rate of firing of fuel* thus increasing the rate of heat release.

4. Due to compactness less floor space is required.

5. The tendency of scale formation is eliminated due to high velocity of water through the tubes.

6. All the parts are uniformly heated, therefore, the danger of overheating is reduced and thermal stress problem is simplified.

7. The differential expansion is reduced due to uniform temperature and this reduces the possibility of gas and air leakages.

8. The components can be arranged horizontally as high head required for natural circulation is eliminated using forced circulation. There is a greater flexibility in the components arrangement.

9. The steam can be raised quickly to meet the variable load requirements without the use of complicated control devices.

10. The *efficiency of plant is increased upto 40 to 42 percent* by using high pressure and high temperature steam.

11. A very rapid start from cold is possible if an external supply of power is available. Hence the boiler can be used for carrying peak loads or standby purposes with hydraulic station.

12. Use of high pressure and high temperature steam is economical.

12.11.4   LaMont Boiler

This boiler *works on a forced circulation and the circulation is maintained by a centrifugal pump, driven by a steam turbine using steam from the boiler.* For emergency an electrically driven pump is also fitted.

Fig. 12.22 shows a LaMont steam boiler. The feed water passes through the economiser to the drum from which it is drawn to the circulation pump. The pump delivers the feed water to the tube evaporating section which in turn sends a mixture of steam and water to the drum. The steam in the drum is then drawn through the superheated. The superheated steam so obtained is then supplied to the primemover.

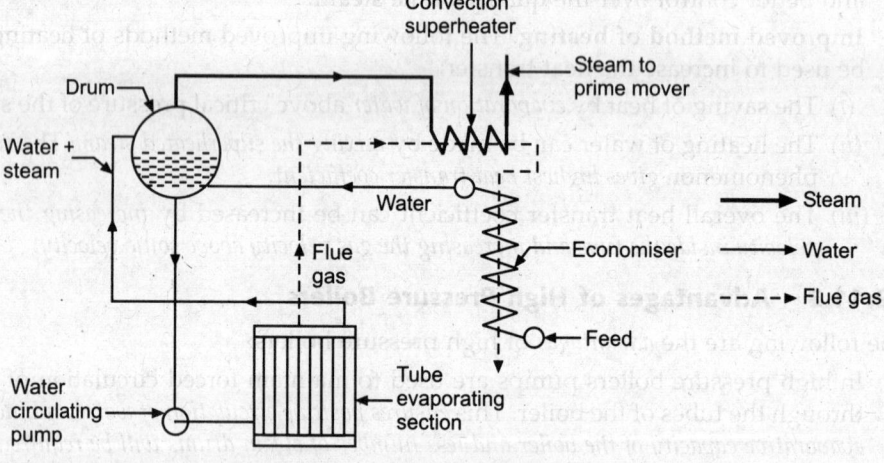

Fig. 12.22. LaMont boiler.

These boilers have been built to generate of 45 to 50 tonnes of superheated steam at a pressure of 130 bar and at a temperature of 500°C.

12.11.5   Loeffler Boiler

In a LaMont boiler the major difficulty experienced is the deposition of salt and sediment on the inner surfaces of the water tubes. The deposition reduces the heat transfer and ultimately the generating capacity. This further increases the danger of overheating the tubes due to salt deposition as it has high thermal resistance. This difficulty was solved in Loeffler boiler by *preventing the flow of water into the boiler tubes*.

This boiler also makes use of forced circulation. Its novel principle is the *evaporating of the feed water by means of superheated steam from the superheater, the hot gases from the furnace being primarily used for superheating purposes*.

Fig. 12.23 shows a diagrammatic view of a Loeffler boiler. The high pressure feed pump draws water through the economiser (or feed water heater) and delivers it into the evaporating drum. The steam circulating pump draws saturated steam from the evaporating drum and passes it through radiant and convective superheaters where steam is heated to required temperature. *From the superheater about one-third of the superheated steam passes to the prime mover (turbine), the remaining two-thirds passing through the water in the evaporating drum in order to evaporate feed water*.

This boiler can carry higher salt concentrations than any other type and is more compact than indirectly heated boilers having natural circulation. These qualities fit it for land or sea transport power generation.

Loeffler boilers with generating capacity of 100 tonnes/h and operating at 140 bar are already commissioned.

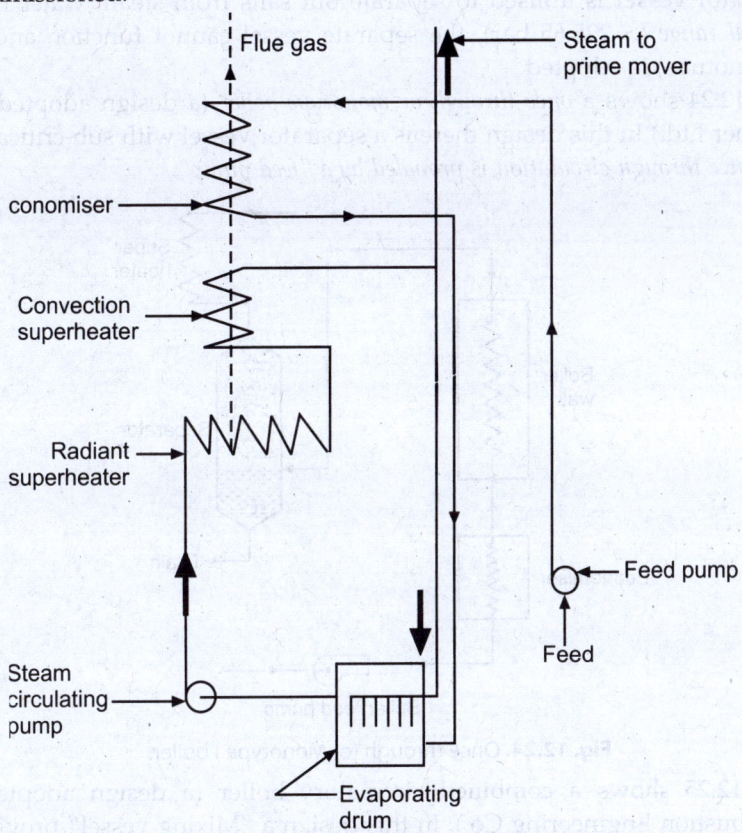

Fig. 12.23. Loeffler boiler.

12.11.6 Once Through Boiler

Owing to the sharply increasing costs of construction and fuel it becomes essential for the designers to economise on the installation cost and to increase fuel efficiency in the new stations by using modern sophisticated technology. Higher size units with higher steam parameters seem a natural choice for economical installation and operation of thermal power plants. The 800 MW units would be designed on supercritical steam pressure with a drumless boiler on *"once through principle"*.

As the steam pressure increases, the *differential* between the specific weight of saturated water in down-comers and specific weight of steam-water mixture in furnace wall tubes, which *causes natural circulation* in boiler, goes on *decreasing. Sluggish circulation causes film boiling.* In film boiling, the tube metal remains in contact with steam bubbles which provide high thermal resistance for heat flow and therefore tube metal sharply deteriorates due to high metal temperature leading to failure of the boiler tubes.

It is *not possible to prevent "film boiling"* in the upper furnace tubes for pressures above *180 bar* with *natural circulation.* Therefore, generally above 160 bar pressure, *controlled circulation* in water walls is used by providing *boiler circulating pump* between down-comers and lower water distributing heaters and the water walls. In such controlled circulation boilers, it is possible to utilize high steam pressure upto 200 bar but beyond this, the effectiveness of boiler drum in separating the saturated steam from water is reduced. Therefore, *beyond 200 bar, drumless boiler is envisaged. In sub-critical range (<225.65 bar),* a **separator vessel** is utilised to separate out salts from steam water mixture, but in *supercritical range (> 225.65 bar),* the separate vessel cannot function and only **once through** (monotube) is adopted.

- Fig. 12.24 shows a *once through or monotype boiler* (a design adopted by Sulzers Brother Ltd.) In this design there is a separator vessel with sub-critical pressures; the *once through circulation is provided by a "feed pump".*

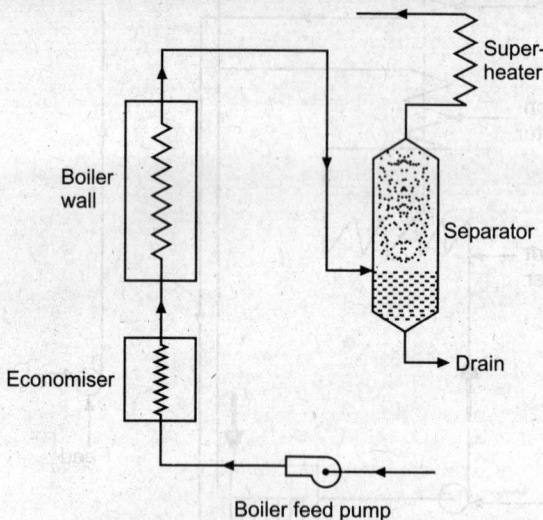

Fig. 12.24. Once through (or Monotype) boiler.

- Fig. 12.25 shows a combined circulatory boiler (a design adopted by M/S Combustion Engineering Co.). In this design a "Mixing vessel" provides suction to boiler circulating pumps at sub-critical pressures and inlet saturated steam to superheater and serves as a receiving header for steam-water mixture from evaporator suction. The boiler circulating pumps are required to function in the start-up or low pressure conditions but when the pressure goes above critical pressure then these are stopped and once through circulation is provided by boiler feed pump.

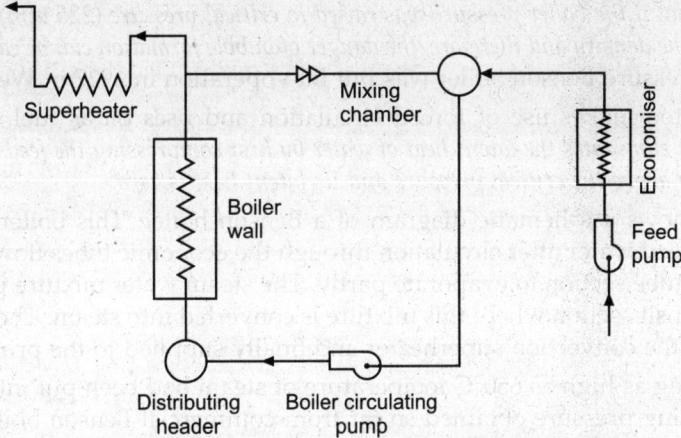

Fig. 12.25. Combined circulatory pump.

Advantages of Once through boilers:

The following are the *advantages of once through boilers for large thermal units*:

1. No higher limit for the higher steam pressure.
2. Full steam temperature can be maintained over a wider range of load.
3. Starting and cooling down of the boiler is fast.
4. No circulation disturbance due to rapid pressure fluctuations.
5. A once through boiler is smaller in size and weighs less in comparison to a natural circulation boiler.
6. Easy control of steam temperature during start-up and shut down (which is very advantageous for start-up of boiler and turbine).
7. Greater freedom in arrangement and location of heating surfaces.
8. Easy to adopt variable pressure operation for better performance at part load operation.
9. Elimination of heavy walled drum decreases the metallurgical sensitivity of boiler against changes in pressure.

Flash Steam Generator:

- A flash steam generator is a special form of boiler having basically a *helix tube fired by down jet combustion of gas or oil*. Water is pumped into the helix and at exit 90 percent of it is in the form of steam, the remaining water fraction being collected in the separator. The *tube helix principle, which eliminates the need for a water space, gives an extremely high heat output in a small area.*
- The *combustion efficiency* is about 80 percent on oil, and 73 percent on gas.
- The *advantages* are *very rapid response* (full steam production within about five minutes) and output ranges upto an evaporation rate of about 1 kg/s with operating steam pressure ranging from 3 to 70 bar.
- This type of boiler is more suitable when the plant is designed to take peak loads.

12.11.7 Benson Boiler

In the LaMont boiler, the main difficult experienced *is the formation and attachment of bubbles on the inner surfaces of the heating tubes. The attached bubbles to the tube surfaces reduce the heat flow and steam generation as it offers high thermal resistance than water film. Benson in 1922 argued that if the boiler pressure was raised to critical pressure (225 atm), the steam and water have the same density and therefore, the danger of bubble formation can be easily eliminated.* The first high pressure Benson boiler was put into operation in 1927 in West Germany.

This boiler too makes use of forced circulation and uses oil as fuel. It chief novel *principle is that it eliminates the latent heat of water by first compressing the feed to a pressure of 235 bar, it is then above the critical pressure and its latent heat is zero.*

Fig. 12.26 shows a schematic diagram of a Benson boiler. This boiler does not use any drum. The feed water after circulation through the economic tubes flows through the radiant parallel tube section to evaporate partly. The steam water mixture produced then moves to the transit section where this mixture is converted into steam. The steam is now passed through the convection superheater and finally supplied to the prime mover.

Boilers having as high as 650°C temperature of steam had been put into service. The maximum working pressure obtained so far from commercial Benson boiler is 500 atm. The Benson boilers of 150 tonnes/h generating capacity are in use.

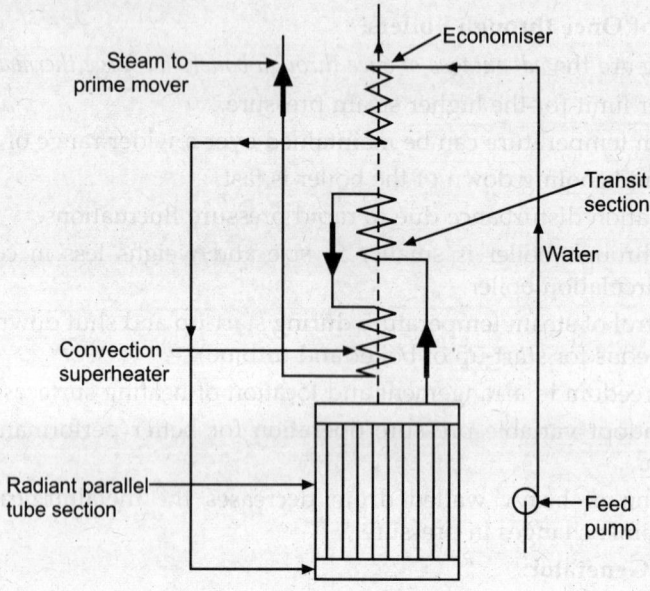

Fig. 12.26. Benson boiler.

Advantages of a Benson boiler:

The Benson boiler possesses the following *advantages:*

1. It can be erected in a comparatively smaller floor area.
2. The total weight of a Benson boiler is 20% less than other boilers, since there are no drums. This also reduces the cost of the boiler.
3. It can be started very quickly because of welded joints.
4. Natural convection boilers require expansion joints but these are not required for Benson boiler as the pipes are welded.
5. The furnace walls of the boiler can be more efficiently protected by using smaller diameter and closed pitched tubes.
6. The transfer of parts of the boiler is easy as no drums are required and majority of the parts are carried to the site without pre-assembly.

7. It can be operated most economically by varying the temperature and pressure at part loads and overloads. The desired temperature can also be maintained constant at any pressure.

8. The blow-down losses of the boiler are hardly 4% of natural circulation boiler of the same capacity.

9. Explosion hazards are not severe as it consists of only tubes of small diameter and has very little storage capacity.

10. The superheater in a Benson boiler is an integral part of forced circulation system, therefore, no special starting arrangement for superheater is required.

12.11.8 Velox Boiler

Refer to fig. 12.27.

It is a well known fact that when the gas velocity exceeds the sound-velocity, the heat is transferred from the gas at a much higher rate than rates achieved with sub-sonic flow. The advantage of this theory is taken to effect the large heat transfer from a smaller surface area in this boiler.

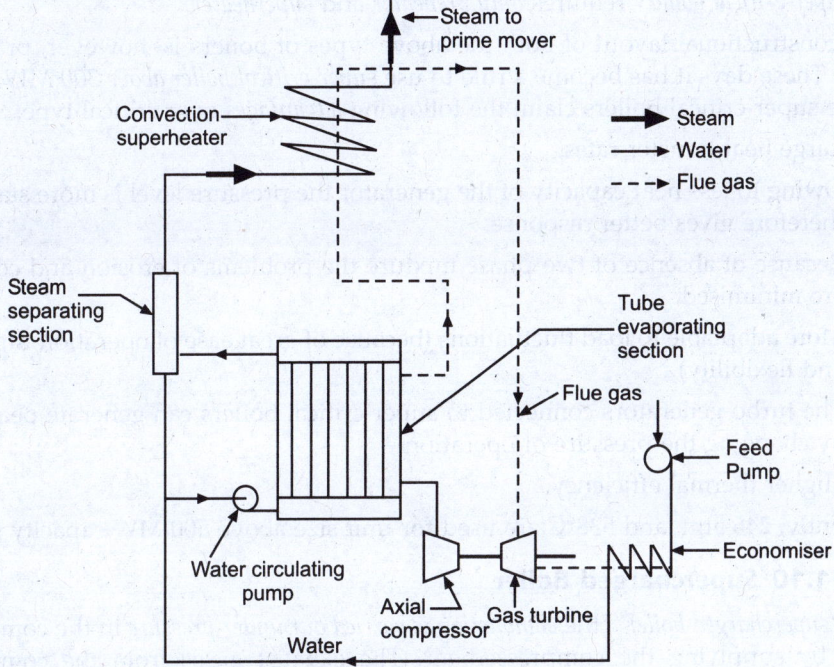

Fig. 12.27. Velox boiler.

This *boiler makes use of* ***pressurised combustion.***

The gas turbine drives the axial flow compressor which raises the incoming air from atmosphere pressure to furnace pressure. The combustion gases after heating the water and steam flow through the gas turbine to the atmosphere. The feed water after passing through the economiser is pumped by a water circulating pump to the tube evaporating section. Steam separated in steam separating section flows to the superheater, from there it moves to the prime mover.

The size of the Velox boiler is limited to 100 tonnes/h because 600 B.H.P. is required to run the air compressor at this output. The power developed by the gas turbine is not sufficient to run the compressor and therefore some power from external source power must be supplied.

Advantages:

1. The boiler is very compact and has greater flexibility.
2. Very high combustion rates are possible.
3. It can be quickly started.
4. Low excess air is required as the pressurised air is used and the problem of draught is simplified.

12.11.9 Super-critical Boilers

A large number of steam generating plants are designed between working ranges of 125 atm. and 510°C to 300 atm. and 660°C; these are basically characterised as *sub-critical and super-critical.*

Usually, a *"sub-critical boiler"* consists of three distinct sections as *preheater* (economiser), *evaporator* and *superheater.*

A *"super-critical boiler"* requires *only preheater* and *superheater.*

The constructional layout of both the above types of boilers is, however, practically identical. These days it has become a rule to use *super-critical boiler above 300 MW capacity units.* The super-critical boilers claim the following *advantages* over critical type:

1. Large heat transfer rates.
2. Owing to less heat capacity of the generator the pressure level is more stable and therefore gives better response.
3. Because of absence of two phase mixture the problems of erosion and corrosion are minimised.
4. More adaptable to load fluctuations (because of great ease of operation, simplicity and flexibility).
5. The turbo-generators connected to super-critical boilers can generate peak loads by changing the pressure of operation.
6. Higher thermal efficiency.

Presently, 246 atm. and 538°C are used for unit size above 500 MW capacity plants.

12.11.10 Supercharged Boiler

In a *"supercharged boiler"*, the *combustion is carried out under pressure* in the combustion chamber by supplying the compressed air. The exhaust gases from the combustion chamber are used to run the gas turbine as they are exhausted to high pressure. The gas turbine runs the air compressor to supply the compressed air to the combustion chamber.

Advantages:

1. Owing to very high overall heat transfer co-efficient the heat transfer surface required is hardly 20 to 25% of the heat transfer surface of a conventional boiler.
2. The part of the gas turbine output can be used to drive other auxiliaries.
3. Small heat storage capacity of the boiler plant gives better response to control.
4. Rapid start of the boiler is possible.
5. Comparatively less number of operators are required.

12.12 CORROSION IN BOILERS AND ITS PREVENTION

For the safety and performance of fossil fired boilers proper selection of tube material is very essential.

It is of paramount importance to keep the tubes clean internally and externally free of deposits that could impair heat transfer and lead to corrosion, ultimately causing tube failures. Corrosion damage is always experienced inside tubes of the boiler, economiser and superheater when water chemistry is not maintained within limit as recommended by the manufacturers.

Materials commonly used for various parts/sections of the boilers are given below:

Parts/Sections	Materials
1. *Furnace walls and economisers*	Carbon steels and ferric alloys with small percentages of chromium (5–10%).
2. *Superheater and reheater tubes*	— Carbon steels ... upto 500°C temp.
	— Alloy steels ... temp. > 500°C.
	— Carbon-molybdenum steel is used at *inlet section*.
	— Ferritic alloy steel with high percentage of chromium is used for downstream section of superheater.
	— Stainless steel and high chromium steels are recommended for *hotter sections* (560–600°C).

Composite tubes (consisting of an outer layer of 50% Cr, 15% Ni steel, metallurgically bonded to inner layer of 800 H) are used when coal-ash attack is severe.

Following are the few important *phenomena which contribute to corrosion*:

(i) Hydrogen induced brittle fracture. ⎫

(ii) Bulk deposit corrosion. ⎬ Water-side problems

(iii) Corrosion fatigue. ⎪

(iv) Stress corrosion cracking. ⎭

(v) Oxidation. ⎫

(vi) Fouling ⎬ Fireside problems

(vii) Slagging ⎭

If inspite of design efforts high temperature corrosion occurs then the problem can be solved by using any one of the following *methods*:

1. Using the fuels having more favourable characteristics.
2. Replacing the damaged tubes with tubes containing high chromium content.
3. Providing stainless steel tube shields at the cost of reduced efficiency.

HIGHLIGHTS

1. A steam power plant converts the chemical energy of the fossil fuels (coal, oil, gas) into mechanical/electrical energy.
2. The layout of a modern steam power plant comprises of the following four circuits:

   (i) Coal and ash circuit   (ii) Air and gas circuit

(*iii*) Feed water and steam flow circuit   (*iv*) Cooing water circuit.

3. The choice of steam conditions depends upon the following:

   (*i*) Price of the coal   (*ii*) Capital cost of the plant

   (*iii*) Time available for erection   (*iv*) Thermal efficiency obtainable

   (*v*) The station 'load factor.'

4. Automatic stokers are classified as follows:

   (*i*) Overfeed stokers   (*ii*) Underfeed stokers

5. Pulverised fuel handling systems are classified as:

   (*i*) Unit system   (*ii*) Central system

6. Pulverised fuel burners are classified as follows:

   (*i*) Long flame burners   (*ii*) Turbulent burners

   (*iii*) Tangential burners   (*iv*) Cyclone burners.

7. Oil burners are classified as follows:

   1. Vapourising oil burners:

      (*a*) Atmospheric pressure atomising burner (*b*) Rotating cup burner

      (*c*) Recirculating burner   (*d*) Wick type burner.

   2. Atomising fuel burners:

      (*a*) Mechanical or oil pressure atomising burner.

      (*b*) Steam or high pressure atomising burner.

      (*c*) Low pressure air atomising burner.

8. Fluidised bed may be defined as the bed of solid particles behaving as a fluid.

9. Ash handling systems may be classified as follows:

   (*i*) Mechanical handling system   (*ii*) Hydraulic system

   (*iii*) Pneumatic system   (*iv*) Steam jet system.

10. The dust collectors may be classified as follows:

    1. Mechanical dust collectors

       (*i*) Wet type (scrubbers):

          (*a*) Spray type   (*b*) Packed type

          (*c*) Impingement type.

       (*ii*) Dry type:

          (*a*) Gravitational separators   (*b*) Cyclone separators.

    2. Electrical dust collectors

       (*i*) Rod type   (*ii*) Plate type.

11. The 'collection efficiency' of a dust collector is the amount of dust removed per unit weight of dust.

12. pH value of water is the logarithm of the reciprocal of hydrogen ion concentration. It is number from 0 to 14 with 7 indicating neutral number.

13. The small pressure difference which causes a flow of gas to take place is termed as a *draught*.

14. The most important classification of steam turbines is as follows:

    (*i*) Impulse turbines   (*ii*) Reaction turbines

    (*iii*) Combination of impulse and reaction turbines.

THEORETICAL QUESTIONS

1. How are steam power plants classified?
2. Give the layout of a modern steam power plant and explain it briefly.
3. What are the essential requirements of steam power station design ?
4. What factors should be taken into consideration while selecting the site for steam power plant?
5. How can the capacity of a steam power plant be determined?
6. On what factors does the choice of steam conditions depend?
7. Enumerate the means by which the coal from coal mines can be transported.
8. What are the requirements of a good coal handling plant?
9. Enumerate and explain the steps involved in handling of the coal.
10. Explain with the help of a neat diagram the arrangement of the Fluidised Bed Combustion (FBC) system.
11. State the characteristics of a good ash handling plant.
12. Enumerate and explain various modern ash-handling systems.
13. How are dust collectors classified?
14. Explain with the help of a diagram the working of a 'cyclone separator'.
15. How do you define the 'collection efficiency' of a dust separator?
16. What are the uses of ash and dust?
17. Give the general layout of ash handling and dust collection system.

CHAPTER

13 DIESEL ENGINE POWER PLANT

13.1. Introduction; 13.2. Advantages and disadvantages of diesel power plants; 13.3. Applications of diesel power plant 13.4. Site selection; 13.5. Heat engines 13.6. Classification of I.C. engines. 13.7 Comparison between a petrol engine and a diesel engine.; 13.8. Essential components of a diesel power plant–Engine-air intake system-exhaust system-fuel system; 13.9. Operation of a diesel power plant; 13.10. Types of diesel engines used for diesel power plants; 13.11. Layout of a diesel engine power plant. *Highlights – Theoretical Questions.*

13.1. INTRODUCTION

- Diesel engine power plants are installed *where supply of coal and water is not available in sufficient quantity or where power is to be generated in small quantity or where standby sets are required for continuity of supply such as in hospitals, telephone exchanges, radio stations and cinemas.* These plants in the range of 2 to 50 MW capacity are used as *central stations* for supply authorities and works and they are universally adopted to supplement hydro-electric or thermal stations where standby generating plants are essential for starting from cold and under emergency conditions.

- In several countries, the demand for diesel power plants is increased for electric power generation because of difficulties experienced in construction of new hydraulic plants and enlargement of old hydro-plants. A long term planning is required for the development of thermo and hydro-plants which cannot keep the pace with many times the increased demand by the people and industries.

- The diesel units used for electric generation are *more reliable* and *long-lived piece of equipment* compared with other types of plants.

13.2. ADVANTAGES AND DISADVANTAGES OF DIESEL POWER PLANTS

The *advantages and disadvantages* of diesel power plants are listed below:

Advantages:

1. Design and installation are very simple.
2. Can respond to varying loads without any difficulty.
3. The standby losses are less.
4. Occupy less space.
5. Can be started and put on load quickly.
6. Require less quantity of water for cooling purposes.
7. Overall capital cost is lesser than that for steam plants.

8. Require less operating and supervising staff as compared to that for steam plants.

9. The efficiency of such plants at part loads does not fall so much as that of a steam plant.

10. The cost of building and civil engineering works is low.

11. Can burn fairly wide range of fuels.

12. These plants can be located very near to the load centres, many times in the heart of the town.

13. No problem of ash handling.

14. The lubrication system is more economical as compared with that of a steam power plant.

15. The diesel power plants are *more efficient than steam power plants* in the range of 150 MW capacity.

Disadvantages:

1. High operating cost.

2. High maintenance and lubrication cost.

3. Diesel units capacity is limited. These cannot be constructed in large size.

4. In a diesel power plant noise is a serious problem.

5. Diesel plants cannot supply overloads continuously whereas steam power plant can work under 25% overload continuously.

6. The diesel power plants are not economical where fuel has to be imported.

7. The life of a diesel power plant is quite small (2 to 5 years or less) as compared to that of a steam power plant (25 to 30 years).

13.3 APPLICATIONS OF DIESEL POWER PLANT

The diesel power plants find wide application in the following fields:

1. Peak load plant
2. Mobile plants
3. Standby units
4. Emergency plant
5. Nursery station
6. Starting stations
7. Central stations—where capacity required is small (5 to 10 MW)
8. Industrial concerns where power requirement is small, say of the order of 500 kW, diesel power plants become more economical due to their higher overall efficiency.

13.4. SITE SELECTION

The following *factors* should be considered while *selecting the site for a diesel power plant:*

1. **Foundation sub-soil condition.** The conditions of sub-soil should be such that a foundation at a reasonable depth should be capable of providing a strong support to the engine.

2. **Access to the site.** The site should be so selected that it is accessible through rail and road.

3. **Distance from the load centre.** The location of the plant should be near the load centre. This reduces the cost of transmission lines and maintenance cost. The power loss is also minimised.

4. **Availability of water.** Sufficient quantity of water should be available at the site selected.

5. **Fuel transportation.** The site selected should be near to the source of fuel supply so that transportation charges are low.

13.5. HEAT ENGINES

Any type of engine or machine which derives heat energy from the combustion of fuel or any other sources and converts this energy into mechanical work is termed as a **heat engine**.

Heat engines may be classified into two main classes as follows:

    1. External Combustion Engines   2. Internal Combustion Engines

1. **External combustion engines (E.C. engines).** In this case, *combustion of fuel takes place outside the cylinder* as in case of *steam engines* where the heat of combustion is employed to generate steam which is used to move a piston in a cylinder. Other examples of external combustion engines are *hot air engines*, *steam turbine* and *closed cycle gas turbine*. These engines are generally used for driving locomotives, ships, generation of electric power etc.

2. **Internal combustion engines (I.C. engines).** In this case, *combustion of the fuel with oxygen of the air occurs within the cylinder of the engine.* The internal combustion engines group includes engines employing mixtures of combustible gases and air, known as *gas engines* those using lighter liquid fuel or spirit known as *petrol engines* and those using heavier liquid fuels, known as *oil compression ignition* or *diesel engines.*

13.6. CLASSIFICATION OF I.C. ENGINES

Internal combustion engines may be *classified* as given below:

1. **According to cycle of operation:**
    (*i*) Two stroke cycle engines
    (*ii*) Four stroke cycle engines.

2. **According to cycle of combustion:**
    (*i*) Otto cycle engine (combustion at constant volume)
    (*ii*) Diesel cycle engine (combustion at constant pressure)
    (*iii*) Dual-combustion or Semi-Diesel cycle engine (combustion partly at constant volume and partly at constant pressure).

3. **According to arrangement of cylinder:**
    (*i*) Horizontal engine   (*ii*) Vertical engine
    (*iii*) V-type engine   (*iv*) Radial engine etc.

4. **According to their uses:**
    (*i*) Stationary engine   (*ii*) Portable engine
    (*iii*) Marine engine   (*iv*) Automobile engine
    (*v*) Aero engine etc.

5. **According to the fuel employed and the method of fuel supply to the engine cylinder:**
    (*i*) Oil engine   (*ii*) Petrol engine
    (*iii*) Gas engine   (*iv*) Kerosene engine etc.
    (*v*) Carburettor, hot bulb, solid injection and air injection engine.

6. **According to the speed of the engine:**
   (i) Low speed engine   (ii) Medium speed engine
   (iii) High speed engine.

7. **According to method of ignition:**
   (i) Spark ignition (S.I.) engine   (ii) Compression ignition (C.I.) engine.

8. **According to method of cooling the cylinder:**
   (i) Air-cooled engine   (ii) Water-cooled engine.

9. **According to method of governing:**
   (i) Hit and miss governed engine   (ii) Quality governed engine
   (iii) Quantity governed engine.

10. **According to valve arrangement:**
    (i) Over head valve engine   (ii) L-head type engine
    (iii) T-head type engine   (iv) F-head type engine.

11. **According to number of cylinders:**
    (i) Single cylinder engine   (ii) Multi-cylinder engine.

13.7. COMPARISON BETWEEN A PETROL ENGINE AND A DIESEL ENGINE

Petrol Engine	Diesel Engine
1. Air petrol mixture is sucked in the engine cyclinder during suction stroke.	1. Only air is sucked during suction stroke.
2. Spark plug is used.	2. Employs an injector.
3. Power is produced by spark ignition.	3. Power is produced by compression ignition.
4. Thermal efficiency up to 25%.	4. Thermal efficiency up to 40%.
5. Occupies less space.	5. Occupies more space.
6. More running cost.	6. Less running cost.
7. Light in weight.	7. Heavy in weight.
8. Fuel (Petrol) costlier.	8. Fuel (Diesel) cheaper.
9. Petrol being volatile is dangerous.	9. Diesel is not dangerous as it is non-volatile.
10. Pre-ignition possible.	10. Pre-ignition not possible.
11. Works on Otto cycle.	11. Works on diesel cycle.
12. Less dependable.	12. More dependable.
13. Used in *cars* and *motor cycles*.	13. Used in heavy duty vehicles like *trucks*, *buses* and *heavy machinery*.

13.8. ESSENTIAL COMPONENTS OF A DIESEL POWER PLANT

Refer to Fig. 13.1. The essential components of a diesel power plant are listed and discussed as follows:

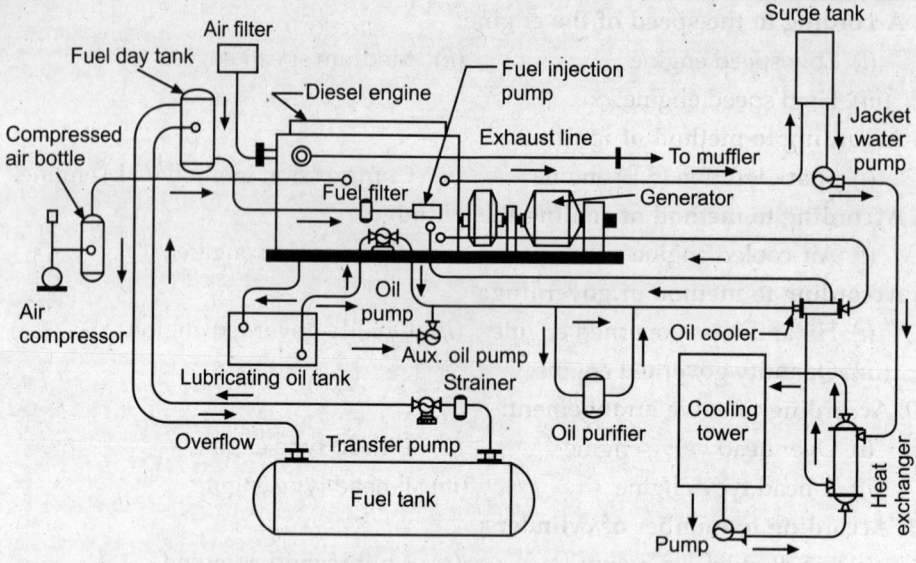

Fig. 13.1. Schematic arrangement of a diesel power plant.

1. Engine
2. Air intake system
3. Exhaust system
4. Fuel system
5. Cooling system
6. Lubrication system
7. Engine starting system
8. Governing system.

13.8.1. Engine

This is the main component of the plant which develops the required power. It is generally directly coupled to the generator as shown in Fig. 13.1.

13.8.2. Air Intake System

The air intake system conveys fresh air through pipes or ducts to: (*i*) Air intake manifold of four stroke engines (*ii*) The scavenging pump inlet of a two stroke engine and (*iii*) The supercharger inlet of a supercharged engine.

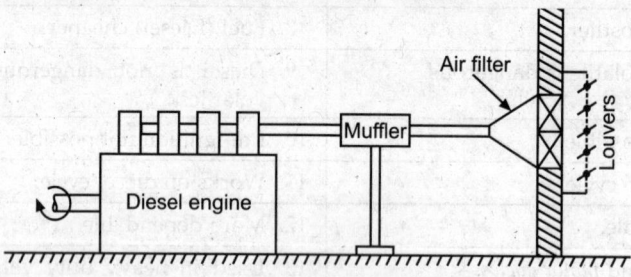

Fig. 13.2. Air intake system.

The air system begins with an intake located outside the building provided with a *filter* to catch dirt which would otherwise cause excessive wear in the engine. The filters may be of *dry or oil bath*. Electrostatic precipitator filters can also be used. *Oil impingement type of filter* consists of a frame filled with metal shavings which are coated with a special oil so that the air in passing through the frame and being broken up into a number of small filaments comes into contact with the oil whose property is to seize and hold any dust particles being carried by the air. The *dry type* of filter is made of cloth, felt, glass wool etc. In case of *oil bath type of filter* the air is swept over or through a pool of oil so that the particles of dust becomes coated. Light weight steel pipe is the material for intake ducts. In some cases, the engine noise may be transmitted back through the air intake system to the outside air. In such cases a silencer is provided between the engine and the intake.

Following *precautions* should be taken while constructing a suitable air intake system:

1. Air intakes may not be located inside the engine room.

2. Air should not be taken from a confined space otherwise air pulsations can cause serious vibration problems.

3. The air-intake line used should neither have too small a diameter nor should be too long, otherwise there may crop up engine starvation problem.

4. Air intake filters may not be located close to the roof of the engine room otherwise pulsating air flow through the filters can cause serious vibrations of the roof.

5. Air intake filters should not be located in an inaccessible location.

13.8.3. Exhaust System

Refer to Fig. 13.3. The purpose of the exhaust system is to discharge the engine exhaust to the atmosphere outside the building. The exhaust manifold connects the engine cylinder exhausts outlets to the exhaust pipe which is provided with a muffler to reduce pressure in the exhaust line and eliminate most of the noise which may result if gases are discharged directly into the atmosphere.

The exhaust pipe leading out of the building should be short in length with minimum number of bends and should have one or two flexible tubing sections which take up the effects of expansion, and isolate the system from the engine vibration. Every engine should be provided with its independent exhaust system.

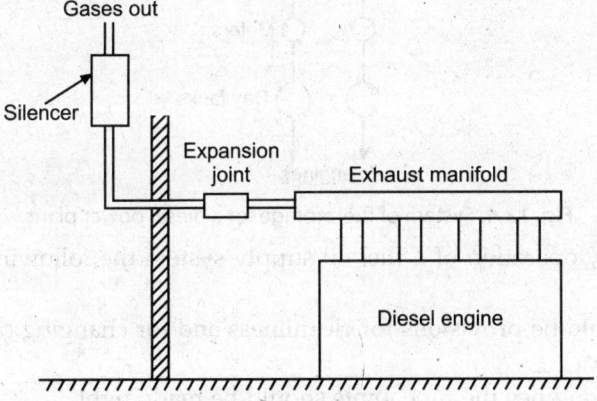

Fig. 13.3. Exhaust system.

The waste heat utilisation in a *diesel-steam* station may be done by providing waste-heat boilers in which most of the heat of exhaust gases from the engine is utilised to raise low pressure steam. Such application is common on *marine plants*. On the *stationary power plant* the heat of exhaust may be utilised to heat water in gas-to-water heat exchangers consisting of a water coil placed in exhaust muffler and using the water in the plant suitably. If air heating is required, the exhaust pipe from the engine is surrounded by the cold air jacket, and transfers the heat of exhaust gases to the air.

13.8.4. Fuel System

Refer to Fig. 13.4.

The fuel oil may be delivered at the plant site by trucks, railroad tank cars or barge and tankers. From tank car or truck the delivery is through the unloading facility to main storage tanks and then by transfer pumps to small service storage tanks known as *engine day tanks*. Large storage capacity allows purchasing fuel when prices are low. The main flow is made workable and practical by arranging the piping equipment with the necessary heaters, by passes, shut-offs, drain lines, relief valves, strainers and filters, flow meters and temperature indicators. The actual flow plans depend on type of fuel, engine equipment, size of the plant etc. The tanks should contain manholes for internal access and repair, fill lines to receive oil, vent lines to discharge vapours, overflow return lines for controlling oil flow and a suction line to withdraw oil. *Coils heated by hot water or steam reduce oil viscosity to lower pumping power needs.*

The minimum storage capacity of at least a month's requirement of oil should be kept in bulk, but where advantage of seasonal fluctuations in cost of oil is to be availed, it may be necessary to provide storage for a few month's requirements. *Day tanks* supply the daily fuel need of engines and may contain a minimum of about 8 hours of oil requirement of the engines. These tanks are usually placed high so that oil may flow to engines under gravity.

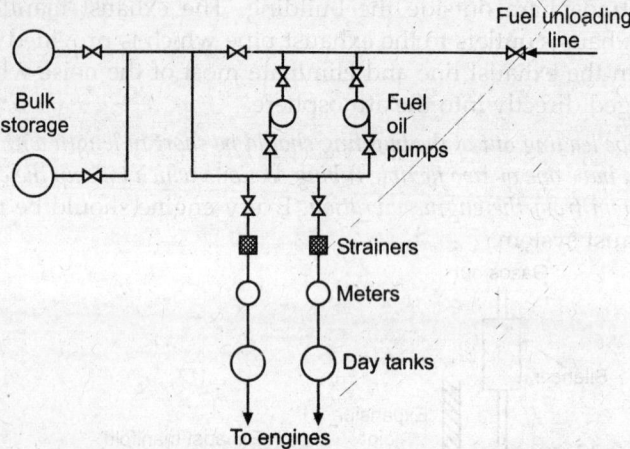

Fig. 13.4. System of fuel storage for a diesel power plant.

For satisfactory operation of a fuel oil supply system the following *points* should be taken care of:

1. There should be provisions for cleanliness and for changing over of lines during emergencies.

2. In all suction lines the pipe joints should be made tight.

3. Before being covered all oil lines should be put under air pressure and the joints tested with soap solution. Small air leaks into the line can be the sources of exasperating operating difficulties and are hard to remedy once the plant is in operation.

4. The piping between filter and the engine should be thoroughly oil flushed before being first placed in service.

5. Considerable importance should be given for cleanliness in handling bulk fuel oil. Dirt particles will ruin the fine lap of injection pumps or plug the injection

nozzle orifices. So *high-grade filters* are of paramount importance to the diesel oil supply system.

13.8.4.1. Fuel Injection System

The mechanical heart of the Diesel engine is the *fuel injection system*. The engine can perform no better than its fuel injection system. A very small quantity of fuel must be measured out, injected, atomised, and mixed with combustion air. The mixing problem becomes more difficult—the larger the cylinder and faster the rotative speed. Fortunately the high-speed engines are the small bore automotive types; however, special combustion arrangements such as precombustion chambers, air cells, etc., are necessary to secure good mixing. *Engines driving electrical generators have lower speeds and simple combustion chambers.*

13.8.4.2. Functions of a fuel injection system

1. Filter the fuel.
2. Meter or measure the correct-quantity of fuel to be injected.
3. Time the fuel injection.
4. Control the rate of fuel injection.
5. Automise or break up the fuel to fine particles.
6. Properly distribute the fuel in the combusion chamber.

The injection system are manufactured with great accuracy, especially the parts that actually meter and inject the fuel. Some of the tolerances between the moving parts are very small of the order of *one micron*. Such closely fitting parts require special attention during manufacture and hence the injection systems are *costly*.

13.8.4.3. Types of fuel injection systems

The following *fuel injection systems* are commonly used in diesel power station:

1. Common-rail injection system.
2. Individual pump injection system.
3. Distributor.

Atomisation of fuel oil has been secured by (*i*) *air blast* and (*ii*) *pressure spray*. Early diesel engines used air fuel injection at about 70 bar. This is sufficient not only to inject the oil, but also to atomise it for a rapid and thorough combustion. The expense of providing an air compressor and tank lead to the development of *"solid"* injection, using a liquid pressure of between 100 and 200 bar which is sufficiently high to atomise the oil it forces through spray nozzles. Great advances have been made in the field of solid injection of the fuel through research and progress in fuel pump, spray nozzles, and combustion chamber design.

13.9. OPERATION OF A DIESEL POWER PLANT

When diesel alternator sets are put in parallel, *"hunting"* or *"phase swinging"* may be produced *due to resonance* unless due care is taken in the design and manufacture of the sets. This condition occurs *due to resonance between the periodic disturbing forces of the engine and natural frequency of the system.* The engine forces results from uneven turning moment on the engine crank which are corrected by the flywheel effect. *"Hunting" results from the tendency of each set trying to pull the other into synchronism* and is characterised by flickering of lights.

To ensure *most economical operation of diesel engines* of different sizes when working together and sharing load it is necessary that they should carry the same percentage of their full load capacity at all times as the fuel consumption would be lowest in this condition. For best operation performance the manufacturer's recommendations should be strictly followed.

In order to get good performance of a diesel power plant the following *points* should be taken care of:

1. It is necessary to maintain the *cooling temperature* within the prescribed range and use of very cold water should be avoided. The cooling water should be free from suspended impurities and suitably treated to be scale and corrosion free. If the ambient temperature approaches freezing point, the cooling water should be drained out of the engine when it is kept idle.

2. During operation the *lubrication system* should work effectively and requisite pressure and temperature maintained. The engine oil should be of the correct specifications and should be in a fit condition to lubricate the different parts. A watch may be kept on the consumption of lubricating oil as this gives as indication of the true internal condition of the engine.

3. *The engine should be periodically run even when not required to be used and should not be allowed to stand idle for more than 7 days.*

4. *Air filter, oil filters and fuel filters* should be periodically serviced or replaced as recommended by the manufacturers or if found in an unsatisfactory condition upon inspection.

5. Periodical checking of engine compression and firing pressures and also exhaust temperatures should be made.

- The engine exhaust usually provides a good indication of satisfactory performance of the engine. *A black smoke in the exhaust is a sign of inadequate combustion or engine overloading.*

- *The loss of compression resulting from wearing out of moving parts lowers the compression ratio causing inadequate combustion.* These defects can be checked by taking *indicator diagrams* of the engine after reasonable intervals.

13.10. TYPES OF DIESEL ENGINES USED FOR DIESEL POWER PLANTS

The diesel engines may be four-stroke or two stroke cycle engines. *The two-stroke cycle engines are favoured for diesel power plants.*

Efforts are being made to use "*dual fuel engines*" in diesel power plants for better economy and proper use of available gaseous fuels in the country. The gas may be a waste product as in the case of sewage treatment installations or oil fuels where the economic advantage is self evident. With the wider availability of natural gas, the dual fuel engines may become an attractive means of utilising gas as fuel at off-peak tariffs for the electric power generation.

Working of Dual Fuel Engines:

The various strokes of a dual fuel engine are as follows:

1. **Suction stroke.** During this stroke *air and gas are drawn* in the engine cylinder.
2. **Compression stroke.** During this stroke the pressure of the mixture drawn is increased. Near the end of this stroke the '*pilot oil*' is injected into the engine cylinder. *The compression heat first ignites the pilot oil* and then gas mixture.

3. **Working/Power stroke.** During this stroke the gases (at high temperature) expand and thus power is obtained.

4. **Exhaust stroke.** The exhaust gases are released to the atmosphere during the stroke.

13.11. LAYOUT OF A DIESEL ENGINE POWER PLANT

Fig. 13.5 shows the layout of a diesel engine power plant.

The most common arrangement for diesel engines is with parallel centre lines, with some room left for extension in future. The repairs and usual maintenance works connected with such engines necessitate sufficient space around the units and consideration should be given to the need for dismantling and removal of large components of the engine generator set. The air intakes and filters as well as the exhaust mufflers are located outside the building or may be separated from the main engine room by a partition wall. The latter arrangement is not vibration free. Adequate space for oil storage and repair shop as well as for office should be provided close to the main engine room. Bulk storage of oil may be outdoor. The engine room should be well ventilated.

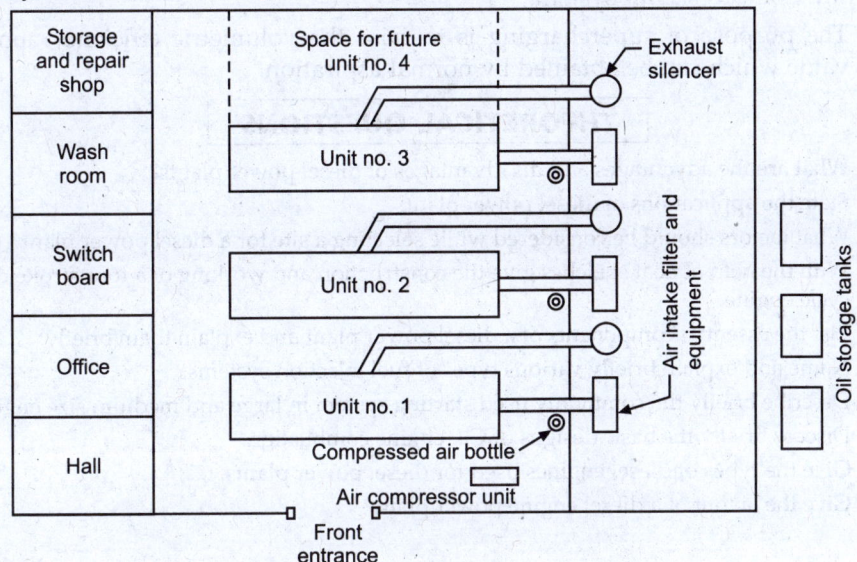

Fig. 13.5. Layout of a diesel engine power plant.

HIGHLIGHTS

1. Any type of engine or machine which derives heat energy from the combustion of fuel or any other source and converts this energy into mechanical work is termed as a *heat engine*.

2. Essential components of a diesel power plant are:

    (*i*) Engine   (*ii*) Air intake system

   (*iii*) Exhaust system   (*iv*) Fuel system

    (*v*) Cooling system   (*vi*) Lubrication system

   (*vii*) Engine starting system   (*viii*) Governing system.

3. Commonly used fuel injection system in a diesel power station:

    (*i*) Common-rail injection system

    (*ii*) Individual pump injection system

    (*iii*) Distribution system.

4. In liquid cooling following methods are used for circulating the water around the cylinder and cylinder head:

    (*i*) Thermo-system cooling   (*ii*) Forced or pump cooling

    (*iii*) Cooling with thermostatic regulator   (*iv*) Pressurised cooling

    (*v*) Evaporative cooling.

5. Various lubrication systems use for I.C. engines are:

    (*i*) Wet sump lubrication system   (*ii*) Dry sump lubrication system

    (*iii*) Mist lubrication system.

6. The following three are the commonly used starting systems in large and medium size engines:

    (*i*) Starting by an auxiliary engine

    (*ii*) Use of electric motors of self starters

    (*iii*) Compressed air system.

7. The purpose of supercharging is to raise the volumetric efficiency above that value which can be obtained by normal aspiration.

THEORETICAL QUESTIONS

1. What are the advantages and disadvantages of diesel power plants?
2. State the applications of diesel power plant.
3. What factors should be considered while selecting a site for a diesel power plant?
4. With the help of neat sketches give the construction and working of a four stroke diesel cycle engine.
5. List the essential components of a diesel power plant and explain them briefly.
6. Name and explain briefly various types of fuel injection systems.
7. Describe briefly the commonly used starting system in large and medium size engines.
8. Discuss briefly the basic designs of C.I. engine combustion.
9. Give the types of diesel engines used for diesel power plants.
10. Give the layout of a diesel engine power plant.

CHAPTER

14 GAS TURBINE POWER PLANTS

14.1 GAS TURBINES—GENERAL ASPECTS

Probably a windmill was the first turbine to produce useful work, wherein there is no precompression and no combustion. The characteristic features of a gas turbine as we think of the name today include a *compression process* and a *heat-addition* (or combustion) process. The *'gas turbine'* represents perhaps the *most satisfactory way of producing very large quantities of power in a self-contained and compact unit.* The gas turbine may have an ample future use in conjunction with the oil engine. For smaller gas turbine units, the inefficiencies in compression and expansion processes become greater and to *improve the thermal efficiency* it is *necessary to use a heat exchanger.* In order that a small gas turbine may compete for economy with the small oil engine or petrol engine it is necessary that a compact effective heat exchanger be used in the gas turbine cycle. The thermal efficiency of the gas turbine alone is still quite modest 20 to 30% compared with that of a modern steam plant 38 to 40%. It is possible to construct *combined plants* whose efficiencies are of the order of 45% or more. Higher efficiencies might be attained in future.

The following are the *major fields of application of gas turbines*:

1. Aviation.
2. Power generation.
3. Oil and gas industry.
4. Marine propulsion.

The efficiency of a gas turbine is not the criteria for the choice of this plant. A gas turbine is used in *aviation* and *marine fields* because *it is self contained, light weight not requiring cooling water and generally fit into the overall shape of the structure.* It is selected for *'power generation'* because of its *simplicity, lack of cooling water, needs quick installation and quick starting*. It is used in *oil and gas industry* because of *cheaper supply of fuel and low installation cost.*

The gas turbines have the following *"limitations"* :

1. They are not self starting.
2. Low efficiencies at part loads.
3. Non-reversibility.
4. Higher rotor speeds.
5. Low overall plant efficiency.

In the last two decides, rapid progress has been observed in the *development and improvement of the gas turbine plants for electric power production.* The major progress has been observed in the following *directions*:

(*i*) Increase in unit capacities of gas turbine units.

(*ii*) Increase in their efficiency.

(*iii*) Drop in capital cost.

14.2 APPLICATIONS OF GAS TURBINE PLANTS

Gas turbine plants for the purpose of power plant engineering find the following *applications* :

1. To drive generators and supply peak loads in steam, diesel or hydroplants.
2. To work as combination plants with conventional steam boilers.
3. To supply mechanical drive for auxiliaries.

- These plants are well suited for *peak load service* since the fuel costs are somewhat higher and initial cost low. Moreover, *peak load operation permits use of water injection which increases turbine work by about 40% with an increase in heat rate of about 20%.* The short duration of increase in heat rate does not prove of any much harm.

- The combination arrangement of gas turbines with conventional boilers may be supercharging or for heat recovery from exhaust gases. In the *supercharging system* air is supplied to the boiler under pressure by a compressor mounted on the common shaft with turbine, and gases formed as result of combination after coming out of the boiler pass through the gas turbine before passing through the economiser and the chimney.

- The application of the gas turbine to drive the auxiliaries is not strictly included under direct electric power generation by the turbines and would not be discussed.

14.3 ADVANTAGES AND DISADVANTAGES OF GAS TURBINE POWER PLANTS OVER DIESEL AND THERMAL POWER PLANTS

A. Advantages over Diesel plants:

1. The work developed per kg of air is large compared with diesel plant.
2. Less vibrations due to perfect balancing.
3. Less space requirements.
4. Capital cost considerably less.
5. Higher mechanical efficiency.

6. The running speed of the turbine (40,000 to 100,000 r.p.m.) is considerably large compared to diesel engine (1000 to 2000 r.p.m.).

7. Lower installation and maintenance costs.

8. The torque characteristics of turbine plants are far better than diesel plants.

9. The ignition and lubrication systems are simpler.

10. The specific fuel consumption does not increase with time in gas turbine plant as rapidly as in diesel plants.

11. Poor quality fuels can be used.

Disadvantages :

1. Poor part load efficiency.

2. Special metals and alloys are required for different components of the plants.

3. Special cooling methods are required for cooling the turbine blades.

4. Short life.

B. Advantages over Steam power plant:

1. No ash handling problem.

2. Low capital cost.

3. The gas turbine plants can be installed at selected load centre as space requirement is considerably less where steam plant cannot be accommodated.

4. Fewer auxiliaries required/used.

5. Gas turbines can be built relatively quicker. They require much less space and civil engineering works and water supply.

6. The gas turbine plant as peak load plant is more preferable as it can be brought on load quickly and surely.

7. The components and circuits of a gas turbine plant can be arranged to give the most economic results in any given circumstances which is not possible in case of steam power plants.

8. For the same pressure and initial temperature conditions the ratio of exhaust to inlet volume would be only 3.95 in case of gas turbine plant as against 250 for steam plant.

9. Above 550°C, the thermal efficiency of the gas turbine plant increases *three times* as fast the steam cycle efficiency for a given top temperature increase.

10. The site of the steam power plant is dictated by the availability of large cooling water whereas an open cycle gas turbine plant can be located near the load centre as no cooling water is required. The cooling water required for closed cycle gas turbine is hardly 10% of the steam power plant.

11. The gas turbine plants can work quite economically for short running hours.

12. Storage of fuel is much smaller and handling is easy.

14.4. SITE SELECTION

While selecting the site for a gas turbine plant the following *points* should be given due consideration :

1. The plant should be located near the load centre to avoid transmission costs and losses.

2. The site should be away from business centres due to noisy operations.

3. Cheap and good quality fuel should be easily available.

4. Availability of labour.

5. Availability of means of transportation.

6. The land should be available at a cheap price.

7. The bearing capacity of the land should be high.

14.5. THE SIMPLE GAS TURBINE PLANT

A gas turbine plant may be defined as one *in which the principal primemover is of the turbine type and the working medium is a permanent gas.*

Refer to Fig. 14.1. A simple gas turbine plant consists of the following:

1. *Turbine.*

2. *A compressor* mounted on the same shaft or coupled to the turbine.

3. *The combustor.*

4. *Auxiliaries* such as starting device, auxiliary lubrication pump, fuel system, oil system and the duct system etc.

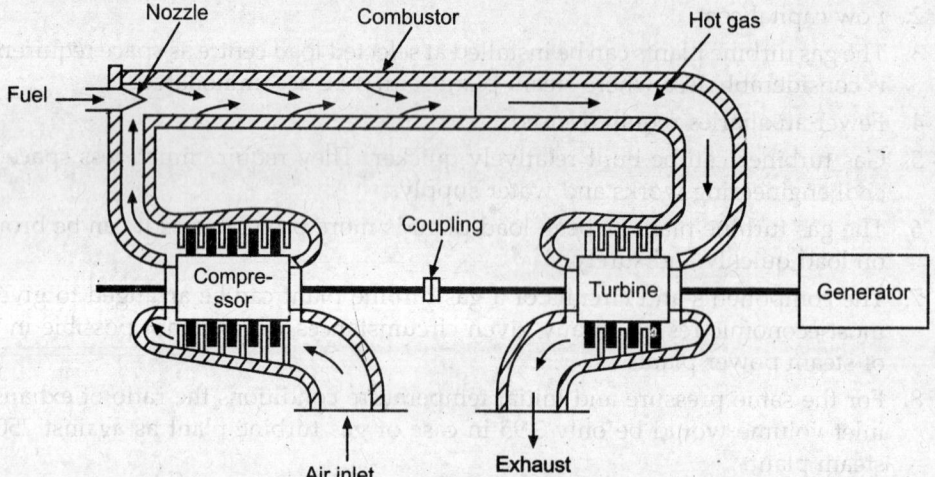

Fig. 14.1. Arrangement of a simple gas turbine plant.

A modified plant may have in addition to above an *intercooler, a regenerator, a reheater* etc.

The working fluid is compressed in a compressor which is generally rotary, multistage type. Heat energy is added to the compressed fluid in the chamber. This high energy fluid, at high temperature and presssure, then expands in the turbine unit thereby generating power. Part of the power generated is consumed in driving the generating compressor and accessories and the rest is utilised in electrical energy. The gas turbines work on open cycle, semi-closed cycle or closed cycle. In order to improve efficiency, compression and expansion of working fluid is carried out in *multistages*.

14.6. ENERGY CYCLE FOR A SIMPLE-CYCLE GAS TURBINE

Fig. 14.2 shows an energy-flow diagram for a simple-cycle gas turbine, the description of which is as follows:

— The air brings in minute amount of energy (measured above 0°C).

— Compressor adds considerable amount of energy.

— Fuel carries major input to cycle.

— Sum of fuel and compressed-air energy leaves combustor to enter turbine.

— In turbine *smallest part of entering energy goes to useful output, largest part leaves in exhaust.*

Shaft energy to drive compressor is about twice as much as the useful shaft output.

Actually the shaft energy keeps circulating in the cycle as long as the turbine runs. The important comparison is *the size of the output with the fuel input.* For the simple-cycle gas turbine the output may run about 20 per cent of the fuel input for pressure and temperature conditions at turbine inlet. This means 80% of the fuel energy is wasted. While the 20% thermal efficiency is not too bad, it can be improved by including *additional heat recovery apparatus.*

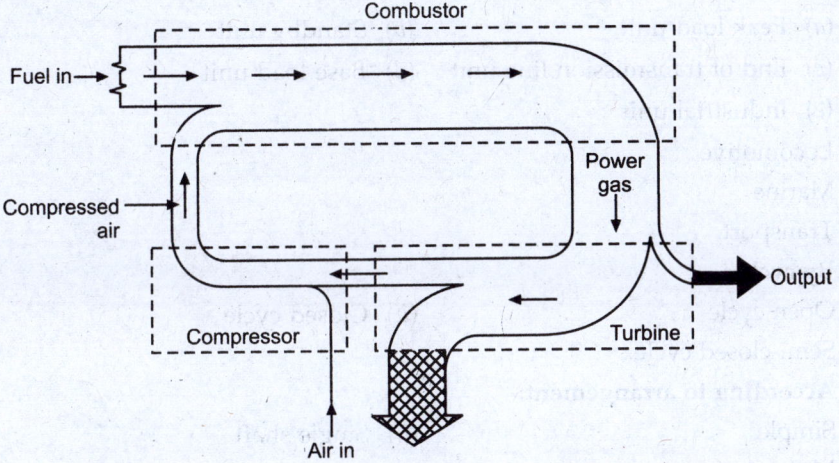

Fig. 14.2. Energy flow diagram for gas-turbine unit.

14.7. PERFORMANCE TERMS

Some of the *important terms used to measure performance of a gas turbine are defined* as follows :

1. **Pressure ratio.** It is the ratio of cycle's highest to its lowest pressure, usually highest-pressure-compressor discharges to the lowest-pressure-compressor inlet pressures.

2. **Work ratio.** It is the ratio of network output to the total work developed in the turbine or turbines.

3. **Air ratio.** Kg of air entering the compressor inlet per unit of cycle net output, for example, kg/kWh.

4. **Compression efficiency.** It is the ratio of work needed for ideal air compression through a given pressure range to work actually used by the compressor.

5. **Engine efficiency.** It is the ratio of the work actually developed by the turbine expanding hot power gas through a given pressure range to that would be yielded for ideal expansion conditions.

6. **Machine efficiency.** It is the collective term meaning both engine efficiency and compressor efficiency of turbine and compressor, respectively.

7. **Combustion efficiency.** It is the ratio of heat actually released by 1 kg of fuel to heat that would be released by complete perfect combustion.

8. **Thermal efficiency.** It is the percentage of total energy input appearing as net work output of the cycle.

14.8. CLASSIFICATION OF GAS TURBINE POWER PLANTS

The gas turbine power plants may be *classified* according to the following criteria :

1. **By application:**
   (i) *In aircraft:*
      (a) Jet propulsion   (b) Prop-jets.
   (ii) *Stationary:*
      (a) Peak load unit   (b) Standby unit
      (c) End of transmission line unit   (d) Base load unit
      (e) Industrial unit.
   (iii) Locomotive
   (iv) Marine
   (v) Transport.

2. **By cycle :**
   (i) Open cycle   (ii) Closed cycle
   (iii) Semi-closed cycle.

3. **According to arrangement:**
   (i) Simple   (ii) Single shaft
   (iii) Multi-shaft   (iv) Intercooled
   (v) Reheat   (vi) Regenerative
   (vii) Combination.

4. **According to combustion :**
   (i) Continuous combustion   (ii) Intermittent combustion.

5. **By fuel:**
   (i) Solid fuel   (ii) Liquid fuel
   (iii) Gaseous fuel.

14.9. CLASSIFICATION OF GAS TURBINES

The gas turbines are mainly divided into *two groups* :

1. **Constant pressure combustion gas turbine:**
   (a) Open cycle constant pressure gas turbine
   (b) Closed cycle constant pressure gas turbine.

2. **Constant volume combustion gas turbine:**

In almost *all the fields open cycle gas turbine plants are used. Closed cycle plants were introduced at one stage because of their ability to burn cheap fuel.* In between their progress remained slow because of availability of cheap oil and natural gas. Because of rising oil prices, now again, the attention is being paid to closed cycle plants.

14.10. COMBINATION GAS TURBINE CYCLES

14.10.1. Combined Turbine and Steam Power Plants

The characteristics of the gas turbine plants render these plants very well suited for use in combination with steam or hydro-plants. These plants can be quickly started for emergency or peak load service. The combination *'gas-turbine-steam cycles' aim at utilising the heat of exhaust gases from the gas turbine and thus, improve the overall plant efficiency.*

Three popular designs of combination cycle comprise of:

1. Heating feed water with exhaust gases.

2. Employing the gases from a supercharged boiler to expand in the gas turbine.

3. Employing the gases as combustion air in the steam boiler.

1. Heating feed water with exhaust gases :

Refer to Fig. 14.3. By employing exhaust gases to heat the feed water it is possible to utilise the entire steam supply to the turbines to expand through entire range of expansion, and thus result in increasing the output of work, since bleeding would not be required.

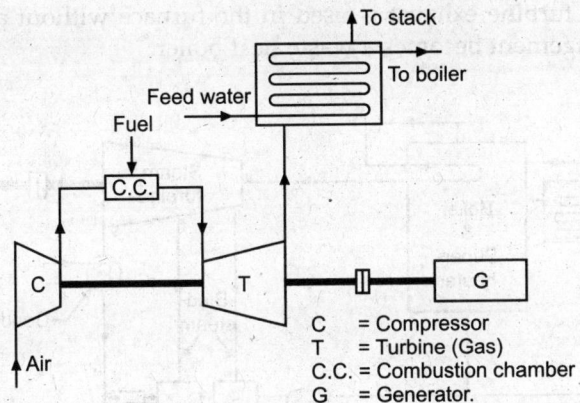

Fig. 14.3. Use of exhaust gases to heat feed water of steam cycle.

2. Employing the gases from a supercharged boiler to expand in the gas turbine :

Refer to Fig. 14.4. The supercharged boiler is good application of the combined gas turbine-steam cycle. In this arrangement the boiler furnace works under a pressure of about 5 bar and the gases are expanded in the gas turbine, its exhaust being used to heat feed water before being discharged through the stack. The *heat transfer rate in this boiler is very high as compared to that in a conventional boiler due to higher pressure of gases; and a smaller size of steam generator is needed* for the same steam raising capacity as of the conventional plant. Furthermore, since the gases in the furnace are already under pressure, *no induced draught or forced draught fans are needed* and there is saving in power consumption which would otherwise be spent in mechanical draught supply. Through this combination an *overall improvement in heat rate is to the extent of above 7 percent.*

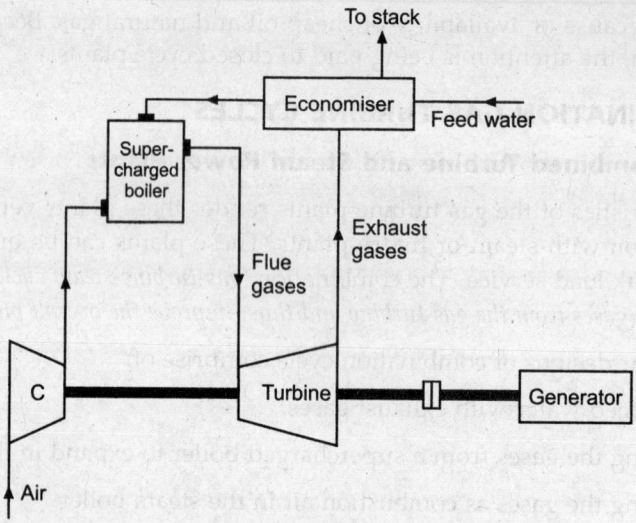

Fig. 14.4. Arrangement of supercharged boiler.

3. **Employing the gases as combustion air in the steam boiler :**

Refer to Fig. 14.5. When exhaust gases are used as preheated air for combustion in the boiler, an improvement of about 5 percent in overall heat rate of the plant results. The boiler is fed with supplementary fuel and air, and is made larger than the conventional furnace. If only the turbine exhaust is used in the furnace without any supplementary fuel firing, the arrangement becomes a waste heat boiler.

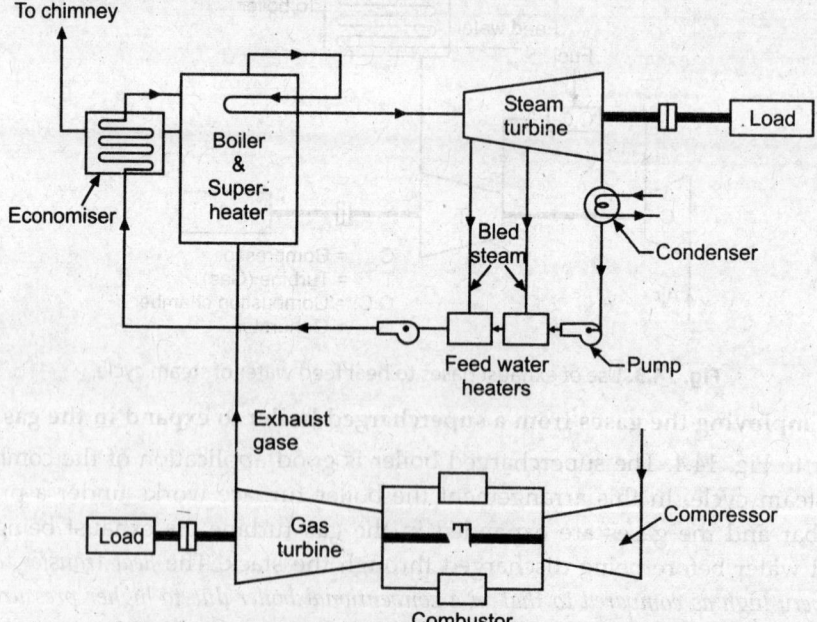

Fig. 14.5. Use of exhaust gases for combustion in the furnace of the steam plant.

- Fig. 14.6 shows the gain in heat rate due to combination cycle.
- Fig. 14.7 shows the comparison of a steam and closed cycle gas plant.

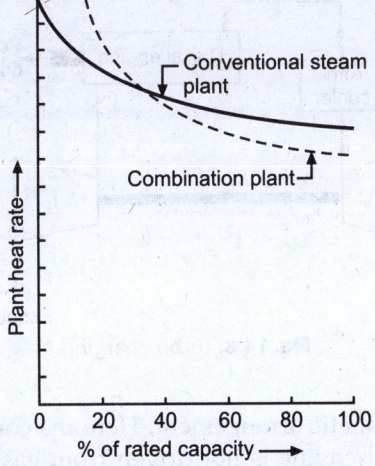

Fig. 14.6

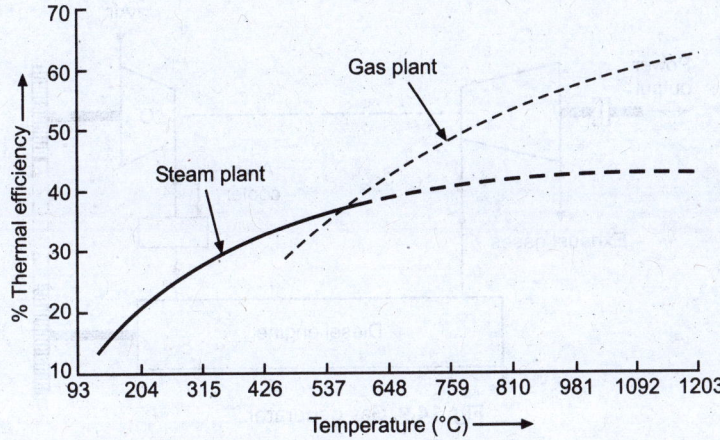

Fig. 14.7. Comparison between steam and closed cycle gas turbine plant.

14.10.2. Combined Gas Turbine and Diesel Power Plants

The performance of a diesel engine can be improved by combining it with an exhaust driven gas turbine. It can be achieved by the following *three combinations* :

1. Turbo-charging.

2. Gas-generator.

3. Compound engine.

1. Turbo-charging :

Refer to Fig. 14.8. This method is known as *supercharging*. Here the exhaust of the diesel engine is expanded in the gas turbine and the work output of the gas turbine is utilised to run a compressor which supplies the pressurised air to the diesel engine to increase its output. The load is coupled to the diesel engine shaft and the output of the gas turbine is just sufficient to run the compressor.

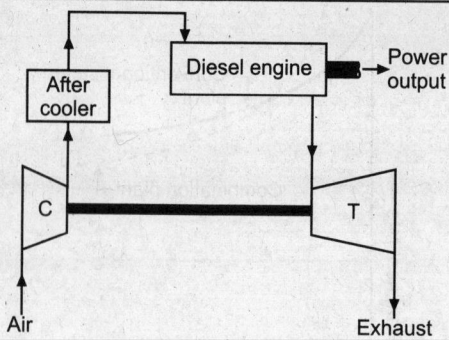

Fig. 14.8. Turbo-charging.

2. Gas-generator :

Fig. 14.9 shows the schematic arrangement. Here the compressor which supplies the compressed air to the diesel engine is not driven from gas turbine but from the diesel engine through some suitable drive. The output of the diesel engine is consumed in driving the air compressor and the gas turbine supplies the power.

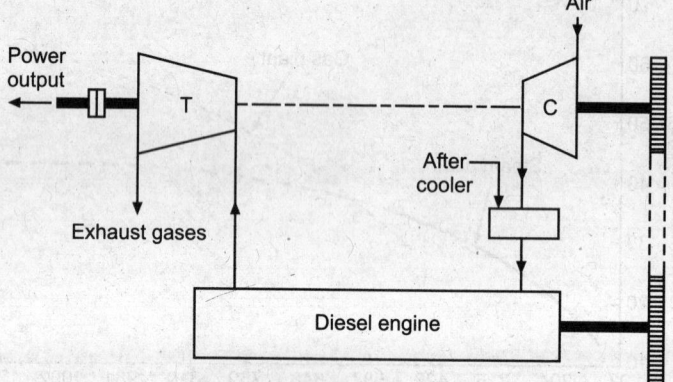

Fig. 14.9. Gas-generator.

3. Compound engine :

Refer to Fig. 14.10. In this arrangement the air compressor is driven from both diesel engine and gas turbine through a suitable gearing and the power output is taken from the diesel engine shaft.

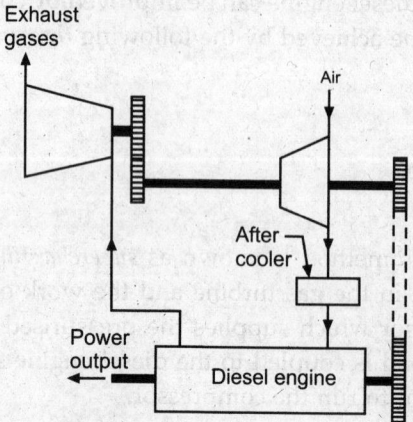

Fig. 14.10. Compound engine.

14.10.3. Advantages of Combined Cycle

The *advantages of combined cycle* are :

1. Improved efficiency.
2. More suitable for rapid start and shut down.
3. Less cooling water requirement.
4. It gives high ratio of power output to occupied ground space.
5. The combined system offers self-sustaining feature.
6. The capital cost of combined plant with supplementary firing is slightly higher than a simple gas turbine plant and much below those of a classical steam plant of the same power capacity.

14.11. OPERATION OF A GAS TURBINE

The *operation of a gas turbine* includes the following :

1. Starting.
2. Shut down.

1. Starting of a gas turbine :

Starting of a gas turbine power plant requires an *auxiliary power source,* till the plant's own compressor inducts air and compresses it to a pressure such that expansion from reasonable temperature will develop enough power to sustain operation. The starter may be (*i*) an I.C. engine (*ii*) a steam turbine (*iii*) an auxiliary electric motor or (*iv*) another gas turbine.

It must be coupled to the turbo-compressor shaft with a disengaging or over-running clutch. A main generator or its direct connected exciter may be pressed into temporary service as a motor.

Starting procedure :

1. Run the unit and induct air.
2. Actuate the combustion ignition system and inject the fuel. The fuel flow is controlled to obtain the necessary warm up.
3. Adjust the speed and voltage and synchronise the alternator.
4. Build up the load on the alternator by governor gear control.

2. Shut down of a gas turbine :

Shut down of a gas turbine occurs very quickly after the fuel is cut off from the combustor. A rapid shut down is desirable since it minimizes the *chilling effect* resulting from passing cold air through the hot turbine. Turning gears are usually employed, or alternatively the starting device may be operated at reduced speed to ensure symmetrical cooling of rotor. This avoids thermal *'kinking'* of the shaft.

14.12. GAS TURBINE POWER PLANT LAYOUT

The layout of a gas turbine plant exercises a very important effect on the overall performance of the plant, as it is possible to incur a loss to the extent of 20% of the power developed in the interconnecting ducts having a considerable number of sharp bends. Therefore ample care should be taken in the design and layout of the air as well as gas circuits. Fig. 14.11 shows a schematic simplified diagram of a gas turbine power plant layout. The following points are *worth noting* :

- In this case the main building is the turbine house in which major portion of the plant as well as auxiliaries are installed. Usually it is similar to the steam plant house in many respects.

- The storage tanks containing fuel oil are arranged outside but adjoining the turbine house. In some installations even heat exchangers are also out of the doors.

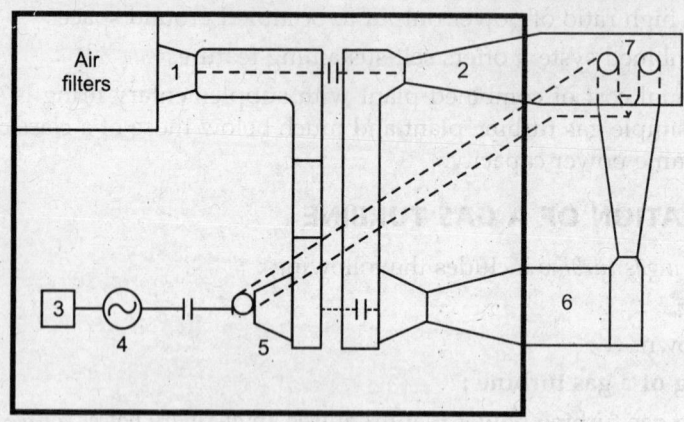

1 = L.P. compressor   2 = L.P. turbine
3 = Starting motor   4 = Alternator
5 = H.P. compressor   6 = Combustion chamber
7 = Heat exchangers

Fig. 14.11. Gas turbine power plant layout.

- Whereas the major portion of the total space is occupied by the intercoolers, combustion chambers, heat exchangers, waste heat boilers and interconnecting duct work, the rotating parts of the plant form a very small part of the total volume of the plant.

14.13 VARIOUS ARRANGEMENTS OF GAS TURBINE POWER PLANTS

The various arrangements of gas turbine plants are shown in Figs. 14.12 to 14.17.

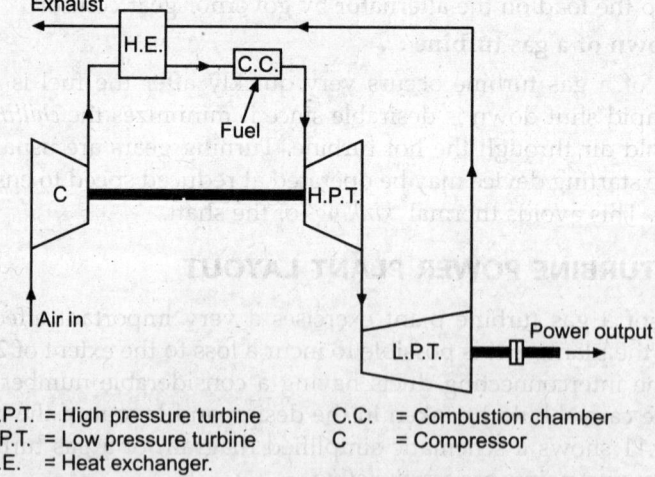

H.P.T. = High pressure turbine   C.C. = Combustion chamber
L.P.T. = Low pressure turbine   C = Compressor
H.E. = Heat exchanger.

Fig. 14.12. Open cycle gas turbine with separate power turbine.

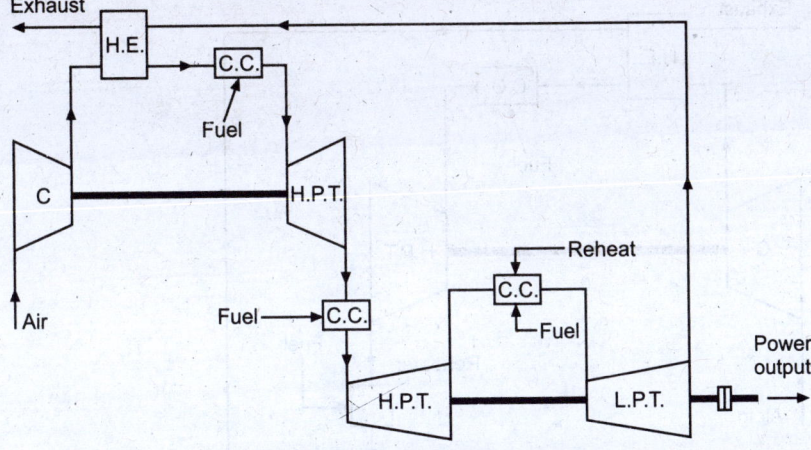

Fig. 14.13. Series flow gas turbine plant.

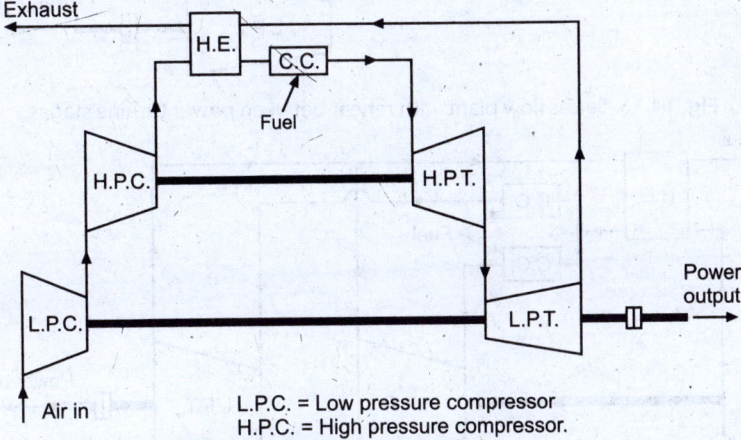

L.P.C. = Low pressure compressor
H.P.C. = High pressure compressor.

Fig. 14.14. Parallel flow gas turbine plant.

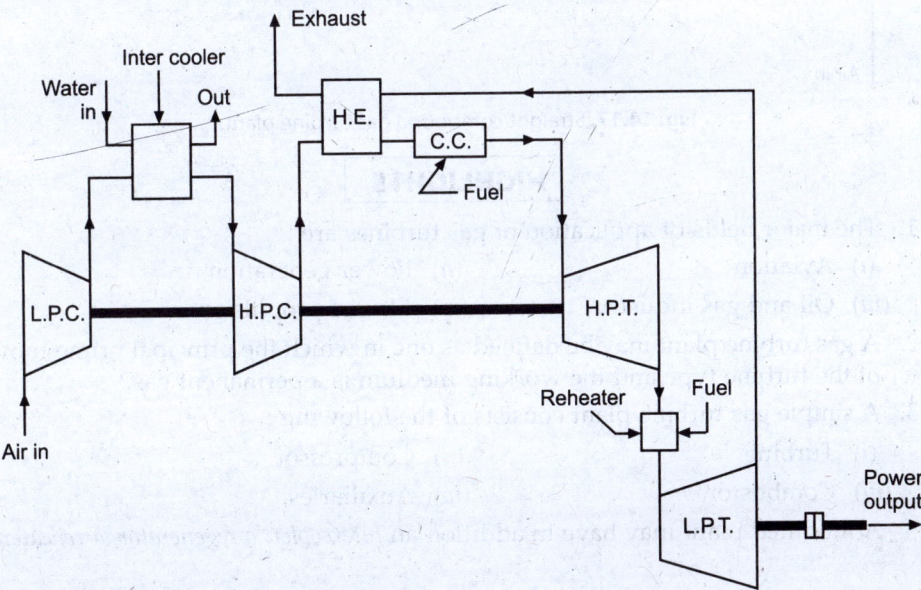

Fig. 14.15. Series flow plant with intercooled compression.

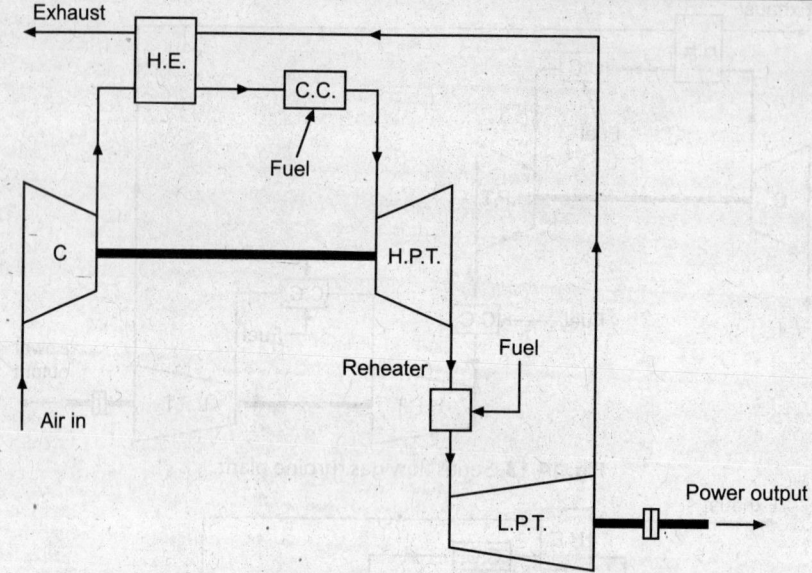

Fig. 14.16. Series flow plant with reheat between power turbine stages.

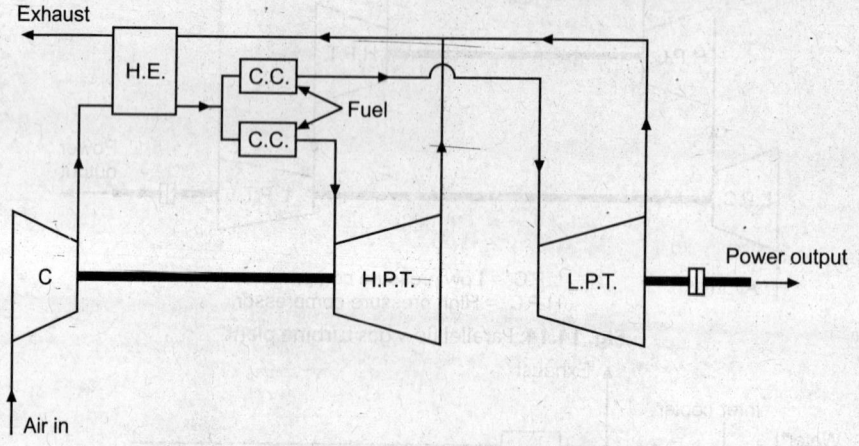

Fig. 14.17. Straight compound gas turbine plant.

HIGHLIGHTS

1. The major fields of application of gas turbines are :
   - (*i*) Aviation
   - (*ii*) Power generation
   - (*iii*) Oil and gas industry
   - (*iv*) Marine propulsion.

2. A gas turbine plant may be defined as one in which the principal prime-mover is of the turbine type and the working medium is a permanent gas.

3. A simple gas turbine plant consists of the following :
   - (*i*) Turbine
   - (*ii*) Compressor
   - (*iii*) Combustor
   - (*iv*) Auxiliaries.

   A modified plant may have in addition an *intercooler, a regenerator, a reheater etc.*

4. Methods for improvement of thermal efficiency of open cycle gas turbine plant are :

   (*i*) Intercooling   (*ii*) Reheating

   (*iii*) Regeneration.

5. *Free-piston engine plants* are the conventional gas turbine plants with the difference that the air compressor and combustion chamber are replaced by a free piston engine.

THEORETICAL QUESTIONS

1. What are the major fields of application of gas turbine ?
2. State the limitations of gas turbines.
3. List the applications of gas turbine plants.
4. State the advantages and disadvantages of gas turbine power plants over diesel and thermal power plants.
5. What factors should be considered while selecting a site for a gas turbine power plant ?
6. Give the description of a simple gas turbine plant.
7. Explain with the help of a neat diagram the energy cycle for a simple-cycle gas turbine.
8. Define the following performance terms :

   Air ratio, Pressure ratio, Work ratio, Compressor efficiency, Engine efficiency, Machine efficiency, Combustion efficiency and Thermal efficiency.
9. How are gas turbine power plants classified ?
10. How are gas turbines classified ?
11. What do you mean by "combination gas turbine cycles". Explain briefly combined gas turbine and steam power plants.
12. List the advantages of 'combined cycle'.
13. How is a gas turbine 'started' and 'shut down' ?
14. Explain with a neat sketch the layout of a gas turbine power plant.
15. Enumerate and explain briefly the components of a gas turbine power plant.

CHAPTER
15

HYDRO-ELECTRIC POWER PLANT

15.1. Introduction; 15.2 Application of Hydro-electric plants; 15.3 Advantages and disadvantages of hydro-electric plants; 15.4 Selection of site for a hydro-electric plant; 15.5 Essential features/elements of hydro-electric power plant; 15.6 Classification of hydro-electric power plants – High head power plants – Medium head power plants – Low head power plants – Base load plants – Peak load plants – Run-of-river plants without pontage – Run-of-river plant with pondage – Storage type plants – Pump storage plants – Mini and Micro–hydel plants; 15.7. Hydraulic turbines – Classification of hydraulic turbines; 15.8. Combined hydro and steam power plants; 15.9 Comparison of hydro-power stations with thermal power stations; 15.10 Calculation of available hydro-power; 15.11 Cost of hydro-power; 15.12 Hydrology – Introduction – The hydrological cycle – Measurement of run-off hydrograph – Flow duration curve – Mass curve. Highlights – Theoretical Questions.

15.1. INTRODUCTION

In hydro-electric plants energy of water is utilised to move the turbines which in turn run the electric generators. The energy of water utilised for power generation may be kinetic or potential. The *kinetic energy* of water is its energy in motion and is a function of mass and velocity, while the *potential energy* is a function of the difference in level/head of water between two points. In either case continuous availability of a water is a basic necessity ; to ensure this, water collected in natural lakes and reservoirs at high altitudes may be utilised or water may be artificially stored by constructing dams across flowing streams. The ideal site is one in which a good system of natural lakes with substantial catchment area, exists at a high altitude. *Rainfall is the primary source of water* and depends upon such factors as temperature, humidity, cloudiness, wind etc. The usefulness of rainfall for power purposes further depends upon several complex factors which include its intensity, time distribution, topography of land etc. However, it has been observed that only a small part of the rainfall can actually be utilised for power generation. A significant part is accounted for by *direct evaporation*, while another similar quantity *seeps* into the soil and forms the underground storage. Some water is also absorbed by vegetation. Thus, only a part of water falling as rain actually flows over the ground surface as direct run off and forms the streams which can be utilised for hydro-schemes.

- First hydro-electric station was probably started in America in 1882 and thereafter development took place very rapidly. In India the first major hydro-electric development of 4.5 MW capacity named as Sivasamudram Scheme in Mysore was commissioned in 1902. In 1914 a hydropower plant named Khopali project of 50 MW capacity was commissioned in Maharashtra. The hydro-power capacity, upto 1947, was nearly 500 MW.

- Hydro (water) power is a conventional renewable source of energy which is clean, free from pollution and generally has a good environmental effect. However the following factors are *major obstacles in the utilisation of hydro-power resources*:

  (*i*) Large investments   (*ii*) Long gestation period

  (*iii*) Increased cost of power transmission.

Next to thermal power, hydro-power is important in regard to power generation. The hydro-electric power plants provide 30 per cent of the total power of the world. The total hydro-potential of the world is about 5000 GW. In some countries (like Norway) almost total power generation is hydrobased.

15.2 APPLICATION OF HYDRO-ELECTRIC PLANTS

Earlier hydro-electric plants have been used as exclusive source of power, but the trend is towards use of hydropower in an *interconnected system with thermal stations*. As a self-contained and independent power source, a hydro-plant is most effective with adequate storage capacity, otherwise the maximum load capacity of the station has to be based on minimum flow of stream and there is a great wastage of water over the dam for greater part of the year. This increases the per unit cost of installation. By interconnecting hydro-power with steam, a great deal of *saving in cost can be effected due to*:

  (*i*) Reduction in necessary reserve capacity,

  (*ii*) Diversity in construction programmes,

  (*iii*) Higher utilisation factors on hydroplants, and

  (*iv*) Higher capacity factors on efficient steam plants.

- In an interconnected system the base load is supplied by hydropower when the maximum flow demand is less than the stream flow while steam supplies the peak. When stream flow is lower than the maximum demand the hydroplant supplies the peak load and steam plant the base load.

15.3 ADVANTAGES AND DISADVANTAGES OF HYDRO-ELECTRIC PLANTS

Advantages of hydro-electric plant:

1. No fuel charges.
2. An hydro-electric plant is highly reliable.
3. Maintenance and operation charges are very low.
4. Running cost of the plant is low.
5. The plant has no stand by losses.
6. The plant efficiency does not change with age.
7. It take a few minutes to run and synchronise the plant.
8. Less supervising staff is required.
9. No fuel transportation problem.
10. No ash problem and atmosphere is not polluted since no smoke is produced in the plant.
11. In addition to power generation these plants are also used for flood control and irrigation purposes.
12. Such a plant has comparatively a long life (100-125 years as against 20-45 years of a thermal plant).
13. The number of operations required is considerably small compared with thermal power plants.

14. The machines used in hydro-electric plants are more robust and generally run at low speeds at 300 to 400 r.p.m. where the machines used in thermal plants run at a speed 3000 to 4000 r.p.m. Therefore, there are no specialised mechanical problems or special alloys required for construction.

15. The cost of land is not a major problem since the hydro-electric stations are situated away from the developed areas.

Disadvantages:

1. The initial cost of the plant is very high.

2. It takes considerable long time for the erection of such plants.

3. Such plants are usually located in hilly areas far away from the load centre and as such they require long transmission lines to deliver power, subsequently the cost of transmission lines and losses in them will be more.

4. Power generation by the hydro-electric plant is only dependent on the quantity of water available which in turn depends on the natural phenomenon of rain. So if the rainfall is in time and proper and the required amount of water can be collected, the plant will function satisfactorily otherwise not.

15.4. SELECTION OF SITE FOR A HYDRO-ELECTRIC PLANT

The following *factors* should be considered while selecting the site for a hydro-electric plant:

1. Availability of water   2. Water storage
3. Water head   4. Accessibility of the site
5. Distance from load centre   6. Type of the land of site.

1. Availability of water:

The most important aspect of hydro-electric plant is the availability of water at the site since all other designs are based on it. Therefore the run-off data at the proposed site must be available before hand. It may not be possible to have run-off data at the proposed site but data concerning the rainfall over the large catchment area is always available. Estimate should be made about the average quantity of water available throughout the year and also about maximum and minimum quantity of water available during the year. These details are necessary to:

(i) decide the capacity of the hydro-electric plant,

(ii) setting up of peak load plant such as steam, diesel or gas turbine plant and to,

(iii) provide adequate spillways or gate relief during the flood period.

2. Water storage:

Since there is a wide variation in rainfall during the year, therefore, it is always necessary to store the water for continuous generation of power. The storage capacity can be calculated with the help of mass curve. Maximum storage should justify the expenditure on the project.

The *two types of storages in use* are:

(i) The storage is so constructed that it can make water available for power generation of one year only. In this case storage becomes full in the beginning of the year and becomes empty at the end of each year.

(ii) The storage is so constructed that water is available in sufficient quantity even during the worst dry periods.

3. Water head:

In order to generate a requisite quantity of power it is necessary that a large quantity of water at a *sufficient head* should be available. An increase in effective head, for a given output, reduces the quantity of water required to be supplied to the turbines.

4. Accessibility of the site:

The site where hydro-electric plant is to be constructed should be easily accessible. This is important if the electric power generated is to be utilised at or near the plant site. The site selected should have transportation facilities of rail and road.

5. Distance from the load centre:

It is of paramount importance that the power plant should be set up *near the load centre*; this will *reduce the cost of erection and maintenance of transmission line*.

6. Type of the land of the site:

The land to be selected for the site should be cheap and rocky. The ideal site will be one where the dam will have largest catchment area to store water at high head and will be economical in construction.

The necessary requirements of the foundation rocks for a masonry dam are as follows:

(*i*) The rock should be strong enough to withstand the stresses transmitted from the dam structure as well as the thrust of the water when the reservoir is full.

(*ii*) The rock in the foundation of the dam should be reasonably impervious.

(*iii*) The rock should remain stable under all conditions.

15.5. ESSENTIAL FEATURES/ELEMENTS OF HYDRO-ELECTRIC POWER PLANT

The following are the *essential elements of hydro-electric power plant:*

1. Catchment area	2. Reservoir	3. Dam
4. Spillways	5. Conduits	6. Surge tanks
7. Prime movers	8. Draft tubes	9. Power-house and equipment.

Fig. 15.1 shows the flow sheet of hydro-electric power plant.

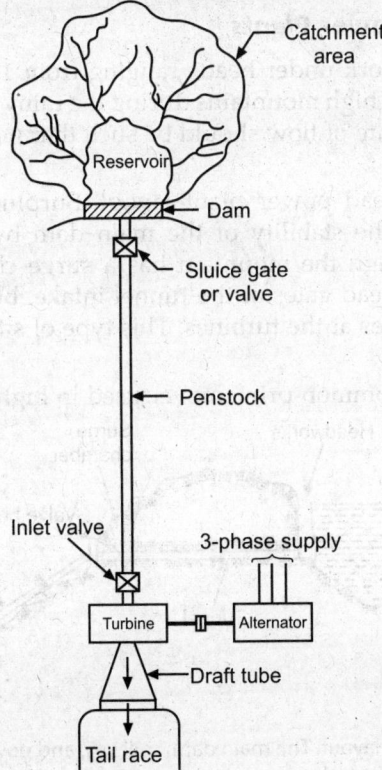

Fig. 15.1. Flow sheet of hydro-electric power plant.

15.6. CLASSIFICATION OF HYDRO-ELECTRIC POWER PLANTS

Hydro-electric power stations may be *classified* as follows:

A. According to availability of head:

1. High head power plants   2. Medium head power plants
3. Low head power plants.

B. According to the nature of load:

1. Base load plants   2. Peak load plants.

C. According to the quantity of water available:

1. Run-of-river plant without pondage
2. Run-of-river plant with pondage
3. Storage type plants
4. Pump storage plants
5. Mini and micro-hydel plants.

A. According to availability of head:

The following figures give a rough idea of the heads under which the various types of plants work:

(*i*) High head power plants	100 m and above
(*ii*) Medium head power plants	30 to 100 m
(*iii*) Low head power plants	25 to 80 m.

Note: It may be noted that figures given above overlap each other. Therefore it is difficult to classify the plants directly on the basis of head alone. The basis, therefore, technically adopted is the *specific speed* of the turbine used for a particular plant.

15.6.1. High Head Power Plants

These types of plants work under heads ranging from 100 to 2000 metres. Water is usually stored up in lakes on high mountains during the rainy season or during the season when the snow melts. The rate of flow should be such that water can last throughout the year.

Fig. 15.2 shows high head power plant layout. Surplus water discharged by the spillway cannot endanger the stability of the main dam by erosion because they are separated. The tunnel through the mountain has a surge chamber excavated near the exit. Flow is controlled by head gates at the tunnel intake, butterfly valves at the top of the penstocks, and gate valves at the turbines. This type of site might also be suitable for an underground station.

The *Pelton wheel* is the common primemover used in high *head power plants*.

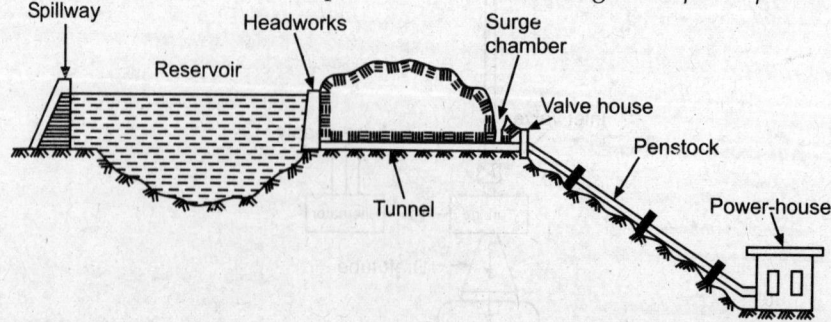

Fig. 15.2. High head power plant layout. The main dam, spillway, and power-house stand at widely separated locations. Water flows from the reservoir through a tunnel and penstocks to the turbines.

15.6.2. Medium Head Power Plants

Refer to Fig. 15.3. When the operating head of water lies between 30 to 100 metres, the power plant is known as medium head power plant. This type of plant commonly uses *Francis turbines*. The forebay provided at the beginning of the penstock serves as the water reservoir. In such plants, the water is generally carried in open canals from main reservoir to the forebay and then to the power-house through the penstock. The forebay itself works as a surge tank in this plant.

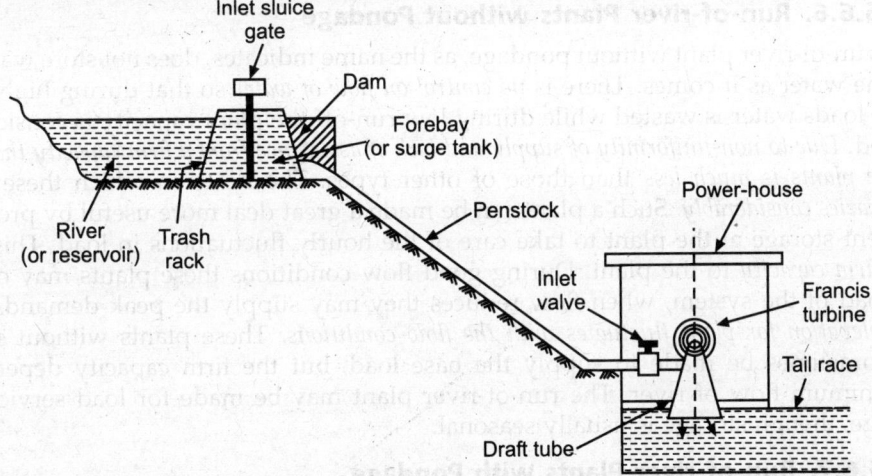

Fig. 15.3. Medium head power plant layout.

15.6.3. Low Head Power Plants

Refer to Fig. 15.4. These plants usually consist of a dam across a river. A side way stream diverges from the river at the dam. Over this stream the power-house is constructed. Later this channel joins the river further downstream. This type of plant uses vertical shaft Francis turbine or Kaplan turbine.

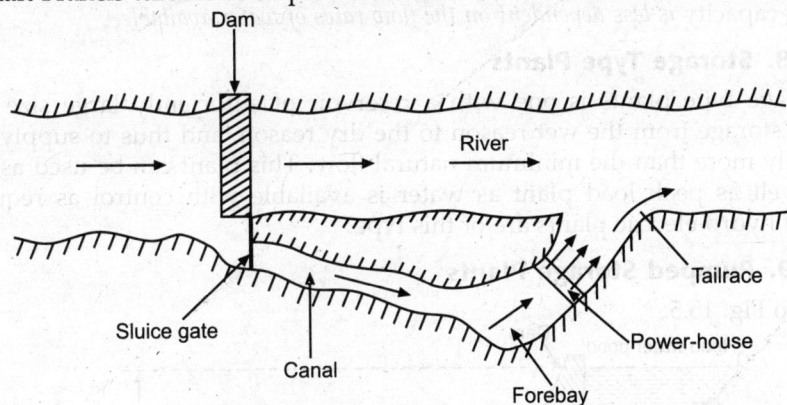

Fig. 15.4. Low head power plant layout.

B. *According to nature of load:*

15.6.4. Base Load Plants

The plants which cater to the base load of the system are called **base load plants.** These plants are required to *supply a constant power when connected to the grid.* Thus they *run without stop* and are often remote-controlled with which *least staff is required for such* plants. Run-of-river plants without pondage may sometimes work as base load plant, but the firm capacity in such cases, will be much less.

15.6.5. Peak Load Plants

The plants which can supply the power during peak loads are known as **peak load plants.** Some of such plants supply the power during average load but also supply peak load as and when it is there ; whereas other peak load plants are required to work during peak load hours only. The run-of-river plants may be made for the peak load by providing pondage.

C.   According to the quantity of water available:

15.6.6. Run-of-river Plants without Pondage

A run-of-river plant without pondage, as the name indicates, does not store water and uses the water as it comes. There is *no control on flow of water* so that during high floods or low loads water is wasted while during low run-off the plant capacity is considerably reduced. *Due to non-uniformity of supply and lack of assistance from a firm capacity the utility of these plants is much less* than those of other types. The head on which these plants work *varies considerably.* Such a plant can be made a great deal more useful by providing sufficient storage at the plant to take care of the hourly fluctuations in load. This lends some *firm capacity* to the plant. During good flow conditions these plants may cater to base load of the system, when flow reduces they may supply the peak demands. *Head water elevation for plant fluctuates with the flow conditions.* These plants without storage may sometimes be made to supply the base load, but the firm capacity depends on the minimum flow of river. The run-of-river plant may be made for load service with pondage, though storage is usually seasonal.

15.6.7. Run-of-river Plants with Pondage

Pondage usually refers to the collection of water behind a dam at the plant and increases the stream capacity for a short period, say a week. *Storage* means collection of water in up stream reservoirs and this increases the capacity of the stream over an extended period of several months. Storage plants may work satisfactorily as base load and peak load plants.

This type of plant, as compared to that without pondage, is *more reliable* and its generating capacity *is less dependent on the flow rates of water available.*

15.6.8. Storage Type Plants

A storage type plants is one with a reservoir of sufficiently large size to permit carry-over storage from the wet reason to the dry reason, and thus to supply firm flow substantially more than the minimum natural flow. This plant can be used as base load plant as well as peak load plant as water is available with control as required. The majority of hydro-electric plants are of this type.

15.6.9. Pumped Storage Plants

Refer to Fig. 15.5.

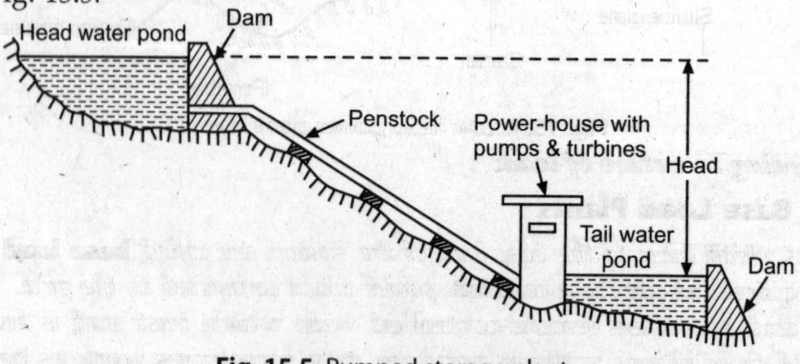

Fig. 15.5. Pumped storage plant.

Pumped storage plants are employed at the places where the quantity of water available for power generation is *inadequate*. Here the water passing through the turbines is stored in *'tail race pond'*. During low load periods this water is pumped back to the head reservoir using the extra energy available. This water can be again used for generating power during peak load periods. Pumping of water may be done seasonally or daily depending upon the conditions of the site and the nature of the load on the plant.

Such plants are *usually interconnected* with steam or diesel engine plants so that off peak capacity of interconnecting stations is used in pumping water and the same is used during peak load periods. Of course, the energy available from the quantity of water pumped by the plant is *less* than the energy input during pumped operation. Again while using pumped water the *power available is reduced* on account of losses occuring in primemovers.

Advantages:

The pump storage plants entail the following *advantages:*

1. There is substantial increase in peak load capacity of the plant at comparatively low capital cost.
2. Due to load comparable to rated load on the plant, the operating efficiency of the plant is high.
3. There is an improvement in the load factor of the plant.
4. The energy available during peak load periods is higher than that of during off peak periods so that inspite of losses incurred in pumping there is *overall gain*.
5. Load on the hydro-electric plant remains uniform.
6. The hydro-electric plant becomes partly independent of the stream flow conditions.

Under pump storage projects almost 70 percent power used in pumping the water can be recovered. In this field the use of *"Reversible Turbine Pump"* units is also worth noting. These units can be used as turbine while generating power and as pump while pumping water to storage. The generator in this case works as motor during reverse operation. The efficiency in such case is high and almost the same in both the operations. With the use of reversible turbine pump sets, additional capital investment on pump and its motor can be saved and the scheme can be worked more economically.

15.6.10. Mini and Micro-hydel Plants

In order to meet with the present energy crisis partly, a solution is to develop *mini* (5 m to 20 m head) and *micro* (less than 5 m head) hydel potential in our country. The low head hydro-potential is scattered in this contry and estimated potential from such sites could be as much as 20,000 MW.

By proper planning and implementation, it is possible to commission a small hydro-generating set up of 5 MW within a period of one and half year against the period of a decade or two for large capacity power plants. Several such sets upto 1000 kW each have been already installed in Himachal Pradesh, U.P., Arunachal Pradesh, West Bengal and Bhutan.

To reduce the cost of micro-hydel stations than that of the cost of conventional installation the following *considerations are kep in view:*

1. The civil engineering work needs to be kept to a *minimum* and designed to fit in with already existing structures *e.g.,* irrigation, channels, locks, small dams etc.
2. The machines need to be manufactured in a small range of size of simplified design, allowing the use of unified tools and aimed at reducing the cost of manufacture.

3. These installations must be automatically controlled to reduce attending personnel.

4. The equipment must be simple and robust, with easy accessibility to essential parts for maintenance.

5. The units must be light and adequately subassembled for ease of handling and transport and to keep down erection and dismantling costs.

Micro-hydel plants (micro-stations) *make use of standardised bulb sets with unit output ranging from 100 to 1000 kW working under heads between 1.5 to 10 metres.*

15.7. HYDRAULIC TURBINES

A hydraulic turbine converts the potential energy of water into mechanical energy which in turn is utilised to run an electric generator to get electric energy.

15.7.1. Classification of Hydraulic Turbines

The hydraulic turbines are *classified* as follows:

(*i*) According to the head and quantity of water available.

(*ii*) According to the name of the originator.

(*iii*) According to the action of water on the moving blades.

(*iv*) According to the direction of flow of water in the runner.

(*v*) According to the disposition of the turbine shaft.

(*vi*) According to the specific speed N_s.

1. According to the head and quantity of water available:

(*i*) *Impulse turbine*—requires *high head and small quantity of flow.*

(*ii*) *Reaction turbine*—requires *low head and high rate of flow.*

Actually there are two types of reaction turbines, one for medium head and medium flow and the other for low head and large flow.

2. According to the name of the originator:

(*i*) *Pelton turbine*–named after Lester Allen Pelton of California (USA). It is an impulse type of turbine and is used for *high head* and *low discharge.*

(*ii*) *Francis turbine*–named after James Bichens Francis. It is a reaction type of turbine for *medium high to medium low heads* and *medium small to medium large quantities of water.*

(*iii*) *Kaplan turbine*–named after Dr. Victor Kaplan. It is a *reaction type of turbine for low heads and large quantities of flow.*

3. According to action of water on the moving blades:

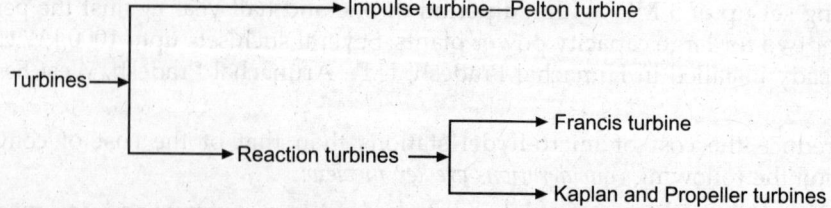

4. According to direction of flow of water in the runner:

(*i*) Tangential flow turbine (Pelton turbine)

(*ii*) Radial flow turbine (no more used)

(*iii*) Axial flow turbine (Kaplan turbine)

(*iv*) Mixed (radial and axial) flow turbine (Francis turbine).

In *tangential flow* turbine of Pelton type the water strikes the runner tangential to the path of rotation.

In *axial flow* turbine water flows parallel to the axis of the turbine shaft. Kaplan turbine is an axial flow turbine. In Kaplan turbine the runner blades are *adjustable and can be rotated* about pivots fixed to the boss of the runner. If the runner blades of the axial flow turbines are *fixed, these are called* "Propeller turbines".

In *mixed flow* turbines the water enters the blades radially and comes out axially, parallel to the turbine shaft. Modern Francis turbines have mixed flow runners.

5. According to the disposition of the turbine shaft:

Turbine shaft may be either vertical or horizontal. In modern turbine practics, Pelton turbines usually have horizontal shafts whereas the rest, especially the large units, have vertical shafts.

6. According to specific speed:

The *specific speed* of a turbine is defined as the speed of a geometrically similar turbine that would develop *one brake horsepower* under the *head of one metre*. All geometrically similar turbines (irrespective of their sizes) will have the same specific speed when operating under the same conditions of head and flow.

Specific speed $\qquad N_s = \dfrac{N\sqrt{P_t}}{H^{5/4}}$

where, $\qquad\qquad\quad N =$ The normal working speed in r.p.m.,

$\qquad\qquad\qquad P_t =$ Power output of the turbine, and

$\qquad\qquad\qquad H =$ The net or effective head in metres.

The specific speeds for the various types of runners are given below:

Type of turbine	Type of runner		Specific speed (N_s)
Pelton	⌈ Slow		10 to 20
	Normal		20 to 28
	⌊ Fast		28 to 35
			60 to 120
Francis	⌈ Slow		120 to 180
	Normal		180 to 300
	⌊ Fast		300 to 1000
Kaplan	—		

Turbines with low specific speeds work under a high head and low discharge condition, while high specific speed turbines work under low head and high discharge conditions.

15.7.2. Description of Various Types of Turbines

15.7.2.1. Impulse turbines

Pelton wheel, among the various impulse turbines that have been designed and utilized, is by far the important. The Pelton wheel or Pelton turbine is a *tangential flow impulse turbine*. It consists of a rotor, at the periphery of which are mounted equally spaced double-hemispherical or double-ellipsoidal buckets. Water is transferred from a high head source through penstock pipes. A branch pipe from each penstock pipe ends in a nozzle, through which the water flows out as a high speed jet. A needle or spear moving inside the nozzle controls the water flow through the nozzle and at the same time, provides a smooth flow with negligible energy loss. All the available *potential energy is thus converted into kinetic energy* before the jet strikes the buckets. The pressure all over the wheel is *constant and equal to atmosphere, so that energy transfer occurs due to purely impulse action.*

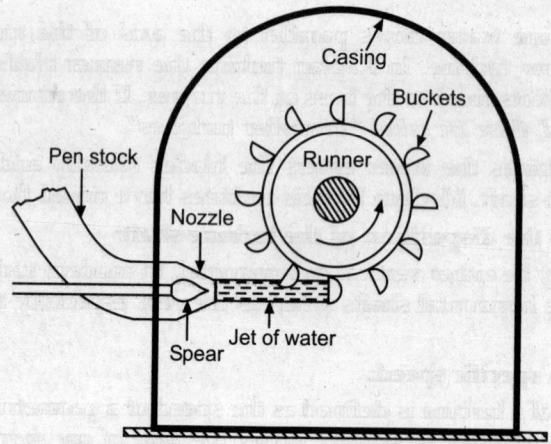

Fig. 15.6. Pelton wheel.

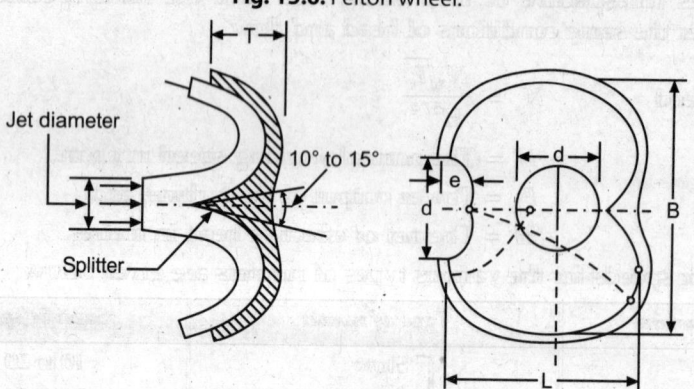

$\dfrac{L}{d}$ to 2.5 to 2.8, $\dfrac{D}{d} = 14$ to 16, $\dfrac{T}{d} = 0.95$, Notch (width) $= 1.1d + 5$ mm

Fig. 15.7. The bucket dimensions.

Fig. 15.6 shows a schematic diagram of a Pelton wheel, while Fig. 15.7 shows two views of its bucket.

The jet emerging from the nozzle hits the splitter symmetrically and is equally distributed into the two halves of hemispherical bucket as shown. *The bucket centre-line cannot be made exactly like a mathematical cusp, partly because of manufacturing difficulties and partly because the jet striking the cusp invariably carries particles of sand and other abrasive materials which tend to wear it down.* The inlet angle of the jet is therefore between 1° and 3°, but it is always assumed to be zero in all calculations. Theoretically, if the bucket were exactly hemispherical, it would deflect the jet through 180°. Then, the relative velocity of the jet leaving the bucket, C_{r2}, would be opposite in direction to the relative velocity of the entering jet C_{r1}. This cannot be achieved in practice since the jet leaving the bucket would then strike the *back of the succeeding bucket to cause splashing and interference so that the overall turbine efficiency would fall to low values.* Hence, in practice, the angular deflection of the jet in the bucket is limited to about 165° or 170°, and the *bucket is therefore slightly smaller than a hemisphere in size.*

15.7.2.2. Reaction turbines

In reaction turbines, the *runner utilizes both potential and kinetic energies*. As the water flows through the stationary parts of the turbine, whole of its pressure energy is not transferred to kinetic energy and when the water flows through the moving parts, there is a change both in the pressure and in the direction and velocity of flow of water. As the water gives up its energy to the runner, both its pressure and absolute velocity get reduced. The water which acts on the runner blades is under *a pressure above atmospheric and the runner passages are always completely filled with water.*

A. Francis turbines:

Refer to Fig. 15.8 (*a*), (*b*). The modern Francis water turbine is an *inward mixed flow reaction turbine i.e.,* the water under pressure, *enters the runner from the guide vanes towards the centre in radial direction and discharges out of the runner axially.* The Francis turbine operates under *medium heads* and also requires *medium quantity* of water. It is employed in the medium head power plants. This type of turbine covers a wide range of heads. Water is brought down to the turbine and directed to a number of stationary orifices fixed all around the circumference of the runner. These stationary orifices are commonly termed as *guide vanes or wicket gates.*

The head action on the turbine is partly transformed into kinetic energy and the rest remains as pressure head. There is a difference of pressure between the guide vanes and the runner which is called the *reaction pressure* and is responsible for the motion of the runner. That is why a Francis turbine is also known as *reaction turbine.*

In Francis turbine the pressure at the inlet is more than that at the outlet. This means that the water in the turbine must flow in a closed conduit. Unlike the Pelton type, where the water strikes only a few of the runner buckets at a time, in the Francis turbine the *runner is always full of water. The movement of the runner is affected by the change of both the potential and the kinetic energies of water.* After doing its work the water is discharged to the tail race through closed tube of gradually enlarging section. This tube is known as *draft tube.* It does not allow water to fall freely to tail race level as in the Pelton turbine. The free end of the draft tube is submerged deep in the tail water making, thus, the entire water passage, right from the head race upto the tail race, totally enclosed.

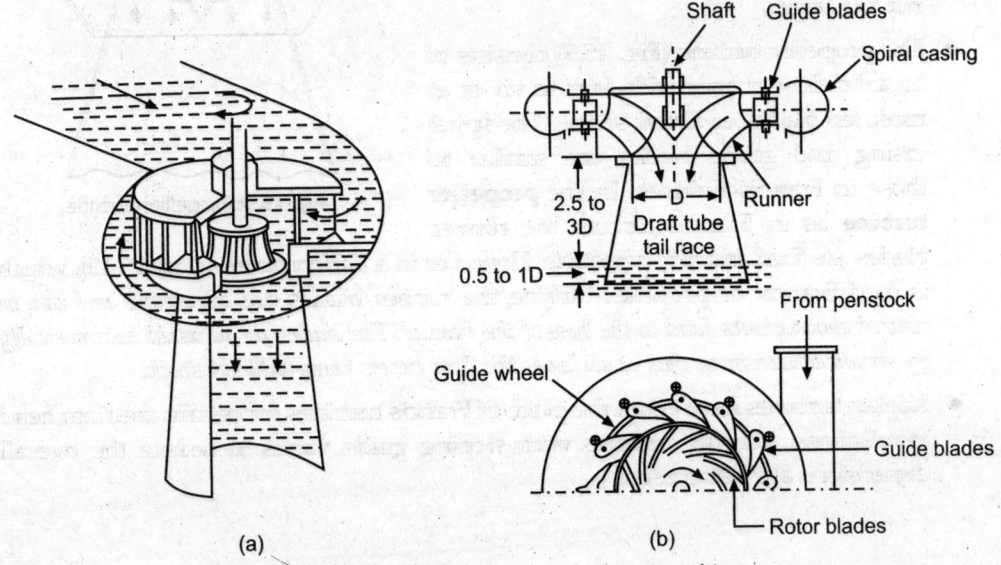

Fig. 15.8: Schematic diagram of a Francis turbine.

Draft tube:

The draft tube serves the following two purposes:

1. *It allows the turbine to be set above tail-water level, without loss of head, to facilitate inspection and maintenance.*

2. *It regains, by diffuser action, the major portion of the kinetic energy delivered to it from the runner.*

At rated load the velocity at the upstream end of the tube for modern units ranges from 7 to 9 m/s, representing from 2.7 to 4.8 m head. As the specific speed (it is the speed of a geometrically similar turbine running under a unit head and producing unit power) is increased and the head reduced, it becomes increasingly important to have an efficient draft tube. Good practice limits the velocity at the discharge end of the tube to 1.5 to 2.1 m/s, representing less than 0.3 m velocity head loss.

The following two types of *draft tubes* are commonly used:

(*i*) The straight conical or concentric tube.

(*ii*) The elbow type.

Properly designed, the two types are about equally efficient, over 85%.

B. Propeller and Kaplan turbines:

- The need to utilize low heads where large volumes of water are available makes it essential to provide a large flow area and to run the machine at very low speeds. The propeller turbine is a reaction turbine used for heads between 4 m and 80 m, and has a specific speed ranging from 300 to 1000. It is purely *axial-flow* device providing the largest possible flow area that will utilize a large volume of water and still obtain flow velocities which are not too large.

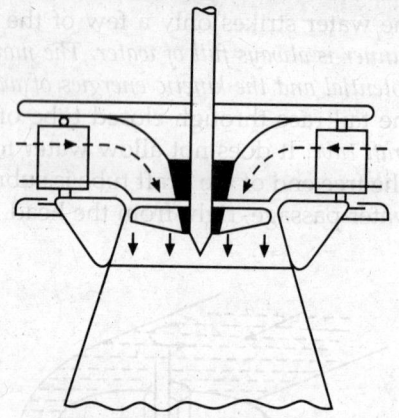

- The propeller turbine (Fig. 15.9) consists of an axial-flow runner with four to six or at most ten blades of airfoil shape. The spiral casing and guide blades are similar to those in Francis turbines. In the **propeller turbine** as in Francis turbine the runner

Fig. 15.9: Propeller turbine.

blades are *fixed* and *non-adjustable*. However in a *Kaplan turbine* (Fig. 15.10), which is modification of propeller turbine the runner blades are *adjustable and can be rotated about pivots fixed to the boss of the runner. The blades are adjusted automatically by servomechanism so that at all loads the flow enters them without shock.*

- Kaplan turbines have taken the place of Francis turbines for certain medium head installations. Kaplan turbines with sloping guide vanes to reduce the overall dimensions are being used.

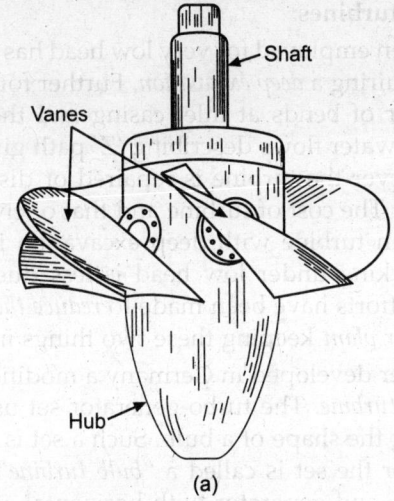

(a)

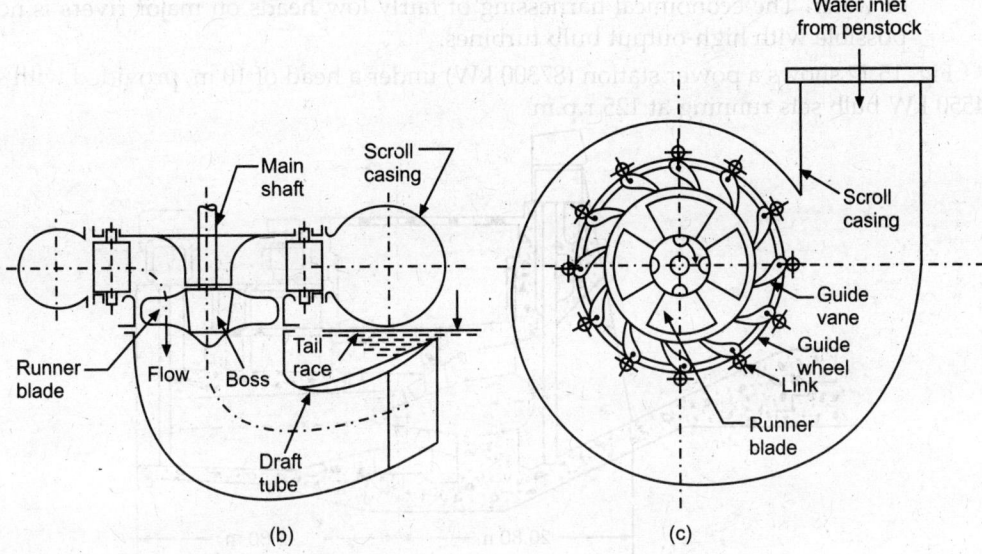

(b)   (c)

Fig. 15.10. Schematic diagram of a Kaplan turbine.

Fig. 15.11 shows the comparison of efficiencies of propeller (fixed blades) and Kaplan turbines.

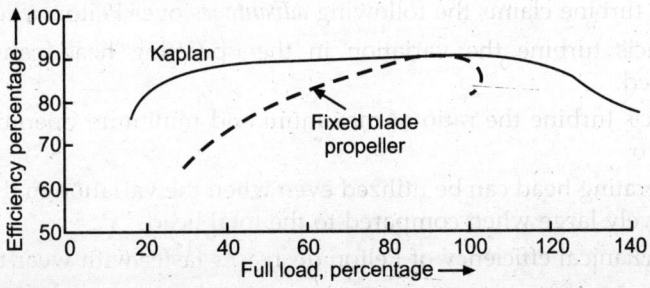

Fig. 15.11. Comparison of efficiencies of propeller (fixed blades) and Kaplan turbines.

C. Tubular (or Bulb) turbines:

- Kaplan turbine when employed for very low head has to be installed below the tail race level, thus requiring a *deep excavation.* Further for Kaplan turbine installation there are a number of bends at inlet casing and the draft tube of elbow type through which the water flows describing 'Z' path giving rise to *continuous losses at the bends.* Whenever the turbine is repaired or dismantled, the generator has to be removed first. The cost of turbine and that of civil engineering works using conventional Kaplan turbine with deep excavation is very high. The efficiency of such plants working under low head is less due to excessive losses at the bends. Therefore, efforts have been made to *reduce the overall cost* and *improve the efficiency of the power plant* keeping these two things in view.

- In 1937 Arno Fischer developed in Germany a modified axial flow turbine which is know as *tubular turbine.* The turbo-generator set using tubular turbine has an outer casing having the shape of a bulb. Such a set is now termed as *bulb set* and the turbine used for the set is called a *"bulb turbine".* The bulb is a water tight assembly of turbine and generator with horizontal axis, submerged in a stream of water. The economical harnessing of fairly low heads on major rivers is now possible with high-output bulb turbines.

Fig. 15.12 shows a power station (87300 kW) under a head of 10 m, provided with six 14550 kW bulb sets running at 125 r.p.m.

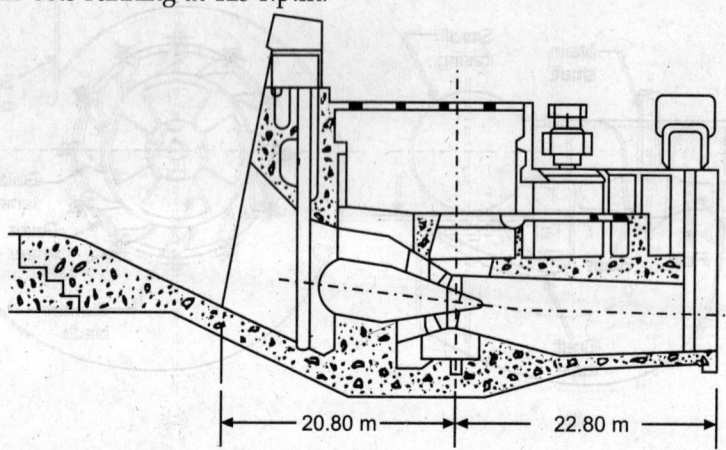

|← 20.80 m →|← 22.80 m →|

Fig. 15.12. Power station—using bulb turbines.

Comparison of hydraulic turbines:

I. *Francis turbine versus Pelton wheel:*

The Francis turbine claims the following *advantages* over Pelton wheel:

1. In Francis turbine the variation in the operating head can be more easily controlled.

2. In Francis turbine the ratio of maximum and minimum operating heads can be even two.

3. The operating head can be utilized even when the variation in the tail water level is relatively large when compared to the total head.

4. The mechanical efficiency of Pelton decreases faster with wear than Francis.

5. The size of the runner, generator and power-house required is small and economical if the Francis turbine is used instead of Pelton wheel for same power generation.

Drawbacks of Francis turbine:

As compared with Pelton wheel, the Francis turbine has the following *drawbacks:*

1. Water which is not clean can cause very rapid wear in high head Francis turbine.
2. The overhaul and inspection is much more difficult comparatively.
3. Cavitiation is an ever-present danger.
4. The water hammer effect is more troublesome with Francis turbine.
5. If Francis turbine is run below 50% head for a long period it will not only decrease its efficiency but also the cavitation danger will become more serious.

II. *Kaplan verses Francis turbine:*

Kaplan turbine claims the following *advantages* over Francis turbine:

1. For the same power developed Kaplan turbine is more compact in construction and smaller in size.
2. Part-load efficiency is considerably high.
3. Low frictional losses (because of small number of blades used).

15.8. COMBINED HYDRO AND STEAM POWER PLANTS

An electrical power system should fulfil the following objectives:

1. To ensure an *adequate and reliable electric power supply* at all loads and at all times.
2. The source of energy should be such as to give the *minimum overall cost of the system as a whole.*

The above objectives (unless a country/region is rich either in abundant supply of cheap fuel or ample water power resources which can be !eveloped at suitable site) can be best realised by a *judicious combination of both hydro and thermal power.* Hydro-power represents a renewable source of energy which enjoys many intrinsic advantages as compared to thermal-power. Although the cost of construction of hydro-power plant is nearly same as that of a coal based steam power plant in terms of investment for MW, but hydro-power plant uses water for power generation which is available in abundance in nature.

It is known that *hydro-plant can meet the demands of load variations more rapidly and easily.* Thus, *when the rate of flow of water is low, the steam plant can work at constant load producing a better efficiency and the hydro-plant will work most effectively as peak load plant and its output can be varied to meet the load fluctuations.* The steam and hydro-plants reverse their functions (*steam plant providing the peak load and the hydro-plant providing base load*) when high rate of water flow is available. But even under this condition, the *steam plant output will remain constant* and the hydro-plant output *will be varied to meet the load fluctuations.*

15.9. COMPARISON OF HYDRO-POWER STATIONS WITH THERMAL-POWER STATIONS

The comparison between hydro-power stations and thermal-power stations is given below:

S. No.	Aspects	Hydro-power station	Thermal-power station
1.	*Raw material consumption*	Nil. Water power is inexhaustible and perpetual and is continuously replenished by the direct agency of sun.	Huge quantity of coal consumed, therby exhausting "fuel reserves".
2.	*Cost of energy*	Cheaper	Costlier

3.	Cost of energy generation	Immune to inflation.	Very much influenced by the increase in the cost of fuel.
4.	Life of plant	Long useful life.	Not so long comparatively. The component parts deteriorate and become obsolete at a faster rate.
5.	Pollution	Free from problems of pollution.	Causes pollution and subsequently create health hazards.
6.	Design, construction and reliability	Simple in design, robust in construction and reliable in operation	Comparatively more complicated in design, less robust in construction and less reliable in operation.
7.	Running below a certain minimum load factor.	Can be run.	Cannot be run.
8.	Reserve capacity and variation in power demands.	Particularly suited to provide reserve capacity as well as meeting the exacting needs of daily variation in power demands.	Comparatively not suited for the mentioned requirements.
9.	Employment potential	More. Affords a relatively high employment potential and better utilization of the available local talent and resources.	Less
10.	Man power required; Labour problem	Small Less	Large More
11.	Foreign exchange requirements for equipment	Less	More
12.	Construction time required	Almost same as thermal-power station	Almost same as hydro-power station.
13.	Overall capital expenditure requirements	Low	High

15.10. CALCULATION OF AVAILABLE HYDRO-POWER

The theoretical power available from falling water can be calculated using the following formula:

$$P_{th.} = \frac{wQH}{1000} \ kW \qquad \qquad ...(15.1)$$

where, P_{th} = Theoretical output in kW,

w = Weight density of water in N/m^3,

Q = Flow through turbine (or quantity of water available for hydro-power generation) in m^3/s, and

H = Head available in *metres*.

The actual useful or *effective output* depends upon the efficiency of the various parts of the installation.

If, η_1 = Efficiency of pipelines, intake etc., and

η_2 = Efficiency of hydraulic turbine,

Then, *overall efficiency* $\eta_0 = \eta_1 \times \eta_2$.

Since the turbine and the generator are directly coupled on common shaft the hydro-electrical power available will be given by the equation:

$$P_{actual} = P_{th.} \times \eta_0 \qquad \qquad ...(15.2)$$

or,

$$P_{actual} = \frac{wQH}{1000} \times \eta_0 \; kW \qquad \qquad ...(15.3)$$

15.11. COST OF HYDRO-POWER

The initial cost of any hydro-plant is very high but the power produced by it is the cheapest.

The following *costs* are included in development of a hydro-plant:

1. Cost of land and riparian rights.
2. Cost of railways and highways required for the construction work.
3. Cost of construction.
4. Cost of engineering supervision of the project.
5. Cost of building etc.
6. Cost of equipment.
7. Cost of equipment used for power transmission.

15.12. HYDROLOGY

15.12.1. Introduction

Hydrology *may be defined as the science which deals with the depletion and replenishment of water resources.* It deals with the surface water as well as the ground water. It is also concerned with the *transportation of water* from one place to another, and from one form to another. It helps us in determining the occurence and availability of water.

Hydrology aims at answering the following major questions:

(*i*) How is the water going to precipitate ?

(*ii*) How is water going to behave ?

(*iii*) What will happen to water after precipitation ?

The basic knowledge of this science is a must for every civil engineer, particularly the one who is engaged in the design, planning or construction of irrigation structures, bridges and highway culverts, or flood control works, etc.

15.12.2. The Hydrologic Cycle

Most of the earth's water sources, such as rivers, lakes, oceans and underground sources, etc. get their supplies from rains, while the rain water itself is the evaporation from these sources. Water is lost to the atmosphere as vapour from earth, which is then precipitated back in the form of rain, snow, hail, dew, sleet or frost, etc. This evaporation and precipitation continues for ever, and thereby, a balance is maintained between the two. This process is known as *Hydrologic Cycle*. It can be represented graphically as shown in Fig. 15.13. Hydrologic equation is expressed as follows:

$$P = R + E \qquad \qquad ...(15.4)$$

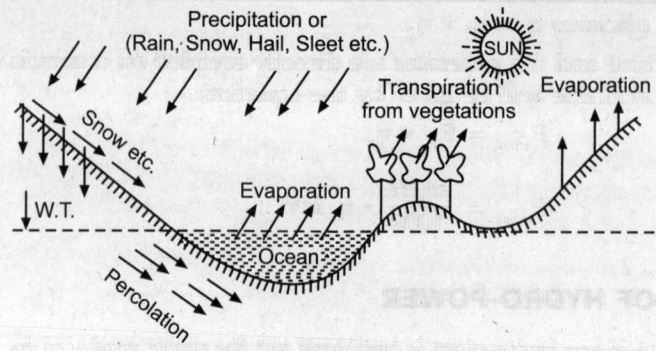

Fig. 15.13. Hydrologic cycle.

where, P = Precipitation,

                          R = Run-off, and

                          E = Evaporation.

Precipitation. It includes all the water that falls from atmosphere to earth surface. Precipitation is of two types:

(*i*) Liquid precipitation (rain fall).

(*ii*) Solid precipitation (snow, hail).

Run-off and surface run-off. Run-off and surface run-off are two different terms and should not be confused. *Run-off* includes all the water flowing in the stream channel at any given section. While the *surface run-off* includes only the water that reaches the stream channel without first percolating down to the water table.

Run-off can, therefore, also be named as Discharge or Stream flow. Rainfall duration, its intensity and a real distribution influence the rate and volume of run-off.

Evaporation. Transfer of water from liquid to vapour state is called *evaporation*.

Transpiration. The process by which water is released to the atmosphere by the plants is called *transpiration*.

15.12.3. Measurement of Run-off

Run-off can be measured daily, monthly, seasonal or yearly. It can be measured by the following *methods:*

1. From rainfall records.

2. Empirical formulae.

3. Run-off curves and tables.

4. Discharge observation method.

1. From rainfall records. In this method consistent rainfall record for a sufficiently long period is taken and then average depth of rainfall over the catchment is determined. Then considering all the factors which affect run-off process, a coefficient is arrived at for that catchment. Now a simple equation can be used to find out the run-off over the catchment.

$$\text{Run-off} = \text{Rainfall} \times \text{coefficient} \qquad\qquad ...(15.5)$$

2. Empirical formulae. In this method an attempt is made to derive a direct relationship between the rainfall and subsequent run-off. For this purpose some constants are established which give fairly accurate result for a specified region. Some important formulae are given below:

(a) **Khosla's formula:**

$$R = P - 4.811\,T$$

where,
R = Annual run-off in mm,
P = Annual rainfall in mm, and
T = Mean temperature in °C.

(b) **Inglis formulae** for hilly and plan areas of Maharashtra:

For *Ghat region*

$$R = 0.88\,P - 304.8$$

For *plain region*

$$R = \frac{(P-177.8)\times P}{2540}$$

(c) **Lacey's formula:**

$$R = \frac{P}{1+\dfrac{3084\,F}{PS}}$$

where,
R = Monsoon run-off in mm,
P = Monsoon rainfall in mm,
S = Catchment area factor, and
F = Monsoon duration factor.

Values of S for various types of catchment are given below:

Type of catchment	Value of S
Flat, cultivated and black cotton soils	0.25
Flat, partly cultivated, various soils	0.6
Average catchment	1.00
Hills and places with little cultivation	1.70
Very hilly and steep, with hardly any cultivation	3.45

Values of F for various durations of monsoon are given below:

Class of monsoon	Value of F
Very short	0.50
Standard length	1.00
Very long	1.50

3. Run-off curves and tables. Each region has its own catchment area and rainfall characteristics. Thus formulae given above and coefficients derived there in cannot be applied universally. However, for the same region the characteristics mostly remain unchanged. Based on this fact the run-off coefficients are derived once for all. Then a graph is plotted in which one axis represents rainfall and the other run-off. The curves obtained are called *run-off curves.* Alternatively a *table* can be prepared to give the run-off for a certain value of rainfall for a particular region.

4. Discharge observation method. By actual measurement of discharge at an outlet of a drainage basin run-off over a catchment can be computed. The complication in this method is that the discharge of the stream at the outlet comprises surface run-off as well as sub-surface flow. To find out the sub-surface run-off it is essential to separate the sub-surface flow from the total flow.

The separation can be done on an approximate basis but with correct analysis.

Factors affecting the run-off

The following factors affect run-off:

1. *Rainfall pattern.*
2. Character of catchment area.
3. Topography.
4. Shape and size of the catchment area.
5. Vegetation.
6. Geology of the area.
7. Weather conditions.

15.12.4. Hydrograph

Hydrograph is defined as a graph showing discharge (run-off) of flowing water with respect to time for a specified time. Discharge graphs are known as flood or run-off graphs. Each hydrograph has a reference to a particular river site. The time period for discharge hydrograph may be hour, day, week or month.

Hydrograph of stream of river will depend on the characteristics of the catchment and precipitation over the catchment. Hydrograph will access the flood flow of rivers hence it is essential that anticipated hydrograph could be drawn for river for a given storm.

Hydrograph indicates the power available from the stream at different times of day, week or year.

Typical hydrographs are shown in Figs. 15.14 and 15.15.

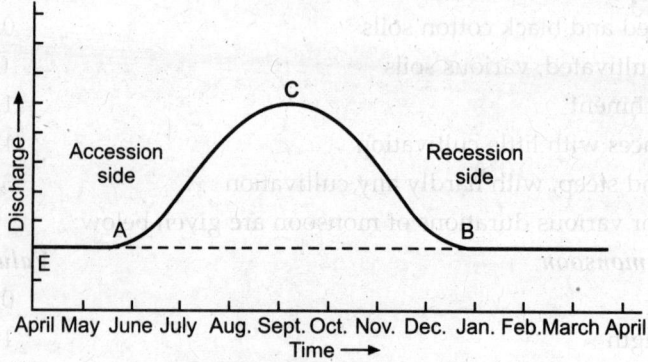

Fig. 15.14. Typical hydrograph.

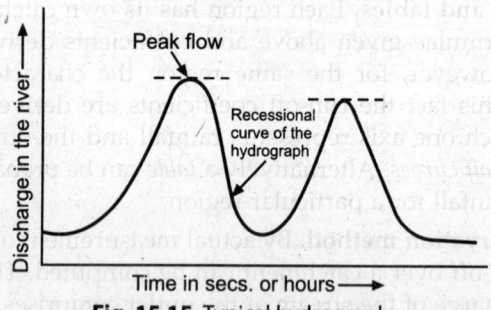

Fig. 15.15. Typical hydrograph.

The Unit hydrograph:

The peak flow represents a momentary value. Therefore the peak flow alone does give sufficient information about the run-off. It is necessary to understand the full hydrograph of flow. Introduction of unit hydrograph theory in 1932 made it possible to predict a run-off hydrograph corresponding to an observed or hypothetical storm. *The basic concept of unit hydrograph is that the hydrographs of run-off from two identical storms would be the same.* In parctice identical storms occur very rarely. The rainfall generally varies in duration, amount and areal distribution. This makes it necessary to construct a typical hydrograph for a basin which could be used as a unit of measurement of run-off.

A Unit hydrograph *may be defined as a hydrograph which represents unit run-off resulted from an intense rainfall of unit duration and specific areal distribution.*

The *following steps* are used for the *construction of unit hydrograph:*

1. Choose an isolated intense rainfall of unit duration *from past records.*
2. Plot the discharge hydrograph for outlet from the rainfall records.
3. Deduct the base flow from stream discharge hydrograph to get hydrograph of surface run-off.
4. Find out the volume of surface run-off and convert this volume into cm of run-off over the catchment area.
5. Measure the ordinates of surface run-off hydrograph.
6. Divide these ordinates by obtained run-off in cm to get ordinates of unit hydrograph.

Thus for any catchment unit hydrograph can be prepared once. Then whenever peak flow is to be found out, *multiply the maximum ordinate of unit hydrograph by the run-off value expressed in cm.* Similarly to obtain run-off hydrograph of the storm of same unit duration multiply the ordinates of the unit hydrograph by the run-off value expressed in cm. If the storm is of longer duration calculate the run-off in each unit duration of the storm. Then super-impose the run-off hydrographs in the same order giving a lag of unit period between each of them. Finally draw a summation hydrograph by adding all the overlapping ordinates. Generally the computations are done in a tabular form before the hydrograph is plotted.

Fig. 15.16 is self explanatory and shows how a run-off hydrograph is constructed from a unit hydrograph.

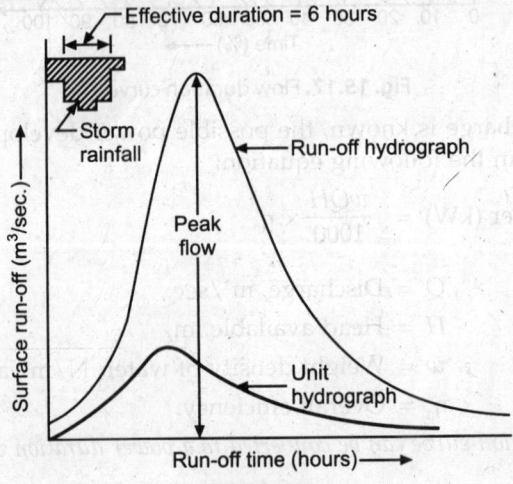

Fig. 15.16.

Limitations to the use of unit hydrographs:

1. Its use is limited to areas about 5000 sq. kilometres since similar rainfall distribution over a large area from storm to storm is rarely possible.

2. The odd-shaped basins (particularly long and narrow) have very uneven rainfall distribution, therefore, unit hydrograph method is not adopted to such basins.

3. In mountain areas, the areal distribution is very uneven, even then unit hyrograph method is used because the distribution pattern remains same from storm to strom.

15.12.5. Flow Duration Curve

Refer to Fig. 15.17. Flow duration curve is another useful form *to represent the run-off data for the given time. This curve is plotted between flow available during a period versus the fraction of time.* If the magnitude on the ordinate is the potential power contained in the stream flow, then the curve is known as *"power duration curve".* This curve is a very useful tool in the analysis for the development of water power.

The flow duration curve is drawn with the help of a hydrograph from the available run-off data and, here it is necessary to find out the length of time duration during which certain flows are available. This information either from run-off data or from hydrograph is tabulated. Now the flow duration curve taking 100 percent time on X-axis and run-off on Y-axis can be drawn.

The area under the flow duration curve (Fig. 15.17) gives the total quantity of run-off during that period as the flow duration curve is representation of graph with its flows arranged in order of descending magnitude.

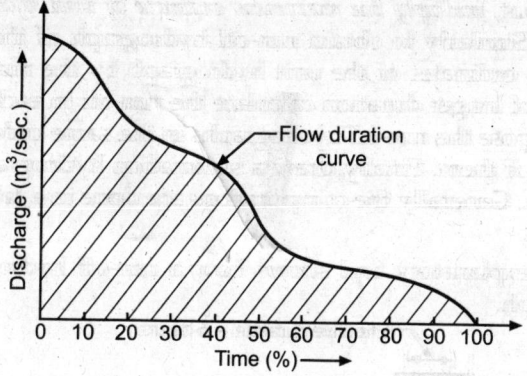

Fig. 15.17. Flow duration curve.

If the head of discharge is known, the possible power developed from water in kW can be determined from the following equation:

$$\text{Power (kW)} = \frac{wQH}{1000} \times \eta_0$$

where,

Q = Discharge, m³/sec.,

H = Head available, m,

w = Weight density of water, N/m³, and

η_0 = Overall efficiency.

Thus the *flow duration curve can be converted to a power duration with some other scale on the same graph.*

Flow duration curve are most useful in the following cases:

(*i*) *For preliminary studies*   (*ii*) *For comparison between streams.*

Uses of flow duration curve:

1. A flow duration curve allows the evaluation of low level flows.

2. It is highly useful in the planning and design of water resources projects. In particular, for hydropower studies, the flow duration curve serves to determine the potential for *firm power generation*. In the case of a run-off-the-river plant, with no storage facilities, the firm power is usually computed on the basis of flow available 90 to 97 percent of the time. The *firm power* is also known as the *primary power*. *Secondary power* is the power generated at the plant utilising water other than that used for the generation of firm power.

3. If a sediment rating curve is available for the given stream, the flow duration curve can be converted into cumulative sediment transport curve by multiplying each flow rate by its rate of sediment transport. The area under this curve represents the total amount of sediment transported.

4. The flow duration curve also finds use in the design of drainage systems and in flood control studies.

5. A flow duration curve plotted on a log-log paper provides a qualitative description of the run-off variability in the stream. If the curve is having steep slope throughout, it indicates a stream with highly variable discharge. This is typical of the conditions where the flow is mainly from surface run-off. A flat slope indicates small variability which is a characteristic of the streams receiving both surface run-off and ground water run-off. A flat portion at the lower end of the curve indicates substantial contribution from ground water run-off, while the flat portion at the upper end of the curve is characteristic of streams with large flood plain storage, such as lakes and swaps, or where the high flows are mainly derived from snowmelt.

6. The shape of the flow duration curve may change with the length of record. This aspect of the flow duration curve can be utilised for extrapolation of short records.

Shortcomings/Defects of flow duration curve:

1. It does not present the flows in natural source of occurrence.

2. It is also not possible to tell from flow duration curve whether the lowest flows occurred in consecutive periods or were scattered throughout the considered period.

15.12.6. Mass Curve

Refer to Fig. 15.18.

A '*mass curve*' *is the graph of the cumulative values of water quantity (run-off) against time.* A mass curve is an *integral curve of the hydrograph* which expresses the area under the hydrograph from one time to another.

It is a convenient device *to determine storage requirement* that is needed to produce a certain dependable flow from fluctuating discharge of a river by a reservoir.

Mass curve can also be used to solve the reserve problem of determining the maximum demand rate that can be maintained by a given storage volume. However, it is a trial and error procedure.

The mass curve *will always have a positive shape* but of a greater or less degree depending upon the variations in the quality of inflow water available. The negative inclination of mass curve would show that the amount of water flowing in the reservoir was less than the loss due to evaporation and seepage.

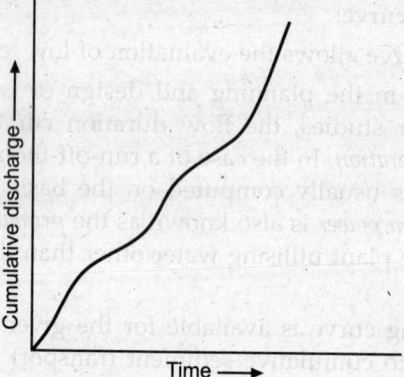

Fig. 15.18. Mass curve.

HIGHLIGHTS

1. A *dam* is a barrier to confine or raise water for storage or diversion to create a hydraulic head.

2. A *canal* is an open waterway excavated in natural ground. A *flume* is an open channel excavated on the surface or supported above ground on a trestle. A *tunnel* is a closed channel excavated through a natural obstruction such as a ridge of higher land between the dam and the powerhouse.

3. A *surge tank* is a small reservoir or tank in which the water level rises or falls to reduce the pressure swings so that they are not transmitted in full to a closed circuit.

4. A *draft tube* serves the following two purposes:
   (*i*) It allows the turbine to be set above tail-water level, without loss of head, to facilitate inspection and maintenance.
   (*ii*) It regains, by diffuser action, the major portion of the kinetic energy delivered to it from the runner.

5. The plants which cater to the base load of the system are called *'base load plants'* whereas the plants which can supply the power during peak loads are known as *peak load plants.*

6. Micro-hydel plants (micro-stations) make use of standardized *bulb sets* with unit output ranging from 100 to 1000 kW working under heads between 1.5 to 10 metres.

7. The *specific speed* of a turbine is defined as the speed of a geometrically similar turbine that would develop one brake horse power under a head of one metre.

8. The *Pelton* turbine is a tangential flow impulse turbine. The pressure over the Pelton wheel is constant and equal to atmosphere, so that energy transfer occurs due to purely impulse action.

9. The modern *Francis* water turbine is an inward mixed flow reaction turbine. It operates under medium heads and also requires medium quantity of water.

10. In the *propeller turbine the runner blades are fixed and non-adjustable.* In *Kaplan turbine,* which is a modification of propeller turbine the runner blades are adjustable and can be rotated about the pivots fixed to the boss of the runner.

11. 'Cavitation' may be defined as the phenomenon which mainfests itself in the pitting of the metallic surfaces of turbine parts because of formation of cavities.

12. 'Hydrology' may be defined as the science which deals with the depletion and replenishment of water resources.

13. Run-off includes all the water flowing in the stream channel at any given section. It can be measured by the following methods:

    (i) From rainfall records.

    (ii) Empirical formulae.

    (iii) Run-off curves and tables.

    (iv) Discharge observation method.

14. Hydrograph is defined as a graph showing discharge (run-off) of flowing water with respect to time for a specified time. It indicates the power available from the stream at different times of day, week or year.

15. Flow duration curve represents the run-off data for the given time. It is plotted between flow available during a period versus the fraction of time.

16. Mass curve is the graph of the cumulative values of water quantity (run-off) against time. It is an integral curve of the hydrograph which expresses the area under the hydrograph from one time to another.

THEORETICAL QUESTIONS

1. Give the application of hydro-electric plants.

2. Enumerate advantages and disadvantages of hydro-plants.

3. Enumerate and explain briefly the factors which should be considered while selecting the site for hydro-electric plant.

4. Enumerate essential elements of hydro-electric power plant.

5. Give the classification of hydro-electric plants.

6. Explain briefly mini and micro-hydro plant.

7. What is a catchment area?

8. What is a reservoir?

9. What is a dam? What are its various types?

10. Explain the advantages of combined operation of hydro-electric station and thermal station.

11 Compare hydro and thermal power plants.

16 CHAPTER

ENVIRONTAL ASPECTS OF ENERGY GENERATION AND POLLUTION CONTROL

16.1. Introduction; 16.2 Pollution from thermal power plants – Gaseous emission and its control – Particulate emission and its control – Solid waste disposal – Thermal pollution; 16.3 Pollution from nuclear power plant; 16.4 Pollution from hydroelectric power plants and solar power generating stations: Highlights – Theoretical Questions.

16.1. INTRODUCTION

All power production plants, invariably, pollute the atmosphere and the resulting imbalance on Ecology has a bad effect. The pollution is inevitable in some cases and has to be minimised to the extent possible. This is being achieved by effective legislations all over the world.

The power plant pollutants of major concern are :

A. From fossil power plants:
   (*i*) Sulphur oxide
   (*iii*) Carbon oxide
   (*v*) Particulate matter.
   (*ii*) Nitrogen oxide
   (*iv*) Thermal pollution

B. From nuclear power plants:
   (*i*) Radioactivity release
   (*iii*) Thermal pollution.
   (*ii*) Radioactive wastes

Besides this, pollutants such as lead and hydrocarbons are contributed by automobiles.

In this chapter, we shall look at the type of pollutions various power plants cause and the method of minimising these bad effects.

16.2. POLLUTION FROM THERMAL-POWER PLANTS

The environment is polluted to a great extent by thermal power plants. The emission from the chimney throws unwanted gases and particles into the atmosphere while the heat is thrown into the atmosphere and rivers. Both these aspects pollute the environment beyond tolerable limits and now are being controlled by appropriate regulations. The types of emissions, effects and methods of minimising these pollutions are discussed below.

The *air pollution* in a large measure is caused by the thermal-power plants burning conventional fuels (coal, oil or gas). The combustible elements of the fuel are converted to gaseous products and non-combustible elements to ash. Thus the *emission can be classified* as follow :

   1. Gaseous emission.
   3. Solid waste emission.
   2. Particulate emission.
   4. Thermal pollution (or waste heat).

16.2.1. Gaseous Emission and its Control

The various gaseous pollutants are :

(i) Sulphur dioxide   (ii) Hydrogen sulphide

(iii) Oxides of nitrogen   (iv) Carbon monoxide etc.

The effects of pollutants on environment are as follows :

S. No.	Pollutant	Effects		
		On man	**On vegitation**	**On materials/animals**
1.	SO_2	Suffocation, irritation of throat and eyes, respiration system.	Destruction of sensitive crops and reduced yield.	Corrosion.
2.	NO_2	Irritation, bronchitis, oedema of lungs.	—	—
3.	H_2S	Bare disease, repiratory diseases.	Destruction of crops.	Flourosis in cattle grazing.
4.	CO	Poisoning, increased accident-liability.	—	—

Removal of sulpher dioxide (SO_2):

SO_2 is removed by wet scrubbers as shown in Fig. 16.1.

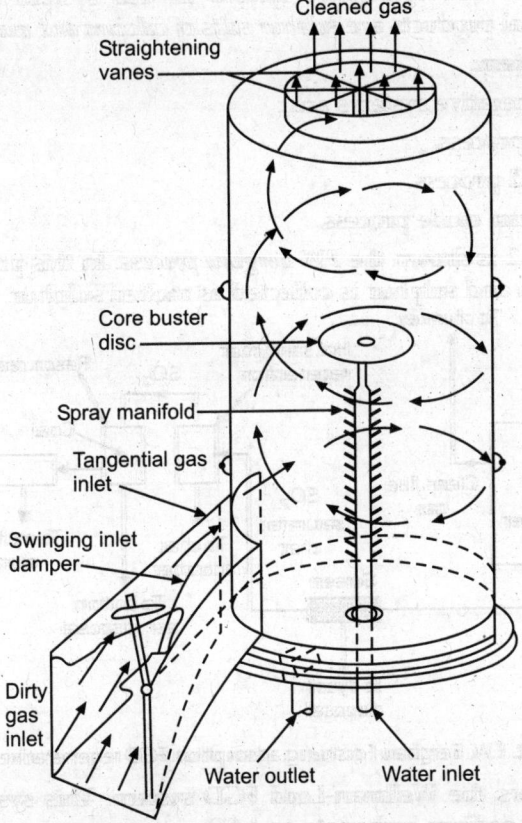

Fig. 16.1. Wet scrubber.

— The gases to be cleaned are admitted tangentially into the *scrubber* which will also help in separating the particulate matters. Water spray absorbs these gases, and particulate matters which collect on the surface of the scrubber, are washed down by the water and this water is further treated, filtered and reused.

— The wet scrubbers also find application in chemical and grain milling industries.

— The *collection efficiency of scrubber is about 90 per cent.*

The following are *disadvantages* of using wet scrubbers :

1. The gases are cooled to such an extent that they must be reheated before being sent to the stack.

2. The pressure drops are very high.

3. Water used, after dissolving sulphur oxides, will contain sulphuric and sulphurous acids which may corrode the piplines and the scrubber itself. This water cannot be let out into the rivers for obvious reasons.

In power plants where high sulphur content coal is the only source available, it is preferable to remove the sulphur from the coal before it is burnt. This is done by coal washing which reduces the flyash as well as some sulphur oxides in the flue gases. But the power plants employ **"Flue-gas desulphurization"** (FGD) system similar to wet scrubber system. **FGD** can be of the following types :

1. *The recovery or regenerative system.*

2. *Throw away or non-regenerative system.* In this system the reactants are not recovered and the final products are *sulphur salts of calcium and magnesium.*

Regenerative system:

Some of the regenerative systems are :

1. FW-Bergbau process.

2. Wellman-Lord process.

3. Wet magnesium oxide process.

— In the Fig. 16.2 is shown the *FW-Bergbau process.* In this process, SO_2 is removed by adsorption and sulphur is collected as molten sulphur.

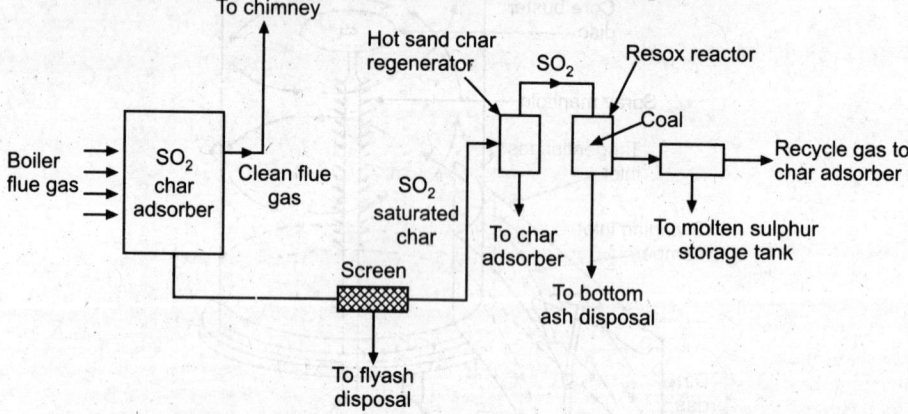

Fig. 16.2. F.W. Bergbau Forshung adsorption FGD regenerative system.

— Fig. 16.3 shows the Wellman-Lord FGD system. This system removes SO_2 by absorption in sodium carbonate and SO_2 is recovered as sulphur or sulphuric acid products.

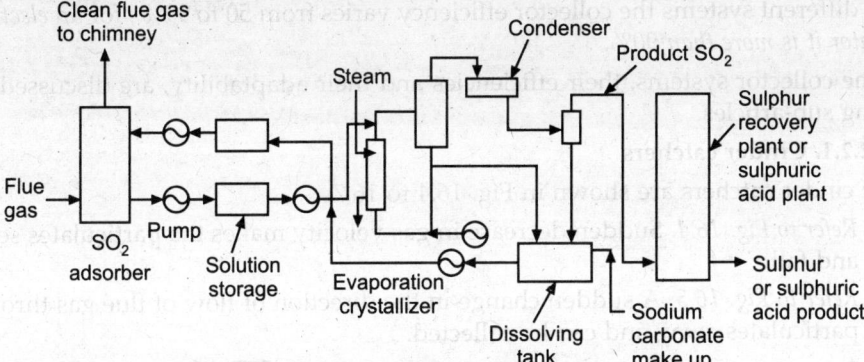

Fig. 16.3. Wellman-Lord absorption FGD regenerative system.

In non-regenerative systems the principal reactant is either lime or limestone. The slurry is made into sludge by adding flyash and other proprietory sludge additives and the sludge is disposed. This method could prove a bit more expensive since no sulphur or sulphur product is recovered and the reactant is not generated as in the case of FW Bergbau porcess.

Emission of NO$_x$:

Nitrigen oxides are compounds of the elements nitrogen and oxygen, both of which are present in air. The combustion of *fossil fuels* in air is accompained by the formation of nitric oxide (*NO*) which is subsequently partly oxidised to nitrogen dioxide (*NO$_2$*). The resulting mixture of variable combustion is represented by the symbol *NO$_x$*, where *x* has a value between 1 and 2. Nitrogen oxides are present in stack gases from coal, oil and gas furnaces (and also in the exhaust gases from internal combustion engines and gas turbines).

The following methods are commonly used to reduce the emission of *NO$_x$* from thermal (and gas turbine) power plants :

1. Reduction of temperature in combustion zone.
2. Reduction of residence period in combustion zone.
3. Increase in equivalence ratio in the combustion zone.

16.2.2. Particulate Emission and its Control

The particulate emission, in power plants using fossil fuels, is easiest to control. Particulate matter can be either *dust* (Particles having a diameter of 1 micron) which do not settle down or *particles* with a diameter of more than 10 microns which settle down to the ground. The particulate emission can be *classified* as follows :

Smoke. It composes of stable suspension of particles that have a diameter of *less than 10 microns* and are visible only in the aggregate.

Fumes. These are very small particles resulting from chemical reactions and are normally composed of metals and metallic oxides.

Flyash. These are ash particles of diameters of *100 microns or less*.

Cinders. These are ash particles of diameters of *100 microns or more*.

The above particulates, in any system of controlling the particulate emission, are to be effectively collected from the flue gases. The performance parameters for any particulate remover is called the *collection efficiency* defined as :

$$\text{Collector efficiency} = \frac{\text{Mass of dust removed}}{\text{Mass of dust present}} \times 100$$

For different systems the collector efficiency varies from 50 to 99% ; for an *electrostatic precipitator it is more than 90%*.

Some collector systems, their efficiencies and their adaptability, are discussed in the following sub-articles.

16.2.2.1. Cinder catchers

The cinder catchers are shown in Fig. 16.4 to 16.7.

— *Refer to Fig. 16.4.* Sudden decrease in gas velocity makes the particulates separate and fall.

— *Refer to Fig. 16.5.* A sudden change in the direction of flow of flue gas throws the particulates away and can be collected.

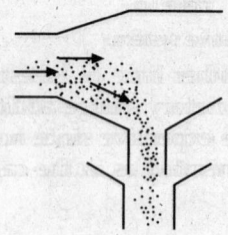

Fig. 16.4. Sudden decrease in gas velocity.

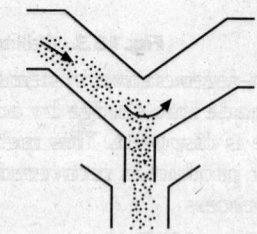

Fig. 16.5. Sudden change in the direction of the flow of flue gas.

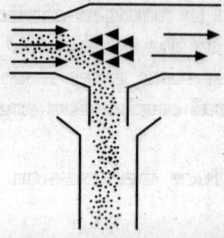

Fig. 16.6. Impingement of flue gases on a series of baffle stops.

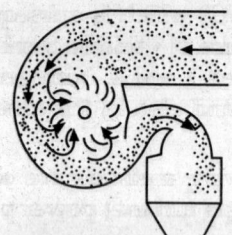

Fig. 16.7. Cinder-vane fan.

— *Refer to Fig. 16.6.* Impingement of flue gases on a series of baffle stops the particulate matter as shown. These are commonly used in stoker and small cyclone furnaces where crushed coal is burned rather than the very fine pulverised coal. The collection efficiencies of cinder catchers are from 50 to 75%.

— *Refer to Fig. 16.7.* **Cinder-vane fan.** The cinder-vane fan uses the fan which imparts centrifugal force to the particulates and they are collected as shown. The efficiency is from 50 to 75%

16.2.2.2. Wet scrubbers

— Wet scrubbers as described for removal of gases can also be used for removal of particulate matters; but the gases will have to be reheated before they are sent to the stack.

— The wet scrubbers are *not commonly used to remove particulate matters*.

16.2.2.3. Electrostatic precipitator

— An electrostatic precipitator is shown in Fig. 16.8. In this device a very high voltage of 30 kV to 60 kV is applied to the wires suspended in a gas-flow passage between two grounded plates.

The particles in the gas stream acquire a charge from the negatively charged wires and are then attracted to the ground plates. The grounded plates are periodically rapped by a steel plug which is raised and dropped by an electromagnet and dust is collected in the hoppers below.

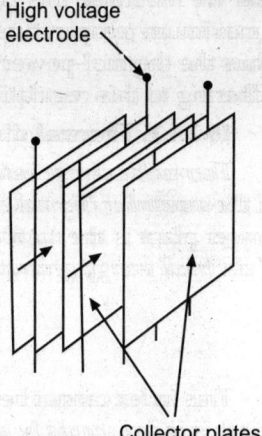

Fig. 16.8. Electrostatic precipitator.

— In this type of collector, *care must be taken to see that large quantity of unburnt gases do not enter the precipitator.* If such a mixture enters, power should be turned off, otherwise there could be *explosion because of constant sparking between* wires and plates.

— The collection efficiency is about 99 per cent.

— Electrostatic precipitators are suitable for power plants where *flyash content is high.* Flyash having high electrical resistivity *does not* separate in the electrostatic precipitator. This problem can be solved by injecting *sulphur trioxide* into the exhaust gas which *improves the conductivity of flyash*. This again poses a problem of discharging objectionable sulphur trioxide into the atmosphere ; this needs a wet scrubber after the electrostatic precipitator.

16.2.2.4. Baghouse filters

Fig. 16.9 shows a baghouse filter. Baghouse filters are found useful in removing the particulate matters where *low sulphur coal is used.*

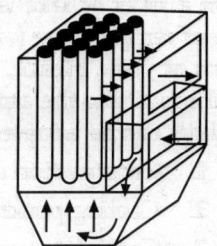

Fig. 16.9. Baghouse filter.

— The cloth filters cost about 20 per cent of installation cost and last for $1\frac{1}{2}$ to 3 years.

— The baghouse filter is usually cleaned by forcing air in the reversed direction. They need large filter areas of about 6.5 m²/MW of power generation. Hence the installation cost could be high.

— Although baghouse filters are expensive, yet they are being widely used in *coal-fired systems.*

16.2.3. Soild Waste Disposal

From the fossil fuel fired power plants considerable amount of solids in the form of ash is discharged. This ash is removed as bottom ash or slug from the furnace. The fossil fuel fired system also discharges solid wastes such as calcium and magnesium salts generated by absorption of SO_2 and SO_3 by reactant like limestone.

16.2.4. Thermal Pollution

Discharge of thermal energy into waters is commonly called *'Thermal pollution'*.

Thermal power stations invariably will have to discharge enormous amounts of energy into water since water is one medium largely used to condense steam. If this heated water from condensers is discharged into lakes or rivers, the water temperature goes up. The ability of water to hold dissolved gases goes down when the temperature increases. *At about 35°C, the dissolved oxygen will be so low that the acquatic life will die.* But in very cold regions, letting out hot water into the lakes or rivers helps in increasing the fish growth. But, in our country, such places are not many and hence, it is necessary that we minimise this thermal pollution of water. One of the regulations stipulates that a maximum temperature of water let out can be 1°C above the atmospheric temperature. Thus the thermal-power plants or any other industry has to resort to various methods of adhering to this regulation.

16.2.4.1. Thermal discharge index (TDI)

Thermal discharge index (TDI) is the term usually used in connection with the estimation of the *amount of thermal energy released to environment from a thermal-power plant.* TDI of any power plant is the *number of thermal energy units discharged to the environment for ever unit of electrical energy generated.*

$$\text{TDI} = \frac{\text{Thermal power discharged to environment in MW}_{\text{thermal}}}{\text{Electrical power output in MW}_{\text{electrical}}}$$

This index cannot be zero or else the plant violates the Second low of thermodynamics; but this index *should be as low as possible to improve the efficiency of the plant as well as to keep the pollution level low.*

The thermal discharge index (TDI) is *strongly dependent on the thermal efficiency of the plant.*

16.2.4.2. How to reduce thermal pollution?

While considering the efficiency of the thermal plant, it is desirable that the water from a river or lake is pumped through the condenser and fed back to the source. The rise of remperature will be about 10° C which is highly objectionable from the pollution point of view. Hence, this waste heat which is removed from the condenser will have to be thrown into the atmosphere and not into the water source, in this direction following *methods* can be adopted :

1. Construction of a separate lake.
2. Cooling pond.
3. Cooling towers.

1. Construction of a separate lake. A sufficiently large water storage in the form of a lake can be built and once-through cooling the condenser can be adopted. It the natural cooling of water from the lake is not sufficient, floating spray pumps can be employed.

This method *improves the thermal efficiency* of the plant but can prove *expensive.* Also, it may not always be possible to have a large enough lake artificially built.

2. Cooling pond. A cooling pond with continuously operating fountains can be adopted for smaller power plants. This will also serve as a beautifying feature of the power plant site.

3. Cooling towers. In order to throw heat into the atmosphere most power stations adopt the cooling towers. The hyperbolic shape given to the tower automatically induces air from the bottom to flow upwards and the water is cooled by coming in direct contact with the air. This is a natural convection cooling and is also called *'wet-cooling tower'.* The overall efficiency of such plants will be lower than those of the plants adopting once-through cooling system. There will be considerable vapour flumes escaping from the cooling towers. Sometime, make-up cooling water may be scarce. In such cases, *dry cooling tower* can be adopted. Dry cooling towers are much more expensive than wet cooling towers.

All cooling towers, whether dry or wet, are *expensive and add to the initial investment of the plant.* Small plants can adopt mechanical-draft systems using induced or forced draft systems. This helps in avoiding height to the cooling towers. Thus, the initial cost is reduced but the maintenance cost of mechanical-draft systems are high.

16.3. POLLUTION FROM NUCLEAR POWER PLANTS

The various types of pollution from nuclear power plants are :

(*i*) Radioactive pollution.

(*ii*) Waste from reactor (solid, liquid, gases).

(*iii*) Thermal pollution.

(*i*) **Radioactive pollution.** This is the most dangerous and serious type of pollution. This is due to radioactive elements and fissionable products in reactor. The best way to abate is the *radioactive shield around the reactor.*

(*ii*) **Waste from reactor.** Due to nuclear reactor reaction nuclear wastes (mixtures of various Beta and Gamma emitting radioactive isotopes with various half lives) are produced which cannot be neutralised by any chemical method. If the waste is discharged in the atmosphere, air and water will be contaminated beyond the tolerable limits. Some methods of storage or disposal of radioactive waste materials are discussed below :

1. Storage tanks. The radioactive wastes can be buried underground (very deep below the surface) in corrosion resistance tanks located in isolated areas. With the passage of time these will become stable isotopes.

2. Dilution. After storing for a short time, low energy wastes are diluted either in liquid or gaseous material. After dilution, they are disposed off in sewer without causing hazard.

3. Sea disposal. This dilution can be used by adequately diluting the wastes and this method is being used by the British.

4. Atmospheric dilution. This method can be used for gaseous radioactive wastes. But solid particles from the gaseous wastes must be filtered out thoroughly since they are the most dangerous with higher half lives.

5. Absorption by the soil. Fission products are disposed off by this method. The radioactive particles are absorbed by the soil particles. But this is *expensive.*

6. Burying in sea. Solid nuclear wastes can be stored is concrete blocks which are buried in the sea. This method is *expensive but no further care is needed.*

16.4. POLLUTION FROM HYDRO-ELECTRIC POWER PLANTS AND SOLAR POWER GENERATING STATIONS

Hydro-electric and Solar Power Generation plants have no polluting effect on the environment. The hydro-electric project does not pollute the atmosphere at all, but it can be argued that the solar power stations in the long run may upset the balance in nature. To extend the argument to the logical end, imagine a very vast area of land is covered by solar collectors of different forms. Then the minimum required sun's rays may not reach the earth's surface. This will certainly kill the vegetation on the earth and also the bacteria which are destroyed by sun's rays may survive giving rise to new types of health problems. Further, the evaporation of water and consequent rains may change their cycles. Added to these, the average temperatures of the earth and ocean may change. This may result in new balances among the living creatures which cannot be easily predicted. Since we do not envisage such a large scale coverage of earth's surface in the near future, we can safely state that the solar energy power plants do not pollute the atmosphere.

HIGHLIGHTS

1. The power plant pollutants of major concern are :
   (a) From fossil power plants :
   - (i) Sulphur oxide
   - (ii) Nitrogen oxide
   - (iii) Carbon oxide
   - (iv) Thermal pollution
   - (v) Particulate matter.

   (b) From nuclear power plants:
   - (i) Radioactivity release
   - (ii) Radioactive wastes
   - (iii) Thermal pollution.

2. The emission may be classified as follows :
   - (i) Gaseous emission
   - (ii) Particulate emission
   - (iii) Solid waste emission
   - (iv) Thermal pollution (or waste heat).

3. Thermal discharge index (TDI) of any power plant is the number of thermal energy units discharged to the environment for every unit of electrical energy generated.

4. Thermal pollution can be reduced by:
   - (i) Constructing a separate lake
   - (ii) Cooling pond
   - (iii) Cooling towers.

5. Some methods of storage or disposal of radioactive waste materials are :
   - (i) Storage tanks
   - (ii) Dilution
   - (iii) Burial
   - (iv) Sea disposal
   - (v) Atmospheric dilution
   - (vi) Absorption by soil
   - (vii) Burying in sea
   - (viii) Projecting into space.

THEORETICAL QUESTIONS

1. Name some power plant pollutants of major concern.
2. How are emissions from thermal power plants classified ?
3. Name important gaseous pollutants discharged by thermal-power plants. How are they controlled ?
4. Describe briefly, with the help of a neat diagram, 'wet-scrubber' used for removing SO_2.
5. What is 'Particulate emission' ? How is it controlled ?
6. Define the following:
   Smoke, Fumes, Fly-ash, Cinders.
7. What is an electrostatic precipitator ?
8. Where are 'Baghouse filters' used ?
9. What do you mean by Thermal Pollution' ?
10. What is "Thermal Discharge Index' (TDI) ?
11. Enumerate and explain briefly various methods of reducing the thermal pollution.
12. Write a short note on 'Pollution from Nuclear power plants'.
13. What are the various methods of storage or disposal of radioactive waste materials ?

SECTION :
SHORT ANSWER QUESTIONS

Q. 1. *Define the terms "Energy and Energy technology".*

Ans. *Energy:* It is the capability to produce motion, force, work, change in shape, change in form etc.

Energy technology: It is the applied part of energy sciences ("*energy science*" focuses attentions on the '*energy*' and 'energy transformations' involved in the various other branches of science to National economy and civilisation) for work and processes, useful to human society, nations and individual.

Q. 2. *What are the characteristics of energy?*

Ans. Energy possesses the following *characteristics*:

1. It can be stored.
2. It can neither be created nor destroyed.
3. It is available is several forms.
4. It does not have absolute value.
5. It is associated with a potential. Free flow of energy takes place only from a higher potential to a lower potential.
6. It can be transported from one system to other system or from one place to another place.
7. The energy is measured in Nm or in joules.

Q. 3. *What is 'Energy intensity'?*

Ans. *Energy intensity:*

The term '*energy intensity*' is defined as *energy consumption per unit of Gross National Product (GNP).*

When the per unit energy consumption for the production of energy intensive raw materials, like steel and aluminium, is reduced, there may be a marginal fall in energy (GNP ratio) with continuation of the downward trend.

Developed countries have reduced '*energy intensity*', resulting in less energy consumption and at the same time achieving higher production.

Q. 4. *What is 'Energy-GDP elasticity:*

Ans. *Energy-GDP elasticity:*

It is defined as the *percentage growth in energy requirement for 1% growth in GDP.*

The *lower* the value of elasticity, the *higher* is the overall efficiency.

- The value of elasticity for the developed countries ranges from 0.8 to 1.0 whereas for India it is about 1.2.

Q. 5. *What do you mean by 'Energy audit'?*

Ans. *Energy audit* is *an official survey/study of energy consumption/processing/supplying aspects related with an organisation, system, process, plant, equipment.*

The *objectives* of the 'Energy Audit' are to recommend "steps" to be taken by the management for:

(*i*) Improving the energy efficiencies,

(*ii*) Reducing the energy costs, and

(*iii*) Improving the productivity without sacrificing quality, standard of living/comforts and environmental balance.

The Energy Audit is officially recommended by the *Management* and is carried out by the Energy Audit Group headed by the Energy Auditor.

Q. 6. *What are 'commercial or conventional energy sources'?*

Ans. *Coal, oil, gas, uranium and hydro* are commonly known as **commercial or conventional energy sources**. These represent about 92% of the total energy used in the world.

Q. 7. *What are 'non-commercial energy sources?*

Ans. *Firewood, animal dung and agricultural waste* etc. are called as **non-commercial energy sources**. These represent about 8% of the total energy used in the world.

Q. 8. *What are 'renewable and non-renewable energy sources?*

Ans. **Renewable:** These sources are being *continuously produced in nature and are inexhaustible.*

Examples: Wood, Wind energy, Biomass, Biogas, Solar energy etc.

Non-renewable: These are *finite and exhaustible.*

Examples: Coal, petroleum etc.

Q. 9. *What are primary and secondary energy sources?*

Ans. **Primary energy sources.** These sources are obtained from environment.

Examples: Fossil fuels, Solar energy, Hydro energy and Tidal energy.

Secondary energy resources. These resources do not occur in nature but are *derived from primary energy resources.*

Q. 10. *What is Magnet-Hydro-Dynamic (MHD) generator?*

Ans. The MHD working principle is based on *Faraday's law of electromagnetic induction which states that change in magnetic field induces an electric field in any conductor located in the magnetic field.* This electric field while acting on the free charges in the conductor causes a current to flow in the conductor. In **MHD generator**, an *ionised gas is used as a conductor. If such ionised gas is passed at a high velocity through a powerful magnetic field, then current is generated* and *can be extracted by placing electrodes in a suitable position in the stream.* It produces *D.C. power directly.*

In MHD generation, all kinds of heat sources like coal, gas, oil, solar etc. can be used. MHD systems are of two types: (*i*) Open cycle system; and (*ii*) Close cycle system.

Q. 11. *What is 'solar radiation'?*

Ans. **Solar radiation** is the *energy radiated by the sun.*

— The *radiated energy received on earth surface is called* **Solar irradiation**.

— *Solar radiation received on a flat horizontal surface on earth* is called **Solar insolation**.

Q. 12. *What are 'Extraterrestrial and terrestrial solar radiations?*

Ans. **Extraterrestrial solar radiation:** The *intensity of sun's radiation outside the earth's atmosphere* is called "*extraterrestrial*" and has no *diffuse components.*

Extraterrestrial radiation is the *measure* of solar radiation that would be received in the *absence of atmosphere.*

Terrestrial solar radiation: The *radiation received on the earth surface* is called "*terrestrial radiation*" and is nearly 70 percent of extraterrestrial radiation.

Q. 13. *Define the terms beam, diffuse and total radiations.*

Ans. Refer to Fig. 1.

Beam (or direct) radiation (I_b). *Solar radiation received on the surface of earth without change in directions* is known as "*beam or direct radiation*".

Diffuse radiation (I_d). The solar *radiation received* from the sun *after its direction has been changed by reflection and scattering by atmosphere* is known as "*diffuse radiation*".

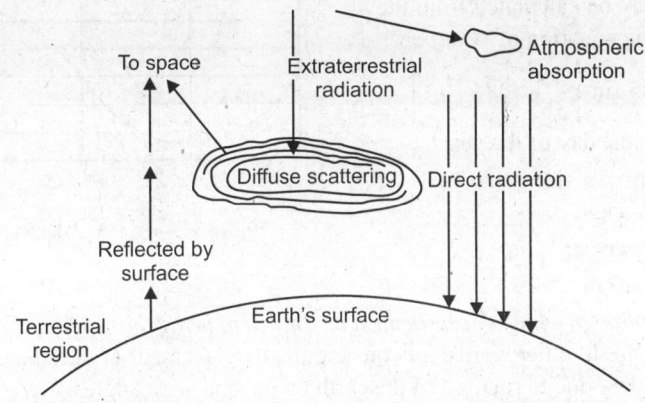

Figure 1. Direct, diffuse and total solar radiations.

Total radiation (I_T). The *sum of beam and diffuse radiations intercepted at the surface of earth per unit area* of *location* is known as "*total radiation*". It is also known as **Insolation**.

Mathematically: $I_T = I_b + I_d$

Q. 14. *What is air mass (m_a)?*

Ans. *Airmass (m_a). It is the path length of radiation through the atmosphere, considering the vertical path at level as unity.*

m_a = 1, when sun is at zenith (*i.e.,* directly above head).

m_a = 2, when zenith angle (θ_z) is 60°.

m_a = sec θ_z, when $m_a > 3$

m_a = 0 just above the earth's atmosphere.

Q. 15. *What are the reasons for variation in solar radiations reaching the earth than received on the outside of the atmosphere?*

Ans. As solar radiations pass through the earth's atmosphere the shortwave '*intraviolet rays*' are '*absorbed*' by *ozone* in atmosphere and the long wave '*infrared waves*' are '*absorbed*' by *carbon dioxide* and *moisture* in the atmosphere. A portion of radiations is '*scattered*' by the components of atmosphere such as water vapour and dust. A portion of this scattered radiation always reaches the earth's surface as '*diffuse radiation*'. Thus radiations finally received at the earth's surface consists partly of beam radiation and partly of diffuse radions.

Q. 16. *What is 'clarity index'?*

Ans. *Clarity Index: The ratio of radiation received on earth's horizontal surface over a given period to radiation on equal surface area beyond earth's atmosphere in direction perpendicular to the beam is called "Clarity index".*

It depends upon the clarity of atmosphere for passage of solar beam radiation. Clarity index can be between 0.1 to 0.7.

Q. 17. *What is 'concentration ratio'?*

Ans. *Concentration ratio: It is the ratio of solar power per unit area of the concentrator surface (kW/m²) to power per unit area on the line focus or point focus (kW/m²).*

Q. 18. *What is declination angle?*

Ans. *Declination angle (δ):*

It is the angle made by the line *joining the centres of the sun and the earth with its projection on the equatorial plane. This angle varies from a maximum value of +23.5° on June 21 to minimum of −23.5° on December 21.*

The declination (in degrees) for any day may be calculated from the approximate equation of "*Cooper*":

$$\delta = 23.45 \sin\left[\frac{360}{365}(284 + n)\right]$$

where, *n* is the day of the year.

Fig. 2. shows the variation of declination angle.

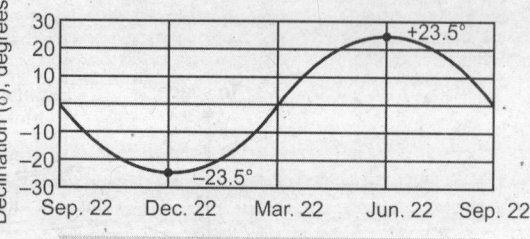

Figure 2. Variation of declination angle.

Q. 19. *What is 'Hour angle (ω)?*

Ans. *Hour angle* **(ω):**

It is angle through which the earth must be rotated to bring the meridian of the plane directly under the sun. In other words, it is the angular displacement of the sun, east or west of the local meridian, due to rotation of the earth on its axis at an angle of 15° per hour.

- At solar noon ω being zero and each hour angle equating 15° of longitude with "morning positive" and "afternoon negative" (*e.g.* ω = +15° for 11.00 and ω = −37.5° for 14.30), hour angle can be expressed as:

$$\omega = 15(12 - LST)$$

Q. 20. *What are the limitations of focusing type collectors*?

Ans. The "*concentrators or focusing type collectors*" can give high temperatures than *flat plate collectors*, but they entail the following *shortcomings/limitations*.

1. *Non-availability and high cost of materials required.* These materials must be easily shapeable, yet have a long life; they must be lightweight and capable of retaining their brightness in tropical weather. Anodised aluminium and stainless steel are two such materials but they are *expensive* and *not readily available in sufficient quantities*.

2. They require *direct light* and are *not operative* when the sun is *even partly covered with clouds*.

3. They *need tracking systems* and reflecting surfaces undergo deterioration with the passage of time.

4. These devices are also subject to *similar vibration and movement problems as radar antenna dishes*.

Q. 21. *What is the difference between neutral and selective collectors?*

Ans. *The collectors for which ratio is equal to unity are called* '**Neutral collectors**' *and those for which the ratio is greater than unity are called* '**Selective collectors**'.

Q. 22. *Define the efficiency of a solar collector.*

Ans. The "efficiency *of a solar collector* (η_c)" is defined as the ratio of *the useful heat output of the collector to the solar energy flux incident on the collector.*

Q. 23. *What are the advantages of 'compound parabolic concentrator' (CPC)?*

Ans. The advantages of CPC are:

1. *High concentration ratio.*

2. *No need of tracking.*

3. *Efficiency* for accepting diffuse radiation is *much larger* that conventional concentrators.

Q. 24. *Define the 'optical efficiency of a concentrating collector'.*

Ans. The **optical efficiency** of a concentrating collector (η_{opt}) is defined as the ratio of the solar radiation absorbed by the absorber to the beam solar radiation on the concentrator.

Q. 25. *What is 'Energy storage system'?*

Ans. An **energy storage system** stores the collected amount of energy in excess of requirement of the demand and supplies this energy when the demand exceeds the supply energy.

Q. 26. *What is a solar pond?*

Ans. A *solar pond* (also called 'salt pond') is an artificially designed pond, filled with salty water, maintaining a definite concentration gradient. It combines solar energy radiation and sensible heat storage, and as such, it is *utilised for collecting and storing solar energy.*

Q. 27. *List some of the commonly used direct solar energy applications.*

Ans. Some of the commonly used direct solar energy applications are:

1. Solar water heating.
2. Solar heating and cooling.
3. Solar distillation.
4. Solar pumping.
5. Solar drying.
6. Solar cooking.
7. Solar Green-house.
8. Solar power plant.
9. Solar photovoltaic system.

Q. 28. *What is solar still?*

Ans. A *solar still* is a device which is used to convert saline water into pure water by using solar energy.

Q. 29. *What is solar air heater?*

Ans. A *conventional solar air heater is essentially a flate-plate collector with an 'absorber plate', a 'transparent cover system' at the top and 'insulation' at the bottom and on the sides.* The whole assembly is encased in a sheet metal container. The working fluid is *air*, though the passage for its flow varies according to the type of air heater.

The 'materials' used for construction of air heaters are similar to those of liquid flat-plate collectors. 'Selective coating' on the absorber plate can be used to improve the collection efficiency but cost effectiveness criterion should be kept in mind.

Q. 30. *What is a 'green-house'? List the advantage of green-houses.*

Ans. A *green-house is an enclosed space which provides the required environment for growth and production of plants under adverse climatic conditions.* Its design depends upon the local climatic conditions and the environment needed for the growth.

Plants *manufacture their food by 'photosynthesis process'* which maintains a balance with respiration.

Advantages:

This type of structure is less expensive to build than a fully insulated structure. It provides the following *advantages*.

1. Inexpensive, good quality food can be grown.
2. An additional heat source (temperature control) is available for the house attached to it.
3. A source of moderator for the humidity (humidity control) in the house.

Fig. 3 shows a schematic diagram of a pipe-framed green-house.

Q. 31. *What is wind? What are its characteristics?*

Ans. **Wind** *is air set in motion by small amount of insolation reaching the upper atmosphere of earth.*

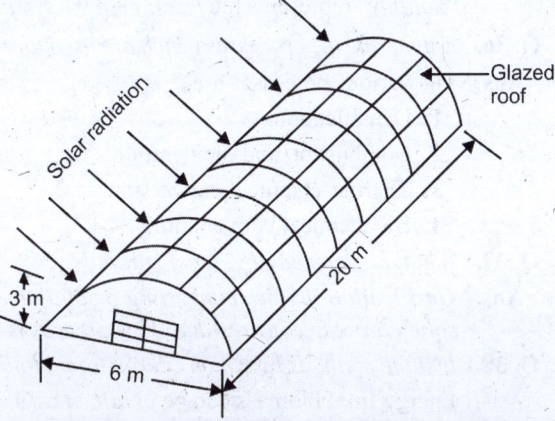

Figure 3. Schematic diagram of pipe-framed greenhouse.

It contains *kinetic energy (K.E.)* which can easily be *converted to electrical energy*. Nature generates about 1.67×10^5 kWh of wind energy annually over land area of earth and 10 times this figure over the entire globe.

The ***main characteristics*** of wind are:

- *Wind speed increases roughly as $\frac{1}{7}$ th power of height. Typical* tower heights are about 20–30 m.

- ***Energy-pattern factor.*** *It is the ratio of the actual energy in varying wind to energy calculated from the cube of mean wind speed.* This factor is always *greater than unity* which means the energy estimates based on mean (hourly) speed are *pessimistic.*

Q. 32. *What is wind energy pattern factor?*

Ans. The ***energy pattern factor (EPF)*** *is the ratio of power from speed distribution to the power from coverage speed of the turbine blades.*

i.e. $$EPF = \frac{\text{Power from speed distribution}}{\text{Power from average speed}}$$

Generally, *EPF* lies between 2 to 5.

Q. 33. *What is an airfoil?*

Ans. An ***airfoil*** is a streamlined air surface designed for air to flow around it in order to produce *low drag and high lift force.*

Q. 34. *What is stalling?*

Ans. The *lift increases as the angle formed at the junction of the airfoil and the air stream (the angle of attack) becomes less and less acute, upto the point where the angle of the air flow on low pressure side becomes excessive.* When this happens, the air flow breaks away from the low pressure side, a *lot of the turbulence ensues, the lift decreases* and the *drag increases quite substantially; this phenomenon is known as **stalling**.*

Q. 35. *What is 'solidity'?*

Ans. ***Solidity (S)*** is *defined as the ratio of the blade area* to *the rotor circumference.* It determines the quantity of blade material required to intercept a certain wind area.

Mathematically, *solidity*, $S = \dfrac{Nb}{\pi D}$ ...(1)

where, N = No. of blades,

                              b = Blade width, and

                              D = Diameter of the circle described by a blade.

"*Solidity*" represents the *fraction of the swept area of the rotor which is covered with metal.*

Q. 36. *Enumerate the processes which are used for the biomass conversion to energy or to biofuels.*

Ans. The various processes are:

1. Densification.
2. Combustion and incineration.
3. Thermo-chemical conversion.
4. Biochemical conversion.

Q. 37. *What do you mean by 'Gasification'?*

Ans. ***Gasification*** *implies converting a solid or liquid into a gaseous fuel without leaving any solid carbonaceous residue.* This process is carried out in a *'gasifer'.*

Q. 38. *List the methods by which energy from biomass can be obtained.*

Ans. Energy from biomass can be *obtained* by using the following *methods* :

1. Combustion;   2. Anaerobic digestion;

   3. Pyrolysis;   **4.** Hydrolysis and ethanol fermentation;   **5.** Gasifier.

Q. 39. *What do you mean by 'Energy plantation'?*

Ans. *When land plants are grown purposely for their fuel value, by capturing solar radiation in them is called Energy plantation. 'Energy plantations' by design are managed and operated to provide substantial amounts of unusable fuel continuously throughout the year at the costs competitive with other fuels. Annual plants, typical of important farm crops, are unsuitable for providing a year-around supply of fuel.*

Q. 40. *What is tide?*

Ans. *The periodic rise and fall of the water level of sea which are carried by the action of the sun and moon on water of the earth is called the tide.*

Q. 41. *What is wave energy?*

Ans. *Wave energy comes from the interaction between the winds and surfaces of oceans. The energy available varies with the size and frequency of waves. It is estimated that about 50 kW of power is available for every metre width of true wave front.*

Ocean wave energy is due to the periodic to-and-fro, up-and-down motion of water particles in the form of progressive waves. The period of ocean waves is the order of a few seconds. Ocean waves are superimposed on ocean water. Ocean water surface level varies with ocean tidal cycle.

Ocean waves possess potential energy (P.E.) and kinetic energy (K.E.). The ocean waves originate in different parts of the ocean surface due to the surface winds. The waves travel in the direction of the wind to the shore. The waves may be due to the local winds or the planetary winds. The *height of the waves depends upon the wind velocities, depth of the ocean, contour of the shore etc.*

The typical ranges of ocean waves are:

   (*i*) *Wind height* = 2 × amplitude = 0.2 m to 4 m

   (*ii*) *Wave period* = 4 s to 12 s.

Very dangerous and destructive waves occur during storms and gusts. They may reach heights of 10 *m* and may topple ships and damage the ocean energy plants.

Q. 42. *What are the various types of OTCE plants.*

Ans. *The plants which employ carnot-type process to generate power between the two steady temperatures are called Ocean Thermal Energy Conversion (OTEC) plants.*

The two basic types of OTEC systems:

   1. Closed cycle (or Anderson cycle) system.

   2. Open cycle (or Claude cycle) system.

Q. 43. *What are 'Direct energy conversion (DEC) devices'?*

Ans. *The devices which convert naturally available energy into electricity, without an intermediate conversion, into mechanical energy; energy source may be thermal, solar or chemical are called Direct energy conversion (DEC) devices.*

Q. 44. *What is a fuel cell?*

Ans. *A fuel cell is an electrochemical device in which the chemical energy of a conventional fuel is converted directly and efficiently into low voltage, direct current electrical energy.*

   ● Fuel cell systems generally *operate on pure hydrogen and air to produce electricity.*

Q. 45. *What are the desirable characteristics of a fuel cell?*

Ans. The *fuel cell* should have the following *characteristics:*

   1. It should have high *energy conversion efficiency.*

   2. It should produce *low chemical pollution.*

   3. It should be *flexible to choose any fuel*.

   4. It should have cogeneration capability and *rapid load response*.

Q. 46. *Name a few commonly used fuel cells*.

 Ans. *A few commonly used fuel cells are*:

   1. Hydrogen-oxygen fuel cell (Hydrox cell) – Alkine fuel cell (AFC)

   2. Phosphoric Acid Fuel Cell (PAFC)

   3. Polymer Electrolyte Membrane Cell (PEMFC)

   4. Molten Carbonate Fuel Cell (MCFC)

   5. Solid Oxide Fuel Cell (SOFC)

   6. Regenerative Cell.

Q. 47. *What are the advantages of hydrogen as fuel?*

 Ans. Following are the *advantages* of hydrogen as fuel:

   1. *Very high energy content*.

   2. Burning is *non-polluting*.

   3. Hydrogen produced from biomass and supplied to the consumers in the transport sector *costs only 50%* compared to hydrogen produced electrolytically.

   4. For fuel-cell operated bus, hydrogen produced from biomass can *compete well* with gasoline-operated vehicles.

   5. It is a *superior fuel for turbojet aircraft* due to *greater economy*, *lower noise level* and *little pollution*.

   6. Hydrogen as a velicular fuel can reduce dependence on fossil fuel which is increasing in cost every year.

   7. Hydrogen can easily be transported and distributed through pipelines.

   8. Hydrogen being a high density fuel, its low transport cost compensates for its high product cost to make it an *economically viable fuel*.

   9. Hydrogen can be used for generating electricity for domestic appliances, in domestic cooking as a fuel, in automobiles etc.

Q. 48. *What are the applications of hydrogen (energy)?*

 Ans. Following are the *applications* of hydrogen energy:

   1. It can be used for H_2–O_2 *fuel cell* for production of electrical energy.

   2. Hydrogen is used as fuel in *aircrafts and rockets in liquid form*.

   3. *Used in cooking*, *water heaters* and *air-conditioning*.

   4. Can also be used in *furnaces*.

   5. Can also be used in *generators*.

   6. Widely used in petroleum refining.

   7. Widely used in manufacture of vanaspati, fertilizers and alcohols.

   • Hydrogen-based vehicles have been developed by Mazda Motor Corporation, BMW Germany, Toyota hybrid highlander and Taiwanese scooter. In India, Tatas are working on the modification of I.C. engines in the present vehicles that can be run on hydrogen fuel.

   • National Hydrogen Energy Board (NHEB) has prepared a workable plan to make hydrogen as a *commercial fuel*. *In the near future, large amounts of hydrogen could be produced in remote wind farms, solar stations and ocean power plants and stored underground*.

INDEX

NOTES

NOTES

NOTES

NOTES